AF882101

Recent Advances in Medicinal Chemistry
(Volume 1)

Editors

Atta-ur-Rahman
Honorary Life Fellow
Kings College
University of Cambridge
UK

Muhammad Iqbal Choudhary
H.E.J. Research Institute of Chemistry
International Center for Chemical and Biological Sciences
University of Karachi
Karachi
Pakistan

&

George Perry
Department of Biology
University of Texas
San Antonio
Texas
USA

CONTENTS

CHAPTERS

FOREWORD

The 1st volume of "Recent Adances in Medicinal Chemistry" offers a very rich panorama of authoritative reviews covering old , as well as new, unexpected issues. Once established, drugs rarely become undisputed therapeutic weapons against one specific kind of disease and reevaluations of such drugs needs to be carried out regularly. The present volume does indeed provide a number of such reevaluations, as, for instance, the aureolic acid group of anti-cancer drugs and the non-steroidal anti-inflammatory drugs (Chakraborty *et al.*). In the ever-expanding collection of anti-cancer drugs, the aromatase inhibitor tamoxifen has long been the gold standard to treat hormone-receptor positive, early forms of breast cancer. However, new generation aromatase inhibitors show definitive advantages over tamoxifen and these advantages are discussed by Colozza *et al.* On the same subject, the precise identification of estrogen receptors is becoming of prime importance in deciding which inhibitor to use (Nagai M. and Brentani M.M.).

The already widespread utilization of statins may further extend to neoplastic diseases, as statins were shown to inhibit Rho-GTPases (often hyperexpressed in tumors) by inhibiting the synthesis of their prenyl groups (C. Riganti).

New molecules such as plant glycosides (saponins) display anti-tumor activity that could be utilized in combination with existing anti-tumor compounds (C. Bachran *et al.*).

Chemoprevention undoubtedly offers effective means to delay or altogether prevent the development of most diseases. Macáková *et al.,* discuss extensively the protective effects of tannins, a group of natural polyphenols with anti-oxidant activity. On the same line, Young *et al.,* consider the activity of carotenoids in the chemoprevention of prostate cancer.

Delivery of pharmacologically active compounds and drugs was always of importance to ascertain that active molecules preferentially reach their target cells or tissues, and Murakami *et al.,* review the recent advances in nanoparticular vehicles.

Antimicrobials are acutely needed to treat bacterial diseases, especially those caused by micro-organisms resistant to existing antibiotics. Pérez-Castillo *et al.,* review how the 3-D structure of the targeted bacterial enzyme may be utilized to design new antibiotics.

Enzymatic activities are essential to carry out biological functions and their perturbations are often caused by disturbances of the intracellular redox potential. In this respect, Nagahara reports on the mechanisms to maintain the cellular redox equilibrium. The plasma membrane redox system is an important property of plasma membranes. In particular, Avigliano *et al.,* review the plasma membrane redox system of blood platelets and the role of anti-oxidant vitamins in this context.

The acute spinal cord injuries (traumatic) represent an unsolved problem in medicine. H. Mestre *et al.,* elaborate on the ongoing efforts made to develop a pharmacological way to ensure the neuroprotection of the injured spine.

The precise assessement of the levels of pharmacologically active molecules in patients needs to be met and Samanidou *et al.,* review how HPLC (high-performance liquid chromatography) can be utilized to monitor the levels of tricyclic anti-depressants.

Lastly, Santucci *et al.,* review chosen instances where misfolded proteins cause pathologies, namely how non-native forms of cytochrome c are involved in apoptosis triggering and participate in amyloid formation.

By the variety of subjects reviewed dealing with chemoprevention and treatment of common diseases, the changing scope of action of medicinal chemistry is illustrated in this volume. It is indeed quite possible that our physiology and pathologies also evolve comcomittantly with the drugs we take to prevent or treat diseases.

Daniel C. Hoessli

Panjwani Center for Molecular Medicine and Drug Research
International Center for Chemical and Biological Sciences
University of Karachi, Karachi-75270
Pakistan

PREFACE

Consolidation, failure of high throughput approaches and rationale design to deliver as promised, and uncertainty of future finances have drawn industry away from innovation to reformulation and repackaging. In this new world, academic research plays an even greater role in drug development, whether built on natural products or novel insights. Blurring of the lines between the traditional fields of chemistry and biology offers opportunities for new discoveries. In this milieu "Recent Advances in Medicinal Chemistry" will find a welcome home.

George Perry

College of Sciences
University of Texas at San Antonio
San Antonio
Texas
USA

List of Contributors

Amalia Bosia

Department of Oncology, Biology and Biochemistry, Research Center on Experimental Medicine (CeRMS), University of Torino, Via Santena 5/bis, 10126 Torino, Italy

Ann Nowé

Computational Modeling Lab (CoMo), Department of Computer Sciences, Faculty of Sciences, VrijeUniversiteit Brussel, Pleinlaan 2, B-1050 Brussel, Belgium

Anna Hošťálková

Department of Pharmaceutical Botany and Ecology, ADINACO Research Group, Faculty of Pharmacy, Charles University, Hradec Kralove, Czech Republic and Centro de Investigación del Proyecto CAMINA A.C., Mexico City, Mexico

Antonio Ibarra

Facultad de Ciencias de la Salud, Universidad Anáhuac México Norte, Huixquilucan, Estado de México, Mexico

Charles Y.F. Young

Departments of Urology, Biochemistry and Molecular Biology, Mayo Clinic College of Medicine, Mayo Clinic, Rochester, Minnesot, USA

Chiara Riganti

Department of Oncology, Biology and Biochemistry, Research Center on Experimental Medicine (CeRMS), University of Torino, Via Santena 5/bis, 10126 Torino, Italy

Christopher Bachran

Institut für Laboratoriumsmedizin, Klinische Chemie und Pathobiochemie, Charité – Universitätsmedizin Berlin, Campus Benjamin Franklin, Hindenburgdamm 30, D-12200 Berlin, Germany

D. Del Principe

Department of Experimental Medicine & Surgery, University of Rome 'Tor Vergata', Italy

Daniel Jun

Centre of Advanced Studies, Faculty of Military Health Sciences, University of Defence, Hradec Kralove, Czech Republic

Dario Ghigo

Department of Oncology, Biology and Biochemistry, Research Center on Experimental Medicine (CeRMS), University of Torino, Via Santena 5/bis, 10126 Torino, Italy

Diana Bachran

Institut für Laboratoriumsmedizin, Klinische Chemie und Pathobiochemie, Charité – Universitätsmedizin Berlin, Campus Benjamin Franklin, Hindenburgdamm 30, D-12200 Berlin, Germany

Dipak Dasgupta

Biophysics Division, Saha Institute of Nuclear Physics, 1/AF Bidhannagar, Kolkata 700 064, India

E. De Azambuja

Medical Oncology Clinic, Jules Bordet Institute, Boulevard de Waterloo, 125, 1000 Brussels, Belgium

E. Minenza

S.C. Oncologia Medica, Ospedale Civile, Via Santa Lucia 2, 62100 Macerata, Italy

Elisabetta Aldieri

Department of Oncology, Biology and Biochemistry, Research Center on Experimental Medicine (CeRMS), University of Torino, Via Santena 5/bis, 10126 Torino, Italy

Fabio Polticelli

Dipartimento di Biologia, Università Roma Tre, V.le Marconi 446, 00146 Rome, Italy; Dipartimento di Biologia,Università Roma Tre, V.le Marconi 446, 00146 Rome, Italy and Istituto Nazionale di Fisica Nucleare, Sezione Rome Tre, 00146 Rome, Italy

Federica Sinibaldi

Dipartimento di Scienze Cliniche e Medicina Traslazionale, Università di Roma 'Tor Vergata', Via Montpellier 1, 00133 , Rome, Italy

George Perry

College of Sciences, University of Texas at San Antonio, San Antonio, Texas, USA

H.-Q. Yuan

Institute of Biochemistry and Molecular Biology, Medical College, Shandong University , Jinan, People's Republic of China

Hendrik Fuchs

Institut für Laboratoriumsmedizin, Klinische Chemie und Pathobiochemie, Charité – Universitätsmedizin Berlin, Campus Benjamin Franklin, Hindenburgdamm 30, D-12200 Berlin, Germany

Hirak Chakraborty

Department of Biochemistry and Biophysics, University of North Carolina at Chapel Hill, Chapel Hill, NC 27599, USA

Humberto Mestre

Facultad de Ciencias de la Salud, Universidad Anáhuac México Norte, Huixquilucan, Estado de México, Mexico

I. Savini

Department of Experimental Medicine & Surgery, University of Rome 'Tor Vergata', Italy

I.N. Papadoyannis

Laboratory of Analytical Chemistry, Department of Chemistry, Aristotle University of Thessaloniki, Thessaloniki, 541 24, Greece

Ivana Campia

Department of Oncology, Biology and Biochemistry, Research Center on Experimental Medicine (CeRMS), University of Torino, Via Santena 5/bis, 10126 Torino, Italy

J.-Y. Zhang

Institute of Biochemistry and Molecular Biology, Medical College, Shandong University , Jinan, People's Republic of China

Jakub Chlebek

Department of Pharmaceutical Botany and Ecology, ADINACO Research Group, Faculty of Pharmacy, Charles University, Hradec Kralove, Czech Republic

K.V. Donkena

Departments of Urology and Biochemistry and Molecular Biology, Mayo Clinic College of Medicine, Mayo Clinic, Rochester, Minnesota, USA

Kamil Kuča

Centre of Advanced Studies, Faculty of Military Health Sciences, University of Defence, Hradec Kralove, Czech Republic

Kateřina Macáková

Department of Pharmaceutical Botany and Ecology, ADINACO Research Group, Faculty of Pharmacy, Charles University, Hradec Kralove, Czech Republic

Kunihiro Tsuchida

Division for Therapies against Intractable Diseases, Institute for Comprehensive Medical Science (ICMS), Fujita Health University, Toyoake, Aichi 470-1192, Japan

Luciana Avigliano

Department of Experimental Medicine & Surgery, University of Rome 'Tor Vergata', Italy

Laura Fiorucci

Dipartimento di Scienze Cliniche e Medicina Traslazionale, Università di Roma 'Tor Vergata', Via Montpellier 1, 00133 , Rome, Italy

Lubomír Opletal

Department of Pharmaceutical Botany and Ecology, ADINACO Research Group, Faculty of Pharmacy, Charles University, Hradec Kralove, Czech Republic

Lucie Cahlíková

Department of Pharmaceutical Botany and Ecology, ADINACO Research Group, Faculty of Pharmacy, Charles University, Hradec Kralove, Czech Republic

M. Nunzi

S.C. Oncologia Medica, Azienda Ospedaliera, Via Tristano di Joannuccio 1, 05100 Terni, Italy

Maria Aparecida Nagai

Disciplina de Oncologia, Departamento de Radiologia e Oncologia da Faculdade de Medicina da Universidade de São Paulo, Av. Dr. Arnaldo, 455, 4°andar, CEP 01246-903, São Paulo, Brazil and Laboratório de Genética Molecular do Centro de Investigação Translacional em Oncologia, Av. Dr. Arnaldo, 251, 8 andar, CEP 01246-000, São Paulo, Brazil

M.K. Nika

Laboratory of Analytical Chemistry, Department of Chemistry, Aristotle University of Thessaloniki, Thessaloniki, 541 24, Greece

M.-L. He

Institute of Cancer Research, Life Science School, Tongji University, Shanghai, People's Republic of China

M.M. Brentani

Disciplina de Oncologia, Departamento de Radiologia e Oncologia da Faculdade de Medicina da Universidade de São Paulo, Av. Dr. Arnaldo, 455, 4°andar, CEP 01246-903, São Paulo, Brazil

M.V. Catani

Department of Experimental Medicine & Surgery, University of Rome 'Tor Vergata', Italy

Mariantonietta Colozza

S.C. Oncologia Medica, Azienda Ospedaliera, Via Tristano di Joannuccio 1, 05100 Terni, Italy

Mark Sutherland

University of Bradford, Institute of Cancer Therapeutics, Bradford, West Yorkshire, United Kingdom

Masako Yudasaka

National Institute of Advanced Industrial Science and Technology (AIST), Nanotube Research Center, Central 5, 1-1-1 Higashi, Tsukuba 305-8565, Japan

Matheus Froeyen

Laboratory for Medicinal Chemistry, Rega Institute for Medical Research, Katholieke Universiteit Leuven, Minderbroedersstraat 10, B-3000 Leuven, Belgium

Miguel Ángel Cabrera-Pérez

Molecular Simulation & Drug Design Group, Centro de BioactivosQuímicos, Universidad Central de Las Villas, Santa Clara, 54830, Villa Clara, Cuba, USA and Engineering Department, Pharmacy and Pharmaceutical Technology Area, Faculty of Pharmacy, University Miguel Hermandez, Plicante D3550, Spain

Munna Sarkar

Chemical Sciences Division, Biophysics Division, Saha Institute of Nuclear Physics, 1/AF Bidhannagar, Kolkata 700 064, India

Noriyuki Nagahara

Department of Environmental Medicine, Nippon Medical School, 1-1-5 Sendagi Bunkyo-ku, Tokyo 113-8602, Japan

P. Dinh

Medical Oncology Clinic, Jules Bordet Institute, Boulevard de Waterloo, 125, 1000 Brussels, Belgium

Pukhrambam Grihanjali Devi

Department of Chemistry, Imphal College, Imphal, , India

R. Califano

Department of Medical Oncology, The Christie NHS Foundation Trust, Wilmslow Road, Manchester, M20 4BX, UK

Ricardo Balanza

Facultad de Ciencias de la Salud, Universidad Anáhuac México Norte, Huixquilucan, Estado de México, Mexico

Roberto Santucci

Dipartimento di Scienze Cliniche e Medicina Traslazionale, Università di Roma 'Tor Vergata', Via Montpellier 1, 00133 , Rome, Italy

S. Sabatini

S.C. Oncologia Medica, Azienda Ospedaliera, Via Tristano di Joannuccio 1, 05100 Terni, Italy

Silke Bachran

Institut für Laboratoriumsmedizin, Klinische Chemie und Pathobiochemie, Charité – Universitätsmedizin Berlin, Campus Benjamin Franklin, Hindenburgdamm 30, D-12200 Berlin, Germany

Sophie Doublier

Department of Oncology, Biology and Biochemistry, Research Center on Experimental Medicine (CeRMS), University of Torino, Via Santena 5/bis, 10126 Torino, Italy

Sumio Iijima

National Institute of Advanced Industrial Science and Technology (AIST), Nanotube Research Center, Central 5, 1-1-1 Higashi, Tsukuba 305-8565, Japan

Tatsuya Murakami

Division for Therapies against Intractable Diseases, Institute for Comprehensive Medical Science (ICMS), Fujita Health University, Toyoake, Aichi 470-1192, Japan

V.F. Samanidou

Laboratory of Analytical Chemistry, Department of Chemistry, Aristotle University of Thessaloniki, Thessaloniki, 541 24, Greece

Vít Kolečkář

Centre of Advanced Studies, Faculty of Military Health Sciences, University of Defence, Hradec Kralove, Czech Republic

Yunierkis Pérez-Castillo

Molecular Simulation & Drug Design Group. Centro de Bioactivos Químicos. Universidad Central de Las Villas, Santa Clara, 54830, Villa Clara, Cuba; Laboratory for Medicinal Chemistry, Rega Institute for Medical Research, Katholieke Universiteit Leuven, Minderbroedersstraat 10, B-3000 Leuven, Belgium and Computational Modeling Lab (CoMo), Department of Computer Sciences, Faculty of Sciences, VrijeUniversiteit Brussel, Pleinlaan 2, B-1050 Brussel, Belgium

CHAPTER 1

New Functions of Old Drugs: Aureolic Acid Group of Anti-Cancer Antibiotics and Non-Steroidal Anti-Inflammatory Drugs

Hirak Chakraborty[1,§], Pukhrambam Grihanjali Devi[2,§], Munna Sarkar[3,*] and Dipak Dasgupta[3,*]

[1]Department of Biochemistry and Biophysics, University of North Carolina at Chapel Hill, Chapel Hill, NC 27599 USA; [2]Department of Chemistry, Imphal College, Imphal, India and [3]Chemical Sciences Division, Biophysics Division, Saha Institute of Nuclear Physics, 1/AF, Block-AF, Bidhannagar, Kolkata-700 064, India

Abstract: Non-steroidal anti-inflammatory drugs and aureolic acid group of anti-cancer drugs belong to the class of generic drugs. Research with some members of these two groups of drugs in different laboratories has unveiled functions other than those for which they were primarily developed as drugs. Here we have reviewed the molecular mechanism behind the multiple functions of these drugs that might lead to employ them for treatment of diseases in addition to those they are presently employed. The distinct advantage of using old drugs for alternate functions lies in their well-studied Absorption Distribution Metabolism Excretion and Toxicity (ADMET) profile.

Keywords: Alternate functions, alzheimer disease, anti-cancer, aureolic acid group, COX dependent and COX independent pathways, Drug repositioning, metal chelation, non steroidal anti-inflammatory drugs, painkillers.

1. INTRODUCTION

'Old drugs' that have been in the market for long constitutes a large pool of compounds available for further research. Many of these drugs have outlived their patents, allowing them to be legally produced as generic drugs. Generic drugs

***Address correspondence to Munna Sarkar and Dipak Dasgupta:** Chemical Sciences Division, Biophysics Division, Saha Institute of Nuclear Physics, 1/AF Bidhannagar, Kolkata 700 064, India; Tel: +91 33 23375345; E-mail: munna.sarkar@saha.ac.in and dipak.dasgupta@saha.ac.in
[§]Both authors contributed equally in writing the chapter

have the same chemical ingredients as their brand name counterpart and show the same benefits and risks. Since they are 'off patented' the cost of production is low keeping their pricing much cheaper than their brand name counterparts. Many of these drugs in the market show unconventional functions, which are quite distinct from the function other than the intended one. Understanding the molecular mechanism behind these unconventional functions would allow the utilization of these 'old drugs' for new disease targets. Recently, phenotype and molecular target based screening of generic drugs against multiple targets have become an important strategy in accelerating rational drug design [1]. On the other hand, repositioning approach typically uses an interesting side effect of an approved medication to develop it for its new function [2]. An example is that of 'Propecia', a drug now used in the treatment of hair loss was developed to act against benign enlarged prostate gland. The advantage of using old medication for novel application is that their doses, *in vivo* pharmacokinetics and ADMET (Adsorption, Distribution, Metabolism. Excretion, Toxicity) profiles are well studied since they have passed the necessary clinical trials for their conventional use. Another conceptually interesting approach, aimed at reducing side effects of a drug, is based on targeted drug delivery using tumor specific peptides capable of translocating drugs across cell membranes [3]. This allows better internalization of the drugs allowing delivery of the dosage required for tumor elimination. Drug resistance in cancer therapy is a common problem. Recently, drug resistance against common cancer drug cisplatin, has been overcome using gene therapy. Infection with a recombinant adenovirus expressing the human retinoblastoma tumor suppressor gene is sufficient to impart lethality in tumor cells in absence of cisplatin by triggering cell cycle arrest in the G1 phase [4]. Even though rapid screening against multiple targets allows identification of novel hits, understanding the molecular mechanism behind the multiple functions of generic drugs might lead to a better usage of these drugs. The approach helps to develop a new class of drugs based on the chemical platform of old drugs but aimed at a specific function. In this chapter we will discuss the multiple functions of two classes of generic drugs, *viz.* synthetically produced non-steroidal anti-inflammatory drugs (NSAIDs) and aureolic acid group of anti-cancer antibiotic obtained from bacterial sources.

The conventional use of NSAIDs is to control pain and inflammation. This class of drugs has been in the market for a very long time with aspirin being the oldest drug that was marketed more than hundred years back. The principal targets for their conventional function are cyclooxygenases (COX) enzymes, which are membrane-associated proteins. There are two isoforms of COX, *viz*. COX-1 and COX-2 [5]. Over the past one decade research have shown that these NSAIDs can have several other functions which include chemoprevention [6, 7] and chemosuppression [8, 9] against several types of cancers, protection against neurodegenerative diseases like Alzheimer disease (AD) [10-12], UV-sensitizer [13-15], UV-protector [16, 17] *etc*. Till date, there seems to be no general consensus as to the exact mechanism behind these novel functions. Several targets have been implicated in almost all the different functions. Not all NSAIDs show same kind of behavior towards a specific function and a great variation exist in the extent of their efficacies. Since NSAIDs encompasses several chemical motifs, a closer look at the chemical basis of their novel function could be a good approach for future drug discovery/designing.

The aureolic acid anticancer antibiotics, chromomycin A_3 (CHR) and mithramycin (MTR), (Fig. (**1**)) are the two naturally occurring antibiotics first reported from *Streptomyces gresius*, and *Streptomyces plicatus*, respectively, in the 1950s and 1960s [18]. Since then a number of structurally related antibiotics have been reported from different bacterial sources. They are glycosylated aromatic polyketides with an intense yellow color and fluoresce under UV light, which is the genesis for the name of the family. With the exception of chromocyclomycin, which is a tetracyclic compound, the aglycons of this family consist of a tricyclic ring system fused to a unique dihydroxy-methoxy-oxo-pentyl aliphatic side chain attached at C-3.

In some compounds, a small alkyl residue (methyl, isobutyl) is attached at position C-7. Each consists of the chromomycinone moiety, the aglycon ring either side of which is linked to six-membered sugar residues *via* O-glycoside linkages [19]. It is interesting to note that anionic MTR (pH 8.0) undergoes concentration dependent self-association, whereas neither the neutral form (pH 3.5) nor the dimer complex with Mg^{2+} aggregates under similar condition [20].

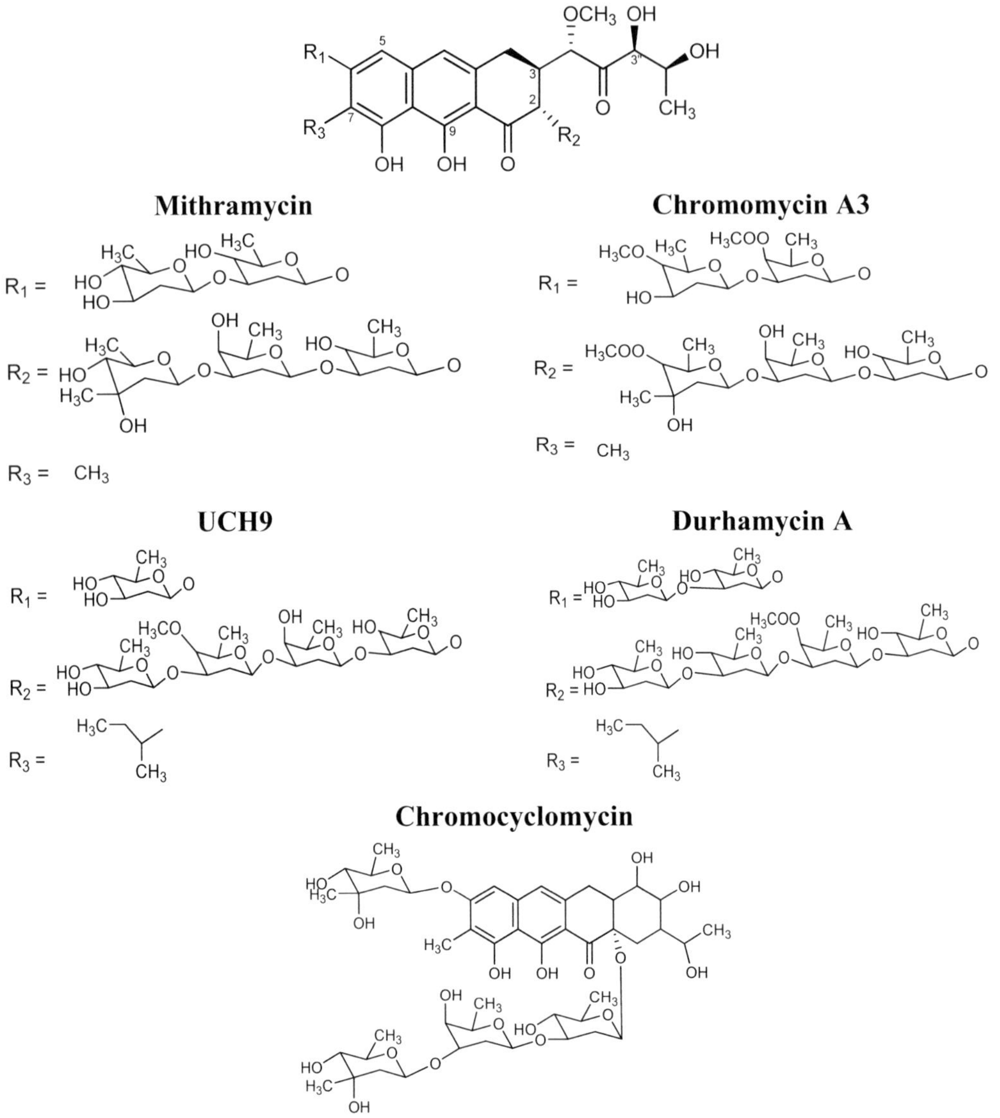

Figure 1: Structures of aureolic acid antibiotics.

Aureolic acids, MTR and CHR, were initially meant to be used for their antibiotic activity against gram-positive bacteria but now have clinical applications because of its anti-cancer property. Earlier studies with the antibiotics have proposed that they act by inhibiting DNA-dependent RNA synthesis both *in vivo* and *in vitro via* reversible interaction with (G.C)-rich DNA [18, 19, 21, 22] in the presence of bivalent metal ion, like Mg^{2+}. They inhibit the expression of genes with (G.C) rich promoter. Even though their role as anticancer drugs has been studied well, there has been resurgence in the study of these antibiotics to examine their therapeutic potential for the treatment of human disorders other than cancer. Extensive work of isolation and characterization has been done on the biosynthesis gene clusters of the two antibiotics [23, 24]. The knowledge about the biosynthetic pathway of the antibiotics along with the identification of the associated gene cluster have opened the prospect of employing genetically modified structural analogues for therapeutic purpose. Thus, engineering of the MTM biosynthetic pathway has produced thc 3-sidc-chainmodificd analogs MTM SK (SK) and MTM SDK (SDK), with enhanced anticancer activity and improved therapeutic index. Major limitations of therapy with mithramycin are low bioavailability, short plasma retention time, and low tumor accumulation. Keeping in view of these shortcomings, a recent study of two nanoparticulate formulations, poly(ethylene glycol)-poly(aspartate hydrazide) self-assembled and cross-linked micelles, were currently reported for investigations with regard to the ability to load and pH dependently release of the antibiotic [25].

In this chapter we shall briefly highlight the alternate functions of NSAIDs as well as that of aureolic acid anticancer antibiotics. An approach to understand the chemical basis of unconventional use of these drugs will also be discussed. It should be mentioned that NSAIDs encompasses several chemical motifs, whereas the principal chemical motif of the aureolic acid anti-cancer antibiotics are same with small changes in the sugar moieties. It is therefore expected that for the NSAIDs, the chemical structure should play a more determining role in the manifestation of their new functions. However, as will be shown later, even small changes in the sugar moieties of the aureolic acid antibiotics can affect the extent of their various functions.

2. PRINCIPAL FUNCTIONS OF NSAIDS AND AUREOLIC ACID ANTICANCER ANTIBIOTICS AND THEIR MECHANISM OF ACTION

Non-steroidal anti-inflammatory drugs (NSAIDs) have been commonly used to reduce pain and inflammation in different arthritic and post-operative conditions [5, 26]. NSAIDs are also used as anti-pyretic, analgesic and uricosuric agents [27]. Their anti-inflammatory effect is mainly due to their ability to inhibit the activities of cyclooxygenases (COX) enzymes those mediate the production of prostaglandins from arachidonic acid, which is a dietary fatty acid (Fig. (**2a**)).

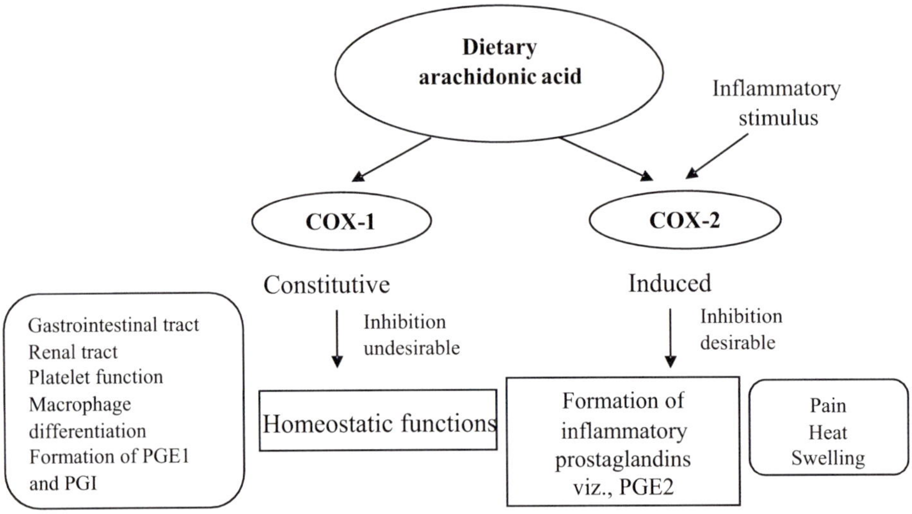

Figure 2(a): Mechanism of action of NSAIDs: The Cyclooxygenase Pathway functions of the human body and is not desirable. Designing COX-2 specific inhibitors is an important strategy in controlling inflammatory processes.

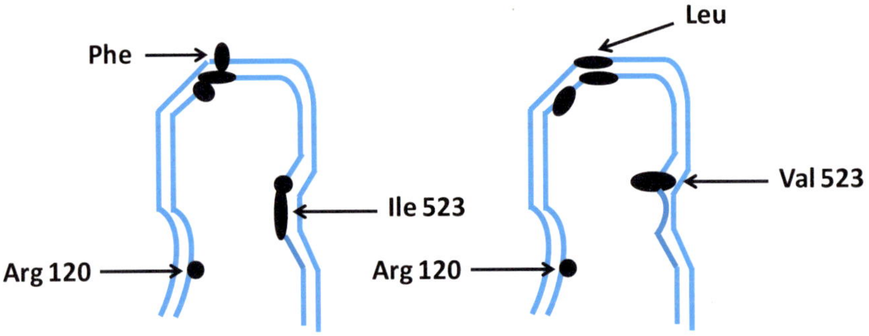

Figure 2(b): Schematic Diagram of COX-1 and COX-2.

There are two isoforms of cyclooxygenases *viz.*, cyclooxygenase-1 (COX-1) and cyclooxygenase-2 (COX-2) [5, 28].

Prostaglandins are powerful signaling agents in the human body. Some prostaglandins, mainly synthesized by the COX-2 isoform are substantially involved in bringing about and maintaining inflammatory processes by increasing vascular permeabilities and amplifying the effects of other inflammatory mediators such as kinins, serotonin and histamin [29]. Hence reducing and controlling the formation of these prostaglandins can reduce the swelling, heat and pain of inflammation. However, not all prostaglandins are harmful for the human body. Some of them, synthesized by COX-1, are important in protecting the stomach lining, promoting clotting of blood, regulating salt and fluid balance maintaining blood flow to the kidneys *etc.* [30]. Hence inhibition of COX-1 will lead to loss of many prostaglandins important for the homeostatic functions of the human body and is not desirable. Designing COX-2 specific inhibitors is an important strategy in controlling inflammatory processes.

X-ray crystallography of the 3-D structures of COX-1 and COX-2 has done much to illuminate how COX-2 specific NSAIDs can be designed. The two isoforms, COX-1 and COX-2 have very similar 3-D structures consisting of a long narrow channel with a hairpin bend at the end [5]. The isoforms are membrane-associated so arachidonic acid released from damaged membranes adjacent to the opening of the enzyme channel, which is largely hydrophobic, is sucked in, twisted around the hairpin bend, two oxygens are inserted and a free radical extracted, resulting in the five-carbon ring that characterizes prostaglandins [5]. From fluorescence quenching study of arachidonic acid it is found that older NSAIDs block both COX-1 and COX-2 about halfway down the channel. X-ray crystallography suggested that this blocking occurs by hydrogen bonding to the polar arginine at position 120 (Fig. (**2b**)) leading to non-specific inhibition of the two isoforms.

A single amino acid difference is critical for the selectivity of many drugs. At position 523 there is an isoleucine molecule in COX-1 and a valine (smaller by a single methyl group) in COX-2. The smaller valine molecule in COX-2 leaves a gap in the wall of the channel (Fig. (**2b**)), giving access to a side pocket, which is thought to be the site of binding of many COX-2 selective drugs. The bulkier isoleucine at 523 in COX-1 is large enough to block access to the side pocket. So targeted single amino acid substitution of valine for isoleucine is sufficient to turn COX-1 into a protein that can be inhibited by COX-2 selective inhibitors [31, 32]. Various NSAIDs have been designed with differential specificity towards COX-1 and COX-2 using different chemical templates. NSAIDs can be classified according to their chemical structure.

Different chemical templates that are being used as NSAIDs have differential efficiency against COX-1 and COX-2. For a particular group of NSAIDs, different drugs are synthesized by the method of isosteric substitution considering a particular drug as the mother template [32]. For example, in the oxicam group of NSAIDs, piroxicam is the mother compound and meloxicam, tenoxicam, lornoxicam *etc.* are synthesized by small changes in the piroxicam chemical template. Generally NSAIDs are broadly divided into six major classes as shown in Table **1**. Of these, aspirin, belonging to the salicylic acid group is the oldest NSAID in the market. NSAIDs, designed after the discovery of the structural differences between COX-1 and COX-2 isoforms, show better selectivity towards COX-2. Coxibs (rofecoxib, celecoxib) and NS-398 are highly selective towards COX-2 whereas flurbiprofen, ketoprofen *etc.* show high level of selectivity towards COX-1. Dichlofenac, etodolac, meloxicam, nimesulide *etc.* are relatively COX-2 selective where as aspirin, ibuprofen, indomethacin have equal affinity towards COX-1 and COX-2. Though it was assumed that COX-2 inhibition is the most effective pathway in controlling pain and inflammation but there are several other functions of COX-2, which are important for homeostasis in health. COX-2 is expressed constitutively in kidney particularly in macula densa, cyclical induction of COX-2 has an important role in ovulation, uterine COX-2 induced at the end of pregnancy, where it is important for the onset of labor and COX-2 inhibitors cause fluid retention [5]. It is because of these homeostatic functions of COX-2, COX-2 specific inhibitors *viz.*, rofecoxib and celecoxib show side effects leading to myocardial infarction and hence rofecoxib has been withdrawn from the market. Hence use of COX-2 specific NSAIDs may land up into more complex situation of side effects than gastrointestinal ulceration caused by older non-specific COX-inhibitors. So in controlling pain and inflammation the preferential COX-1 inhibitors or older NSAIDs may be much more effective because their side effects are well studied and can be managed by using combination drug treatment.

Despite the above problems it may not be wise to discard the COX-2 selective NSAIDs because they have immense potential against various diseases other than to control pain and inflammation. In the past few years it has been shown that COX-2 is a key player in many biochemical processes like apoptosis, angiogenesis, amyloidoses, *etc.*, which will be discussed in the subsequent sections of this chapter. So modification may be made using the present day

COX-2 inhibitors as the main chemical templates to develop drugs against various diseases other than their traditional use, which will be more economic and time saving.

As has been mentioned before, the aureolic acid anticancer antibiotics chromomycin A_3 (CHR) and mithramycin (MTR) (Fig. (**1**)) act *via* inhibition of DNA-dependent RNA synthesis both *in vivo* and *in vitro*. The presence of bivalent metal ion, like Mg^{2+}, is an obligatory factor for the transcription inhibitory property at physiological pH. The anionic antibiotic binds to the metal ion [33, 34] and the resulting complex(es) is(are) the DNA binding ligand(s) at and above physiological pH. They bind to DNA *via* minor groove [35-37]. It was established in our laboratory from spectroscopic and thermodynamic studies that the modes of binding of the two ligands with natural DNA, polynucleotides and oligomeric duplexes are different [33-37]. We also illustrated the role of DNA minor groove size and the accessibility of the 2-amino group in the minor groove of guanosine in drug-DNA interaction using designed nucleotide sequences [37-39]. Detailed NMR studies from other laboratories have helped to understand how the bulky complex of the type $[(drug)_2Mg^{2+}]$ is accommodated at the cost of a considerable widening of the minor groove in B-DNA type structure [40, 41]. In our laboratory we have shown from a detailed thermodynamic analysis of the association of the dimer complex with different DNAs, natural and synthetic, with defined sequences that B to A type transition in the groove leads to a positive change in enthalpy. This is compensated by a positive change in entropy arising from the release of bound water in the minor groove. Sugars present in the antibiotics play a significant role during the association with nucleic acids [35-39]. Absence of substituents like acetoxy group in the sugar moieties of mithramycin imparts conformational flexibility to greater degree than chromomycin. Therefore, the drug dimer of mithramycin has been found to have a better conformational plasticity than chromomycin when it binds to the minor groove of DNA. Its strong antitumor activity against a number of cancer cell lines has been ascribed to the DNA-binding property drug-metal complex(es).

As these antibiotics specifically bind to GC-rich regions, they are employed as strong inhibitors of specific promoter regions like *c-myc* [42] and *c-src* [43] thus preventing the association of regulatory proteins and transcription factors like Sp1 and the resulting formation of the transcription initiation complex. It has an adverse effect upon the general processes such as transcription elongation. DNase I foot printing has identified that MTR binds to the P1 and P2 promoter regions of the *c-myc* gene [44]. This also explains their antiviral property of deactivating the HIV-I provirus [45]. Treatment of different cancer cell lines with MTR was found to facilitate different apoptotic pathways such as TNF [46], tumor necrosis factor-alpha-related apoptosis-inducing ligand (TRAIL) [47] and Fas [48]. These properties give MTR (trade name Plicamycin) its clinical application as an FDA approved drug. MTR is clinically employed for the treatment of neoplastic diseases like chronic myelogenous leukemia, testicular carcinoma and Paget's disease [18].

Table 1: Different chemical groups of NSAIDs

Class	Example	Template
Acetyl salicylic acid	Aspirin	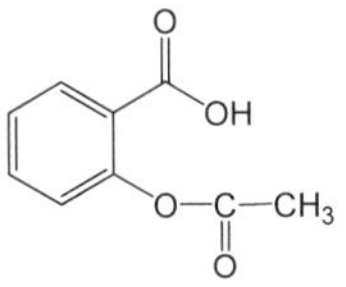 Aspirin
Acetic acid	Diclofenac, Indomethacin Ketorolac, Nabumetone, Sulindac, Tolmetin	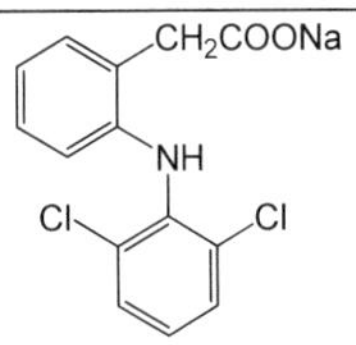 Diclofenac sodium salt
Fenamates	Meclofenamate, Mefenamic acid	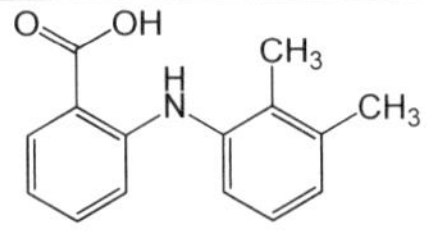 Mefenamic acid
Oxicam	Piroxicam, Meloxicam Tenoxicam, Lornoxicam	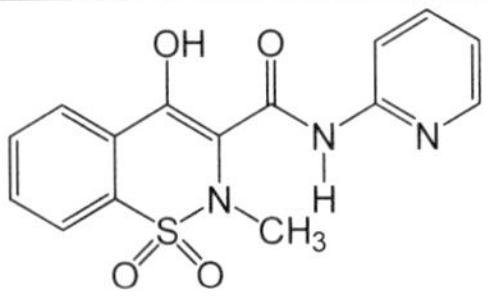 Piroxicam
Propionic acid	Ibuprofen, Ketoprofen, Naproxane, Oxaprozin	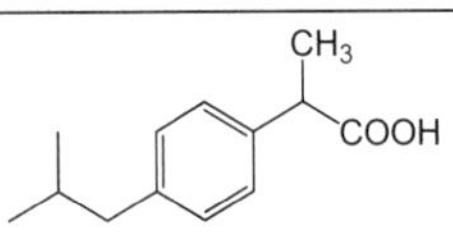 Ibuprofen

Table 1: contd...

Coxib	Celecoxib, Rofecoxib Valdecoxib	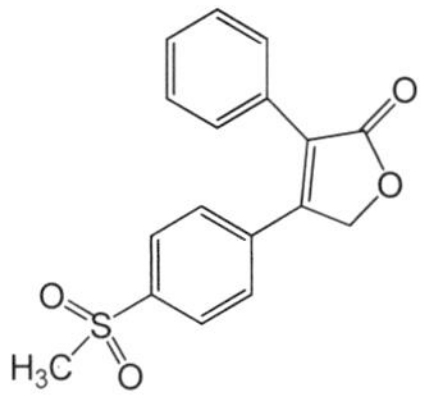 Rofecoxib

Since the above promoters are part of the chromatin, we have studied the effects of these drugs upon the chromatin structure [49-53]. Spectroscopic studies such as absorbance, fluorescence and CD have demonstrated directly the association of the above complexes with chromatin and its components under different conditions [49, 53]. The reduced binding affinity of the antibiotic: Mg^{2+} complexes to nucleosome or chromatin might be a consequence of bending of double helix or, additionally, of unusual DNA conformations induced by the histone binding [51, 52]. Presence of histones might also reduce the accessibility of the minor groove to this class of groove binders. Alternatively, one can say that histone-DNA contacts and N-terminal tail domains of individual core proteins in nucleosome core particle reduce the accessibility of nucleosomal DNA to antibiotic: Mg^{2+} complexes [52]. In the chromatin, presence of linker H1 further reduces the binding potential of the ligand. These drugs also induce instability in nucleosome leading to DNA release.

These antibiotics also show other functions quite diverse in nature. They are briefly described below.

3. OTHER FUNCTIONS OF THESE DRUGS AND MECHANISM OF ACTIONS

3.1. Other Functions of NSAIDs

3.1.1. Chemoprevention and Chemosuppression

Numerous experimental, epidemiological and clinical studies suggest that non-steroidal anti-inflammatory drugs (NSAIDs) have a great potential as anticancer agents [6, 8, 54-59]. Again nonrandomized epidemiological studies have found that long-term users of aspirin or other NSAIDs have a lower risk of colorectal adenomatous polyposis and colorectal cancer than non-users [60]. There exists a wealth of data which show that NSAIDs have both chemoprevention and chemosuppression ability. Two schools of thoughts exist regarding the molecular mechanism of these drugs as chemopreventive and chemosuppressive agent. There is enough literature, which supports the COX-dependent mechanism whereas many data exist in literature, which opposes the COX-dependent pathway. We will present some key studies that demonstrate both schools of thoughts.

3.1.1.1. COX-Dependent Pathway

Apart from producing different prostaglandins from arachidonic acid multiple lines of compelling evidences support that COX-2 plays a crucial role in carcinogenesis (Scheme **1**). Molecular studies on the relationship between poly-unsaturated fatty acid metabolism and carcinogenesis have revealed novel molecular targets for cancer prevention and treatment [61, 62]. Literature data exists that show over-expression of COX-2 in tumor cells of colon carcinoma [63], colorectal carcinoma [64], esophagaeal carcinoma [65], pancreatic carcinoma [66], malignancies of breast, skin, cervix, ovary, bladder, head, neck, *etc.* [67]. In addition to the finding that COX-2 is commonly over-expressed in premalignant and malignant tissues, there exists considerable evidence that links COX-2 to the development of cancer. The most specific data that support a cause-and-effect connection between COX-2 and tumorogenesis come from genetic studies. Multiparous female transgenic mice are engineered to overexpress human COX-2 in mammary glands, develop focal mammary gland hyperplasia, dysplasia and metastatic tumors [68]. These findings are consistent with the idea that under some conditions, increased expression of COX-2 induces tumor formation. In a related study, transgenic mice that overexpress COX-2 in skin develop epidermal hyperplasia and dysplasia. Consistent with these studies, knocking out COX-2 markedly reduces the development of intestinal tumors and skin papillomas.

Colon cancers are thought to arise as the result of a series of histopathlogic and molecular changes that transform normal colonic epithelial cells into a colorectal carcinoma, with an adenomatous polyp as an intermediate step in the process.

Analysis of COX-2 expression shows that it is elevated in up to 90% of sporadic colon carcinomas and 40% of colonic adenomas but is not elevated in the normal

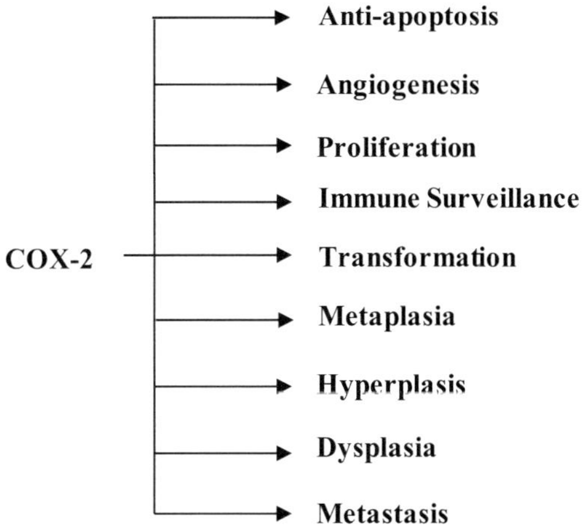

Scheme 1: Role of COX-2 in different carcinogenesis.

colonic epithelial cells [69]. Increased level of COX-2, prostaglandins or both are found in adenomas of the patients with familial adenomatous polyposis (FAP) and in experimentally induced colon tumors in rodent models [70]. COX-2 overexpression in normal alveolar type II cells may be directly involved in increasing the sensitivity of these cells to the effects of carcinogens and enhancing tumor development after initiation [71]. The Ras/ERK signaling pathway appears to play a role in the regulation of COX-2 expression. Human non-small cell lung cancer cell lines with mutations in Ki-Ras have high expression levels of COX-2, and inhibition of Ras activity in these cell lines decreases COX-2 expression [72]. It has also been found that gall bladder cancer cell growth can be stimulated by mitogens and that mitogens can decrease apoptosis. A specific COX-2 inhibitor can decrease the mitogenic stimulus, and the COX-2 inhibition can increase apoptosis. The decreased mitogenesis and increased apoptosis produced by the COX-2 inhibitors were associated with decreased PGE_2 formation [6]. Eberhart *et al.* [73] and others demonstrated that COX-2 enzyme is overexpressed in human colorectal tumors compared with adjacent normal colonic mucosa [73, 74]. Employing a specific COX-2 inhibitor, NS-398, Tsuji *et al.* [75] demonstrated that proliferation of a gastric cancer cell lines and a colon cancer cell lines could be inhibited by the COX-2 enzyme inhibitor [75, 76]. Others have subsequently demonstrated that in a variety of epithelial cell lines, specific COX-2 inhibitors decrease mitogenesis [77, 78]. The studies in animal models also confirm that the polyposis is directly related to the COX-2 expression, Mice containing mutations in both the adenomatous polyposis coli (APC) gene and the COX-2 gene developed fewer intestinal polyps than mice with a functional COX-2 gene [79]. A new selective COX-2 inhibitor, SC58125, suppressed tumor growth *in vivo* and induced apoptosis *in vitro* in cell lines that express high levels of COX-2, but the COX-2 inhibitor was ineffective in HCT-116 cells, which have undetectable level of COX-2 expression [78], These studies place COX-2 in a central position of colon carcinogenesis and suggest that selective COX-2 inhibition may be a useful approach for chemoprevention or even treatment of the cancers, particularly those with high levels of COX-2 expression.

There are evidences that COX-2 inhibitors can also act as anti-angiogenic agents. The link between the COX-2 activity and vascular endothelial growth factor (VEGF) production and action has been established. The disruption of COX-2 gene in mice dramatically suppressed VEGF production in fibroblasts [80] and tumor cells [81]. Again COX-2 inhibitors prevented VEGF-induced MAPK activation in endothelial cells. Ruegg *et al.* [82] have established a link between COX-2 and integrin αVβ3-mediated endothelial cell migration and angiogenesis [82-84]. Inhibition of COX-2 activity in endothelial cells by NSAIDs suppressed αVβ3-dependent endothelial cell spreading and migration *in vitro* and FGF-2 induced angiogenesis *in vivo* [85-87]. Exogenous PGE_2 rescued endothelial cell spreading and migration in the presence of COX-2 inhibitors [83, 88]. The effect of NSAIDs was due to the inhibition of αVβ3-dependent activation of Cdc42 and Rac, two members of Rho family of GTPases that regulate cytoskeletal organization and cell migration. Besides promoting Rac activation and cell spreading, the COX-2 metabolite PGE_2 also accelerates αVβ3-mediated endothelial cell adhesion [88]. The important role of Rac in angiogenesis was also demonstrated by VEGF-required Rac activation [84, 89] and the inhibition of the Rac effector p21-activated kinase (PAK)-1, suppressed endothelial cell tube formation *in vitro* and angiogenesis in the chick CAM assay *in vivo* [90, 91].

So the multiple lines of evidences indicate that COX-2 is an important pharmacological target for anti-cancer therapy. Epidemiological studies show that use of NSAIDs, prototypic inhibitors of COX-2, is associated with a reduced risk of several malignancies, including colorectal cancers. Consistent with this, tumor formation and growth are reduced in animals that are engineered to be COX-2 deficient or treated with a selective COX-2 inhibitor. In the clinical trial, it has been found that treatment with celecoxib, a selective COX-2 inhibitor, reduced the number of colorectal polyps in patients with familial adenomas polyposis (FAP) [92, 93]. Based on these findings many clinical trials are under way to assess the potential efficacy of selective COX-2 inhibitors in preventing and treating human cancers.

3.1.1.2. COX-Independent Pathway

In the previous section we have presented evidences of COX-2 as being an important target for the chemopreventive and chemosuppressive functions of NSAIDs. However, the precise mechanisms by which various NSAIDs exert their antiproliferative effects on cancer cells are still controversial. Emerging evidences suggest that these effects can, at least in some cases, be exerted through COX-2 independent pathways. In this section of the chapter, we will discuss the recent progress in understanding the different COX independent pathways that lead to chemoprevention and chemosuppression by the NSAIDs.

Several independent studies have shown that various NSAIDs can show apoptotic effect in cell lines irrespective of their level of expression of COX-1 and COX-2. For example, indomethacin, a non selective COX-inhibitor, induced apoptosis in both Seg-1 (COX-1/2 positive) and Flo-1 (COX-1/2 negative) esophageal adenocarcinoma cells [94]. Sulindac sulfide and sulindac sulfone induced apoptosis in malignant melanoma cell lines independent of COX-2 expression [95]. Using cell lines with controlled COX-2 expression, they were unable to detect any differences between COX-2 expressing and COX-2 deficient Caco-2 cell clones in the ability of celecoxib to inhibit the cell cycle [96]. Combination of statins and NSAIDs has been proposed to produce synergistic effect in their role in chemoprevention. In colon cancer cell lines HCT116 and HT29, combined action of Atrovastatin and celecoxib in inducing apoptosis is much more than seen in case when the drugs are treated individually [97]. Indomethacin and NS398 had antiproliferative activity on both COX-2 positive cell line (HT29 and HCA7) and COX-2 negative cell line (SW480 and HCT116) [98]. Sulindac sulfide and piroxicam induced apoptosis in both COX-2 expressing HT29 human colon cancer cell lines and COX-2 deficient HCT15 cells [99]. Furthermore, though the COX-2 inhibiting ability of rofecoxib and celecoxib is similar, but celecoxib has a much higher antiproliferative activity in COX-2 positive A549 epithelial cells and COX-2 negative BALL1 hematopoietic cells than rofecoxib [100]. NS398, a COX-2 selective inhibitor, induced apoptosis in HT29 (COX-2 positive) and S/KS (COX negative) human colorectal carcinoma cell lines with comparable IC_{50} [101]. In addition to these studies with a spectrum of cancer cell lines it has also been demonstrated that cells genetically engineered to lack expression of COX-1 and COX-2 or both can remain sensitive to the antiproliferative effects of NSAIDs indicating that NSAIDs can bypass COX to exert their anti-cancer effect.

The heterozygote Min/+ mouse model, like patients with FAP, carries a nonsense mutation in the APC gene that results in the spontaneous development of intestinal adenomas (100% incidence). Administration of sulindac to Min/+ mice reduced the tumor number but did not alter the level of PGE_2 and leukotriene B_4 in intestinal tissues [102]. Furthermore, increasing PGE_2 and interleukine B_4 levels with dietary arachidonic acid supplementation had no effect on tumor number or size [102]. Similarly, when PGE_2 is given to rats concomitantly with indomethacin does not reverse the tumor reducing effect of indomethacin in these animals [103]. In support, it has been shown that celecoxib has an antitumorigenic effect in COX-2 deficient tumors in the nude mice model and also induces apoptosis in the cells, which do not express COX-2 [104]. Again some NSAID derivatives that do not inhibit COX activity retain their chemopreventive activity in the Min/+ mouse model of intestinal polyposis [105]. R-flurbiprofen induces cell cycle blocking and apoptosis in human colon carcinoma cell lines HCT116 by activating C-Jun-N-terminal Kinase (JNK) and down-regulating cyclin D1 expression [106]. Sulindac sulfone, the oxidative metabolite of sulindac, is completely devoid of COX-inhibitory activity but inhibits growth and induces apoptosis in variety of human cancer derived cell lines [107].

Additional studies indicate that sulindac sulfone and its derivatives CP248 and CP461, which activate PKG, lead to rapid and sustained activation of JNK1, a kinase known to play a role in the induction of apoptosis by other cellular stress related events. Mechanistic studies indicate the existence of a novel PKG-

MEKK1-SEK1-JNK1 pathway for the induction of apoptosis by sulindac sulfone [108]. Once activated JNK1 plays a role in apoptotic signaling pathways, JNK1 can phosphorylate and inactivate the anti-apoptotic proteins Bcl-2 and Bcl-XL [29] and it can include the expression of pro-apoptotic proteins (Bad and Bim) through activation of the transcription factor AP-1. MEK/ERK signaling may regulate mitochondrial events that lead to activation of caspases.

Akt plays a key role in tumorigenesis and cancer progression by stimulating cell proliferation and inhibiting apoptosis [109]. The Akt is composed of a -NH_2 terminal plackstrin homology domain and a –COOH terminal kinase catalytic domain. It is activated by a dual regulatory mechanism that requires both translocation to the plasma membrane and phosphorylation. Recently, Wu *et al.* have demonstrated that celecoxib regulates the phosphorylation of Akt and inhibited PDK1 and PTEN phosphorylation in cholangiocarcinoma cells [110]. The anti-sense depletion of COX-2 failed to alter the level of phospho-Akt, which indicates the existence of COX-2 independent effect. This result is supported by the studies from other investigations showing that celecoxib induces apoptosis *via* COX-2 independent mechanism in other human cancer cell lines. In a separate study Lai *et al.* [111] showed that in suppression of rat cholangiocarcinoma (cultured C611B cells) and neu-transformed WB344 rat liver epithelial stem-like cells (WBneu cells), concentration of celecoxib needed to suppress growth and induce apoptosis was markedly higher than that needed for effective inhibition of PG production by these malignant cell types. Studies also show that celecoxib reduces neointimal hyperplasia after angioplasty through inhibition of Akt signaling in a COX-independent manner. Results suggested that celecoxib affects the Akt/GSK signaling axis, leading to vascular smooth muscle cells (VSMC) proliferation and an increase in VSMC apoptosis [112]. Several reviews have been devoted to highlight the detail effect of different NSAID and/or COX-2 inhibitors on different cancers. Different COX-independent targets have been indicated that might play a crucial role for the NSAID to exert their anticancer effect. There are many established COX-independent pathways, which includes cell surface death receptor-mediated pathway [113, 114]. This pathway is initiated by extracellular hormones or agonists that belong to the tumor necrosis factor (TNF) super family including TNFα, Fas/CD95 ligand and Apo2 ligand. These agonists recognize and activate their corresponding receptors, members of TNF/NGF receptor family, such as TNFR1, Fas/CD95 and Apo2. Another important target in the COX-independent pathway is nuclear factor kappa B (NF-κB) [107]. In vertebrates Rel/NF-κB homodimers and heterodimers bind to DNA target sites, collectively called κB sites and directly regulate gene transcription. Many NSAIDs inhibit the NF-κB followed by induction of apoptosis [115]. NSAIDs like salicylate was shown to inhibit the activation of p70S6 kinase, which results the down regulation of c-myc, cyclin D1, cyclin A and might contribute to salicylate induced growth arrest [116]. NS-398 and piroxicam block JNK phosphorylation and inhibit AP-1 activity, resulting in induction of apoptosis [117]. Peroxisome proliferation-activated receptors α, γ and δ (PPAR α,γ and δ) are members of a class of nuclear hormone receptors involved in controlling the transcription of various genes that regulate energy metabolism, cell differentiation, apoptosis and inflammation [118]. Some of the NSAIDs activate the PPAR α, γ and δ, which enhances the apoptosis. Fig. (**3**) shows COX-independent targets of NSAID. Up regulation or down regulation by NSAIDs is indicated by color code. Possible NSAIDs affecting a particular target are also indicated.

All these studies make it clear that NSAIDs can have multiple targets to exert their anticancer effects. What is important from literature is that not all NSAIDs act equally well, rather a selective group of NSAIDs work on a specific target. To understand this preference one needs to look for common chemical templates involved in the interaction with a specific target. However there exists very little literature where this approach has been made and some of them will be discussed later.

3.1.2. Beneficiary Effects on Alzheimer Disease (AD)

Alzheimer disease (AD) is a neurodegenerative disorder characterized by impairment in memory and cognition. The pathogenesis of AD is characterized by cerebral deposits of amyloid β-peptides (Aβ) as amyloid plaques and neurofibrillary tangles (NFTs), which are surrounded by inflammatory cells. Plaque material mainly consists of extra-cellular aggregates of Aβ peptides. Misfolding of the soluble native peptide leads to self association to form

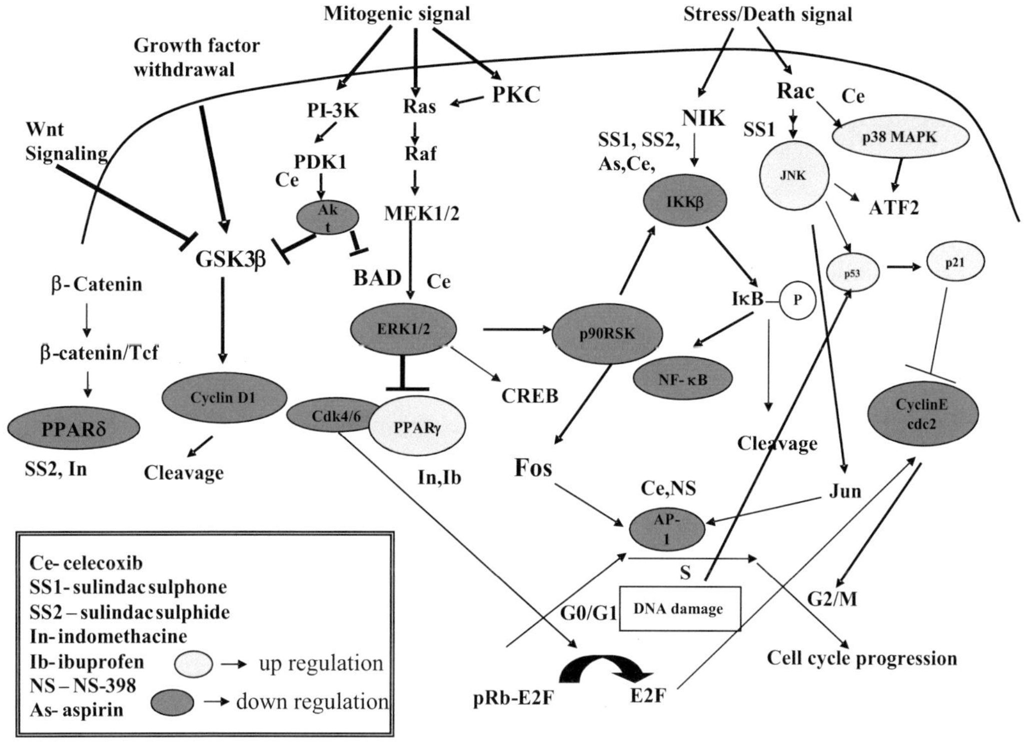

Figure 3: Different COX-2 independent targets of NSAIDs.

oligomers, protofibrils or other intermediates in the fibril formation pathway. Oligomers of Aβ can be detected *in vitro* [119] in cell culture transgenic mouse model of AD [120-122] and also in postmortem of AD patients' brain specimens [123]. NFTs mostly consist of intra-cellular aggregation of phosphorylated tau protein. Severe inflammatory response develops around the Aβ deposition [124, 125], which is initiated by the activation of microglia and the recruitment of astrocytes. These cells secrete many inflammatory cytokines and chemokines that may contribute to neural degeneration and cell death by various mechanisms [126, 127]. It is still controversial as to which event is the key player in AD pathogenesis, whether it is the formation of amyloid peptides (Aβ) by the neurons that mediate neurodegeneration [128] or the inflammatory response associated with the presence of neuritic plaques that cause the neurotoxicity [126, 129]. Even though there exists a wealth of experimental data, there is still neither a direct correlation nor do we understand the mechanism by which amyloidosis or neuroinflammation mediate neurodegeneration. Epidemiological studies have found strong correlation between long term use of NSAIDs with reduced risk for developing AD and delay in the onset of the disease [130]. Since NSAID group of drugs are primarily used to control pain and inflammation, it is an obvious expectation that they would target neuroinflammation to exert their beneficiary effects on AD. However, the picture is not so clear and the mechanism behind the role of NSAID in AD pathogenesis is controversial and riddled with several hypotheses.

Selective NSAIDs *viz.*, Sulindac sulfide, ibuprofen, indomethacin and flurbiprofen reduce Aβ levels in cultured cells from peripheral, glial and neuronal origins [12, 131-134]. Recently, reexamination of large scale clinical trials showed that when patient with preexisting conditions are removed from the trial set, naproxen reduces AD risks by 67% [135]. Interference with the formation of Aβ oligomers have been proposed as a possible mechanism [136]. Other NSAIDs like, acetaminophen, aspirin and celecoxib showed no effect on amyloid pathway. This effect is proposed to be by a mechanism independent of COX-pathway by directly affecting amyloid pathology in the brain that reduces Aβ 42 peptide levels. This is achieved by subtly modulating γ-secretase activity [12] without perturbing Amyloid Precursor Protein (APP) or Notch processing. Among the Aβ-effective NSAIDs, flurbiprofen is particularly important [132]. The R-enantiomer of flurbiprofen does not inhibit COX but does reduce Aβ–42 levels *in vitro* and *in vivo* supporting the fact that this NSAID does indeed reduce Aβ 42 levels by COX independent mechanism [131]. NSAIDs like nimesulide, ibuprofen and indomethacin have been shown to favor nonamyloidogenic APP processing by enhancing α-secretase activity thereby reducing the formation of amyloidogenic derivatives [126]. NSAIDs have also been implicated to target different components of neuroinflammation. Neuroinflmmation is secondary to neuritic plaques. Activated microglia and reactive astocytes surrounding extracellular deposits of Aβ protein initiate an inflammatory response [10, 127, 130]. Microglial COX expression is considered to be important in the pathogenesis of AD [10]. However, in adult human microglia *in vitro*, COX-1 is constitutively expressed but not COX-2, on exposure to Aβ or plaque associated cytokines. So COX-1 is said to be a better target than COX-2 [130]. This could explain the failure of COX-2 specific inhibitors like celecoxib to produce any beneficiary effect in AD pathogenesis. Pilot trial with therapeutic dose (dose of COX inhibition) of traditional NSAIDs showed promise but higher dose may be required.

Another target of neuroinflammation *i.e.*, peroxisome proliferator activated receptor-γ (PPAR- γ) have been implicated in the mechanism of action of NSAIDs [127]. PPAR- γ belongs to a family of nuclear receptors that is able to regulate the transcription of proinflammatory molecules. NSAIDs have been hypothesized to activate PPAR- γ thereby reducing the inflammatory response.

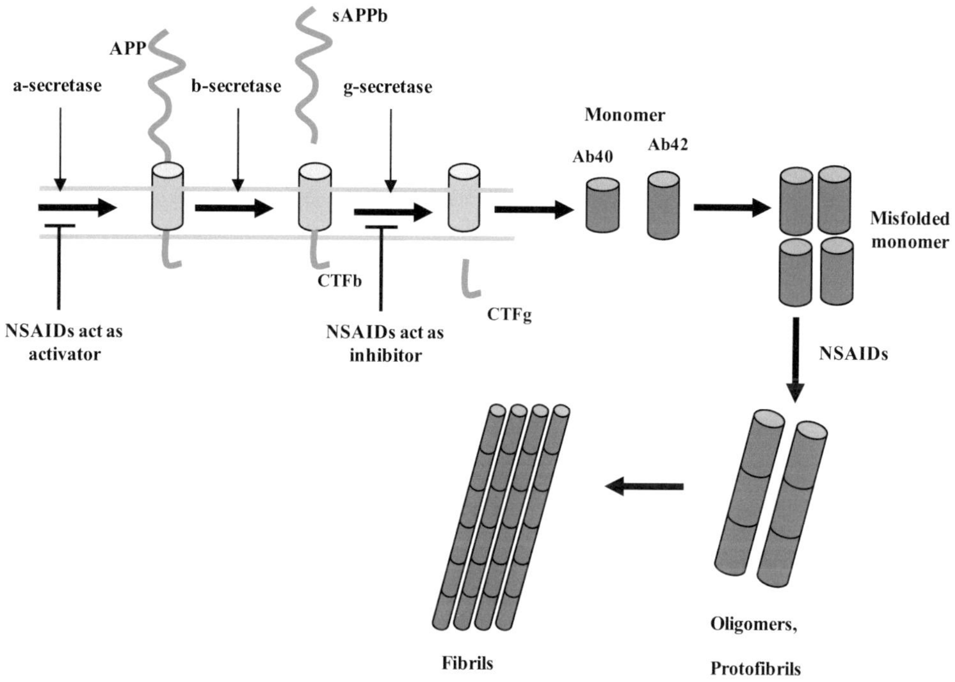

Figure 4: Different APP processing pathways leading to formation of Aβ–42 peptide, which leads to oligomerization and fibril formation.

In Fig. (**4**), we show the various stages of APP processing that leads to formation of Aβ–peptides, which then misfolds, oligomerizes to form fibrils. The possible steps that can be affected by NSAIDs in the entire pathway are also indicated. It is obvious that the NSAIDs can exert their effects on AD pathogenesis by various mechanisms. Recently, Aβ oligomers have been implicated as the primary cytotoxic agents. Even the small Aβ dimers can affect synaptic functions [137, 138]. Selective amyloid lowering agent R-flurbiprofen which is COX inactive were used in a clinical trial which failed in 2008 due to lack of efficacy [139, 140]. This has been attributed mainly due to lack of bioavailability which could be a result of poor blood brain barrier crossing. To overcome this difficulty, hybrid nitrates as NO-donor NSAID (NO-NSAID) are being designed as selective amyloid lowering agents [141]. It is now an established fact that only selected NSAIDs show beneficiary effects against AD. Till date, there has been no study devoted to understand the chemical basis for this preference for a few selective NSAIDs. This could be an important approach, which could lead to specific target identification in the AD pathogenesis. Inhibitory effects of NSAIDs on Aβ fibril formation span NSAIDs having different chemical templates which contradict the importance of chemical motifs in determining the mechanism. It is therefore important to look for the chemical basis of action of NSAID on AD pathogenesis.

3.1.3. Consequences of NSAID Membrane Interaction: Perturbation/Fluidity/Fusion

The principal targets for the primary functions of NSAIDs are COX isoenzymes that are membrane bound. To reach their targets, these drugs first need to interact with the membrane. Hence membrane interaction might be a decisive factor in their clinical outcome. The major constituent of membranes is phospholipids which are diverse in nature, having varied types of head groups and hydrophobic tail regions. These guide the microenvironment of the membrane interior and surface that in turn are expected to modulate the drug-membrane interaction. NSAIDs of oxicam family (piroxicam, meloxicam, tenoxicam, *etc.*) and other chemical groups (nimesulide, indomethacin, ibuprofen, *etc.*) are known to interact with the phospholipids. The interaction changes the mechanical properties of the membrane, which are typically quantified by the change in fluidity, bending modulus, *etc.* Using neutron spin-echo measurement it has been shown that ibuprofen reduces the bending modulus of dimyristoylphosphatidylcholine (DMPC) membrane [142]. Bending modulus is a key determinant for cell division, fusion, shape change, adhesion, and permeability [143, 144]. Several NSAIDs lower the fluidity of the mouse splenocyte membrane [145]. Many reports indicated that this direct NSAID-phospholipid interaction is the potential cause for gastric injury promoted by these drugs [146]. Indomethacin and naproxen have the ability to attenuate the phospholipid-related hydrophobic properties of the gastric mucosa by more than 80-85% in a dose dependent manner when they were administered to rats. The hydrophobicity of the luminal surface of the stomach wall was assessed by contact angle analysis [147]. Ibuprofen interacts with red cell membranes and changes their shapes at a very low concentration (as low as 10 μM) [148]. Ibuprofen also induces a significant increase in the generalized polarization of Large Unilamellar Vesicles (LUV) of DMPC, hence indicating that ibuprofen molecules are located in the polar head group region of DMPC. Study of surface pressure *versus* specific molecular area isotherms of Langmuir monolayers of DMPC on pure water in absence and presence of piroxicam, meloxicam and tenoxicam in the sub phase revealed that they interact with the lipid monolayer and the location of the drugs are different in the monolayer depending on their chemical and physical properties [149]. This suggests NSAIDs not only interact with the assembly of lipids or membranes but also are capable of interacting with the lipid monolayers, thereby, pointing at the chemical affinity of the NSAID molecules towards the phospholipids, especially zwitterioninc phospholipids. The development of novel NSAIDs showing less serious side effects during medical applications will also depend on the understanding of the processes initiating and promoting gastric injury. Such mechanisms are complex, and the cascade of events leading to mucosal damage must therefore be characterized and can also be related to the topical irritancy of NSAIDs. Evidence of the direct superficial damaging effects of several drugs that are members of the NSAID family have been subsequently provided by many investigators who showed histological, biochemical, and permeability changes in the gastric mucosa [147, 150]. However, the "barrier breaking" activity of the drugs has not been established on a molecular basis. Although it is clear that the GI side-effects of NSAIDs are in part attributable to their ability to inhibit the biosynthesis of gastro-protective prostaglandins, a significant amount of evidence exists that NSAIDs can act directly on local mucosa to induce GI ulcers and bleeding by prostaglandins independent mechanism [151, 152]. They may chemically associate with phospholipids and destabilize them from the mucus gel layer. Such a transition would increase the wettability of the stomach and result in an increase in the back-diffusion of luminal acid into the mucosa; consequently, the development of erosions must be expected [152].

Recently, membrane fusion, a new and alternate function of the NSAIDs has been identified. It has been shown that NSAIDs from oxicam group are capable of inducing fusion of small unilamellar vesicles (SUVs) at physiologically relevant concentration [153-155]. Data showed that all three oxicam NSAIDs, namely, meloxicam, piroxicam and tenoxicam have differential rates of content mixing and leakage though they are of the same genre, with meloxicam showing the maximum rate and extent for content mixing with tenoxicam showing the lowest. For all three oxicam NSAIDs, fusion increases with concentration of the drugs (Drug/Lipid (D/L) ratio) and reaches a maximum value at a particular threshold D/L ratio, which is different for the three drugs. Beyond this threshold concentration of the drugs, fusion decreases because drug induced leakage from the vesicles overwhelms the fusion event [156]. The enhanced leakage at concentration beyond the threshold is indicative of increased permeabilization of the membrane by the drugs. Membrane fusion induced by small drug molecules at physiologically relevant concentration is a rare event. Even among NSAIDs, this property is shown only by the oxicams and is not shared by drugs from other chemical groups (indomethacin, ibuprofen, *etc.*) [155]. It should be mentioned that the reason why small drug molecules cannot induce and complete membrane fusion, lies in their inability to impart enough energy by conformational change to overcome the barriers of intermediates of the fusion event [157]. Large molecules like proteins and peptides share this advantage hence *in vivo* they constitute the most common group of fusogenic agents [158, 159]. One of the consequences of the fusogenic property of the oxicams is reflected in the ability of piroxicam to induce fusion and rupture of mitochondrial outer membrane. This leads to the release of cytochorme *c* in the cytosol of V79 cell lines from chinese hamster, which in turn leads to the activation of proapototic caspase-3 in a dose dependent manner [160]. Activation of mitochondria dependent apoptotic pathway is a good strategy in cancer therapy [161]. Besides the pro-apoptotic caspase activation, effect of piroxicam on the membrane morphology of isolated mitochondria leading to fusion and rupture was directly imaged by Scanning Electron Microscope (SEM). [160]. Hence, this fusogenic property of NSAIDs might also be a putative cause of gastric ulcer, which needs further attention. Understanding the mechanism behind NSAID induced membrane fusion will also open the path to apply small drugs to induce fusion in biotechnological and biomedical procedures where membrane fusion plays an integral role.

Based on the detailed information on how NSAIDs interact with the membranes, a strategy could be made to reduce the side-effects on GI-tract. Studies revealed that instead of using the bare drugs, drugs chemically associated with the zwitterionic phospholipid (like dipalmitoylphosphatidylcholine, DPPC) reduces the side-effects of these drugs as demonstrated in animal models of acute chronic NSAID injury [152]. Also the anti-pyretic and anti-inflammatory activity of aspirin appeared to be consistently enhanced when associated with zwitterionic phospholipids. This unexpected finding may be attributable to the increase in lipid permeability and solubility of aspirin complex, which should promote movement of the NSAIDs across membranes and/or barriers and into target cells, to allow its therapeutic actions to be manifested. This suggests the importance of detail understanding of the NSAIDs-phospholipid interaction to develop better antipyretic and anti-inflammatory drugs with minimum side effect on GI-tract. Also, understanding their effects on membranes of cells and cell organelles will help to elucidate the mechanism behind their alternate functions.

3.2. Other Functions of Aureolic Acid Group of Antibiotics

3.2.1. Inducer of Erythroid Differentiation and Fetal Hemoglobin Production

Mithramycin is a potent inducer of γ-globin mRNA accumulation and fetal hemoglobin (HbF) production in erythroid cells from healthy human subjects and β-thalassemia patients [162, 163]. Results from the study suggest potential clinical application of MTR for induction of HbF in patients affected by β-thalassemia or sickle cell disease. The authors proposed that the mechanism of action involves alteration in the pattern of protein binding to the γ-globin promoter, leading to transcriptional activation. They did not rule out the possibility of a direct effect of the antibiotic on other genes involved in the activation of erythroid differentiation.

3.2.2. Prolongation of Survival in Mouse Model of Huntington's Disease (HD)

Pharmacological treatment of a transgenic mouse model of HD (R6/2) with mithramycin extends survival by 29.1%, greater than any single agent reported to date. Improved motor performance and markedly delayed neuropathological sequelae are the important recoveries. The functional mechanism for the healthy effects of mithramycin was linked to the prevention of the increase in H3 methylation observed in R6/2 mice. The enhanced survival and neuroprotection might be ascribed to the alleviation of repressed gene expression important to neuronal function and survival. These findings are also significant because they demonstrate the penetrability of mithramycin across the blood– brain barrier to exert some of its beneficial effects. These protective effects appear to require GC-DNA specificity, because an AT-DNA binding antibiotic, distamycin, has no effect on either R6/2 transgenic mouse survival or rotorod performance. However, the pathway of action is not clear. While displacement of the Sp1 family of transcriptional activators from their canonical GC DNA binding sites is a plausible pathway, the transcriptional activation of a hitherto unknown gene/s involved in promoting neuronal death could be another possibility. Sp1 could be acting directly or indirectly to repress transcription of a prosurvival gene(s). Mithramycin treatment might also work by decreasing htt transgene expression in R6/2 mice. Indeed, the demonstration that mithramycin can reduce striatal lesions induced by the mitochondrial toxin 3-NP also suggests that the protective action of mithramycin is downstream of mutant htt [164-166].

3.2.3. Potential Therapy for Neurological Diseases Associated with Aberrant Activation of Apoptosis

Mithramycin A and its structural analog chromomycin A_3 are reported to be effective inhibitors of neuronal apoptosis induced by glutathione depletion-induced oxidative stress or the DNA-damaging agent camptothecin. It indicates the potential of mithramycin A and its structural analogs as effective agents for the treatment of neurological diseases associated with aberrant activation of apoptosis which can be induced by many pathological stimuli. The displacement of the transcription factors Sp1 and Sp3 from binding at their cognate GC box was attributed the reason behind the protective effects of mithramycin A [167].

3.2.4. Downregulation of Proinflammatory Cytokine Induced MMP Gene Expression

Osteoarthritis (OA) and rheumatoid arthritis patients show articular cartilage degeneration as a major pathological manifestation with increased levels of several matrix metalloproteinases (MMPs) in cartilage, synovial membrane and synovial fluid. MMP-3 and MMP-13 cleave collagens and aggrecan of cartilage extracellular matrix. Proinflammatory cytokines, interleukin-1 (IL-1), IL-17 and tumor necrosis factor (TNF)-α are also increased in arthritic joints thereby inducing catabolic pathways leading to an enhanced expression of MMPs. Reduction in the damage in arthritic tissues can be attained by an inhibition of these proteases. Mithramycin downregulates MMP-3 and MMP-13 gene expression induced by IL-1β, TNF-α and IL-17 in human chondrosarcoma SW1353 cells and in primary human and bovine femoral head chondrocytes. Treatment with the drug also leads to a suppression of constitutive and IL-1-stimulated MMP-13 levels in bovine and human cartilage explants. Interference of Sp1 binding by MTR is proposed as one of the possible mechanisms behind the above function of the drug and its potential to reduce cartilage degeneration. However, the authors did not rule out the possibility of unidentified mechanisms of action [168].

3.2.5. Activation of Fas Death Pathway in Leukemic Cell Lines

MTR induces apoptosis in Fas sensitive Jurkat cells and Fas resistant KGa1 cell lines. The cell lines showed morphological changes with Fas aggregation, DISC (Death Inducing Signaling Complex) formation and caspase processing when treated with MTR ending in chromatin condensation and cell apoptosis. The authors speculate that cytoxicity and severe side effects of MTR such as sepsis and toxic hepatitis could be a result of activation of Fas apoptotic pathway in these cells. On the other hand, sublethal doses of MTR sensitize the Fas-resistant cells to other chemotherapeutic agents such as Aracytin and Mitoxantrone thus advocating the use of MTR in combination therapy [48].

3.2.6. Other Functions of Aureolic Acids

The above reports aim at exploration of the additional therapeutic potential of the aureolic acid antibiotic, mithramycin. However, these studies propose that promoter binding ability culminating in the alteration of the gene expression is the predominant pathway for the therapeutic property of the drug. We want to summarize those results from our laboratory which emphasize an alternate site and pathway of its action. We have earlier reported the association of the antibiotics with the cytoskeletal protein, spectrin both in the absence and presence of Mg^{2+}. The association is as strong as its spectrin-ankyrin interaction [169]. The observation suggests the cytoskeletal proteins as another class of potential target inside the cell. During our in-depth analysis of the energetic of binding of the antibiotics with different levels of chromatin structure, it has not escaped our notice that some of the results indicate interaction of the antibiotics with histones. Therefore, it might be conjectured that these antibiotics might either interact with histone(s) or modulate the histone modification. Effect upon the histone modification would culminate into alteration of gene expression [170]. One report has suggested that MTR might interfere the acetylation of a core histone like H3. ESET/SETDB1 gene expression and histone H3 (K9) trimethylation in Huntington's disease is inhibited by a combination of drugs, mithramycin and cystamine [166]. A recent report has shown that mithramycin exerts multiple points of control in the regulation of Hdm2 protein synthesis. The oncoprotein plays a key role in the regulation and activation of p53 protein. Firstly, the constitutive hdm2-P1 promoter is inhibited by MTR. Secondly, MTR induces p53-dependent transcription from the hdm2-P2 promoter. Thirdly, and critically, MTR also inhibits Hdm2 synthesis at the post-transcriptional level, with negative effects on hdm2 mRNA nuclear export and translation [171].

3.2.7. Metal Chelation Property of the Aureolic Acids as an Alternate Mode of Action

An interesting property of these antibiotics that has been overlooked for a potential therapeutic value is its metal chelating property. The importance of the metal chelating property of the aureolic acid antibiotics and their intracellular action on bivalent metal-dependent enzymes and proteins, *e.g.* Zn-dependent proteins like transcription factors and matrix metalloproteases (MMPs) cannot be overlooked. This property and its consequence have not been adequately pursued to evaluate its therapeutic potential in the treatment of diseases relating to metal dyshomeostasis. In the absence of DNA, MTR and CHR bind to many bivalent metal ions like Mg^{2+} [33], Zn^{2+} [172] and Cu^{2+} forming antibiotic-metal complexes. The metal binding potential of these antibiotics have also been demonstrated for Fe^{2+}, Mn^{2+}, Co^{2+}, Cd^{2+} and Ni^{2+} [173-175]. It has also been shown that MTR also forms complexes with monovalent and trivalent metal ions like Li^{+}, Na^{+}, K^{+}, Tb^{3+} and Gd^{3+} but these cations were not able to promote MTR-DNA interaction [176]. Fluoresence microscopic studies also showed the presence of metal-drug complex in the cellular nucleus [174]. Studies from our laboratory have shown that MTR and CHR form two different types of complexes with Mg^{2+}. The stoichiometries of these complexes are 1:1 (complex I) and 2:1 (complex II) in terms of antibiotic:Mg^{2+} [33-36]. In contrast, it forms only one type of complex of type II with other metal ions. Recently, we have made a detailed report based on the biophysical characterization of the complex formed upon the association of the antibiotics with Zn^{2+} and Mg^{2+} ions [146]. In another study from our laboratory, we have also characterized the formation of the complex between Cu^{2+} and the drugs. Like the zinc ion, they form only 2:1 type complex with Cu^{2+} (unpublished data). Table **2** summarizes the affinity parameters for the association of these antibiotics with the metal ions. We have observed a unique property of the Cu^{2+} complexes of the antibiotics, MTR and CHR. The $[(CHR/MTR)_2{:}Cu^{2+}]$ complex do not interact with double stranded B-form DNA [177] even (G,C)- base containing oligonucleotides sequences. The inability of the Cu^{2+} complex to bind DNA has been attributed to its distorted octahedral structure. Owing to the d^9 electronic configuration of Cu^{2+}, its complexes possess non-zero crystal field stabilization energy (CFSE) and hence transition to more energetically favorable conformations is unlikely once the complex adopts a particular conformation. It is also interesting to note that CHR/MTR preferentially binds to Cu^{2+} in presence of equimolar amount of Zn^{2+}, as is observed from monoisotopic mass spectrometric studies [177].

Mg^{2+} and Zn^{2+} ions show similarities in many physical properties [178], however, there are major differences also. The 100-fold slower exchange of water from the $[Mg(H_2O)_6]^{2+}$ complex as compared to that of $[Zn(H_2O)_6]^{2+}$ might be a plausible reason for the formation of two types of complexes with Mg^{2+} during the association with the antibiotics. There are similarities in the spectral changes of the antibiotics upon 2:1 complex formation with the two metal ions [35, 37, 172]. The structure of the $[(D)_2Mg^{2+}]$ ('D' is the antibiotic) complex has been proposed to be octahedral [176, 178-181] with Mg^{2+} having a co-ordination number of six of which four are satisfied by two each from the carbonyl oxygen (C1-O) and the negatively charged oxygen of hydroxyl group (C9-OH) of the chromomycinone ring (see Fig. (**1**)). The other two coordination sites are satisfied by water as ligands. However, as Zn^{2+} ion has zero crystal field stabilization energy as a result of its filled d-orbital, it shows no preference for a particular geometry. Thus, tetra coordination is a plausible option. This is supported from the results of ^{1}H NMR spectroscopy that favors the possibility of the complex, $[(D)_2Zn^{2+}]$, containing tetra-coordinated Zn^{2+} without the participation of water as ligand [172].

Hou and co-workers showed the stabilization of the $[(CHR)_2Mg^{2+}]$ complex, derived from X-ray crystal structure where the metal ion has octahedral coordination, is due to mutual stacking of the aromatic part of the chromophore of one CHR molecule with the C-D glycosidic linkage of the other molecule. They also stated from the NMR structure of the same complex that the metal ion has a tetrahedral geometry [182]. The formation of the antibiotic-metal complex is always entropy driven with the following kinetics [172]. In general, the rate of the reaction and the product formation decreased with the increase in pre-incubation concentration of the antibiotic. The enzymatic inhibition of ADH by the antibiotic(s) follows non-competitive inhibition.

Table 2: Binding constants and stoichiometry for the formation of antibiotic-metal complexes

$$D + Zn^{2+} \underset{k_{-1}}{\overset{\text{fast},\ k_1}{\leftrightarrows}} (D)Zn^{2+} \xrightarrow{\text{slow},\ k_2} (D)Zn^{2+*} \xrightarrow{\text{fast},\ +D} [(D)_2Zn^{2+}] \quad (D = \text{antibiotic})$$

Complex	**Antibiotic**	**Stoichiometry (Antibiotic:Metal)**	**Dissociation Constant (M^2)**
Mg^{2+}	MTR	1:1 (Complex I)	5.5×10^{-5} (M)
		2:1(ComplexII)	3.5×10^{-8}
	CHR	1:1 (Complex I)	5.3×10^{-5}(M)
		2:1(ComplexII)	9.1×10^{-8}
Zn^{2+}	MTR	2:1	10.55×10^{-10}
	CHR	2:1	3.20×10^{-10}
Cu^{2+}	MTR	2:1	0.67×10^{-10}
	CHR	2:1	0.15×10^{-10}

The ability of the bivalent metal ions to bind with the drug(s) at physiological pH coupled with the their intracellular concentrations and the stability parameters of the drug(s)-metal complex imply that the mode of action of these drugs inside the cell need not be limited to the DNA binding property of the drug-dimer complex only. In addition, their clinical use may be extended for the purpose where diseases occur due to misbalance of metal homeostasis inside the cell.

Recently, we have also shown the negative effect for the preincubation of Zn(II)-containing alcohol dehydrogenase (ADH) with MTR (& CHR) upon their enzymatic activity which is a consequence of induced structural alteration as a sequel to its complex formation with the Zn^{2+} ions present in the enzyme [183]. Alcohol dehydrogenase (E.C. 1.1.1.1) is a tetrameric enzyme containing two zinc ions per subunit, one each at the catalytic and structural sites, was studied. A negative effect for the pre-incubation of ADH with MTR/CHR upon its enzymatic activity was observed which was characterized by a decrease in the rate and the extent of NADH formed. In Our studies have opened up the possibility to exploit the metal binding potential of the aureolic acid antibiotics for therapeutic purpose as discussed above.

3.2.7.1. Zn(II)-Containing Proteins and Enzymes as Potential Targets

Zinc is the second most abundant trace metal in higher animals with the average human body containing 2-4 g of zinc. Zn(II)-containing proteins are known to have functions extending from catalysis of metabolic pathways and macromolecular synthesis to the regulation of gene expression. With the demonstrated ability of MTR and CHR to inhibit the function of the Zn(II)-containing enzyme, alcohol dehydrogenase [183] and the possibility of interaction of MTR with MMP during its inhibition of MMP gene expression [168], Zn(II)-containing enzymes and proteins might be explored as an alternative therapeutic target as these proteins are involved in most pathological conditions. It is known that proteases make up nearly 2% of the human genome and represent 5-10% of all known potential drug targets. In particular, there are nearly 200 distinct metalloproteases whose functions are linked to a number of clinically relevant conditions, most notably cancer [184]. All MMPs share a common active-site motif: in which three histidine residues bind to a zinc ion. Many of the inhibitors currently undergoing clinical testing are small peptide mimics that chelate the zinc ion and block the function of the enzymes. As new and valuable information to understanding the complex functional roles of metalloproteases in disease progression come up, clinical trials with MTR as matrix metalloprotease inhibitor could be considered for therapeutic gain.

3.2.7.2. Potential Agent in Chelation Therapy in Neurodegenerative Diseases Showing Metal Dyshomeostasis such as Alzeimer Disease (AD), Wilson's Disease (WD) and Prion Diseases

Alzheimer disease (AD) is a devastating neurodegenerative disease with progressive and irreversible damage to thought, memory, and language. The etiology of AD is not well understood but accumulating evidence supports oxidative stress generated by various mechanisms to be a major risk factor that initiates and promotes neurodegeneration. Many studies show the elevated concentrations of metal like iron, copper, zinc and aluminum in the brain of AD patients which indicate that the environmental conditions in AD, exacerbated by imbalances in several metals, has the potential for catalyzing and stimulating free radical formation and enhancing neuron degeneration thus opening the door for metal chelation therapy. A metal chelator with affinity for multiple metals such as aluminum, copper, and zinc turn out to be useful rather than damaging effects since various metals are implicated as oxidative instigators. This may be the reason behind the demonstrated therapeutic benefits of metal chelators like desferioxamine (DFO) in patients with AD [185]. Wilson's Disease (WD) is a disorder in copper metabolism involving mutation in the ATP7B gene and is manifested by copper overload in the cells. Metal chelator like penicillamine has been used to treat patients with WD [186]. As the aureolic acids like MTR and CHR are metal chelators with affinity for multiple metals like iron, zinc and copper, they have potential as therapeutics in metal chelation therapy for the treatment of neurodegerative diseases involving metal dyshomeostasis like AD and WD.

It was reported that prion proteins (PrP) might play a crucial role in zinc homeostasis within the brain, either through binding the metal ion at the synapse and being involved in a re-uptake mechanism, or as part of a protective sequestering system and/or a cellular sensor, thus preventing cellular toxicity of excess metal. However, in prion disease, when the protein undergoes a conformational change to the infectious form, this function of PrP in zinc homeostasis might be compromised thereby contributing to disease pathogenesis and changes in zinc binding. Studies with scrapie infected mice showed zinc content being reduced to 20% and the copper content reduced to 35–50% of the levels in uninfected mice while the blood content of zinc increased significantly. These data suggest that in prion disease, zinc is displaced from its normal localization and displaced in the blood [187]. As metal chelators have proved to be efficient therapeutics in other neurodegenerative diseases, it opens up the potential for the aureolic acids as potential therapeutic in prion diseases also.

3.2.8. Potential Therapeutic Functions of the Modified Antibiotics

It is well known that chromatin-interacting small molecules are of exceptional importance in medicine, accounting for a significant portion of all anticancer drugs. Majority of the clinical anticancer drugs introduced through 2002 are natural products or natural product derivatives, and exert their effects by acting on genomic DNA and/or histone(s). Although very few minor groove binders are clinically used, they are thought to hold the most promise for *de novo* design [188]. This holds great promise for the aureolic acid antibiotics MTR and CHR, which are minor groove binders. Therefore, we have included a section to give a brief overview of the potential leads for better and novel aureolic acid antibiotics with improved therapeutic index. The biosynthetic pathways of the parent antibiotics have been extensively studied and there have been reports of new structurally modified antibiotics. All leads are however based upon their potential as transcription inhibitors *via* reversible association with (G.C) rich DNA. Mithramycin SK, a novel antitumor drug with improved therapeutic index, Mithramycin SA, and demycarosyl-mithramycin SK are three new products generated in the mithramycin producer Streptomyces argillaceus through combinatorial biosynthesis [189]. Other new DNA-binding ligands were designed to mimic Chromomycin A3 (CHR), which contains a hydroxylated tetrahydroanthracene chromophore substituted with di and trisaccharides. In these new model compounds, a simple alkyl group attached to the chromophore part mimics the trisaccharide part of CHR. These have been successfully demonstrated for their Mg^{2+}-coordinated dimer complexes to exhibit DNA-binding affinity [190]. Recently, there was another report on a new analog, Mithramycin SDK, obtained by targeted gene inactivation of the ketoreductase MtrW catalyzing the last step in MTR biosynthesis. SDK exhibits greater activity as transcriptional inhibitor compared to MTR with a high degree of selectivity toward GC-rich DNA-binding transcription factors. SDK inhibited proliferation, inducing apoptosis in ovarian cancer cells with minimal effects on normal cells viability. The new MTR derivative SDK could be an effective agent for treatment of cancer and other diseases originating from abnormal expression or activity of GC-rich DNA-binding transcription factors [191].

4. CHEMICAL DISSECTION OF THE DRUGS TO UNDERSTAND ALTERNATE FUNCTIONS

Efforts to understand the molecular mechanism behind multiple functions of NSAIDs have resulted in controversies and no general consensus exists as to the actual mechanism behind any one of the unconventional functions of NSAID. Multiple targets and molecular pathways have been implicated for various functions. However, it is now clear that only selective NSAIDs exhibit a specific function and that too not at an equal level. It is therefore important to look at the chemical basis behind the unconventional functions of NSAID to identify the structural features required for a particular function. This would help to develop new classes of drugs aimed at a specific function using the chemical templates of NSAIDs. Such an approach was taken by Chen and co-workers [109]. Celecoxib, a newer generation COX-2 inhibitor was approved for FAP, an inherited predisposition to colorectal cancer in 1999. Its chemopreventive effect is achieved by sensitizing cancer cells to apoptotic signals. Equally powerful COX-2 inhibitor rofecoxib is at least two orders of magnitude less potent, which implied that the anti-tumor effect of celecoxib is independent of its COX-2 inhibition.

Chen looked at the structural difference between rofecoxib and celecoxib and used a systematic approach to modify the structures of both the drugs to produce 50 compounds. These compounds were then tested for their ability to induce apoptosis in human prostate cancer cells. The pathways through which apoptosis acts were monitored and molecular models were used to identify the key structural features involved in apoptosis. The structural requirements for the induction of apoptosis were found to be distinct from those that mediate COX-2 inhibition. Apoptosis required bulky terminal phenyl ring, a heterocyclic system with negative electrostatic potential and a benzene sulfonamide or benzene carboxamide moiety (Fig. (**5**)). To prove their observation, Chen at. al, modified the structure of rofecoxib to create four compounds to mimic the surface electrostatic potential of celecoxib, one of which showed substantial increase in pro-apoptotic activity. They found that apoptosis was mediated by down regulating the serine/threonine kinaseAkt and extracellular signal regulated kinase-2 (ERK2). They therefore demonstrate how understanding the chemical basis behind the apoptotic function of a NSAID, *viz.* celecoxib can help in designing a new class of compounds that induce apoptosis by targeting the Akt ERK2 signaling pathways in human prostate cancer cells using the molecular template of the NSAID. Efforts were also given to understand the different functions of similar drug molecules. Our group is working on the oxicam group of NSAIDs, which consist of piroxicam (Px), meloxicam (Mx), tenoxicam (Tx) *etc.* These molecules are structurally very similar to one another. Mx, Tx can be synthesized by the method of isosteric substitution considering Px as the mother template (Fig. (**6**)). The advantage of using these kinds of established drugs over the newly synthesized compounds is that the biochemical properties and the ADMET profile of the known drugs are well studied. From Fig. (**6**), it is evident that the structural difference between tenoxicam and that of meloxicam and piroxicam is small.

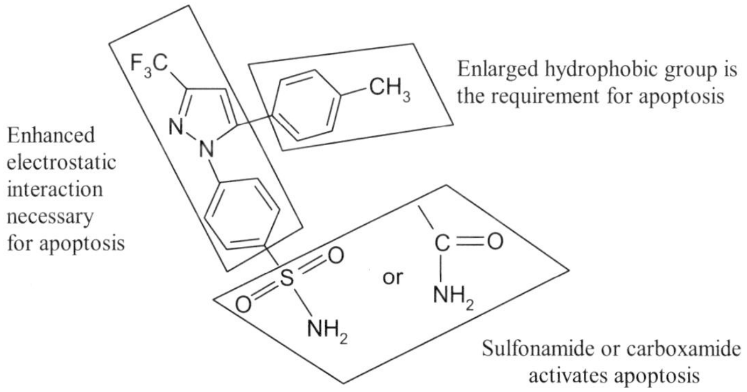

Figure 5: Chemical basis of pro-apoptotic functions of celecoxib.

Even such small changes in chemical structure can lead to completely different chemical/biological property. Oxicam group of NSAIDs, are not equally efficient against different cancer though they are structurally very similar. For example even though Px and Mx show anticancer properties, Tx shows no such effects. In the *in vitro* study it was observed that the Cu(II) complex of piroxicam and meloxicam bind with the calf thymus DNA resulting in alteration in the DNA backbone whereas tenoxicam does not interact with the DNA backbone at all [192, 193]. Another important property of oxicam NSAIDs is their structural dynamism. Structural dynamism is reflected in the ease with which the drugs can 'switchover' or convert from one prototropic form to the other guided by their environment [194]. In the interaction of these drugs with membranes/membrane mimetic systems it has been found that one prototropic form of the drug is converted to the other, depending on the nature of the membrane mimetic systems/membranes [195]. This switchover equilibrium is fine tuned by the electrostatics of the membrane surface, hydrophobicity of the membrane core, presence of counterion and steric effects guiding the water penetration within the membrane [195-200]. Thus the diverse nature of biomembrane *in vivo*, characterized by their different membrane parameters, should be the decisive factor in choosing which structural form of the drug will be finally presented to its target cyclooxygenases. This raises the intriguing possibility that the reason behind the different functions of these drugs could be due to their structural flexibility. However, further studies are required to establish this hypothesis.

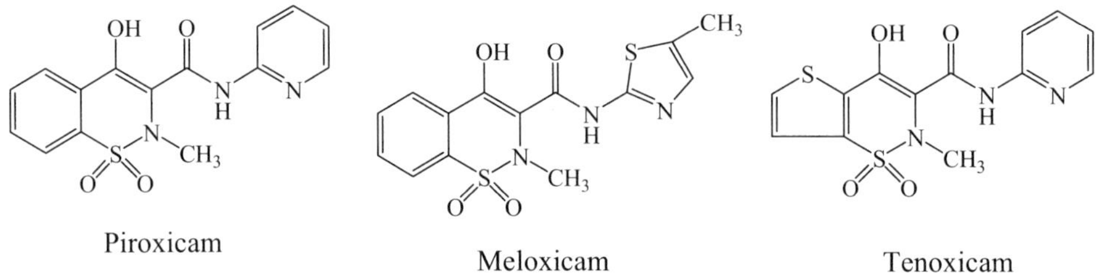

Figure 6: Chemical structure of piroxicam, meloxicam and tenoxicam.

Even in case of membrane fusion, a new function of NSAIDs recently identified by our group, chemical nature plays a crucial role. As has been mentioned before, membrane fusogenic property at physiologically relevant concentration is exhibited by the oxicam chemical group and not shared by NSAIDs belonging to other groups under similar experimental condition. Even among the oxicams, there exists differences in the extent and the rates of fusion process with Mx showing the highest rate followed by Px and Tx. Partial dissection of the reason behind these differences among oxicam induced fusion has been made. Differences in the extent of partitioning, location of the drugs in lipid vesicles and ability to permeabilize lipid bilayers are the primary reasons behind the differential fusogenic behavior of Mx, Px and Tx. The clinical implications of this new function of oxicam NSAIDs are yet to be established. As a preliminary study it has been shown that the fusogenic property of Px is responsible for altering mitochondrial membrane morphology leading to fusion and rupture of mitochondrial membrane [160]. This has been implicated as a possible cause for the release of cytochrome *c* from mitochondria into the cytosol, triggering pro-apoptotic events like caspase-3 activation. Px is also capable of reducing cytochorme *c* with small changes in protein structure. Enhanced GI toxicity could be another possible clinical outcome even though a direct link has not been established.

Aureolic acids, MTR and CHR, on the other hand, are structurally very similar compounds having the same pharmacore, the chromomycinone moity, with little differences in the sugar residues. The presence of the acetoxy group on the B-ring in CHR is believed to be a plausible reason behind its greater toxicity than MTR. It is also noticed that the modes of action of the two antibiotics on metal chelation and DNA binding are the same but the extent of their activity differs. The absence of substituents like the acetoxy group in MTR imparts conformational flexibility to a greater extent than CHR when the antibiotic:metal complex(es) binds to the minor groove of DNA. It was also demonstrated from Biospecific Interaction Analysis (BIA) and Surface Plasma Resonance (SPR) analysis that the $(MTR)_2$ Mg^{2+} -DNA complexes are highly unstable as compared to that of the $(CHR)_2Mg^{2+}$-DNA complexes which might explain its lower toxicity [201]. The rate of the association of CHR with bivalent metal ions like Zn^{2+} is also higher than that of MTR [172]. Both CHR and MTR have the tricyclic ring system fused to a unique dihydroxy-methoxy-oxo-pentyl aliphatic side chain attached at C-3. It has been shown that chemical modification of this 3-aliphatic side chain can generate novel aureolic acid compounds with greater therapeutic index such as MTR SDK, which has a diketo group in the C-2″ and C-3″ aliphatic side chain. The greater activity of SDK as transcriptional inhibitor in cells with tighter DNA binding, efficient protein binding and rapid accumulation in cells as compared to MTR might be a direct consequence of the structural modification [191].

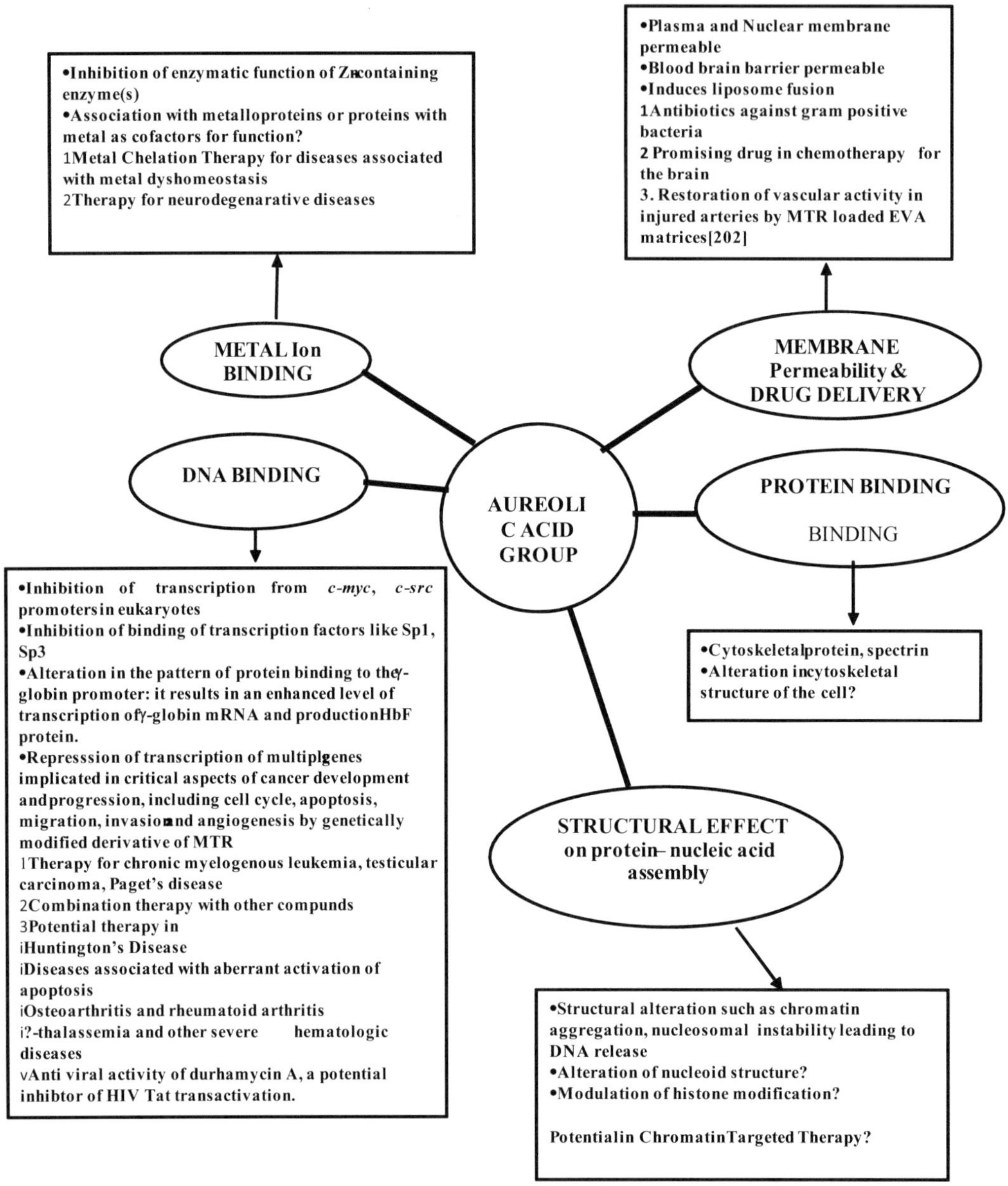

Scheme 2: Sites of action and therapeutic potential. of the aureolic acid group of antibiotics

5. CONCLUSION

In this chapter we have given an overview of function based therapeutic potential of two classes of drugs. NSAIDs have larger market in comparison to aureolic

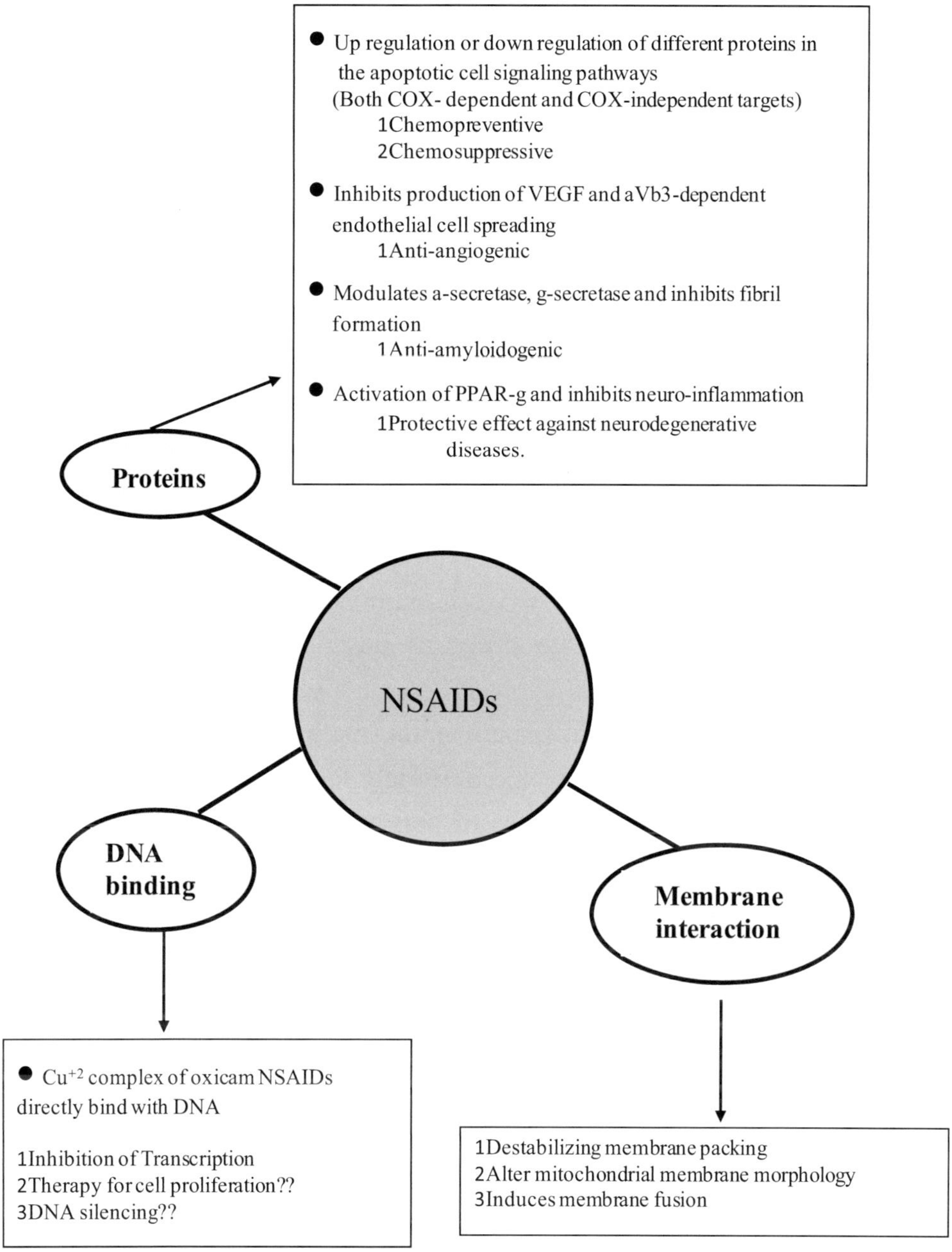

Scheme 3: Sites of action and therapeutic potential of NSAIDs.

acid group of drugs whose use is at present is confined to several types of cancer and as protective agents against neurodegenerative diseases like Alzheimer. In the two cartoons given below, we have catalogued the targets reported till date for the two classes of drugs. We have also indicated the pharmacological uses of the drugs that might employ these targets. An examination of the Schemes **2** and **3** clearly suggests that potential of the two classes of the drugs, as therapeutics, has not been explored to the maximum. For the aureolic acid group of antibiotics, recently there has been a plethora of reports aiming to explore the alternate potential. The identification of the gene cluster for the biosynthesis of the aureolic acid group of antibiotics has further opened up the vista towards a series of modified antibiotics with less toxicity, and improved potential as drug.

The diverse milieu of the potential interaction sites in a cell suggest that it may be an uphill task to predict its potential as a drug for the diseases mentioned in the schemes. In a multicellular organism, there is another degree of complexity arising from the differences in the cellular function and the underlying mechanisms. Notwithstanding the above limitations, it may be worth looking for the other therapeutic uses of the drugs as described above. Trial and error experiments will be the tool of choice. System biology approach based on an integration of a high throughput experimental, mathematical and computational science in an iterative approach would definitely lend further help. In both cases an incisive knowledge about the sites of action leading to alternate therapeutic potential of the drugs is an important input parameter.

ACKNOWLEDGEMENTS

Declared none.

CONFLICT OF INTEREST

The authors confirm that this chapter contents have no conflict of interest.

DISCLOSURE

The chapter submitted for eBook Series entitled: "**Recent Advances in Medicinal Chemistry, Volume 1**" is an update of our article published in **Mini-Reviews in Medicinal Chemistry, Volume 4, pp. 331 to 349**, with additional text and references.

REFERENCES

[1] O'Connor KA, Roth BL. Finding new tricks for old drugs: an efficient route for public-sector drug discovery. Nat Rev Drug Discov. 2005 Dec;4(12):1005-14.

[2] Ashburn TT, Thor KB. Drug repositioning: identifying and developing new uses for existing drugs. Nat Rev Drug Discov. 2004 Aug;3(8):673-83.

[3] Hong FD, Clayman GL. Isolation of a peptide for targeted drug delivery into human head and neck solid tumors. Cancer Res. 2000 Dec 1;60(23):6551-6.

[4] Un F. G1 arrest induction represents a critical determinant for cisplatin cytotoxicity in G1 checkpoint-retaining human cancers. Anticancer Drugs. 2007 Apr;18(4):411-7.

[5] Hawkey CJ. COX-2 inhibitors. Lancet. 1999 Jan 23;353(9149):307-14.

[6] Ritland SR, Gendler SJ. Chemoprevention of intestinal adenomas in the ApcMin mouse by piroxicam: kinetics, strain effects and resistance to chemosuppression. Carcinogenesis. 1999 Jan;20(1):51-8.

[7] Sporn MB, Suh N. Chemoprevention of cancer. Carcinogenesis. 2000 Mar;21(3):525-30.

[8] Goldman AP, Williams CS, Sheng H, Lamps LW, Williams VP, Pairet M, *et al.* Meloxicam inhibits the growth of colorectal cancer cells. Carcinogenesis. 1998 Dec;19(12):2195-9.

[9] Grossman EM, Longo WE, Panesar N, Mazuski JE, Kaminski DL. The role of cyclooxygenase enzymes in the growth of human gall bladder cancer cells. Carcinogenesis. 2000 Jul;21(7):1403-9.

[10] Hoozemans JJ, Veerhuis R, Janssen I, van Elk EJ, Rozemuller AJ, Eikelenboom P. The role of cyclo-oxygenase 1 and 2 activity in prostaglandin E(2) secretion by cultured human adult microglia: implications for Alzheimer's disease. Brain Res. 2002 Oct 4;951(2):218-26.

[11] Sagi SA, Weggen S, Eriksen J, Golde TE, Koo EH. The non-cyclooxygenase targets of non-steroidal anti-inflammatory drugs, lipoxygenases, peroxisome proliferator-activated receptor, inhibitor of kappa B kinase, and NF kappa B, do not reduce amyloid beta 42 production. J Biol Chem. 2003 Aug 22;278(34):31825-30.

[12] Weggen S, Eriksen JL, Das P, Sagi SA, Wang R, Pietrzik CU, *et al.* A subset of NSAIDs lower amyloidogenic Abeta42 independently of cyclooxygenase activity. Nature. 2001 Nov 8;414(6860):212-6.

[13] Anderson R, Eftychis HA, Weiner A, Findlay GH. An *in vivo* and *in vitro* investigation of the phototoxic potential of tenoxicam, a new non-steroidal anti-inflammatory agent. Dermatologica. 1987;175(5):229-34.

[14] Gebhardt M, Wollina U. [Cutaneous side-effects of nonsteroidal anti-inflammatory drugs (NSAID)]. Z Rheumatol. 1995 Nov-Dec;54(6):405-12.

[15] Serrano G, Fortea JM, Latasa JM, SanMartin O, Bonillo J, Miranda MA. Oxicam-induced photosensitivity. Patch and photopatch testing studies with tenoxicam and piroxicam photoproducts in normal subjects and in piroxicam-droxicam photosensitive patients. J Am Acad Dermatol. 1992 Apr;26(4):545-8.

[16] Bayerl C, Pagung R, Jung EG. Meloxicam in acute UV dermatitis--a pilot study. Photodermatol Photoimmunol Photomed. 1998 Oct-Dec;14(5-6):167-9.

[17] Pitarresi G, Cavallaro G, Giammona G, De GG, Salemi MG, Sortino S. New hydrogel matrices containing an anti-inflammatory agent. Evaluation of *in vitro* release and photoprotective activity. Biomaterials. 2002 Jan;23(2):537-50.

[18] Calabresi GC, B.A. In Goodman and Gilman's The Pharmacological Basis of Therapeutics: Chemotherapy of Neoplastic Diseases. Macmillan, New York, 9th Edition. 1991:1225-69.

[19] Wohlert SE, Kunzel E, Machinek R, Mendez C, Salas JA, Rohr J. The structure of mithramycin reinvestigated. J Nat Prod. 1999 Jan;62(1):119-21.

[20] Lahiri S, Devi PG, Majumder P, Das S, Dasgupta D. Self-association of the anionic form of the DNA-binding anticancer drug mithramycin. J Phys Chem B. 2008 Mar 13;112(10):3251-8.

[21] Dimaraco AA, F.; Zunino, F. In Antibiotics, Mechanism of Action of Antimicrobial & Antitumor Agents. Springer, Berlin Heidelberg New York. 1975:101-28.

[22] Goldberg IH, Friedman PA. Antibiotics and nucleic acids. Annu Rev Biochem. 1971;40:775-810.

[23] Menendez N, Nur-e-Alam M, Brana AF, Rohr J, Salas JA, Mendez C. Biosynthesis of the antitumor chromomycin A3 in Streptomyces griseus: analysis of the gene cluster and rational design of novel chromomycin analogs. Chem Biol. 2004 Jan;11(1):21-32.

[24] Lombo F, Menendez N, Salas JA, Mendez C. The aureolic acid family of antitumor compounds: structure, mode of action, biosynthesis, and novel derivatives. Appl Microbiol Biotechnol. 2006 Nov;73(1):1-14.

[25] Scott D, Rohr J, Bae Y. Nanoparticulate formulations of mithramycin analogs for enhanced cytotoxicity. Int J Nanomedicine. 2011;6:2757-67.

[26] Otterness IG, Larson DL, Lombardino JG. An analysis of piroxicam in rodent models of arthritis. Agents Actions. 1982 Jul;12(3):308-12.

[27] Miller RLI, P.A.; Melmon, K.L. Clinical Pharmacology. Macmillan Publishing Co, New York, 2nd Edition. 1978.

[28] Fu JY, Masferrer JL, Seibert K, Raz A, Needleman P. The induction and suppression of prostaglandin H2 synthase (cyclooxygenase) in human monocytes. J Biol Chem. 1990 Oct 5;265(28):16737-40.

[29] Maundrell K, Antonsson B, Magnenat E, Camps M, Muda M, Chabert C, *et al.* Bcl-2 undergoes phosphorylation by c-Jun N-terminal kinase/stress-activated protein kinases in the presence of the constitutively active GTP-binding protein Rac1. J Biol Chem. 1997 Oct 3;272(40):25238-42.

[30] Cashman JN. The mechanisms of action of NSAIDs in analgesia. Drugs. 1996;52 Suppl 5:13-23.

[31] Gierse JK, McDonald JJ, Hauser SD, Rangwala SH, Koboldt CM, Seibert K. A single amino acid difference between cyclooxygenase-1 (COX-1) and -2 (COX-2) reverses the selectivity of COX-2 specific inhibitors. J Biol Chem. 1996 Jun 28;271(26):15810-4.

[32] Luong C, Miller A, Barnett J, Chow J, Ramesha C, Browner MF. Flexibility of the NSAID binding site in the structure of human cyclooxygenase-2. Nat Struct Biol. 1996 Nov;3(11):927-33.

[33] Aich P, Dasgupta D. Role of Mg++ in the mithramycin-DNA interaction: evidence for two types of mithramycin-Mg++ complex. Biochem Biophys Res Commun. 1990 Dec 14;173(2):689-96.

[34] Aich P, Sen R, Dasgupta D. Interaction between antitumor antibiotic chromomycin A3 and Mg2+. I. Evidence for the formation of two types of chromomycin A3-Mg2+ complexes. Chem Biol Interact. 1992 Jun 15;83(1):23-33.

[35] Aich P, Dasgupta D. Role of magnesium ion in mithramycin-DNA interaction: binding of mithramycin-Mg2+ complexes with DNA. Biochemistry. 1995 Jan 31;34(4):1376-85.

[36] Aich P, Sen R, Dasgupta D. Role of magnesium ion in the interaction between chromomycin A3 and DNA: binding of chromomycin A3-Mg2+ complexes with DNA. Biochemistry. 1992 Mar 24;31(11):2988-97.

[37] Majee S, Sen R, Guha S, Bhattacharyya D, Dasgupta D. Differential interactions of the Mg2+ complexes of chromomycin A3 and mithramycin with poly(dG-dC) x poly(dC-dG) and poly(dG) x poly(dC). Biochemistry. 1997 Feb 25;36(8):2291-9.

[38] Chakrabarti S, Bhattacharyya D, Dasgupta D. Structural basis of DNA recognition by anticancer antibiotics, chromomycin A(3), and mithramycin: roles of minor groove width and ligand flexibility. Biopolymers. 2000;56(2):85-95.

[39] Chakrabarti SB, B.; Dasgupta, D. Interaction of Mithramycin and Chromomycin A3 with d(TAGCTAGCTA)2: Role of Sugars in Antibiotic−DNA Recognition. J Phys Chem B. 2002;106(27):6947-53.

[40] Keniry MA, Banville DL, Simmonds PM, Shafer R. Nuclear magnetic resonance comparison of the binding sites of mithramycin and chromomycin on the self-complementary oligonucleotide d(ACCCGGGT)2. Evidence that the saccharide chains have a role in sequence specificity. J Mol Biol. 1993 Jun 5;231(3):753-67.

[41] Sastry M, Fiala R, Patel DJ. Solution structure of mithramycin dimers bound to partially overlapping sites on DNA. J Mol Biol. 1995 Sep 1;251(5):674-89.

[42] Hardenbol P, Van Dyke MW. *In vitro* inhibition of c-myc transcription by mithramycin. Biochem Biophys Res Commun. 1992 Jun 15;185(2):553-8.

[43] Remsing LL, Bahadori HR, Carbone GM, McGuffie EM, Catapano CV, Rohr J. Inhibition of c-src transcription by mithramycin: structure-activity relationships of biosynthetically produced mithramycin analogues using the c-src promoter as target. Biochemistry. 2003 Jul 15;42(27):8313-24.

[44] Snyder RC, Ray R, Blume S, Miller DM. Mithramycin blocks transcriptional initiation of the c-myc P1 and P2 promoters. Biochemistry. 1991 Apr 30;30(17):4290-7.

[45] Bianchi N, Rutigliano C, Passadore M, Tomassetti M, Pippo L, Mischiati C, *et al.* Targeting of the HIV-1 long terminal repeat with chromomycin potentiates the inhibitory effects of a triplex-forming oligonucleotide on Sp1-DNA interactions and *in vitro* transcription. Biochem J. 1997 Sep 15;326 (Pt 3):919-27.

[46] Duverger V, Murphy AM, Sheehan D, England K, Cotter TG, Hayes I, *et al.* The anticancer drug mithramycin A sensitises tumour cells to apoptosis induced by tumour necrosis factor (TNF). Br J Cancer. 2004 May 17;90(10):2025-31.

[47] Lee TJ, Jung EM, Lee JT, Kim S, Park JW, Choi KS, *et al.* Mithramycin A sensitizes cancer cells to TRAIL-mediated apoptosis by down-regulation of XIAP gene promoter through Sp1 sites. Mol Cancer Ther. 2006 Nov;5(11):2737-46.

[48] Leroy I, Laurent G, Quillet-Mary A. Mithramycin A activates Fas death pathway in leukemic cell lines. Apoptosis. 2006 Jan;11(1):113-9.

[49] Das S, Dasgupta D. Binding of (MTR)2Zn2+ complex to chromatin: a comparison with (MTR)2Mg2+ complex. J Inorg Biochem. 2005 Mar;99(3):707-15.

[50] Mir MA, Das S, Dasgupta D. N-terminal tail domains of core histones in nucleosome block the access of anticancer drugs, mithramycin and daunomycin, to the nucleosomal DNA. Biophys Chem. 2004 Apr 1;109(1):121-35.

[51] Mir MA, Dasgupta D. Association of the anticancer antibiotic chromomycin A(3) with the nucleosome: role of core histone tail domains in the binding process. Biochemistry. 2001 Sep 25;40(38):11578-85.

[52] Mir MA, Dasgupta D. Association of anticancer drug mithramycin with H1-depleted chromatin: a comparison with native chromatin. J Inorg Biochem. 2003 Feb 1;94(1-2):72-7.

[53] Mir MA, Majee S, Das S, Dasgupta D. Association of chromatin with anticancer antibiotics, mithramycin and chromomycin A3. Bioorg Med Chem. 2003 Jul 3;11(13):2791-801.

[54] Fulzele SV, Shaik MS, Chatterjee A, Singh M. Anti-cancer effect of celecoxib and aerosolized docetaxel against human non-small cell lung cancer cell line, A549. J Pharm Pharmacol. 2006 Mar;58(3):327-36.

[55] Jendrossek V, Handrick R, Belka C. Celecoxib activates a novel mitochondrial apoptosis signaling pathway. Faseb J. 2003 Aug;17(11):1547-9.

[56] Noguchi M, Earashi M, Minami M, Miyazaki I, Tanaka M, Sasaki T. Effects of piroxicam and esculetin on the MDA-MB-231 human breast cancer cell line. Prostaglandins Leukot Essent Fatty Acids. 1995 Nov;53(5):325-9.

[57] Reddy BS, Wang CX, Kong AN, Khor TO, Zheng X, Steele VE, *et al.* Prevention of azoxymethane-induced colon cancer by combination of low doses of atorvastatin, aspirin, and celecoxib in F 344 rats. Cancer Res. 2006 Apr 15;66(8):4542-6.

[58] Shiff SJ, Koutsos MI, Qiao L, Rigas B. Nonsteroidal antiinflammatory drugs inhibit the proliferation of colon adenocarcinoma cells: effects on cell cycle and apoptosis. Exp Cell Res. 1996 Jan 10;222(1):179-88.

[59] Park C, Choi BT, Kang KI, Kwon TK, Cheong J, Lee WH, *et al.* Induction of apoptosis and inhibition of cyclooxygenase-2 expression by N-methyl-N'-nitro-N-nitrosoguanidine in human leukemia cells. Anticancer Drugs. 2005 Jun;16(5):507-13.

[60] Sinicrope FA. Targeting cyclooxygenase-2 for prevention and therapy of colorectal cancer. Mol Carcinog. 2006 Jun;45(6):447-54.

[61] Pettus BJ, Bielawski J, Porcelli AM, Reames DL, Johnson KR, Morrow J, *et al.* The sphingosine kinase 1/sphingosine-1-phosphate pathway mediates COX-2 induction and PGE2 production in response to TNF-alpha. Faseb J. 2003 Aug;17(11):1411-21.

[62] Spiegel S, Merrill AH, Jr. Sphingolipid metabolism and cell growth regulation. Faseb J. 1996 Oct;10(12):1388-97.

[63] Nzeako UC, Guicciardi ME, Yoon JH, Bronk SF, Gores GJ. COX-2 inhibits Fas-mediated apoptosis in cholangiocarcinoma cells. Hepatology. 2002 Mar;35(3):552-9.

[64] Fujita T, Matsui M, Takaku K, Uetake H, Ichikawa W, Taketo MM, *et al.* Size- and invasion-dependent increase in cyclooxygenase 2 levels in human colorectal carcinomas. Cancer Res. 1998 Nov 1;58(21):4823-6.

[65] Zimmermann KC, Sarbia M, Weber AA, Borchard F, Gabbert HE, Schror K. Cyclooxygenase-2 expression in human esophageal carcinoma. Cancer Res. 1999 Jan 1;59(1):198-204.

[66] Molina MA, Sitja-Arnau M, Lemoine MG, Frazier ML, Sinicrope FA. Increased cyclooxygenase-2 expression in human pancreatic carcinomas and cell lines: growth inhibition by nonsteroidal anti-inflammatory drugs. Cancer Res. 1999 Sep 1;59(17):4356-62.

[67] Dannenberg AJ, Altorki NK, Boyle JO, Dang C, Howe LR, Weksler BB, *et al.* Cyclo-oxygenase 2: a pharmacological target for the prevention of cancer. Lancet Oncol. 2001 Sep;2(9):544-51.

[68] Liu CH, Chang SH, Narko K, Trifan OC, Wu MT, Smith E, *et al.* Overexpression of cyclooxygenase-2 is sufficient to induce tumorigenesis in transgenic mice. J Biol Chem. 2001 May 25;276(21):18563-9.

[69] Subbaramaiah K, Telang N, Ramonetti JT, Araki R, DeVito B, Weksler BB, *et al.* Transcription of cyclooxygenase-2 is enhanced in transformed mammary epithelial cells. Cancer Res. 1996 Oct 1;56(19):4424-9.

[70] Thun MJ. NSAID use and decreased risk of gastrointestinal cancers. Gastroenterol Clin North Am. 1996 Jun;25(2):333-48.

[71] Wardlaw SA, March TH, Belinsky SA. Cyclooxygenase-2 expression is abundant in alveolar type II cells in lung cancer-sensitive mouse strains and in premalignant lesions. Carcinogenesis. 2000 Jul;21(7):1371-7.

[72] Heasley LE, Thaler S, Nicks M, Price B, Skorecki K, Nemenoff RA. Induction of cytosolic phospholipase A2 by oncogenic Ras in human non-small cell lung cancer. J Biol Chem. 1997 Jun 6;272(23):14501-4.

[73] Eberhart CE, Coffey RJ, Radhika A, Giardiello FM, Ferrenbach S, DuBois RN. Up-regulation of cyclooxygenase 2 gene expression in human colorectal adenomas and adenocarcinomas. Gastroenterology. 1994 Oct;107(4):1183-8.

[74] Kargman SL, O'Neill GP, Vickers PJ, Evans JF, Mancini JA, Jothy S. Expression of prostaglandin G/H synthase-1 and -2 protein in human colon cancer. Cancer Res. 1995 Jun 15;55(12):2556-9.

[75] Tsuji S, Kawano S, Sawaoka H, Takei Y, Kobayashi I, Nagano K, *et al.* Evidences for involvement of cyclooxygenase-2 in proliferation of two gastrointestinal cancer cell lines. Prostaglandins Leukot Essent Fatty Acids. 1996 Sep;55(3):179-83.

[76] Sawaoka H, Kawano S, Tsuji S, Tsujii M, Murata H, Hori M. Effects of NSAIDs on proliferation of gastric cancer cells *in vitro*: possible implication of cyclooxygenase-2 in cancer development. J Clin Gastroenterol. 1998;27 Suppl 1:S47-52.

[77] Sheng GG, Shao J, Sheng H, Hooton EB, Isakson PC, Morrow JD, *et al.* A selective cyclooxygenase 2 inhibitor suppresses the growth of H-ras-transformed rat intestinal epithelial cells. Gastroenterology. 1997 Dec;113(6):1883-91.

[78] Sheng H, Shao J, Kirkland SC, Isakson P, Coffey RJ, Morrow J, *et al.* Inhibition of human colon cancer cell growth by selective inhibition of cyclooxygenase-2. J Clin Invest. 1997 May 1;99(9):2254-9.

[79] Jacoby RF, Seibert K, Cole CE, Kelloff G, Lubet RA. The cyclooxygenase-2 inhibitor celecoxib is a potent preventive and therapeutic agent in the min mouse model of adenomatous polyposis. Cancer Res. 2000 Sep 15;60(18):5040-4.

[80] Williams CS, Tsujii M, Reese J, Dey SK, DuBois RN. Host cyclooxygenase-2 modulates carcinoma growth. J Clin Invest. 2000 Jun;105(11):1589-94.

[81] Tsujii M, Kawano S, Tsuji S, Sawaoka H, Hori M, DuBois RN. Cyclooxygenase regulates angiogenesis induced by colon cancer cells. Cell. 1998 May 29;93(5):705-16.

[82] Ruegg C, Dormond O, Mariotti A. Endothelial cell integrins and COX-2: mediators and therapeutic targets of tumor angiogenesis. Biochim Biophys Acta. 2004 Mar 4;1654(1):51-67.

[83] Dormond O, Foletti A, Paroz C, Ruegg C. NSAIDs inhibit alpha V beta 3 integrin-mediated and Cdc42/Rac-dependent endothelial-cell spreading, migration and angiogenesis. Nat Med. 2001 Sep;7(9):1041-7.

[84] Soga N, Namba N, McAllister S, Cornelius L, Teitelbaum SL, Dowdy SF, *et al.* Rho family GTPases regulate VEGF-stimulated endothelial cell motility. Exp Cell Res. 2001 Sep 10;269(1):73-87.

[85] Hasegawa K, Ohashi Y, Ishikawa K, Yasue A, Kato R, Achiwa Y, *et al.* Expression of cyclooxygenase-2 in uterine endometrial cancer and anti-tumor effects of a selective COX-2 inhibitor. Int J Oncol. 2005 May;26(5):1419-28.

[86] Iniguez MA, Rodriguez A, Volpert OV, Fresno M, Redondo JM. Cyclooxygenase-2: a therapeutic target in angiogenesis. Trends Mol Med. 2003 Feb;9(2):73-8.

[87] Leahy KM, Ornberg RL, Wang Y, Zweifel BS, Koki AT, Masferrer JL. Cyclooxygenase-2 inhibition by celecoxib reduces proliferation and induces apoptosis in angiogenic endothelial cells *in vivo*. Cancer Res. 2002 Feb 1;62(3):625-31.

[88] Dormond O, Bezzi M, Mariotti A, Ruegg C. Prostaglandin E2 promotes integrin alpha Vbeta 3-dependent endothelial cell adhesion, rac-activation, and spreading through cAMP/PKA-dependent signaling. J Biol Chem. 2002 Nov 29;277(48):45838-46.

[89] Soga N, Connolly JO, Chellaiah M, Kawamura J, Hruska KA. Rac regulates vascular endothelial growth factor stimulated motility. Cell Commun Adhes. 2001;8(1):1-13.

[90] Kiosses WB, Hood J, Yang S, Gerritsen ME, Cheresh DA, Alderson N, *et al.* A dominant-negative p65 PAK peptide inhibits angiogenesis. Circ Res. 2002 Apr 5;90(6):697-702.

[91] Risau W, Flamme I. Vasculogenesis. Annu Rev Cell Dev Biol. 1995;11:73-91.

[92] Hawk ET, Viner JL, Dannenberg A, DuBois RN. COX-2 in cancer--a player that's defining the rules. J Natl Cancer Inst. 2002 Apr 17;94(8):545-6.

[93] Piazza GA, Rahm AL, Krutzsch M, Sperl G, Paranka NS, Gross PH, *et al.* Antineoplastic drugs sulindac sulfide and sulfone inhibit cell growth by inducing apoptosis. Cancer Res. 1995 Jul 15;55(14):3110-6.

[94] Aggarwal S, Taneja N, Lin L, Orringer MB, Rehemtulla A, Beer DG. Indomethacin-induced apoptosis in esophageal adenocarcinoma cells involves upregulation of Bax and translocation of mitochondrial cytochrome C independent of COX-2 expression. Neoplasia. 2000 Jul-Aug;2(4):346-56.

[95] Vogt T, McClelland M, Jung B, Popova S, Bogenrieder T, Becker B, *et al.* Progression and NSAID-induced apoptosis in malignant melanomas are independent of cyclooxygenase II. Melanoma Res. 2001 Dec;11(6):587-99.

[96] Maier TJ, Schilling K, Schmidt R, Geisslinger G, Grosch S. Cyclooxygenase-2 (COX-2)-dependent and -independent anticarcinogenic effects of celecoxib in human colon carcinoma cells. Biochem Pharmacol. 2004 Apr 15;67(8):1469-78.

[97] Xiao H, Yang CS. Combination regimen with statins and NSAIDs: a promising strategy for cancer chemoprevention. Int J Cancer. 2008 Sep 1;123(5):983-90.

[98] Smith ML, Hawcroft G, Hull MA. The effect of non-steroidal anti-inflammatory drugs on human colorectal cancer cells: evidence of different mechanisms of action. Eur J Cancer. 2000 Mar;36(5):664-74.

[99] Hanif R, Pittas A, Feng Y, Koutsos MI, Qiao L, Staiano-Coico L, *et al.* Effects of nonsteroidal anti-inflammatory drugs on proliferation and on induction of apoptosis in colon cancer cells by a prostaglandin-independent pathway. Biochem Pharmacol. 1996 Jul 26;52(2):237-45.

[100] Waskewich C, Blumenthal RD, Li H, Stein R, Goldenberg DM, Burton J. Celecoxib exhibits the greatest potency amongst cyclooxygenase (COX) inhibitors for growth inhibition of COX-2-negative hematopoietic and epithelial cell lines. Cancer Res. 2002 Apr 1;62(7):2029-33.

[101] Elder DJ, Halton DE, Hague A, Paraskeva C. Induction of apoptotic cell death in human colorectal carcinoma cell lines by a cyclooxygenase-2 (COX-2)-selective nonsteroidal anti-

inflammatory drug: independence from COX-2 protein expression. Clin Cancer Res. 1997 Oct;3(10):1679-83.

[102] Chiu CH, McEntee MF, Whelan J. Sulindac causes rapid regression of preexisting tumors in Min/+ mice independent of prostaglandin biosynthesis. Cancer Res. 1997 Oct 1;57(19):4267-73.

[103] Narisawa T, Hermanek P, Habs M, Schmahl D. Reduction of carcinogenicity of N-nitrosomethylurea by indomethacin and failure of resuming effect of prostaglandin E2 (PGE2) against indomethacin. J Cancer Res Clin Oncol. 1984;108(2):239-42.

[104] Tegeder I, Pfeilschifter J, Geisslinger G. Cyclooxygenase-independent actions of cyclooxygenase inhibitors. Faseb J. 2001 Oct;15(12):2057-72.

[105] Wechter WJ, Kantoci D, Murray ED, Jr., Quiggle DD, Leipold DD, Gibson KM, *et al.* R-flurbiprofen chemoprevention and treatment of intestinal adenomas in the APC(Min)/+ mouse model: implications for prophylaxis and treatment of colon cancer. Cancer Res. 1997 Oct 1;57(19):4316-24.

[106] Grosch S, Schilling K, Janssen A, Maier TJ, Niederberger E, Geisslinger G. Induction of apoptosis by R-flurbiprofen in human colon carcinoma cells: involvement of p53. Biochem Pharmacol. 2005 Mar 1;69(5):831-9.

[107] Soh JW, Weinstein IB. Role of COX-independent targets of NSAIDs and related compounds in cancer prevention and treatment. Prog Exp Tumor Res. 2003;37:261-85.

[108] Soh JW, Mao Y, Kim MG, Pamukcu R, Li H, Piazza GA, *et al.* Cyclic GMP mediates apoptosis induced by sulindac derivatives *via* activation of c-Jun NH2-terminal kinase 1. Clin Cancer Res. 2000 Oct;6(10):4136-41.

[109] Hsu AL, Ching TT, Wang DS, Song X, Rangnekar VM, Chen CS. The cyclooxygenase-2 inhibitor celecoxib induces apoptosis by blocking Akt activation in human prostate cancer cells independently of Bcl-2. J Biol Chem. 2000 Apr 14;275(15):11397-403.

[110] Wu T, Leng J, Han C, Demetris AJ. The cyclooxygenase-2 inhibitor celecoxib blocks phosphorylation of Akt and induces apoptosis in human cholangiocarcinoma cells. Mol Cancer Ther. 2004 Mar;3(3):299-307.

[111] Lai GH, Zhang Z, Sirica AE. Celecoxib acts in a cyclooxygenase-2-independent manner and in synergy with emodin to suppress rat cholangiocarcinoma growth *in vitro* through a mechanism involving enhanced Akt inactivation and increased activation of caspases-9 and -3. Mol Cancer Ther. 2003 Mar;2(3):265-71.

[112] Yang HM, Kim HS, Park KW, You HJ, Jeon SI, Youn SW, *et al.* Celecoxib, a cyclooxygenase-2 inhibitor, reduces neointimal hyperplasia through inhibition of Akt signaling. Circulation. 2004 Jul 20;110(3):301-8.

[113] de Vries EG, Timmer T, Mulder NH, van Geelen CM, van der Graaf WT, Spierings DC, *et al.* Modulation of death receptor pathways in oncology. Drugs Today (Barc). 2003;39 Suppl C:95-109.

[114] Sethi G, Ahn KS, Sandur SK, Lin X, Chaturvedi MM, Aggarwal BB. Indirubin enhances tumor necrosis factor-induced apoptosis through modulation of nuclear factor-kappa B signaling pathway. J Biol Chem. 2006 Aug 18;281(33):23425-35.

[115] Katerinaki E, Haycock JW, Lalla R, Carlson KE, Yang Y, Hill RP, *et al.* Sodium salicylate inhibits TNF-alpha-induced NF-kappaB activation, cell migration, invasion and ICAM-1 expression in human melanoma cells. Melanoma Res. 2006 Feb;16(1):11-22.

[116] Law BK, Waltner-Law ME, Entingh AJ, Chytil A, Aakre ME, Norgaard P, *et al.* Salicylate-induced growth arrest is associated with inhibition of p70s6k and down-regulation of c-

myc, cyclin D1, cyclin A, and proliferating cell nuclear antigen. J Biol Chem. 2000 Dec 8;275(49):38261-7.
[117] Liu G, Ma WY, Bode AM, Zhang Y, Dong Z. NS-398 and piroxicam suppress UVB-induced activator protein 1 activity by mechanisms independent of cyclooxygenase-2. J Biol Chem. 2003 Jan 24;278(4):2124-30.
[118] Houseknecht KL, Cole BM, Steele PJ. Peroxisome proliferator-activated receptor gamma (PPARgamma) and its ligands: a review. Domest Anim Endocrinol. 2002 Mar;22(1):1-23.
[119] Huang TH, Yang DS, Plaskos NP, Go S, Yip CM, Fraser PE, *et al.* Structural studies of soluble oligomers of the Alzheimer beta-amyloid peptide. J Mol Biol. 2000 Mar 17;297(1):73-87.
[120] Podlisny MB, Walsh DM, Amarante P, Ostaszewski BL, Stimson ER, Maggio JE, *et al.* Oligomerization of endogenous and synthetic amyloid beta-protein at nanomolar levels in cell culture and stabilization of monomer by Congo red. Biochemistry. 1998 Mar 17;37(11):3602-11.
[121] Takahashi RH, Almeida CG, Kearney PF, Yu F, Lin MT, Milner TA, *et al.* Oligomerization of Alzheimer's beta-amyloid within processes and synapses of cultured neurons and brain. J Neurosci. 2004 Apr 7;24(14):3592-9.
[122] Walsh DM, Tseng BP, Rydel RE, Podlisny MB, Selkoe DJ. The oligomerization of amyloid beta-protein begins intracellularly in cells derived from human brain. Biochemistry. 2000 Sep 5;39(35):10831-9.
[123] Kayed R, Head E, Thompson JL, McIntire TM, Milton SC, Cotman CW, *et al.* Common structure of soluble amyloid oligomers implies common mechanism of pathogenesis. Science. 2003 Apr 18;300(5618):486-9.
[124] Bales KR, Du Y, Holtzman D, Cordell B, Paul SM. Neuroinflammation and Alzheimer's disease: critical roles for cytokine/Abeta-induced glial activation, NF-kappaB, and apolipoprotein E. Neurobiol Aging. 2000 May-Jun;21(3):427-32; discussion 51-3.
[125] Hoozemans JJ, Veerhuis R, Rozemuller AJ, Eikelenboom P. Non-steroidal anti-inflammatory drugs and cyclooxygenase in Alzheimer's disease. Curr Drug Targets. 2003 Aug;4(6):461-8.
[126] McGeer EG, McGeer PL. Inflammatory processes in Alzheimer's disease. Prog Neuropsychopharmacol Biol Psychiatry. 2003 Aug;27(5):741-9.
[127] Sastre M, Klockgether T, Heneka MT. Contribution of inflammatory processes to Alzheimer's disease: molecular mechanisms. Int J Dev Neurosci. 2006 Apr-May;24(2-3):167-76.
[128] Walsh DM, Selkoe DJ. Deciphering the molecular basis of memory failure in Alzheimer's disease. Neuron. 2004 Sep 30;44(1):181-93.
[129] Skovronsky DM, Lee VM, Trojanowski JQ. Neurodegenerative diseases: new concepts of pathogenesis and their therapeutic implications. Annu Rev Pathol. 2006;1:151-70.
[130] McGeer PL, McGeer EG. NSAIDs and Alzheimer disease: epidemiological, animal model and clinical studies. Neurobiol Aging. 2007 May;28(5):639-47.
[131] Eriksen JL, Sagi SA, Smith TE, Weggen S, Das P, McLendon DC, *et al.* NSAIDs and enantiomers of flurbiprofen target gamma-secretase and lower Abeta 42 *in vivo*. J Clin Invest. 2003 Aug;112(3):440-9.
[132] Gasparini L, Ongini E, Wilcock D, Morgan D. Activity of flurbiprofen and chemically related anti-inflammatory drugs in models of Alzheimer's disease. Brain Res Brain Res Rev. 2005 Apr;48(2):400-8.

[133] Gasparini L, Rusconi L, Xu H, del Soldato P, Ongini E. Modulation of beta-amyloid metabolism by non-steroidal anti-inflammatory drugs in neuronal cell cultures. J Neurochem. 2004 Jan;88(2):337-48.

[134] Morihara T, Chu T, Ubeda O, Beech W, Cole GM. Selective inhibition of Abeta42 production by NSAID R-enantiomers. J Neurochem. 2002 Nov;83(4):1009-12.

[135] Cole GM, Frautschy SA. Mechanisms of action of non-steroidal anti-inflammatory drugs for the prevention of Alzheimer's disease. CNS Neurol Disord Drug Targets. 2010 Apr;9(2):140-8.

[136] Kim S, Chang WE, Kumar R, Klimov DK. Naproxen interferes with the assembly of Abeta oligomers implicated in Alzheimer's disease. Biophys J. 2011 Apr 20;100(8):2024-32.

[137] Shankar GM, Li S, Mehta TH, Garcia-Munoz A, Shepardson NE, Smith I, *et al.* Amyloid-beta protein dimers isolated directly from Alzheimer's brains impair synaptic plasticity and memory. Nat Med. 2008 Aug;14(8):837-42.

[138] Gandy S, Simon AJ, Steele JW, Lublin AL, Lah JJ, Walker LC, *et al.* Days to criterion as an indicator of toxicity associated with human Alzheimer amyloid-beta oligomers. Ann Neurol. 2010 Aug;68(2):220-30.

[139] Marder K. Tarenflurbil in patients with mild Alzheimer's disease. Curr Neurol Neurosci Rep. 2010 Sep;10(5):336-7.

[140] Wilcock GK, Black SE, Hendrix SB, Zavitz KH, Swabb EA, Laughlin MA. Efficacy and safety of tarenflurbil in mild to moderate Alzheimer's disease: a randomised phase II trial. Lancet Neurol. 2008 Jun;7(6):483-93.

[141] Schiefer IT, Abdul-Hay S, Wang H, Vanni M, Qin Z, Thatcher GR. Inhibition of amyloidogenesis by nonsteroidal anti-inflammatory drugs and their hybrid nitrates. J Med Chem. 2011 Apr 14;54(7):2293-306.

[142] Boggara MB, Faraone A, Krishnamoorti R. Effect of pH and ibuprofen on the phospholipid bilayer bending modulus. J Phys Chem B. 2010 Jun 24;114(24):8061-6.

[143] Zhou Y, Raphael RM. Effect of salicylate on the elasticity, bending stiffness, and strength of SOPC membranes. Biophys J. 2005 Sep;89(3):1789-801.

[144] Tristram-Nagle S, Nagle JF. HIV-1 fusion peptide decreases bending energy and promotes curved fusion intermediates. Biophys J. 2007 Sep 15;93(6):2048-55.

[145] Ferreira H, Lucio M, Lima JL, Cordeiro-da-Silva A, Tavares J, Reis S. Effect of anti-inflammatory drugs on splenocyte membrane fluidity. Anal Biochem. 2005 Apr 1;339(1):144-9.

[146] Lucio M, Bringezu F, Reis S, Lima JL, Brezesinski G. Binding of nonsteroidal anti-inflammatory drugs to DPPC: structure and thermodynamic aspects. Langmuir. 2008 Apr 15;24(8):4132-9.

[147] Giraud MN, Motta C, Romero JJ, Bommelaer G, Lichtenberger LM. Interaction of indomethacin and naproxen with gastric surface-active phospholipids: a possible mechanism for the gastric toxicity of nonsteroidal anti-inflammatory drugs (NSAIDs). Biochem Pharmacol. 1999 Feb 1;57(3):247-54.

[148] Manrique-Moreno M, Villena F, Sotomayor CP, Edwards AM, Munoz MA, Garidel P, *et al.* Human cells and cell membrane molecular models are affected *in vitro* by the nonsteroidal anti-inflammatory drug ibuprofen. Biochim Biophys Acta. 2011 Nov;1808(11):2656-64.

[149] Kundu S, Chakraborty H, Sarkar M, Datta A. Interaction of Oxicam NSAIDs with lipid monolayer: anomalous dependence on drug concentration. Colloids Surf B Biointerfaces. 2009 Apr 1;70(1):157-61.

[150] Wallace JL. Nonsteroidal anti-inflammatory drugs and gastroenteropathy: the second hundred years. Gastroenterology. 1997 Mar;112(3):1000-16.

[151] Whittle BJ, Higgs GA, Eakins KE, Moncada S, Vane JR. Selective inhibition of prostaglandin production in inflammatory exudates and gastric mucosa. Nature. 1980 Mar 20;284(5753):271-3.

[152] Lichtenberger LM, Wang ZM, Romero JJ, Ulloa C, Perez JC, Giraud MN, *et al.* Non-steroidal anti-inflammatory drugs (NSAIDs) associate with zwitterionic phospholipids: insight into the mechanism and reversal of NSAID-induced gastrointestinal injury. Nat Med. 1995 Feb;1(2):154-8.

[153] Roy SM, Bansode AS, Sarkar M. Effect of increase in orientational order of lipid chains and head group spacing on non steroidal anti-inflammatory drug induced membrane fusion. Langmuir. 2010 Dec 21;26(24):18967-75.

[154] Mondal Roy S, Sarkar M. Effect of lipid molecule headgroup mismatch on non steroidal anti-inflammatory drugs induced membrane fusion. Langmuir. 2011 Dec 20;27(24):15054-64.

[155] Chakraborty H, Mondal S, Sarkar M. Membrane fusion: a new function of non steroidal anti-inflammatory drugs. Biophys Chem. 2008 Sep;137(1):28-34.

[156] Mondal S, Sarkar M. Non-steroidal anti-inflammatory drug induced membrane fusion: concentration and temperature effects. J Phys Chem B. 2009 Dec 24;113(51):16323-31.

[157] Mondal Roy S, Sarkar M. Membrane fusion induced by small molecules and ions. J Lipids. 2011;2011:528784.

[158] Haque ME, Chakraborty H, Koklic T, Komatsu H, Axelsen PH, Lentz BR. Hemagglutinin fusion peptide mutants in model membranes: structural properties, membrane physical properties, and PEG-mediated fusion. Biophys J. 2011 Sep 7;101(5):1095-104.

[159] Wiley DC, Skehel JJ. The structure and function of the hemagglutinin membrane glycoprotein of influenza virus. Annu Rev Biochem. 1987;56:365-94.

[160] Chakraborty H, Chakraborty PK, Raha S, Mandal PC, Sarkar M. Interaction of piroxicam with mitochondrial membrane and cytochrome c. Biochim Biophys Acta. 2007 May;1768(5):1138-46.

[161] Fulda S, Galluzzi L, Kroemer G. Targeting mitochondria for cancer therapy. Nat Rev Drug Discov. 2010 Jun;9(6):447-64.

[162] Bianchi N, Osti F, Rutigliano C, Corradini FG, Borsetti E, Tomassetti M, *et al.* The DNA-binding drugs mithramycin and chromomycin are powerful inducers of erythroid differentiation of human K562 cells. Br J Haematol. 1999 Feb;104(2):258-65.

[163] Fibach E, Bianchi N, Borgatti M, Prus E, Gambari R. Mithramycin induces fetal hemoglobin production in normal and thalassemic human erythroid precursor cells. Blood. 2003 Aug 15;102(4):1276-81.

[164] Ferrante RJ, Ryu H, Kubilus JK, D'Mello S, Sugars KL, Lee J, *et al.* Chemotherapy for the brain: the antitumor antibiotic mithramycin prolongs survival in a mouse model of Huntington's disease. J Neurosci. 2004 Nov 17;24(46):10335-42.

[165] Qiu Z, Norflus F, Singh B, Swindell MK, Buzescu R, Bejarano M, *et al.* Sp1 is up-regulated in cellular and transgenic models of Huntington disease, and its reduction is neuroprotective. J Biol Chem. 2006 Jun 16;281(24):16672-80.

[166] Ryu H, Lee J, Hagerty SW, Soh BY, McAlpin SE, Cormier KA, *et al.* ESET/SETDB1 gene expression and histone H3 (K9) trimethylation in Huntington's disease. Proc Natl Acad Sci U S A. 2006 Dec 12;103(50):19176-81.

[167] Chatterjee S, Zaman K, Ryu H, Conforto A, Ratan RR. Sequence-selective DNA binding drugs mithramycin A and chromomycin A3 are potent inhibitors of neuronal apoptosis induced by oxidative stress and DNA damage in cortical neurons. Ann Neurol. 2001 Mar;49(3):345-54.

[168] Liacini A, Sylvester J, Li WQ, Zafarullah M. Mithramycin downregulates proinflammatory cytokine-induced matrix metalloproteinase gene expression in articular chondrocytes. Arthritis Res Ther. 2005;7(4):R777-83.

[169] Majee S, Dasgupta D, Chakrabarti A. Interaction of the DNA-binding antitumor antibiotics, chromomycin and mithramycin with erythroid spectrin. Eur J Biochem. 1999 Mar;260(3):619-26.

[170] Majumder P, Pradhan SK, Devi PG, Pal S, Dasgupta D. Chromatin as a target for the DNA-binding anticancer drugs. Subcell Biochem. 2007;41:145-89.

[171] Phillips A, Darley M, Blaydes JP. GC-selective DNA-binding antibiotic, mithramycin A, reveals multiple points of control in the regulation of Hdm2 protein synthesis. Oncogene. 2006 Jul 13;25(30):4183-93.

[172] Devi PG, Pal S, Banerjee R, Dasgupta D. Association of antitumor antibiotics, mithramycin and chromomycin, with Zn(II). J Inorg Biochem. 2007 Jan;101(1):127-37.

[173] Cons BM, Fox KR. High resolution hydroxyl radical footprinting of the binding of mithramycin and related antibiotics to DNA. Nucleic Acids Res. 1989 Jul 25;17(14):5447-59.

[174] Hou MH, Wang AH. Mithramycin forms a stable dimeric complex by chelating with Fe(II): DNA-interacting characteristics, cellular permeation and cytotoxicity. Nucleic Acids Res. 2005;33(4):1352-61.

[175] Reyzer ML, Brodbelt JS, Kerwin SM, Kumar D. Evaluation of complexation of metal-mediated DNA-binding drugs to oligonucleotides *via* electrospray ionization mass spectrometry. Nucleic Acids Res. 2001 Nov 1;29(21):E103-3.

[176] Demicheli C, Garnier-Suillerot A. Mithramycin: a very strong metal chelating agent. Biochim Biophys Acta. 1993 Aug 20;1158(1):59-64.

[177] Lahiri S, Takao T, Devi PG, Ghosh S, Ghosh A, Dasgupta A, *et al.* Association of aureolic acid antibiotic, chromomycin A3 with Cu2+ and its negative effect upon DNA binding property of the antibiotic. Biometals. 2012 Apr;25(2):435-50.

[178] Feig AL, Panek M, Horrocks WD, Jr., Uhlenbeck OC. Probing the binding of Tb(III) and Eu(III) to the hammerhead ribozyme using luminescence spectroscopy. Chem Biol. 1999 Nov;6(11):801-10.

[179] Gao XL, Patel DJ. Chromomycin dimer-DNA oligomer complexes. Sequence selectivity and divalent cation specificity. Biochemistry. 1990 Dec 11;29(49):10940-56.

[180] Sastry M, Patel DJ. Solution structure of the mithramycin dimer-DNA complex. Biochemistry. 1993 Jul 6;32(26):6588-604.

[181] Silva DJ, Goodnow R, Jr., Kahne D. The sugars in chromomycin A3 stabilize the Mg(2+)-dimer complex. Biochemistry. 1993 Jan 19;32(2):463-71.

[182] Hou MH, Robinson H, Gao YG, Wang AH. Crystal structure of the [Mg2+-(chromomycin A3)2]-d(TTGGCCAA)2 complex reveals GGCC binding specificity of the drug dimer chelated by a metal ion. Nucleic Acids Res. 2004;32(7):2214-22.

[183] Das S, Devi PG, Pal S, Dasgupta D. Effect of complex formation between Zn2+ ions and the anticancer drug mithramycin upon enzymatic activity of zinc(II)-dependent alcohol dehydrogenase. J Biol Inorg Chem. 2005 Jan;10(1):25-32.

[184] Bogyo M. Metalloproteases see the light. Nat Chem Biol. 2006 May;2(5):229-30.
[185] Liu G, Garrett MR, Men P, Zhu X, Perry G, Smith MA. Nanoparticle and other metal chelation therapeutics in Alzheimer disease. Biochim Biophys Acta. 2005 Sep 25;1741(3):246-52.
[186] Brewer GJ, Askari FK. Wilson's disease: clinical management and therapy. J Hepatol. 2005;42 Suppl(1):S13-21.
[187] Watt NT, Hooper NM. The prion protein and neuronal zinc homeostasis. Trends Biochem Sci. 2003 Aug;28(8):406-10.
[188] Tse WC, Boger DL. Sequence-selective DNA recognition: natural products and nature's lessons. Chem Biol. 2004 Dec;11(12):1607-17.
[189] Remsing LL, Gonzalez AM, Nur-e-Alam M, Fernandez-Lozano MJ, Brana AF, Rix U, *et al.* Mithramycin SK, a novel antitumor drug with improved therapeutic index, mithramycin SA, and demycarosyl-mithramycin SK: three new products generated in the mithramycin producer Streptomyces argillaceus through combinatorial biosynthesis. J Am Chem Soc. 2003 May 14;125(19):5745-53.
[190] Imoto S, Haruta Y, Watanabe K, Sasaki S. New DNA binding ligands as a model of chromomycin A3. Bioorg Med Chem Lett. 2004 Oct 4;14(19):4855-9.
[191] Albertini V, Jain A, Vignati S, Napoli S, Rinaldi A, Kwee I, *et al.* Novel GC-rich DNA-binding compound produced by a genetically engineered mutant of the mithramycin producer Streptomyces argillaceus exhibits improved transcriptional repressor activity: implications for cancer therapy. Nucleic Acids Res. 2006;34(6):1721-34.
[192] Chakraborty S, Sehanobish E, Sarkar M. Binding of Cu(II) complexes of oxicam NSAIDs to alternating AT and homopolymeric AT sequences: differential response to variation in backbone structure. J Biol Inorg Chem. 2012 Mar;17(3):475-87.
[193] Roy S, Banerjee R, Sarkar M. Direct binding of Cu(II)-complexes of oxicam NSAIDs with DNA backbone. J Inorg Biochem. 2006 Aug;100(8):1320-31.
[194] Bnerjee RS, M. Spectroscopic studies of microenvironment dictated structural forms of piroxicam and meloxicam. Journal of Luminescence. 2002;99(3):255-63.
[195] Chakraborty H, Banerjee R, Sarkar M. Incorporation of NSAIDs in micelles: implication of structural switchover in drug-membrane interaction. Biophys Chem. 2003 May 1;104(1):315-25.
[196] Chakraborty H, Roy S, Sarkar M. Interaction of oxicam NSAIDs with DMPC vesicles: differential partitioning of drugs. Chem Phys Lipids. 2005 Dec;138(1-2):20-8.
[197] Chakraborty H, Sarkar M. Optical spectroscopic and TEM studies of catanionic micelles of CTAB/SDS and their interaction with a NSAID. Langmuir. 2004 Apr 27;20(9):3551-8.
[198] Chakraborty H, Sarkar M. Effect of counterion on the structural switchover and binding of piroxicam with sodium dodecyl sulfate (SDS) micelles. J Colloid Interface Sci. 2005 Dec 1;292(1):265-70.
[199] Chakraborty H, Sarkar M. Interaction of piroxicam with micelles: effect of hydrophobic chain length on structural switchover. Biophys Chem. 2005 Aug 22;117(1):79-85.
[200] Chakraborty H, Sarkar M. Interaction of piroxicam and meloxicam with DMPG/DMPC mixed vesicles: anomalous partitioning behavior. Biophys Chem. 2007 Feb;125(2-3):306-13.
[201] Gambari R, Feriotto G, Rutigliano C, Bianchi N, Mischiati C. Biospecific interaction analysis (BIA) of low-molecular weight DNA-binding drugs. J Pharmacol Exp Ther. 2000 Jul;294(1):370-7.

[202] Fishbein I, Brauner R, Chorny M, Gao J, Chen X, Laks H, *et al.* Local delivery of mithramycin restores vascular reactivity and inhibits neointimal formation in injured arteries and vascular grafts. J Control Release. 2001 Dec 13;77(3):167-81.

Send Orders for Reprints to reprints@benthamscience.net
56 *Recent Advances in Medicinal Chemistry, Vol. 1*, **2014**, 56-72

CHAPTER 2

Updated Report on a Novel Mercaptopyruvate Sulfurtransferase Thioredoxin-Dependent Redox-Sensing Molecular Switch: A Mechanism for the Maintenance of Cellular Redox Equilibrium

Noriyuki Nagahara*

Department of Environmental Medicine, Nippon Medical School, 1-1-5 Sendagi Bunkyo-ku, Tokyo 113-8602, Japan

Abstract: 3-Mercaptopyruvate sulfurtransferase (MST, EC.2.8.1.2) has two thioredoxin-dependent redox-sensing switches for the regulation of the enzymatic activity. One is an intermolecular disulfide bond formed between two subunits: A cysteine residue on the surface of each subunit was oxidized to form an intersubunit disulfide bond so as to decrease MST activity, and thioredoxin-specific conversion of a dimer to a monomer increased MST activity. Another switch is a catalytic site cysteine, which reversibly forms a low redox potential sulfenate so as to inhibit MST, and thioredoxin-dependent reduction of the sulfenate restored the MST activity. Concludingly, MST partly contributes to the maintenance of cellular redox homeostasis *via* exerting control over cysteine catabolism. This report is an updated version of the previous review [1] with small modifications.

Keywords: Atmospheric oxygen, antioxidative stress, intermolecular disulfide bond, mercaptolactate-cysteine disulfiduria, mercaptopyruvate sulfurtransferase, molecular evolution, redox-sensing switch, thioredoxin.

INTRODUCTION

3-Mercaptopyruvate sulfurtransferase (MST, EC.2.8.1.2) is a 33 kDa simple protein enzyme which is widely distributed in prokaryotes and eukaryotes [2]. Eukaryotic MST is localized in the cytosol and mitochondria [3, 4], and catalyzes a transsulfuration from mercaptopyruvate to pyruvate in a step of degradation process of cysteine. MST possesses thiosulfate sulfurtransferase (TST, EC.2.8.1.1) activity (a transsulfuration from thiosulfate to sulfinate) [2, 5], and detoxifies environmental cyanide *via* a conversion of cyanide to thiocyanate [6-8].

***Address correspondence to Noriyuki Nagahara:** Department of Environmental Medicine, Nippon Medical School, 1-1-5 Sendagi Bunkyo-ku, Tokyo 113-8602, Japan; Tel: +81 3 3822 2131; Fax: + 81 3 5685 3065; E-mail: noriyuki@nms.ac.jp

Our previous study using protein engineering revealed that reciprocal conversion of catalytic properties between MST and TST was successful, which provided strong evidence that MST and TST were evolutionarily related enzymes [2, 5]. MST and TST consist of an N-terminal catalytically inactive domain and a C-terminal catalytically active domain; molecular evolution of the N-terminal domain has been extensively discussed [9]. A primitive TST molecule is a potential precursor of MST [10].

Recently we found that mammalian MST has a novel intermolecular disulfide bond between the dimer, which serves as a redox-sensing switch for the regulation of MST activity, and contributes to the maintenance of cellular redox equilibrium *via* control of cysteine catabolism [11]. Interestingly, mammalian TST has also evidently acquired the switch. From the point of view of molecular evolution and phylogemetics, the switch evolved in MST and TST during and after the increase of the oxygen concentration in the Earth's atmosphere [10]. Therefore, MST and TST serve as an antioxidant protein and/or an environmental adaptation protein [11-13].

Congenital insufficiency or deficiency of MST activity causes an inherited metabolic disorder, mercaptolactate-cysteine disulfiduria (MCDU), with symptoms of oversecretion of mercaptolactate-cysteine disulfide in urine, with or without mental retardation [14-20]. However, the specific pathogenesis of mental retardation has not been clarified. The results of our previous studies suggest the hypothesis that MST can effectively function as an antioxidant in the developing fetal organs, especially the brain, with its increasing oxygen concentrations in the course of development.

In this review, I focus on the intermolecular thioredoxin-dependent redox-sensing switch of MST in comparison with the classical redox-sensing switches which act *via* an intramolecular disulfide bond [21-29].

A REDOX-SENSING MOLECULAR SWITCH FORMED *VIA* AN INTERMOLECULAR DISULFIDE BOND

Behind the Discovery

Rat MST was activated with dithiothreitol (DTT) or reduced thioredoxin; DTT-treated MST was activated with reduced thioredoxin to approximately 2-fold the the activity level, but thioredoxin-treated MST was not activated with DTT [11]. The two different modes of the enzymatic activation indicate the presence of two different sites of action for thioredoxin. The first site is a catalytic site cysteine, which oxidants easily oxidize to inhibit MST *via* formation of a sulfenate, and thioredoxin reduces this to restore the activity [11] (described in the latter section). Furthermore, MST exhibits a monomer-dimer equilibrium *via* an intermolecular disulfide bond, and a change in the equilibrium depending on the redox status. Thus, we hypothesized that the second site was an intermolecular disulfide bond, which served as a redox-sensing switch for regulation of MST activity.

Structural and Functional Properties of the Redox-Sensing Molecular Switch

Mammalian MSTs and TSTs contain two or three exposed and two buried cysteines; one of the two or three exposed cysteines is a catalytic Cys^{247} (rat and human MST) in the active site (Figs. **1** and **2**). The exposed cysteine residue on the surface corresponding to the rat (human) Cys^{263} is conserved among mammalian MSTs, but rat Cys^{154} is unique. From the point of view of molecular evolution of rhodanese family the increase of oxygen in the Earth's atmosphere was critical for the emergence and maintenance of the redox-sensing molecular switch in eukaryotic MSTs (Fig. **3**) [10].

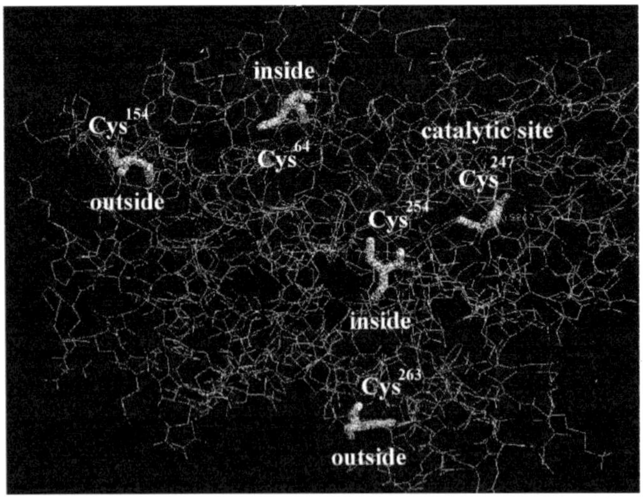

Figure 1: Five cysteines in rat MST. The ternary structure of rat MST was estimated with a computer, based on the nucleic acid sequence data of rat MST [2], and the x-ray structure of bovine rhodasene [42].

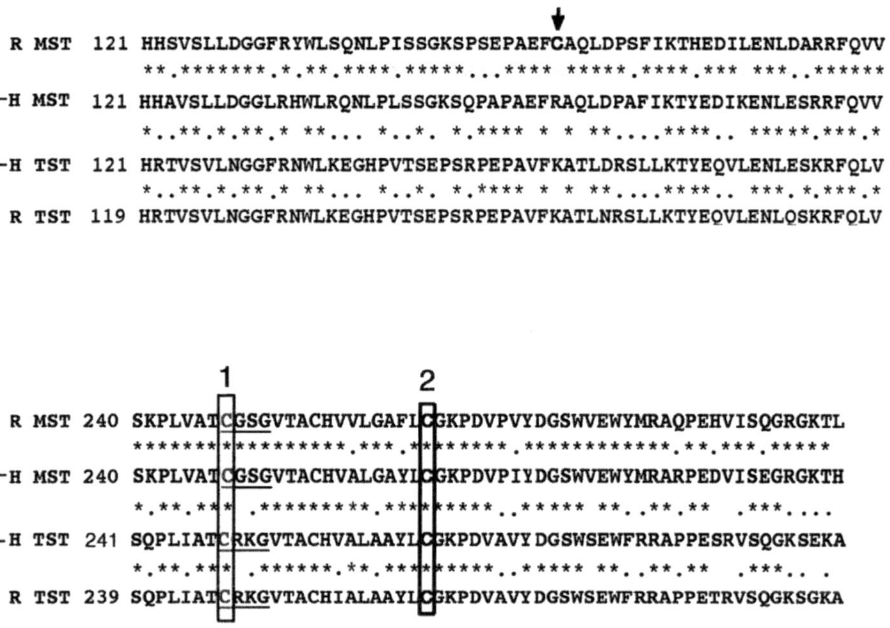

Figure 2: The cysteine(s) composing a redox-sensing molecular switch. Comparison of the amino acid sequences among the rat and human rhodanese family enzymes. H, *Homo sapiens* (BC009450 for MST, D87292 for TST); R, *Rattus norvegicus* (D50564 for MST, BC088449 for TST). Box1, the catalytic site; Box2 and arrow, the redox-sensing switch;, a related amino acid residue to that of human MST; *, an identical amino acid residue to that of human MST.

Rat authentic hepatic and recombinant MSTs exhibit a monomer-dimer equilibrium, and the ratios of the monomer to the dimer were approximately 2 to 1, and 9 to 1, respectively, in 0.2 M potassium phosphate buffer, pH 7.0, containing no reductants under air-saturated conditions [11, 30], which means that an unusual oxidant other than oxygen *per se* reacts with the exposed cysteine residue *in vivo*. HPLC analysis of mutant enzymes (C254S and C263S) strongly suggested that a symmetrical or asymmetrical dimer was formed *via* an intersubunit disulfide bond between Cys^{154} and Cys^{154}, Cys^{154} and Cys^{236}, or Cys^{236} and Cys^{236} [11] (Fig. **4**). Reduced thioredoxin or a reducing system containing thioredoxin, thioredoxin reductase, and NADPH cleaved the intersubunit disulfide bond to activate MST to approximately 2-fold the level of activity of DTT-treated MST (Fig. **5**). On the other hand, reduced glutathione and DTT affected the switch less. As the mid-redox potential of DTT is lower than that of reduced thioredoxin, the reaction specificity is determined by the structural

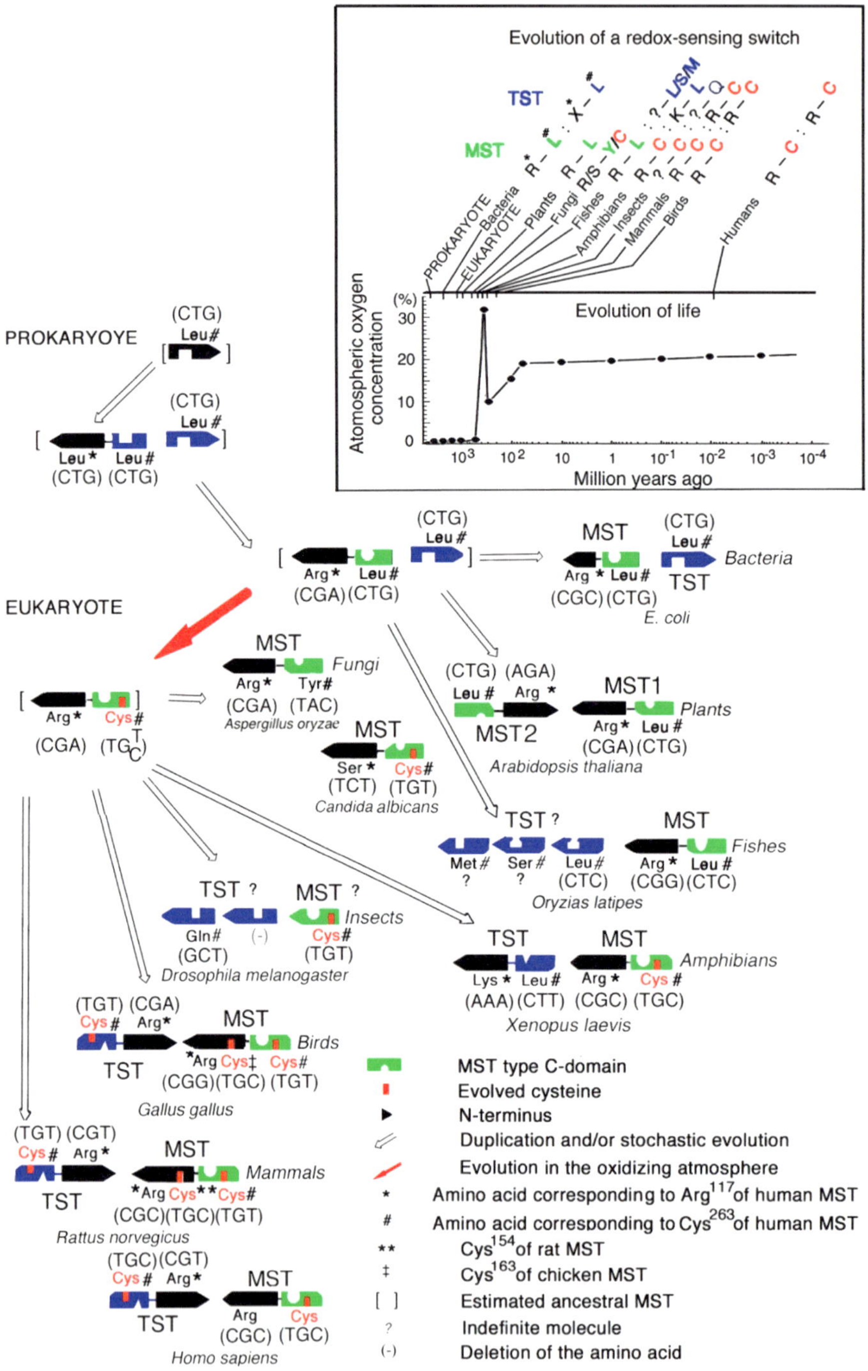

Figure 3: Phylogram of a rhodanese family. Data for codons and deduced amino acid residues were based on cDNA or incomplete genomic DNA data: *Aspergillus oryzae* (AP007175 for MST);

Candida albicans (XM_709437); *Drosophila melanogaster* (BLAST data form FlyBase; National Center for Biotechnology Information; and Berkeley Drosophila Genome Project); *Gallus gallus* (D50564 or XM_001231690 for MST, P25324 or XP_416284 for TST); *Oryzias latipes* (BLAST data form Medaka Expressed Sequence Tags data, the National Bio Resource Project Medaka Genome Project, and National Institute of Genetics, DNA sequencing center); *Xenopus laevis* (BC08421 for MST and BC084422 for TST). Sequence identity was analyzed using GENTYX. Inset, Molecular evolution of a redox-sensing switch in a rhodanese family with changes in oxygen concentration in the Earth's atmosphere. C, Cysteine, which consists of a redox-sensing molecular switch; *, Amino acid corresponding to Arg^{117} of rat MST; #, Amino acid corresponding to Cys^{263} of rat MST. (Reproduced Fig. **2** from Nagahara, N. In: Mohan RM. Ed. Research Advances in Biological Chemistry. Global Research Network, Kerala, 2007; 22).

properties of rat MST, in which the amino acid sequences around Cys^{154} and Cys^{263} are related or identical to parts of the rat and / or *E. coli* thioredoxin reductase sequence [11] (Fig. **6**).

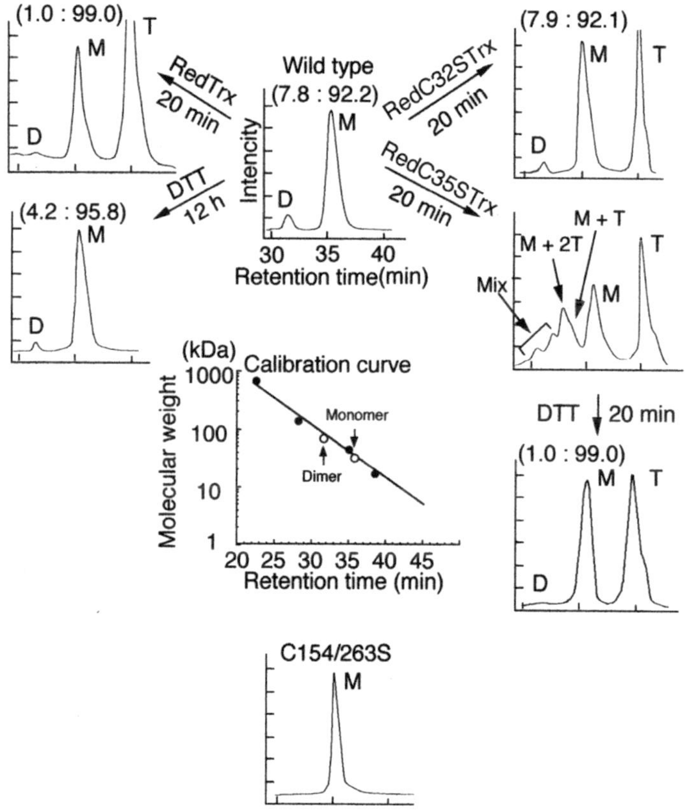

Figure 4: HPLC data for untreated and treated wild-type MST with thioredoxin. An arrow labeled "RedTrx 20 min" (the upper left) means that after *E. coli* reduced thioredoxin (1.5 nmol) was added to the wild-type MST (0.3 nmol), and the mixture was incubated on ice for 20 min. Another

arrow labeled "RedC35STrx 20 min" (lower right) implies that *E. coli* reduced C35S thioredoxin instead was added. The arrow points to a seriously altered profile, suggesting formation of MST-thioredoxin adducts. An arrow proceeding downward labeled "DTT 20 min" implies that DTT (1.5 mM) instead was added, and the mixture was incubated in ice for 20 min. That arrow points to a profile indicating disappearance of the adducts. Another arrow labeled "RedC32STrx 20 min" (upper right) implies that *E. coli* reduced C32S thioredoxin instead was added. The profile was not changed. In another experiment, an arrow labeled "DTT 12 h" (lower left), implying that DTT (1.5 mM) was added, and the mixture was incubated on ice for 12 h. In the middle graph, a calibration curve shows that retention times of standard proteins [thyroglobulin (670 kDa), bovine gamma globulin (158 kDa), chicken ovalbumin (44 kDa), and equine myoglobin (17 kDa)] for the calibration are 22.2, 28.2, 33.3, and 38.3 min, respectively. D, a dimer MST (66.6 kDa, 30.9 min); M, a monomer MST (32.8 kDa, 32.0 min); T, *E. coli* thioredoxin (11.8 kDa, 40.5 min); Mix, mixture of thioredoxin-MST complexes (D, D + T, D + 2T, D + 3T, D + 4T, and M + 2T). Elution profile for the C154/253S MST containing no dimer is shown at the bottom. (Reproduced Fig. **3** from Nagahara, *et al.*, J Biol Chem 2007; 282: 1566).

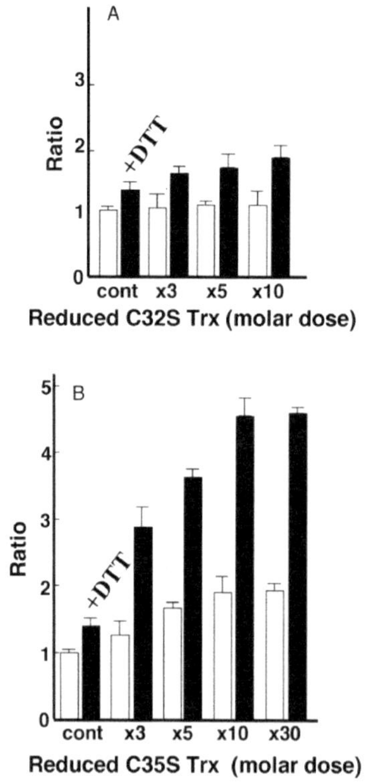

Figure 5: Contribution of Cys^{32} of *E. coli* thioredoxin to MST activation. Wild-type MST (12 μM) was incubated with (A) various concentration of *E. coli* reduced C32S thioredoxin (0, 36, 60, and 120 μM), or (B) various concentration of *E. coli* reduced S35S thioredoxin (0, 36, 60, 120, and 360 μM) in 10 μl of 20 mM potassium phosphate buffer, pH 7.4 on ice for 30 min, a 5-μl aliquot with or without 1.2 mM DTT was used for the rhodanese activity. Data are shown as the mean $\pm$ SE (bar), (n = 3). (Reproduced Fig. **3** from Nagahara, *et al.*, J Biol Chem 2007; 282: 1567).

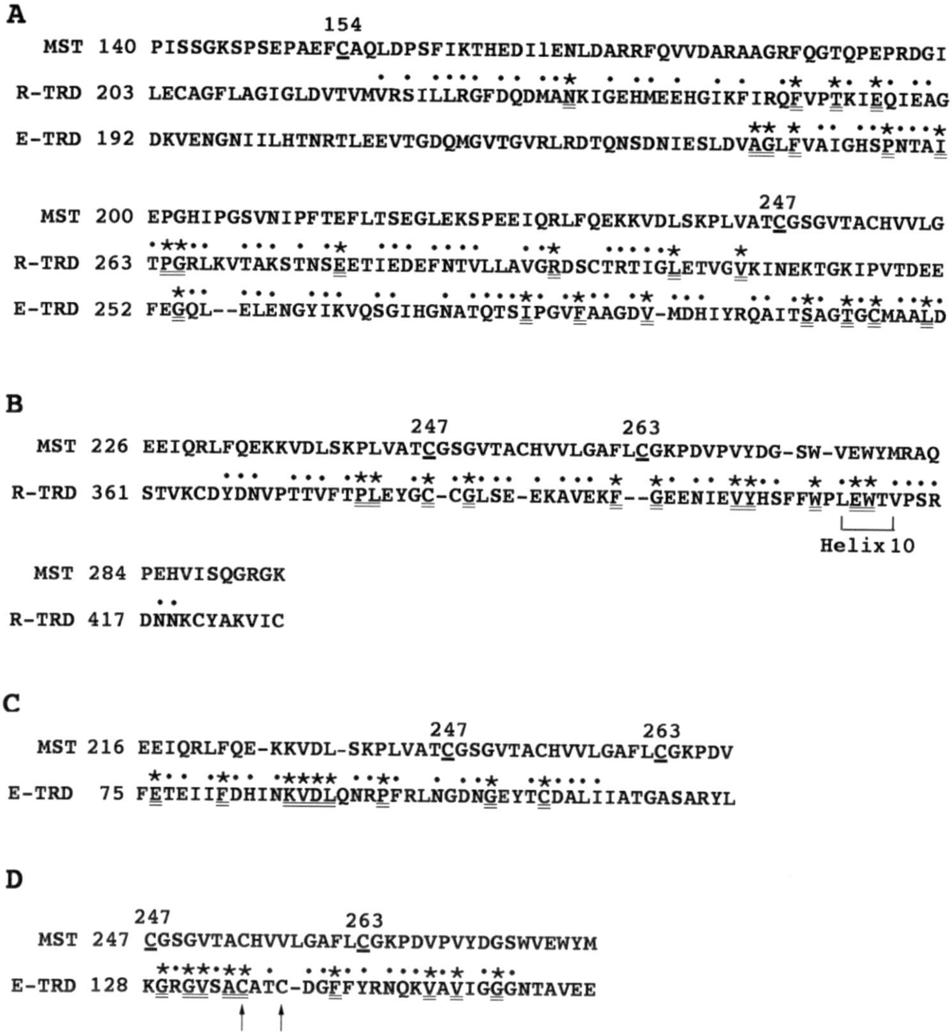

Figure 6: Sequence homology of rat MST to rat TRD and *E. coli* TRD. The sequence around Cys^{154} or Cys^{236} of MST (D50564) is highly related or identical to the sequence of rat TRD (R-TRD) (AF108213) and *E. coli* TRD (E-TRD) (J03762). Helix 10, a proposed Trx binding domain. ·, related amino acid residue; *, identical amino acid residue; ↑, redox active cysteine. (Reproduced Fig. 7 from Nagahara, *et al.*, J Biol Chem 2007; 282: 1568).

In the enzymatic activation experiment, I recognized that the ratio of the concentrations of MST, thioredoxin, thioredoxin reductase, and NDPH in the reducing system was critical for the enzymatic activation of MST (Fig. 7). The ratio in the most effective reducing system was 1 : 1 : 0.05: 12.5 for [rat MST] : [rat thioredoxin] : [rat thioredoxin reductase] : [NADPH], and was 1 : 5 : 0.02 :

12.5 for [rat MST] : [*E. coli* thioredoxin] : [*E. coli* thioredoxin reductase] : [NADPH] [11]. The results suggest two questions.

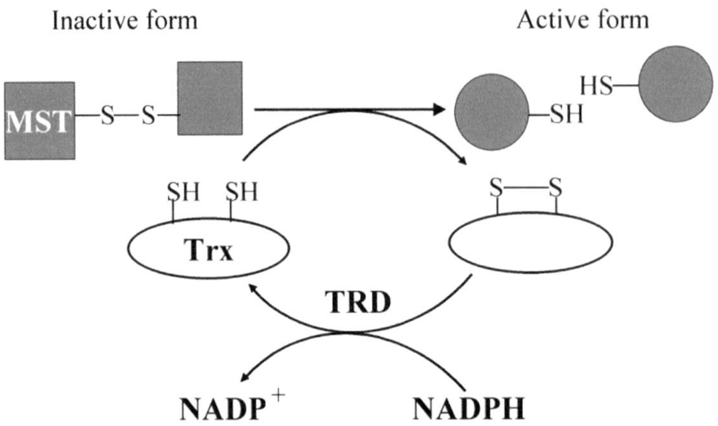

Figure 7: Regulation of the redox-sensing molecular switch. Redox change turns the switch on and off. The thioredoxin-dependent MST activation was regulated by a reducing system containing thioredoxin, thioredoxin reductase and NADPH. In the reaction mixture, MST and thioredoxin reductase may compete in binding thioredoxin. Trx. thioredoxin; TRD, thioredoxin reductase.

First, why is the effective concentration of thioredoxin reductase for the activation of MST low? It is estimated that MST and thioredoxin reductase compete in the binding of thioredoxin, and the affinity of MST to thioredoxin is much lower than that of thioredoxin reductase. Therefore, the concentration of thioredoxin reductase is sufficiently low not to deprive MST of thioredoxin.

Second, why are the effective concentrations of thioredoxin and thioredoxin reductase different between the rat and *E. coli* reducing systems? A reducing system consisting of either a combination of rat thioredoxin and *E. coli* thioredoxin reductase, or *E. coli* thioredoxin and rat thioredoxin reductase, was not available for the activation of MST. Eukaryotic thioredoxin reductase effectively catalyzes the reduction of eukaryotic thioredoxin, but does not effectively catalyze the reduction of prokaryotic thioredoxin (Fig. **8**). On the other hand, rat MST well reacts with both the rat and *E. coli* thioredoxins.

It is concluded that redox status turns the redox-sensing switch on and off to regulate MST activity at the enzymatic level, and MST contributes to maintain the cellular redox homeostasis *via* regulation of cysteine catabolism [11, 14, 30].

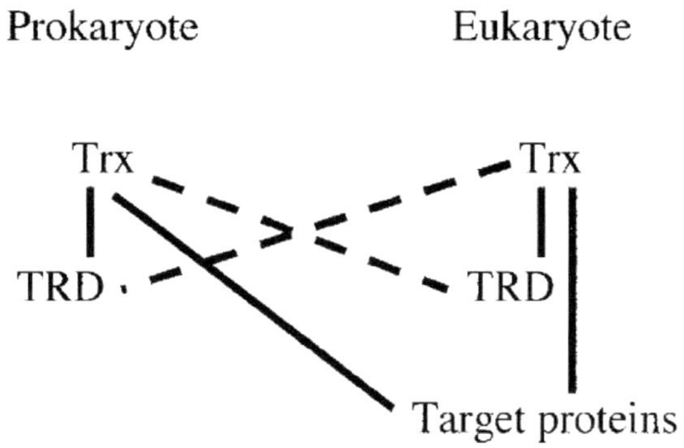

Figure 8: Species specificity for a combination between thioredoxin and thioredoxin reductase. A solid line, a good combination between the two molecules; a dotted line, an incompatible combination. Trx. thioredoxin; TRD, thioredoxin reductase.

Regulatory Mechanism of the Switch by Thioredoxin

Thioredoxin contains two redox-active cysteine residues (Cys^{32} and Cys^{35} in *E. coli* thioredoxin, and Cys^{31} and Cys^{34} in rat thioredoxin), which are involved in the reduction of the oxidized counterparts. *E. coli* reduced C32S thioredoxin did not activate MST [11] (Fig. **5**). On the other hand, *E. coli* reduced C35S thioredoxin dose-dependently activated MST [11] (Figs. **5** and **9**). HPLC analysis revealed that *E. coli* reduced C32S thioredoxin did not affect the dimer, but *E. coli* reduced C35S thioredoxin did [11] (Fig. **4**). An MST-C35S thioredoxin complex (44.6 kD), an MST-2xC35S thioredoxin complex (56.4 kDa), an MST-3xC35S thioredoxin complex (68.2 kDa), a 2xMST-C35S thioredoxin complex (77.4 kDa), a 2xMST-2xC35S thioredoxin complex (89.2 kDa, overlapped), and a 2xMST-3xC35S thioredoxin complex (101 kDa) were formed [11] (Fig. **4**), although the chromatogram did not clearly distinguish all of these MST-thioredoxin complexes. These findings suggest that Cys^{32} of thioredoxin reacts with an intersubunit disulfide bond to form thioredoxin -MST complexes as intermediates.

Other Intermolecular Redox-Sensing Switches

I summarized this issue in a previous review article [31]. In a recent report, plant malate dehydogenese also was found to contain an intermolecular disulfide bond, which either thioredoxin-*h*1 or DTT cleaves to activate the enzymatic activity [32]. This machinary is similar to that of MST. Among the other reported proteins possessing an intermolecular redox-sensing switch, three transcriptional regulators; CprK (H_2O_2 / DTT) (effective treatment with an "oxidant / reductant" for this experiment) [33], ArcB sensor kinase (H_2O_2 / DTT or glutathione) [34], and PpsR1 (H_2O_2 / DTT or ferricyanide) [35] are well characterized. In an exceptional case, the hetero-oligomer ATP synthase is inhibited *via* the formation of an intersubunit disulfide bond between the *bc*' and γ subunits due to a mechanical standstill of the molecular motor ($CuCl_2$ / DTT) [36].

Interestingly, human thioredoxin forms a homo-dimer *via* an intermolecular disulfide bond between the Cys^{72} of each molecule [37, 38], but the physiological significance remains unclear. The fact that glutathionylation of Cys^{72} of human thioredoxin abolished the thioredoxin activity [39] is consistent with the fact that the dimer is an inactive form [38] (Fig. **10**).

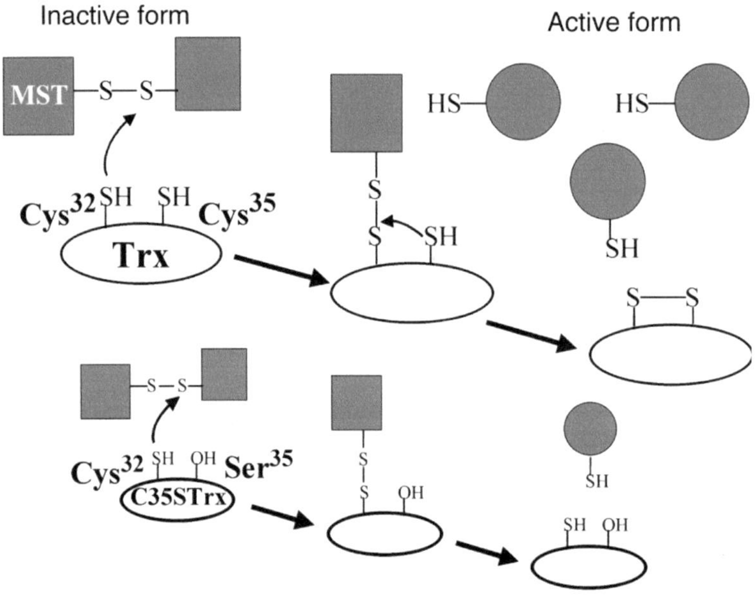

Figure 9: Attack mode of thioredoxin on MST. Trx. thioredoxin. Details are provided in the text.

Intramolecular Redox-Sensing Switches

I summarized this issue in a previous review article [40]. The physiological function of a intramolecular disulfide bond serving as a redox-sensing switch includes regulation of enzymatic activity (phosphatase Cdc25B [21], fructose-1,6-

bisphosphatase [22, 24], AhpF (NADH:peroxiredoxin oxidoreductase) [37], and acetyl-CoA carboxylase [26]), or translational regulation (transcription factor OxyR [28] and NPH1(protein kinase) [29]).

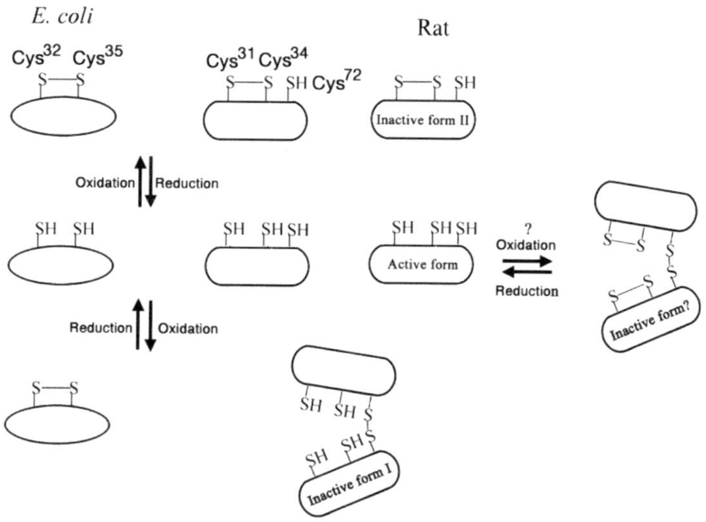

Figure 10: A possible rat thiorodoxin activation induced by redox change. In *E. coli* thioredoxin (NP_418228), exposed cysteines are redox-active Cys^{32} and Cys^{35} alone. In rat thioredoxin (NP_446252), redox active Cys^{61} and Cys^{68} are omitted. Active form, a reduced thioredoxin with full activity; Inactive form I, a dimer form *via* a disulfide bond between Cys^{72} of each monomer without thioredoxin activity due to a steric hindrance; Inactive form II, an oxidized thioredoxin without the activity.

LOW REDOX POTENTIAL SULFENATE FORMATION AT A CATALYTIC SITE CYSTEINE: ANOTHER MECHANISM BY WHICH MST SERVES AS AN ANTI-OXIDANT PROTEIN

In the C-terminal catalytically active domain, a catalytic site cysteine is conserved in the rhodanese family (Fig. **2**). A catalytic site Cys^{247} in rat MST is a target of the oxidants, which is supported by the results of MALDI-TOF mass spectrometric analysis, and also protein chemical study using iodoacetate [12, 30]. A stoichiometric concentration of hydrogen peroxide is easily oxidized to inhibit the enzyme [12, 41] (Fig. **11**). The activity of MST can be completely restored by DTT, reduced thioredoxin or thioredoxin with a reducing system containing thioredoxin reductase and NADPH, but reduced glutathione does not restore the activity [12] (Fig. **10**).

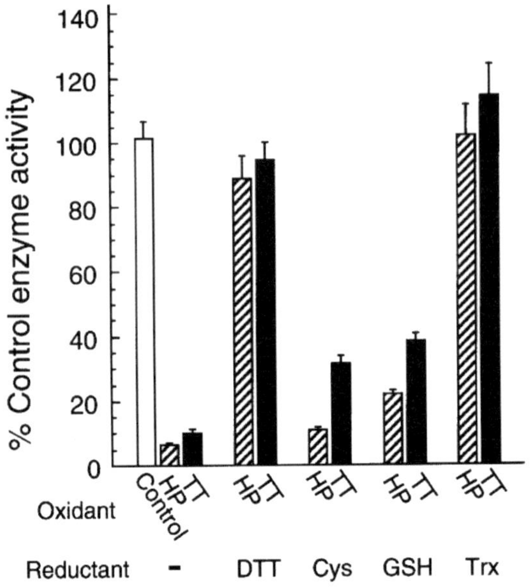

Figure 11: Comparative study for oxidant-inhibited wild type MST and reactivation by reductants. After MST was inhibited by a stoichiometric concentration of hydrogen peroxide (HP) (shaded box) or tetrathionate (TT) (solid black box), free oxidants were removed from each sample and the mixture was concentrated. Each a mixture was incubated with DTT or cysteine.

In the experiment using recombinant *E. coli* reduced thioredoxin (Trx) or yeast reduced glutathione (GSH), the mixture taken from the concentrated sample was added to the reducing system, containing NADPH, thioredoxin or glutathione, and recombinant *E. coli* thioredoxin reductase or glutathione reductase. The mixture was incubated on ice for 20 min. After gel filtration of each mixture, the enzyme-containing fractions were collected and concentrated. Each rhodanese activity of MST was assayed and the data are shown as a percentage of the inactivator-free control rhodanese activity. Data are shown as the mean $\pm$ S.E. (bar), (n = 3). (Reproduced Fig. **2** from Nagahara, *et al.*, J Biol Chem 2005; 280: 34573).

In the reduction process, thioredoxin peroxidase activity was detected [12]. Thus, mild oxidation of rat MST resulted in the formation of a sulfenate (SO^-) at Cys^{247} *via* a donation of one electron to the oxidants, *i.e.*, an antioxidant function. It is noteworthy that the redox potential of the sulfenate is lower than that of reduced glutathione. On the other hand, an excess molar dose of hydrogen peroxide oxidized MST to form a sulfinate or a sulfonate at the catalytic site cysteine, resulting in inactivation of MST.

CONCLUSIONS

1) Oxidants inhibit MST *via* oxidation of a catalytic site cysteine to form a low redox potential sulfenate.

2) At the same time, an inactive form, a dimer is formed *via* an intermolecular disulfide bond.

3) Oxidative stress also inhibits methionine synthase (EC 2.1.1.13) [43, 44], and activates cystathionine β-synthase (EC 4.2.1.22) [45, 46] and glutamate-cysteine ligase (EC 6.3.2.2) [47]. Thus, oxidative stress decreases cysteine degradation, and increases of cellular reductants such as glutathione and thioredoxin.

4) Then, thioredoxin reduces the sulfenate and the disulfide in MST(s) to restore the MST activity.

5) In this process, MST can partly contribute the maintenance of cellular redox equilibrium and the intermolecular disulfide serves as a redox-sensing switch for the regulation of MST activity.

6) These facts suggest that mental retardation in MCDU can be caused by redox imbalance due to congenital defect of MST.

ACKNOWLEDGEMENTS

Declared none.

CONFLICT OF INTEREST

The authors confirm that this chapter contents have no conflict of interest.

DISCLOSURE

The chapter submitted for eBook Series entitled: "**Recent Advances in Medicinal Chemistry, Volume 1**" is an update of our article published in **Mini-Reviews in Medicinal Chemistry, Volume 8, Number 6, pp. 585 to 589**, with additional text and references.

REFERENCES

[1] Nagahara N. A novel mercaptopyruvate sulfurtransferase thioredoxin-dependent redox-sensing molecular switch: a mechanism for the maintenance of cellular redox equilibrium. Mini Rev Med Chem **2008**; 8: 585-9.
[2] Nagahara N, Nishino T. Role of amino acid residues in the active site of rat liver mercaptopyruvate sulfurtransferase. cDNA cloning, overexpression, and site-directed mutagenesis. J Biol Chem **1996**; 271: 27395-401.
[3] Nagahara N, Ito T, Kitamura H, Nishino T. Tissue and subcellular distribution of mercaptopyruvate sulfurtransferase in the rat: confocal laser fluorescence and immunoelectron microscopic studies combined with biochemical analysis. Histochem Cell Biol **1998**; 110: 243-50.
[4] Nakamura T, Yamaguchi Y, Sano H. Plant mercaptopyruvate sulfurtransferases: molecular cloning, subcellular localization and enzymatic activities. Eur J Biochem **2000**; 267: 5621-30.
[5] Nagahara N, Okazaki T, Nishino T. Cytosolic mercaptopyruvate sulfurtransferase is evolutionarily related to mitochondrial rhodanese. Striking similarity in active site amino acid sequence and the increase in the mercaptopyruvate sulfurtransferase activity of rhodanese by site-directed mutagenesis. J Biol Chem **1995**; 270: 16230-5.
[6] Nagahara N, Ito T, Minami M. Mercaptopyruvate sulfurtransferase as a defense against cyanide toxication: molecular properties and mode of detoxification. Histol Histopathol **1999**; 14: 1277-86.
[7] Nagahara N, Li Q, Sawada N. Do antidotes for acute cyanide poisoning act on mercaptopyruvate sulfurtransferase to facilitate detoxification? Curr Drug Targets Immune Endocr Metabol Disord **2003**; 3: 198-204.
[8] Sylvester DM. Hayton W, Morgan RL, Way JL. Effects of thiosulfate on cyanide pharmacokinetics in dogs. Toxicol Appl Pharmacol **1983**; 69: 265-71.
[9] Bordo D, Bork P. The rhodanese/Cdc25 phosphatase superfamily. Sequence-structure-function relations. EMBO Rep **2002**; 3: 741-6.
[10] Nagahara N. Molecular evolution of a thioredoxin-dependent redox-signaling switch in mercaptopyruvate sulfurtransferase. In: Mohan RM, Ed. Research Advances in Biological Chemistry, Vol. 1. Global Research Network: Kerala, **2007**; pp. 19-25.
[11] Nagahara N, Yoshii T, Abe Y, Matsumura T. Thioredoxin-dependent enzymatic activation of mercaptopyruvate sulfurtransferase. An intersubunit disulfide bond serves as a redox switch for activation. J Biol Chem **2007**; 282: 1561-9.
[12] Nagahara N, Yoshii T. An important role of sulfane sulfur at a catalytic cysteine of rat mercaptopyruvate sulfurtransferase in defense against oxidative stress. Amino Acids **2005**; 29: 7.
[13] Nagahara N, Katayama A. Post-translational regulation of mercaptopyruvate sulfurtransferase *via* a low redox potential cysteine-sulfenate in the maintenance of redox homeostasis. J Biol Chem **2005**; 280: 34569-76.
[14] Ampola MG, Efron ML, Bixby EM, Meshorer E. Mental deficiency and a new aminoaciduria. Am J Dis Child **1969**; 117: 66-70.
[15] Crawhall JC, Parker R, Young EP, *et al.* Beta mercaptolactate-cysteine disulfide: analog of cystine in the urine of a mentally retarded patient. Science **1968**; 160: 419-20.
[16] Crawhall JC, Parker R, Sneddon W, Young EP. Beta-mercaptolactate-cysteine disulfide in the urine of a mentally retarded patient. Am J Dis Child **1969**; 117: 71-82.

[17] Crawhall JC, Bir K, Purkiss P, Stanbury JB. Sulfur amino acids as precursors of beta-mercaptolactate-cysteine disulfide in human subjects. Biochem Med **1971**; 5: 109-15.

[18] Crawhall J. Beta-mercaptolactate-cysteine disulfiduria. In: Stanbury JB, Wyngaarden JB, Fredrickson DS, Eds. Metabolic basis of inherited disease, 4th ed. McGraw-Hill, New York, **1978**; pp. 504-13.

[19] Hannestad U, Martensson J, Sjodahl R, Sorbo B. 3-mercaptolactate cysteine disulfiduria: biochemical studies on affected and unaffected members of a family. Biochem Med **1981**; 26: 106-14.

[20] Niederwieser A, Giliberti P, Baerlocher K. beta-Mercaptolactate cysteine disulfiduria in two normal sisters. Isolation and characterization of beta-mercaptolactate cysteine disulfide. Clin Chim Acta **1973**; 43: 405-16.

[21] Buhrman G, Parker B, Sohn J, Rudolph J, Mattos C. Structural mechanism of oxidative regulation of the phosphatase Cdc25B *via* an intramolecular disulfide bond. Biochemistry **2005**; 445: 307-16.

[22] Clancey CJ, Gilbert HF. Thiol/disulfide exchange in the thioredoxin-catalyzed reductive activation of spinach chloroplast fructose-1,6-bisphosphatase. Kinetics and thermodynamics. J Biol Chem **1987**; 262: 13545-49.

[23] Hawkins HC, Blackburn EC, Freedman RB. Comparison of the activities of protein disulphide-isomerase and thioredoxin in catalysing disulphide isomerization in a protein substrate. Biochem J **1991**; 275: 349–53.

[24] Mora-Garcia S, Rodriguez-Suarez R, Wolosiuk RA. Role of electrostatic interactions on the affinity of thioredoxin for target proteins. Recognition of chloroplast fructose-1, 6-bisphosphatase by mutant Escherichia coli thioredoxins. J Biol Chem **1998**; 273: 16273-80.

[25] Reynolds CM, Poole LB. Activity of one of two engineered heterodimers of AhpF, the NADH:peroxiredoxin oxidoreductase from Salmonella typhimurium, reveals intrasubunit electron transfer between domains. Biochemistry **2001**; 40: 3912-9.

[26] Sasaki Y, Kozaki A, Hatano M. Link between light and fatty acid synthesis: thioredoxin-linked reductive activation of plastidic acetyl-CoA carboxylase. Proc Natl Acad Sci USA **1997**; 94: 11096-101.

[27] Sevier CS, Kaiser CA. Disulfide transfer between two conserved cysteine pairs imparts selectivity to protein oxidation by Ero1. Mol Biol Cell **2006**; 17: 2256-66.

[28] Zheng M, Aslund F, Storz G. Activation of the OxyR transcription factor by reversible disulfide bond formation. Science **1998**; 279: 1718-21.

[29] Huala E, Oeller PW, Liscum E, Han I, Larsen E. Science 1997; 278: 2120-3.

[30] Nagahara N, Sawada N. Arabidopsis NPH1: a protein kinase with a putative redox-sensing domain. Curr Med Chem **2006**; 13: 1219-30.

[31] Nagahara N. Redox-sensing cysteine dependent-molecular switches, intermollecular disulfide formation.d In: Taylor JC, Ed. Advances in Chemistry Research. Vol. 12. Nova Science Publishers, Inc., New York, USA, **2012**; pp 243-56.

[32] Hara S, Motohashi K, Arisaka F, *et al.* Thioredoxin-h1 reduces and reactivates the oxidized cytosolic malate dehydrogenase dimer in higher plants. J Biol Chem **2006**; 281: 32065-71.

[33] Pop SM, Gupta N, Raza AS, Ragsdale SW. Transcriptional activation of dehalorespiration. Identification of redox-active cysteines regulating dimerization and DNA binding. J Biol Chem **2006**; 281: 26382-9.

[34] Malpica R, Franco B, Rodriguez C, Kwon O, Georgellis D. Identification of a quinone-sensitive redox switch in the ArcB sensor kinase. Proc Natl Acad Sci USA **2004**; 101: 13318-23.

[35] Jaubert M, Zappa S, Fardoux J. *et al.* Light and redox control of photosynthesis gene expression in Bradyrhizobium: dual roles of two PpsR. Biol Chem **2004**; 279: 44407-16.

[36] Suzuki T, Suzuki J, Mitome N, Ueno H, Yoshida M. Second stalk of ATP synthase. Cross-linking of gamma subunit in F1 to truncated Fob subunit prevents ATP hydrolysis. J Biol Chem **2000**; 275: 37902-9.

[37] Gasdaska JR, Kirkpatrick DL, Montfort, W. *et al.* Oxidative inactivation of thioredoxin as a cellular growth factor and protection by a Cys73->Ser mutation. Biochem Pharmacol **1996**; 52: 1741-7.

[38] Weichsel A, Gasdaska JR, Powis G, Montfort WR. Crystal structures of reduced, oxidized, and mutated human thioredoxins: evidence for a regulatory homodimer. Structure **1996**; 4: 735-51.

[39] Casagrande S, Bonetto V, Fratelli M, *et al.* Glutathionylation of human thioredoxin: a possible crosstalk between the glutathione and thioredoxin systems. Proc Natl Acad Sci USA **2002**; 99: 9745-9.

[40] Nagahara N. Intermolecular disulfide bond to modulate protein function as a redox-sensing switch. Amino Acids **2011**; 41: 59-72.

[41] Horowitz PM, Criscimagna NL. Sulfhydryl-directed triggering of conformational changes in the enzyme rhodanese. J Biol Chem **1988**; 263: 10278-83.

[42] Ploegman JH, Drent G, Kalk KH. Russell J. The covalent and tertiary structure of bovine liver rhodanese. Nature 1978; 273: 124-9.

[43] Mosharov E, Cranford MR, Banerjee R. The quantitatively important relationship between homocysteine metabolism and glutathione synthesis by the transsulfuration pathway and its regulation by redox changes. Biochemistry **2000**; 39: 13005-11.

[44] Taoka S, Ohja S, Shan X, Kruger WD, Banerjee R. Evidence for heme-mediated redox regulation of human cystathionine beta-synthase activity. Biol Chem **1998**; 273: 25179-84.

[45] Chen Z, Chakraborty S, Banerjee R. Demonstration that mammalian methionine synthases are predominantly cobalamin-loaded. J Biol Chem **1995**; 270: 19246-9.

[46] Taoka S, Ohja S, Shan X, Kruger WD, Banerjee R. Evidence for heme-mediated redox regulation of human cystathionine beta-synthase activity. J Biol Chem **1998**; 273: 25179-84.

[47] Dormer UH, Westwater J, Stephen DW, Jamieson DJ. Oxidant regulation of the Saccharomyces cerevisiae GSH1 gene. Biochim Biophys Acta **2002**; 1576: 23-9.

Send Orders for Reprints to reprints@benthamscience.net
Recent Advances in Medicinal Chemistry, Vol. 1, **2014, 73-98**

CHAPTER 3

Characterization of Inorganic Nanomaterials as Therapeutic Vehicles

Tatsuya Murakami[1,4], Masako Yudasaka[2], Sumio Iijima[2,3] and Kunihiro Tsuchida[1,*]

[1]Division for Therapies against Intractable Diseases, Institute for Comprehensive Medical Science (ICMS), Fujita Health University, Toyoake, Aichi 470-1192, Japan; [2]National Institute of Advanced Industrial Science and Technology (AIST), Nanotube Research Center. Central 5, 1-1-1 Higashi, Tsukuba 305-8565, Japan; [3]Faculty of Science and Technology, Meijo University, Shiogamaguchi, Tempaku, Nagoya 468-8502, Japan and [4]Institute for Integrated Cell-Material Sciences, Kyoto University, Sakyo-ku, Kyoto 606-8304, Japan

Abstract: For effective drug actions, concentrations of drugs in the target tissues must be sufficient enough with minimal levels of degradation and dilution. It is desirable that drugs are delivered to the target tissues efficiently. It is also preferable that drugs and therapeutic chemicals do not affect normal tissues. Various methods for drug delivery systems to enhance drug efficacy and reduce adverse drug effects, have been devised by the concomitant development of novel nanomaterials. Nanobiotechnology is one of emerging scientific area that has utilized a variety of inorganic and organic nanomaterials. Each inorganic nanomaterial has its own unique characteristics. In this review, we focus on the usefulness of inorganic nanomaterials, including iron oxide nanoparticles and gold nanoparticles. We also feature fullerenes and carbon nanohorns, both of which are composed entirely of carbons, as therapeutic vehicles, and summarize recent advances in this exciting field of nanoscience and its medical applications.

Keywords: Nanomedicine, drug delivery, iron oxide nanoparticle, gold nanoparticle, fullerene, carbon nanohorn.

INTRODUCTION

Applying nanotechnology to the field of medicine, which has been referred to as nanomedicine has attracted great attention. Nanomedicine covers specific

***Address correspondence to Kunihiro Tsuchida:** Division for Therapies against Intractable Diseases Institute for Comprehensive Medical Sciences (ICMS) Fujita Health University Toyoake, Aichi 470-1192, Japan; Tel: +81-562-93-9384; Fax: +81-562-93-5791; E-mail: tsuchida@fujita-hu.ac.jp intervention with devices and structures within the range from one to several hundred nm, which is the scale of most components of biocompatible complexes. Nanomedicine is expected to lead to the development of scientific and medical tools for drug delivery vehicles, molecular imaging and sensing. Research related with the drug delivery and targeting of pharmaceutical, therapeutic, and diagnostic agents *in vivo*, using nanomaterials, is at the forefront of nanomedicine. Indeed, increased knowledge and techniques developed in the field of nanobiotechnology and nanofabrication has had a groundbreaking impact on devising a variety of drug delivery system.

Drug delivery systems are designed to improve the pharmacological and therapeutic properties of drugs and chemicals administered *in vivo*, and need particulate carriers/vehicles/vectors that can function as drug reservoirs [1-3]. Drug-carrier complexes are needed to achieve controlled release and/or targeting of drugs to desired organs/tissues/infection sites, resulting in alteration of the pharmacokinetics and biodistribution of the drugs [4]. Therefore, nanoparticles are intrinsically advantageous over other conventional materials. The pharmacokinetics of particulate drug carriers is affected mainly by spleen and liver. Splenic filtration occurs at the small interendothelical cell slits (< 200 nm) in the walls of venous sinuses [5, 6]. Furthermore, vesicles larger than 100 nm must be designed to prevent surface opsonization processes, since larger particles undergo phagocytic uptake in liver. Taken together, nanoparticles could be good candidates as long-circulatory drug carriers. Based on stability and enhanced permeability and retention (EPR) effects [7, 8], nanoparticles are particularly useful for cancer therapy and anti-inflammatory therapy.

Among the drug carriers reported to data, anti-fungal drug Amphotericin B and anti-cancer drug doxorubicin (DXR) have been conjugated with liposomes [9, 10] and lipid microspheres have been developed for inflammatory diseases [11]. Several nanocarriers will be in clinical use in near future. Polymer micelles are devised for cancer therapies in clinical applications [12]. Most of them are organic nanomaterials consisting of lipids and/or synthetic polymers. In addition, polymeric nanoparticles have received increasing attention [13]. Compared with the rapid progress in the development of drug delivery systems using organic nanoparticles, inorganic nanoparticle-based drug delivery systems has not been developed until recently.

Recent progress in the field of nanobiotechnology and nanofabrication has led to the production of a variety of inorganic nanomaterials as attractive therapeutic vehicles for drug delivery and imaging. In this review we will feature inorganic nanoparticles, including iron nanoparticles, gold nanoparticles, fullerenes and carbon nanohorns (Table **1**). There are several advantages of these inorganic nanoparticles as drug carriers. First, they are easy to prepare with a defined size in large quantities. More interestingly, they often exhibit multiple functions useful in medical applications, for example as exothermic reactors and contrast agents, whereas organic nanoparticles such as liposomes and microspheres usually serve mainly as drug reservoirs. The emerging roles of these inorganic nanoparticles in drug delivery systems are the focus of this review.

Table 1: Pre-clinical studies on inorganic nanoparticles outlined in this review

Nanoparticle	Payload	Pre-Clinical Model	Outcome	Mechanism of Action	Refs.
Iron oxide	None	Renal cell carcinoma in mice	Remission	Hyperthermia	[21]
	Mitoxantrone	Squamous cell carcinoma in rabbits	Remission	Magnetic drug targeting	[24]
Gold	Plasmid DNA	Normal mice	Electric pulse-assisted gene expression	Systemic gene delivery	[47]
	TNF-α	Colon or mammary carcinoma in mice	Regression	Imaging/drug delivery/ hyperthermia	[51, 52]
	None	Colon carcinoma in mice	Resorption	Photothermal therapy	[58]
Fullerene	Substituents (for HIV therapy)	Normal rat	Rapid clearance from plasma		[61]
Carbon nanohorn	Cisplatin Doxorubicin	Non-small cell carcinoma in mice	Regression	Intratumor drug delivery	[87, 88]
	Zinc phthalocyanine and BSA	c-Ha-ras transformed fibroblast in mice	Regression	Photodynamic therapy (PDT) photohyperthermia (PHT)	[94]

Iron Oxide Nanoparticles

In the late 1980s, iron oxide nanoparticles of 50-100 nm in diameter were first developed as contrast agents for imaging analyses using magnetic resonance [14]. These nanoparticles were solubilized by surface coating with dextran. When intravenously administered to patients with hepatocarcinoma, iron oxide nanomaterials rapidly accumulated within liver, especially in macrophages known as Küpffer cells. Since most cancer tissues contain fewer macrophages compared with normal tissues, cancer tissues is able to be negatively visualized by iron oxide nanoparticles [14]. In order to prolong their half-life in circulation and target tissues and prevent a rapid accumulation in the liver or spleen, much smaller iron oxide nanoparticles with a diameter of ~ 5 nm, designated ultrasmall superparamagnetic iron oxide (USPIO) particles, have been produced [15]. Interestingly, when USPIOs were intravenously administered to rats, they accumulated in iliac, celiac, paraaortic, mesenteric, and mediastinal lymph nodes. The concentration of USPIOs in the lymph nodes was 3.62 ± 0.64%, while that in liver and spleen was 6.32 ± 0.22 and 7.12 ± 0.57%, respectively [15]. Furthermore, it was reported that USPIO could detect axillary lymph nodes in patients with breast cancer [16] and mediastinal lymph nodes in patients with primary lung cancer [17]. Therefore, small iron oxide nanomaterials accumulate in the reticuloendothelial systems including lymph nodes, and are useful for therapeutic vehicles for cancer with lymph node metastasis.

Magnetite nanoparticles of around 10 nm in diameter are attractive compounds for hyperthermia and active drug targeting for cancer therapy [18]. Since the heat generation activity of each single magnetite nanoparticle is not high enough, the magnetite nanoparticles have been encapsulated by cationic liposomes [19] or immunoliposomes [20, 21] to enhance heat generation. Magnetite nanoparticle-loaded anti-HER2 immunoliposomes were used as tumor-targeting vehicles, combining antibody therapy with hyperthermia [20].

The feasibility of magnetite nanoparticles for active drug targeting was reported more than 30 years ago [22]. In that report, magnetite nanoparticles were mixed with cottonseed oil, serum albumin and the anticancer drug, DXR, to prepare magnetic microspheres with an average diameter of 1 μm. The magnetic microparticles could be targeted to tumor tissues in rats under magnetic field irradiation. However, there have been concerns about the long-term deposition of aggregated magnetic particles *in vivo*. Furthermore, strong magnets with constant field gradients are needed. Nonetheless, *in vivo* application of magnetite nanoparticles-mediated hyperthermia seems promising [23].

Several groups have tried to overcome possible problems of magnetite containing drug carriers. Since the physical properties of the emulsified magnetic particles are problematic in some occasions, and not suitable as drug carriers *per se*, Lübbe and coworkers produced magnetite nanoparticles that could adsorb drugs directly [24-26]. The diameter of these nanoparticles is <200 nm, and the magnetite core was coated with starch polymers to stabilize the magnetic nanoparticles and allow chemoabsorptive binding of an anticancer drug, mitoxantrone, through ionic interactions between the phosphate groups of the polymers and the amine groups of mitoxantrone (Fig. (**1a**)) [24]. These magnetite nanoparticles conjugated with mitoxantrone could release most of the mitoxantrone they were carrying within 60 min. When injected intraarterially (*via* femoral artery) near a tumor site in rabbits, under magnetic field irradiation (1.7 Tesla), complete remission of the tumor without recurrence was observed with no sign of toxicity. Furthermore, the intratumoral accumulation of the magnetic nanoparticle was visualized both histologically and by magnetic resonance imaging. In the case of intravenous injection through the ear vein, however, the antitumor activity was largely attenuated. This may indicate that there remains a need to improve the stability of these magnetic particles *in vivo* [24].

Recently, a series of effective methods for the surface modification of magnetite nanoparticles with polymers have been reported. Conversion of hydroxyl groups on the surface into amine groups using γ-(aminopropyl)trialkoxysilane was reported [27, 28]. The converted amine groups were easily accessible and reacted with carboxyl groups at the terminus of poly(ethylene glycol) (PEG) [27] or the aldehyde groups of oxidized dextran [28]. However, these surface-modified magnetite nanoparticles are not suitable to adsorb chemotherapeutic drugs onto their surfaces, because the neutral and hydrophilic polymers are poor adsorbents for drugs. To prevent this kind of molecular characteristics, Jain *et al.* developed a two-step coating procedure to endow magnetite nanoparticles with biocompatibility as well as drug-loading capacity [29]. In the first step, each magnetite nanoparticle was emulsified with oleic acid (OA), which was expected to confer dual functions on the nanoparticles. These OA-coated magnetite nanoparticles were further coated with a PEG-polypropylene oxide (PPO)-

PEG block co-polymer. It is expected that the hydrophobic segments of PPO could anchor at the surface of the OA shell around the magnetite nanoparticle, and that the hydrophilic segments of PEG could extend into the aqueous phase (Fig (**1b**)). Importantly, the double-coated nanoparticles were able to entrap an anticancer drug, DXR, in the OA layer, at 8.2 wt % efficiency, and slowly release approximately 62% of the bound DXR more than one week. Their antiproliferative effect on human breast and prostate cancer cells was also confirmed *in vitro* [29].

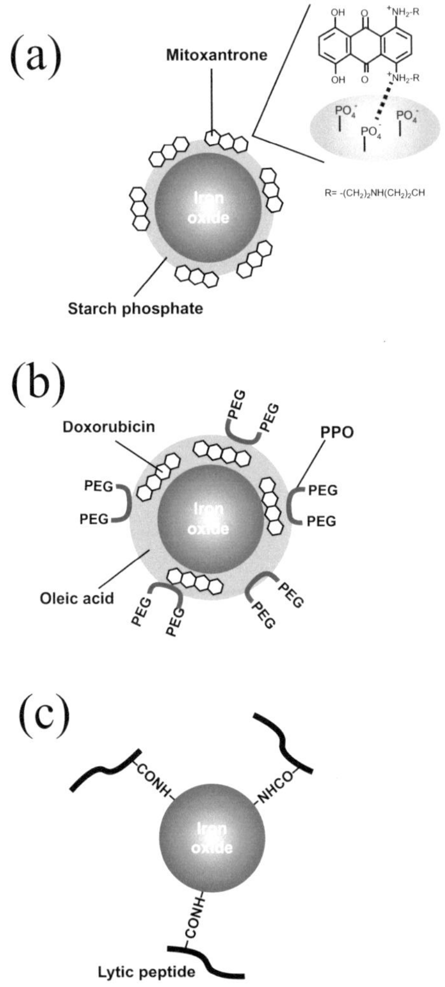

Figure 1: Schematic drawing of various iron oxide nanoparticle-based drug carriers. (a) Iron oxide nanoparticle coated with starch polymers to adsorb mitoxantrone onto the surface through ionic interactions. (b) Iron oxide nanoparticle coated with oleic acids to adsorb doxorubicin through PEG-PPO-PEG block co-polymers. (c) Iron oxide nanoparticle covalently modified with lytic peptides through amide linkage.

The covalent attachment of drugs or chemicals to magnetite nanoparticles is an attractive and alternative approach. Tong *et al.* reported modification of naked magnetite nanoparticles prepared by coprecipitation of ferrous and ferric ions in an ammonia solution with bovine serum albumin, by using carbodiimide [30]. Using this method, lytic peptide-bound magnetite nanoparticles were prepared (Fig. (**1c**)) [31]. The resulting nanomaterials retained superparamagnetism as well as lysis activity toward human breast cancer cells *in vitro*, indicating that lytic peptides also function as hydrophilic and biocompatible polymers. Majority of injected magnetic nanoparticles localized in the liver and spleen, and it decreased over 3 weeks without apparent tissue damages. Therefore, magnetite nanoparticles are biocompatible and can be safely applied for drug delivery and imaging [32]. In summary, magnetite nanoparticles offer attractive drug carriers due to their abilities to achieve hyperthermia and targeted delivery.

Gold Nanoparticles

In the mid-19th century, colloidal gold nanoparticles were first prepared through the chemical reaction of gold chloride with sodium citrate [33]. Gold compounds are known to inhibit cell growth and have low toxicity. They also have reported to decrease cytokine production [34]. It is also of note that several gold compounds, like Auranofin, have been clinically approved and used for the treatment of rheumatoid arthritis and related autoimmune disorders in human [35, 36].

There are a number of reasonable rationales to use gold nanoparticles as vehicles for drug delivery. First of all, gold nanoparticles are relatively easy to synthesize with a defined size [37], and they have low cytotoxicity [38]. Surface modification of gold nanoparticles with multiple polymers and ligands can be achieved in a one-pot synthesis since the gold surface reacts with thiol-groups to form covalent bonds [37]. It should also be of note that gold can be detected at a very low level. Gold nanoparticles can be accurately quantified at a sensitivity as low as 0.001 ppm, by instrumental neutron activation analysis [39].

Like magnetite nanoparticles, gold nanoparticles serve as efficient photothermal agents in therapeutic applications [40]. The absorption maximum of gold nanoparticles is made tunable based on their size and shape. For example, in order to conduct cancer cell-targeted hyperthermia, gold nanoparticles could be conjugated with an anti-epidermal growth factor receptor antibody, which are often overexpressed in plasma membranes of cancer cells to enhance their growth [41]. When the antibody-conjugated gold nanoparticles were incubated with human cancer cells overexpressing epidermal growth factor receptor, and then cells were irradiated at 514 nm with an argon laser, cell death occurred within the irradiated spot. No cell death was observed for benign or normal cells by argon laser irradiation. Unlike magnetite nanoparticles, gold nanoparticles were reported to show anti-angiogenic properties by inhibiting angiogenic cytokines such as vascular endothelial growth factors [42].

Considerable attention has been paid to gold nanoparticles as DNA carriers using "gene gun" technology [43]. DNA-coated gold nanoparticles can be transferred into almost any cell types by the pressure of a compressed gas such as helium or nitrogen [44-46]. In addition, recent reports have shown that gold nanoparticles intrinsically have DNA-delivering capacity. In these reports, gold nanoparticles were reacted with various hetero-bifunctional molecules, such as 2-aminoethanethiol [47], *N,N,N*-trimethyl(11-mercaptoundecyl)ammonium [48], or thiol-modified polyethyleneimine [49], to introduce amine groups onto their surfaces (Fig. (**2a**)). The modified gold nanoparticles could bind to DNA by ionic interactions. Incubating the resulting complexes with medium in cultured cells was sufficient enough to induce DNA expression in mammalian cells. More importantly, the transfection efficiencies of the DNA-gold nanoparticle complexes were several fold higher than those of DNA-polyethyleneimine complexes used as standard transfection reagents [48, 49]. Gold nanoparticles are likely to be applicable even to *in vivo* gene transfer after surface modification with PEG-SH for stabilization [47]. When PEG-modified gold nanoparticles complexed with plasmid DNA were intravenously injected into mice, 5% of the DNA was detected in blood at 5 min after injection, and 20% of the injected gold was detected 120 min after injection, suggesting constant and stable blood circulation of these complexes. In addition, gene expression can be controlled after the intravenous injection of PEG-gold nanoparticles/DNA complexes, by local delivery of electric pulses in the specific area of the tissues and organs. Proinflammatory cytokines were induced by PEG-gold nanoparticles/DNA complexes, however, their level is similar to that of naked DNA, indicating low immunogenicity of gold nanoparticles. However, care must be taken since one report showed that gold nanoparticles could be toxic, even in the absence of electric pulses [49].

In addition, Au/Ni nanorods were synthesized for gene transfer by electrochemical deposition of gold nanoparticles into an Al_2O_3 template of 100 nm in a pore diameter [50]. These nanorods were 100 nm in diameter and 200 nm in length, with 100 nm of gold segments and 100 nm of nickel segments. Each segment of the nanorod was selectively reacted with either thiol- or carboxyl-containing molecules to be further modified with transferrin as a targeting ligand, and plasmid DNAs. The transfection efficiency of DNA-nanorod complexes with targeting ligands in mammalian cells was higher than that of naked DNA or native DNA-nanorod complexes without targeting ligands.

Gold nanoparticles have also been used as therapeutic carriers for polypeptide hormones/growth factors. Tumor necrosis factor (TNF) [51, 52] and insulin [53] have been successfully conjugated with gold nanoparticles, which were prepared by the reduction of $HAuCl_4$. TNF was allowed to react directly with the surface through the formation of covalent bonds, which were cleavable by reduction with 1 μg/ml dithiothreitol treatment. In the case of TNF-immobilized gold nanoparticles, PEG-SH was further introduced to the surface. When these PEG- and TNF-conjugated gold nanoparticles with a mean diameter of ~33 nm were intravenously administered to mice with colon carcinomas, gold nanoparticles accumulated at the tumor site within 3 hours of injection. Although TNF has been shown to have antitumor activity, it evokes toxicity when applied systemically. All mice receiving an injection of native TNF, at a dose of 24 μg per mice, died. However, injection of PEG- and TNF-conjugated gold nanoparticles containing 24 μg of TNF did not cause death. The tumor volume in these mice regressed by more than 90%, indicating that the antitumor action of TNF was effective without toxicity. For insulin binding, the surface was capped with aspartic acid and insulin was adsorbed through hydrogen bonding in a non-covalent manner, because a rapid release of insulin is favorable to decrease blood glucose levels.

Like magnetite nanoparticles, gold nanoparticles are reported to serve as good therapeutic vehicles for drugs with low molecular weights. Firstly, Gu *et al.* prepared

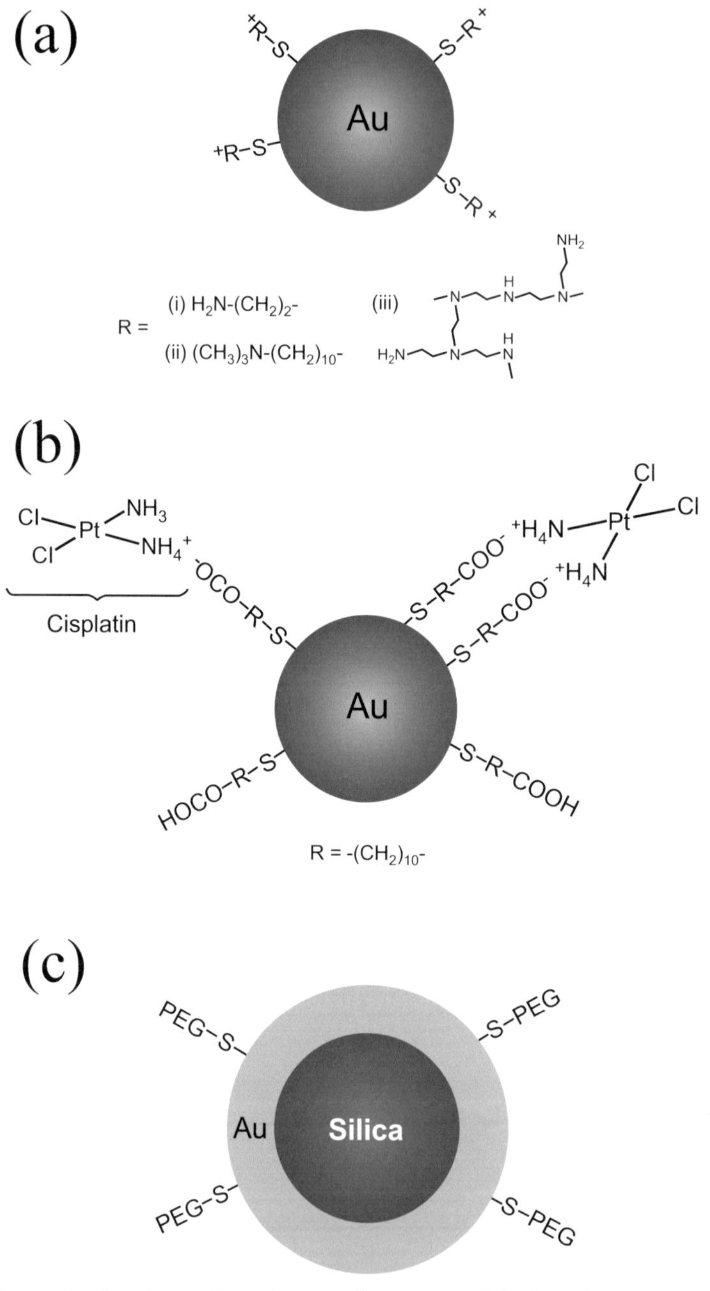

Figure 2: Schematic drawing of various gold nanoparticle-based drug carriers. (a) Gold nanoparticle covalently modified with (i) 2-aminoethanethiol, (ii) *N,N,N*-trimethyl(11-mercaptoundecyl)ammonium, or (iii) thiol-modified polyethyleneimine for gene delivery. (b) Gold nanoparticle covalently modified with 11-mercaptoundecanoic acids to adsorb cisplatin through ionic interaction. (c) Gold nanoparticle with silica core (gold nanoshell). Gold nanoshell is covalently modified with PEG-SH.

multivalent antibiotics conjugated to gold nanoparticles by using thiol-containing vancomycin derivatives [54]. These complexes exhibited sufficient antibiotic activity against vancomycin-resistant enterococci and *E. coli*. The authors speculated that the multivalency of gold nanoparticles and their binding to substrates on the outer membranes of bacteria could play important roles for efficient effects as antibiotics. Secondly, an anticancer drug, cisplatin, was adsorbed onto gold nanoparticles sensitive to near-infrared light, owing to the presence of 11-mercaptoundecanoic acid layers (Fig. (**2b**)) [55]. Based on Fourier transform infrared spectra and ultra violet-visible spectra, cisplatin was likely to bind the carboxylate groups of 11-mercaptoundecanoic acid *via* ionic interaction. The most intriguing characteristics of this complex was that about 90% of the cisplatin in these complexes was released within 1 min of near-infrared irradiation (1064 nm, 100 mJ/pulse, 7 ns per pulse length, 10 Hz repetition rate), whereas only 40% was released after heating to 40°C without irradiation. Therefore, irradiation-induced rapid drug release could be achieved by using gold nanoparticles

Gold nanoshells are a novel class of near-infrared adsorbing nanomaterials (100-150 nm in dimension) consisting of a spherical dielectric silica core surrounded by a thin layer of gold (Fig. (**2c**)) [56]. An intriguing aspect of this type of gold nanoparticle is its possible application to photothermal therapy by using near-infrared lasers, which penetrate into tissues. Gold nanoshells were modified with PEG-SH in the same manner as gold nanoparticles [57, 58]. PEG-gold nanoshells were intravenously administered to mice bearing murine colon carcinomas. When tumors were exposed to near-infrared light (808 nm diode laser, 800 mW at 4 W/cm^2 for 3 min) 6 hours later administration, complete tumor resorption was observed within 10 days [58]. Since gold nanoshell-composite hydrogels were reported to release chemicals and proteins in response to near-infrared irradiation, gold nanoshells could be used as efficient drug carriers [59]. Therefore, gold nanoshells could be powerful therapeutic drug vehicles with near-infrared light-assisted dual functions of heat generation and controlled drug release [60].

Fullerene (C_{60})

Fullerenes and their derivatives have been widely applied in medical fields. Fullerene derivatives have been characterized as inhibitors for human immunodeficiency virus [61, 62], contrast agents for magnetic resonance imaging

[63-65], antioxidants [66, 67], anti-bacterial agents [68, 69]. They are also useful as targeting vectors to mineralized bone [70, 71] and sensitizers for photodynamic therapy [72, 73]. Fullerenes are chemically active nanomaterials. Isobe *et al.* successfully synthesized a series of cationic fullerenes by various spatial arrangements of several amine groups to achieve optimal gene delivery [74, 75]. As a result of efficacy screening based on the transfection efficiency in mammalian cells, a tetraamino fullerene was found to induce gene expression more efficiently than lipofectin, one of the most widely used lipid-based transfection reagents.

Intravenously administered fullerene derivatives exhibited little acute toxicity in mice. However, since fullerenes are retained in the body for long periods, their chronic toxicities remain to be clarified [76]. In order to develop fullerene-based drugs or gene carriers, chemical and/or physical modifications that alter fullerene absorption/excretion profiles will be essential. There are several reports that mention usefulness of fullerenes and their derivatives as potential drug carriers.

Zakharian *et al.* first reported a fullerene-based slow-release system for an anticancer drug, paclitaxel [77]. Both fullerene and paclitaxel were chemically derivatized for conjugation designed to insert an ester linkage between them. This type of paclitaxel conjugate is known to show antitumor activity after being released from fullerene either by cleavage of the ester linkage, or by enzymatic or physicochemical mechanisms. Incubation of the fullerene-paclitaxel conjugate in bovine plasma at 37°C resulted in the slow release of paclitaxel over a period of 4 hours. For aerosol delivery to the lung, the fullerene-paclitaxel conjugates were further embedded in liposomes through the use of the hydrophobic moiety of fullerenes, while maintaining their anticancer activities. Fullerene derivatives not only have intrinsic drug-like actions but also could be engineered to drug-loaded nanomaterials. These unique characteristics are advantageous to establish a multifunctional system for drug and gene delivery [78].

Carbon Nanohorns

Carbon nanotubes (NTs) are carbon allotropes with a cylinder-like nanostructure. NTs have unique chemical, mechanical and electric properties, which are applicable for nanomedicine and material sciences. Single-wall carbon nanohorns

(NHs) are recently discovered aggregates of single-wall NTs with closed ends (Fig. (**3a**)) [79]. NHs are graphic tubules with a diameter of 2-5 nm, larger than the 1.4 nm diameter of typical NTs, and lengths of 40-50 nm. NHs have a mean diameter of 80-100 nm, and have a large inner space compared with NTs. Around the surface of NHs, a large number of horn-shaped tubes can be observed, some of which are kinked (Fig. (**3b**)). These kinked structures and the closed ends of NTs are generated by the presence of pentagonal cells in the graphene sheets consisting of hexagonal cells. The carbon atoms in pentagonal cells are more chemically reactive than those in hexagonal cells. NHs have been reported to be site-specifically modified by reactive oxygen species [80, 81] or sodium amide [82]. In the former, nanometer-sized pores (<2 nm) with various oxygen functionalities at the edges form in the walls of NHs. Small molecules can then penetrate into the interior space of NHs through these pores [83]. Furthermore, synthesis of neither NHs nor oxidized NHs (ox-NHs) requires a metal catalyst. Thus, extremely pure materials can be prepared without potential toxicity derived from metals, which are usually required for NT preparation [84]. These properties of ox-NHs suggest that they may have a potential advantage as novel vehicles in drug delivery systems.

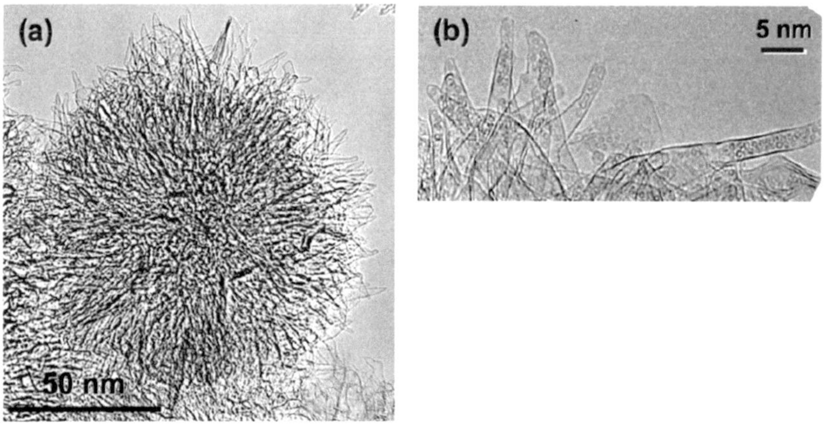

Figure 3: Transmission electron microscopy images of ox-NH. (a) Image of a whole ox-NH. (b) Magnified image showing the surface area of ox-NHs incorporating fullerenes.

The methods for depositing small molecules in the interior spaces of ox-NHs in the liquid phase have been established [85]. The procedure is called as nanoprecipitation and nanoextraction, and it is important to use a solvent with which both guest molecules and ox-NHs are well solubilized. It is reported that ox-NHs can entrap an anti-inflammatory agent, dexamethasone (DEX) [86], and anticancer agents doxorubicin and cisplatin [87, 88], in their surface wall or interior space, using these methods. Fig. (**4a**) shows Langmuir adsorption isotherms showing the adsorption of DEX by NHs, ox-NHs or ox-NHs heat-treated in H_2 at 1200 °C. For all types of NHs, the amount of DEX adsorbed gradually increased in a dose-dependent manner. The amount of DEX adsorbed by ox-NHs in a 0.5 mg/ml of DEX solution was determined to be 200 mg for each gram of ox-NHs, which was approximately six times larger than that obtained for unoxidized NHs. H_2 treatment of ox-NHs, which has been shown to reduce the oxygen functional groups at the pore edges, had only a small effect on the DEX-binding capacity of ox-NHs. These observations strongly suggest that DEX is deposited in the interior space of ox-NHs through the nanometer-sized pores.

Controlled-release of drugs from a drug-carrier complex is one of the essential requirements of drug delivery systems. As shown in Fig. (**4b**), DEX-ox-NH complexes were found to slowly release DEX into phosphate-buffered saline (PBS). This slow release continued for at least two weeks. When DEX-ox-NH complexes were immersed in cell culture media instead of PBS, the initial release rates were significantly increased (Fig. (**4b, inset**)). Various organic compounds present in cell culture medium might be adsorbed to ox-NHs to stimulate release of DEX, or they may enhance the solubility of hydrophobic DEX in media.

It was also shown that DEX released from ox-NH complexes is biologically active. DEX exerts its effects by binding to glucocorticoid receptors in the nucleus, which then activate gene transcription in a DEX-dependent manner [89]. Treatment of mammalian cells transfected with a DEX-dependent reporter plasmid with DEX-ox-NHs activated luciferase expression in a dose-dependent manner, whereas ox-NHs did not have transcriptional activities (Fig. (**5**)). When evaluating the activation level and the DEX-releasing profile during the initial 12 hours, the released DEX was found to retain its biological activity. Prednisolone (PSL), anti-inflammatory glucocorticoid, was also adsorbed on ox-NHs in ethanol-water solvent. PSL was adsorbed on both the inside and outside of the NHs. Locally injected PSL-oxNHs into the tarsal joint of rats suffering from collagen-induced arthritis retarded the progression of the arthritis. Histological analysis indicated the anti-inflammatory effect of PSL-oxNHs *in vivo*. Therefore, PSL-ox-NHs could be useful as therapeutic nanoparticle-based vehicles with anti-inflammatory drugs [90].

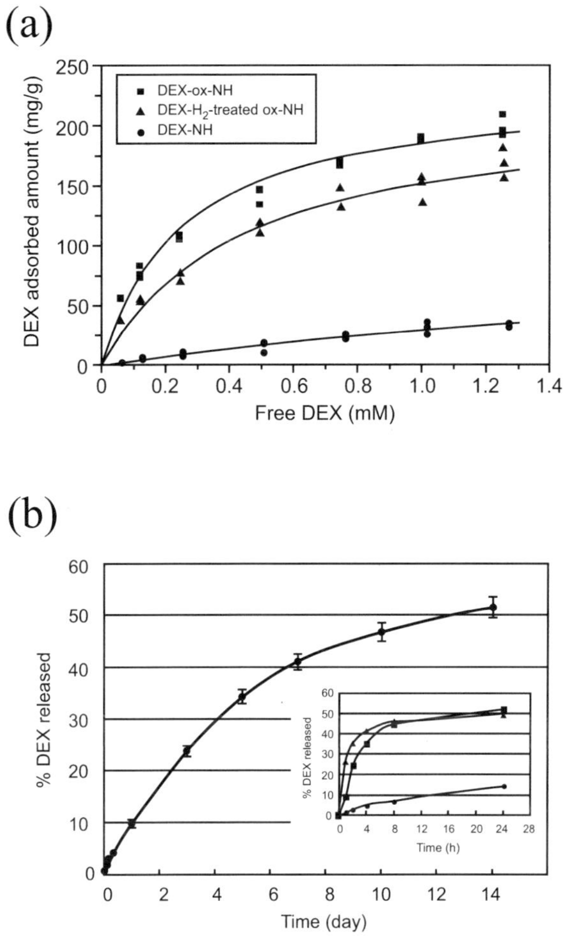

Figure 4: Adsorption and release of DEX by ox-NHs. (a) Langmuir adsorption isotherms showing the adsorption of DEX by ox-NHs, H_2-treated ox-NHs, and NHs: plotted is the amount of DEX adsorbed *vs.* the steady-state drug concentration. (b) Cumulative DEX release profile of ox-NHs in PBS at 37 °C. Total amounts of DEX released up to the indicated times, expressed as percentages of the total DEX bound to DEX-ox-NHs. The inset shows the cumulative release of DEX in PBS (closed circles), RPMI1640 cell culture medium (closed squares), and α-MEM cell culture medium supplemented with 5% FBS (closed triangles) at 37 °C.

Ox-NHs are highly insoluble in aqueous media and readily self-assemble into agglomerates of micrometer size. Particles with a diameter of more than 4 μm may cause vascular occlusion in the human body [91]. Thus, a procedure for dispersing ox-NHs in aqueous solution using an amide-linked polyethylene glycol-doxorubicin (PEG-DXR) conjugate was developed [92]. With its two aromatic rings, DXR is expected to interact with the surfaces of ox-NHs *via* π-π and hydrophobic interactions. When ox-NHs were treated with PEG-DXR, the water solubility increased. PEG-DXR is likely anchored to ox-NHs by its DXR moiety (Fig. (**6a**)).

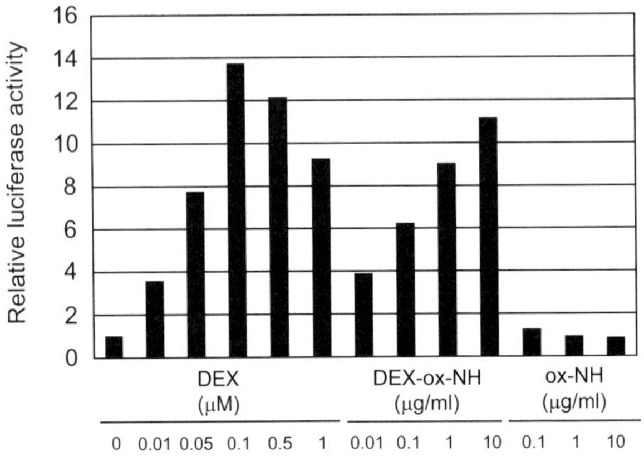

Figure 5: Effects of DEX-ox-NHs and ox-NHs on glucocorticoid receptor-dependent transcriptional activity. Mammalian cells were transfected with a reporter plasmid and treated with DEX, DEX-ox-NHs for 12 hours.

The average diameter of hydrated PEG-DXR-treated ox-NHs was determined to be 174 nm by dynamic light scattering analysis. It should be noted that this diameter is within the range of 120-200 nm that is escapable from being rapidly trapped in liver and spleen [93]. In addition, due to the enhanced permeability and retention (EPR) effect [7, 8], long-circulating nanoparticles of this size have been shown to accumulate within solid tumors. We confirmed the DXR-dependent antiproliferative activity of PEG-DXR-treated ox-NHs on human cancer cells and cancer bearing mice. When PEG-DXR-bound ox-NHs were injected intratumorally to human nonsmall cell lung cancer bearing mice, they significantly retarded tumor growth and prolonged DXR retention in the tumor (Fig. (**6b**)) [88]. Intriguingly, migration of ox-NHs to the adjacent axillary lymph node from the tumor, a major site of metastasis of breast cancer is also observed [88]. Therefore, water-dispersed ox-NHs are expected to be novel drug carriers for chemotherapy against tumors with lymph-node metastasis. Metastasis of cancer in lymph node could be monitored by ox-NHs for imaging analysis.

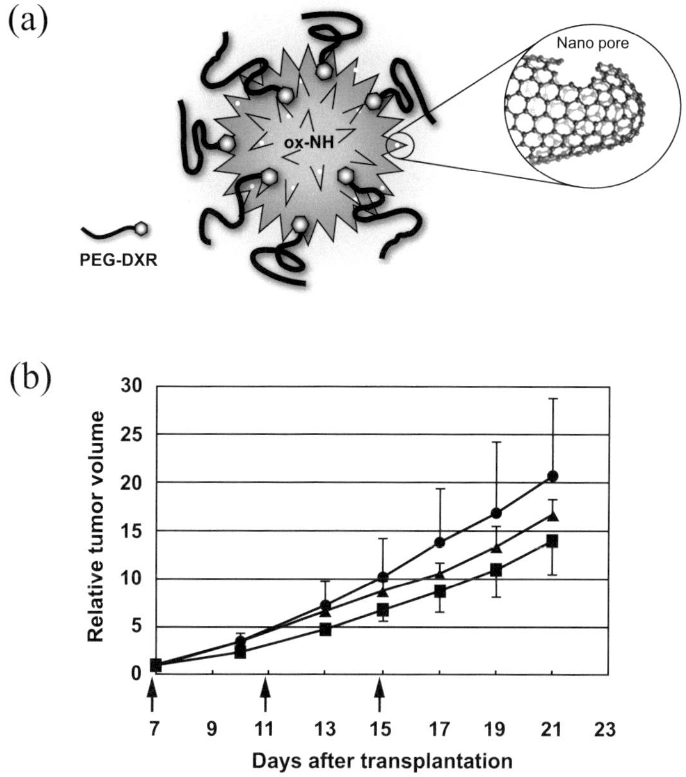

Figure 6: Possible structure and anticancer effects of PEG-DXR-conjugated ox-NH. (a) A possible schematic structure of PEG-DXR-ox-NH with pore. (b) PEG-DXR-ox-NHs showed a higher *in vivo* antitumor activity than PEG-DXR. Human non-small cell lung cancer NCI-H460 cells were implanted subcutaneously in immunodeficient mice. Each drug was intratumorally administered three times on day 7, 11, and 15 (arrows) at the dose of 1.2 mg/kg on a PEG-DXR. Saline (●), PEG-DXR (▲), or PEG-DXR-oxNH (■). The tumor volume is expressed as mean; bar, ±SD. PEG: polyethylene glycol; DXR: doxorubicin.

The solubilization procedure is accompanied by the deposition of the anticancer drug on the outer surface. Furthermore, the interior space of ox-NHs can serve as a carrier for the anticancer drug cisplatin [87]. Therefore, ox-NHs have the potential to function as double reservoirs for two different drugs; one drug on the inside and the other drug on the outside. This multivalency will be a great advantage by devising effective chemotherapies using NHs.

Fabrication of ox-NHs for double photodynamic (PDT) and hyperthermic cancer phototherapy (PHT) is also possible. Zinc phthalocyanine (ZnPc) was loaded onto single-wall carbon nanohorns with opened holes, and bovine serum albumin

(BSA) was attached to the carboxyl groups of ox-NHs. In this system, ZnPc served as the PDT agent, ox-NH was the PHT agent, while BSA enhanced biocompatibility. When ZnPc-ox-NH-BSA was injected into tumors subcutaneously transplanted into mice, the tumors were significantly reduced upon 670-nm laser irradiation [94]. PDT-PHT double therapy system can be devised only with carbon nanotubules such as NHs [94].

In summary, carbon nanohorns are useful carbon nanomaterials for devising multidrug therapy and combining multiple methods for inflammation and cancer treatment.

CONCLUDING REMARKS

In the present review, we focused on the feasibility of inorganic nanoparticles as drug carriers and discussed recent advances in the development of intriguing nanoparticle-derived therapeutic vehicles by using iron oxide nanoparticles, gold nanoparticles, fullerenes, and carbon nanohorns. All of these inorganic nanoparticles have potential advantages for drug targeting and controlled-release, which are the two of the most important characteristics of drug carriers. Such advantages are derived from their structural and/or physicochemical properties, and are hardly observed in other nanoparticles. On the other hand, we should be careful about their potential risk for adverse event as described below.

Recently, increasing concerns have been raised about the toxicities of nanoparticles. Among the nanoparticles featured in this review, however, iron oxide nanoparticles have been clinically approved as contrast agents for magnetic resonance imaging [14]. Gold nanoparticles have been generally recognized to have low toxicity [95], although they may cause tissue-specific cytotoxicity [47]. Fullerenes showed neither cellular toxicity toward mammalian cells [96] nor acute toxicity in mice [76], unless derivatized or modified [97]. Ox-NHs accumulated in spleen and kidney rapidly after intravenous administration to mice, but did not cause any abnormalities at least for two weeks [98]. Future studies should clarify their long-term effects and metabolism of nanoparticles in the body.

In terms of active targeting, superparamagnetic iron oxide nanoparticles are extremely attractive [24, 29]. Although there are many reports regarding immunoliposomes, in which liposomes and antibodies against a specific cell surface antigen are conjugated to achieve active targeting, their therapeutic potential is not as effective as was expected [99, 100]. Therefore, iron oxide nanoparticles could be the good example of clinically practical drug carriers or elements for active drug targeting.

For controlled-release of drugs, inorganic nanoparticles showed unique properties. Gold nanoparticles can achieve controlled release in response to near-infrared irradiation [55, 59]. In most organic drug carriers, including liposomes and lipid microspheres, release of drugs is achieved by drug diffusion, particle erosion, particle degradation, polymer swelling, or diffusion through a membrane or a wall; in other words, it is passively controlled. In this sense, it is safe to say that gold nanoparticles have the potential to exhibit active controlled release induced by exogenous stimuli. Since metal crystals can be deposited in the interior spaces of ox-NHs [98, 101-103], ox-NHs could be endowed with the similar properties to gold nanoparticles as well as iron oxide nanoparticles. In addition, a novel type of drug release by ox-NHs could be developed by introducing "nano caps" for their nanometer-sized pores to control drug release through the pores (Fig. (**6a**)) [104, 105].

Nanomedicine covers multifaceted fields and interdisciplinary cooperation among pharmaceutical and medical sciences, biopolymer and material sciences, tissue engineering and the science of clinical practices. The development of nanomedicine will be partly dependent on the rational design of nano-scaled materials based on a better comprehension of pharmacology and biological processes including metabolism. Since raw inorganic nanoparticles have only low biocompatibility as drug carriers, they should be modified by conjugation with organic materials without impairing their intrinsic properties. Mutual interactions between inorganic and organic material sciences on the nano scale, and interdisciplinary technological cooperation of different research areas, would lead to a breakthrough in the development of drug delivery systems by nanomedicine.

ACKNOWLEDGEMENTS

Declared none.

CONFLICT OF INTEREST

The authors confirm that this chapter contents have no conflict of interest.

DISCLOSURE

The chapter submitted for eBook Series entitled: "**Recent Advances in Medicinal Chemistry, Volume 1**" is an update of our article published in **Mini-Reviews in Medicinal Chemistry, Volume 8, Number 2, pp. 175 to 183**, with additional text and references.

REFERENCES

[1] Barratt, G. Colloidal drug carriers: achievements and perspectives. *Cell. Mol. Life Sci.,* **2003**, *60*, 21-37.

[2] Cuenca, A. G.; Jiang, H.; Hochwald, S. N.; Delano, M.; Cance, W.G.; Grobmyer, S.R. Emerging implications of nanotechnology on cancer diagnostics and therapeutics. *Cancer,* **2006**, *107*, 459-466.

[3] Moghimi, S. M.; Hunter, A. C.; Andresen, T. L. Factors controlling nanoparticle pharmacokinetics: an integrated analysis and perspective. *Annu. Rev. Pharmacol. Toxicol.,* **2012**, *52*, 481-503.

[4] Allen, T. M.; Cullis, P. R. Drug delivery systems: entering the mainstream. *Science,* **2004**, *303*, 1818-1822.

[5] Chen, L. T.; Weiss, L. The role of the sinus wall in the passage of erythrocytes through the spleen. *Blood,* **1973**, 41, 529-537.

[6] Moghimi, S.M.; Porter, C.J.; Muir, I.S.; Illum, L.; Davis, S.S. Non-phagocytic uptake of intravenously injected microspheres in rat spleen: influence of particle size and hydrophilic coating. *Biochem. Biophys. Res. Commun.,* **1991**, *177*, 861-866.

[7] Maeda, H.; Wu, J.; Sawa, T.; Matsumura, Y.; Hori, K. Tumor vascular permeability and the EPR effect in macromolecular therapeutics: a review. *J. Control. Release,* **2000**, *65*, 271-284.

[8] Maeda, H.; Matsumura, Y. EPR effect based drug design and clinical outlook for enhanced cancer chemotherapy. *Adv. Drug Deliv. Rev.,* **2011**, 63, 129-130.

[9] Adler-Moore, J. P.; Proffitt, R. T. Amphotericin B lipid preparations: what are the differences? *Clin. Microbiol. Infect.,* **2008**, *Suppl 4*, 25-36.

[10] Northfelt, D. W.; Martin, F. J.; Working, P.; Volberding, P.A.; Russell, J.; Newman, M.; Amantea, M. A.; Kaplan, L.D. Doxorubicin encapsulated in liposomes containing surface-bound polyethylene glycol: pharmacokinetics, tumor localization, and safety in patients with AIDS-related Kaposi's sarcoma. *J. Clin. Pharmacol.,* **1996**, *36*, 55-63.

[11] Mizushima, Y.; Hoshi, K. DDS for anti-aging and regenerative medicine (review). *J. Drug Target.,* **2006**, *14*, 465-470.

[12] Nishiyama, N.; Morimoto, Y.; Jang, W. D.; Kataoka, K. Design and development of dendrimer photosensitizer-incorporated polymeric micelles for enhanced photodynamic therapy. *Adv. Drug Deliv. Rev.,* **2009**, *61*, 327-338.

[13] Ikuta, Y.; Katayama, N.; Wang, L.; Okugawa, T.; Takahashi, Y.; Schmitt, M.; Gu, X.; Watanabe, M.; Akiyoshi, K.; Nakamura, H.; Kuribayashi, K.; Sunamoto, J.; Shiku, H. Presentation of a major histocompatibility complex class 1-binding peptide by monocyte-derived dendritic cells incorporating hydrophobized polysaccharide-truncated HER2 protein complex: implications for a polyvalent immuno-cell therapy. *Blood,* **2002**, *99*, 3717-3724.

[14] Stark, D. D.; Weissleder, R.; Elizondo, G.; Hahn, P. F.; Saini, S.; Todd, L. E.; Wittenberg, J.; Ferrucci, J. T. Superparamagnetic iron oxide: clinical application as a contrast agent for MR imaging of the liver. *Radiology,* **1988**, *168*, 297-301.

[15] Weissleder, R.; Elizondo, G.; Wittenberg, J.; Rabito, C. A.; Bengele, H. H.; Josephson, L. Ultrasmall superparamagnetic iron oxide: characterization of a new class of contrast agents for MR imaging. *Radiology,* **1990**, *175*, 489-493.

[16] Michel, S. C.; Keller, T. M.; Fröhlich, J. M.; Fink, D.; Caduff, R.; Seifert, B.; Marincek, B.; Kubik-Huch, R. A. Preoperative breast cancer staging: MR imaging of the axilla with ultrasmall superparamagnetic iron oxide enhancement. *Radiology,* **2002**, *225*, 527-536.

[17] Nguyen, B. C.; Stanford, W.; Thompson, B. H.; Rossi, N. P.; Kernstine, K. H.; Kern, J. A.; Robinson, R. A.; Amorosa, J. K.; Mammone, J. F.; Outwater, E. K. Multicenter clinical trial of ultrasmall superparamagnetic iron oxide in the evaluation of mediastinal lymph nodes in patients with primary lung carcinoma. *J. Magn. Reson. Imaging,* **1999**, *10*, 468-473.

[18] Ito, A.; Shinkai, M.; Honda, H.; Kobayashi, T. Medical application of functionalized magnetic nanoparticles. *J. Biosci. Bioeng.* **2005**, *100*, 1-11.

[19] Shinkai, M.; Yanase, M.; Honda, H.; Wakabayashi, T.; Yoshida, J.; Kobayashi, T. Intracellular hyperthermia for cancer using magnetite cationic liposomes: in vitro study. *Jpn. J. Cancer Res.,* **1996**, *87*, 1179-1183.

[20] Ito, A.; Kuga, Y.; Honda, H.; Kikkawa, H.; Horiuchi, A.; Watanabe, Y.; Kobayashi, T. Magnetite nanoparticle-loaded anti-HER2 immunoliposomes for combination of antibody therapy with hyperthermia. *Cancer Lett.,* **2004**, *212*, 167-175.

[21] Shinkai, M.; Le, B.; Honda, H.; Yoshikawa, K.; Shimizu, K.; Saga, S.; Wakabayashi, T.; Yoshida, J.; Kobayashi, T. Targeting hyperthermia for renal cell carcinoma using human MN antigen-specific magnetoliposomes. *Jpn. J. Cancer Res.,* **2001**, *92*, 1138-1145.

[22] Widder, K. J.; Morris, R. M.; Poore, G.; Howard, D. P. Jr.; Senyei, A. E. Tumor remission in Yoshida sarcoma-bearing rts by selective targeting of magnetic albumin microspheres containing doxorubicin. *Proc. Natl. Acad. Sci. USA,* **1981**, *78*, 579-581.

[23] Kobayashi, T. Cancer hyperthermia using magnetic nanoparticles. *Biotechnol. J .,* **2011**, *6*, 1342-1347.

[24] Alexiou, C.; Arnold, W.; Klein, R. J.; Parak, F. G.; Hulin, P.; Bergemann, C.; Erhardt, W.; Wagenpfeil, S.; Lübbe, A. S. Locoregional cancer treatment with magnetic drug targeting. *Cancer Res.,* **2000**, *60*, 6641-6648.

[25] Lübbe, A. S.; Bergemann, C.; Huhnt, W.; Fricke, T.; Riess, H.; Brock, J. W.; Huhn, D. Preclinical experiences with magnetic drug targeting: tolerance and efficacy. *Cancer Res.,* **1996**, *56*, 4694-4701.

[26] Lübbe, A. S.; Alexiou, C.; Bergemann, C. Clinical applications of magnetic drug targeting. *J. Surg. Res.,* **2001**, *95*, 200-206.

[27] Zhang, Y.; Zhang, J. Surface modification of monodisperse magnetite nanoparticles for improved intracellular uptake to breast cancer cells. *J. Colloid. Interface. Sci.,* **2005**, *283*, 352-357.

[28] Sonvico, F.; Mornet, S.; Vasseur, S.; Dubernet, C.; Jaillard, D.; Degrouard, J.; Hoebeke, J.; Duguet, E.; Colombo, P.; Couvreur, P. Folate-conjugated iron oxide nanoparticles for solid tumor targeting as potential specific magnetic hyperthermia mediators: synthesis, physicochemical characterization, and in vitro experiments. *Bioconjug. Chem.,* **2005**, *16*, 1181-1188.

[29] Jain, T. K.; Morales, M. A.; Sahoo, S. K.; Leslie-Pelecky, D. L.; Labhasetwar, V. Iron oxide nanoparticles for sustained delivery of anticancer agents. *Mol. Pharm.,* **2005**, *2*, 194-205.

[30] Tong, X. D.; Xue, B.; Sun, Y. A novel magnetic affinity support for protein adsorption and purification. *Biotechnol. Prog.,* **2001**, *17*, 134-139.

[31] Kumar, C. S.; Leuschner, C.; Doomes, E. E.; Henry, L.; Juban, M.; Hormes, J. Efficacy of lytic peptide-bound magnetite nanoparticles in destroying breast cancer cells. *J. Nanosci. Nanotechnol.,* **2004**, *4*, 245-249.

[32] Jain, T. K.; Reddy, M. K.; Morales, M. A.; Leslie-Pelecky, D. L.; Labhasetwar, V. Biodistribution, clearance, and biocompatibility of iron oxide magnetic nanoparticles in rats. *Mol. Pharm.,* **2008**, *5*, 316-327.

[33] Faraday, M. The Bakerian Lecture: Experimental relations of gold (and other metals) to light. *Philos. Trans. R. Soc. London,* **1857**, *147*, 145-181.

[34] Shukla, R.; Bansal, V.; Chaudhary, M.; Basu, A.; Bhonde, R. R.; Sastry, M. Biocompatibility of gold nanoparticles and their endocytotic fate inside the cellular compartment: a microscopic overview. *Langmuir,* **2005**, *21*, 10644-10654.

[35] Finkelstein, A. E.; Walz, D. T.; Batista, V.; Mizraji, M.; Roisman, F.; Misher, A. Auranofin. New oral gold compound for treatment of rheumatoid arthritis. *Ann. Rheum. Dis.,* **1976**, *35*, 251-257.

[36] Mottram, P. L. Past, present and future drug treatment for rheumatoid arthritis and systemic lupus erythematosus. *Immunol. Cell Biol.,* **2003**, *81*, 350-353.

[37] Daniel, M. C.; Astruc, D. Gold nanoparticles: assembly, supramolecular chemistry, quantum-size-related properties, and applications toward biology, catalysis, and nanotechnology. *Chem. Rev.,* **2004**, *104*, 293-346.

[38] Connor, E. E.; Mwamuka, J.; Gole, A.; Murphy, C. J.; Wyatt, M. D. Gold nanoparticles are taken up by human cells but do not cause acute cytotoxicity. *Small,* **2005**, *1*, 325-327.

[39] Hillyer, J. F.; Albrecht, R. M. Correlative instrumental neutron activation analysis, light microscopy, transmission electron microscopy, and X-ray microanalysis for qualitative and quantitative detection of colloidal gold spheres in biological specimens. *Microsc. Microanal.,* **1998**, *4*, 481-490.

[40] Pitsillides, C. M.; Joe, E. K.; Wei, X.; Anderson, R. R.; Lin, C. P. Selective cell targeting with light-absorbing microparticles and nanoparticles. *Biophys. J.,* **2003**, *84*, 4023-4032.

[41] El-Sayed, I. H.; Huang, X.; El-Sayed, M. A. Selective laser photo-thermal therapy of epithelial carcinoma using anti-EGFR antibody conjugated gold nanoparticles. *Cancer. Lett.,* **2006**, *239*, 129-135.

[42] Mukherjee, P.; Bhattacharya, R.; Wang, P.; Wang, L.; Basu, S.; Nagy, J. A.; Atala, A.; Mukhopadhyay. D.; Soker, S. Antiangiogenic properties of gold nanoparticles. *Clin. Cancer Res.,* **2005**, *11*, 3530-3534.

[43] Lee, P. W.; Peng, S. F.; Su, C. J.; Mi, F. L.; Chen, H. L.; Wei, M. C.; Lin, H. J.; Sung, H. W. The use of biodegradable polymeric nanoparticles in combination with a low-pressure gene gun for transdermal DNA delivery. *Biomaterials,* **2008**, *29*, 742-751.

[44] Larregina, A. T.; Watkins, S. C.; Erdos, G.; Spencer, L.A.; Storkus, W. J.; Berr Stolz, D.; Falo, L. D. Jr. Direct transfection and activation of human cutaneous dendritic cells. *Gene Ther.,* **2001**, *8*, 608-617.
[45] Muangmoonchai, R;. Wong, S. C.; Smirlis, D.; Phillips, I. R.; Shephardl, E. A. Transfection of liver in vivo by biolistic particle delivery: its use in the investigation of cytochrome P450 gene regulation. *Mol. Biotechnol.,* **2002**, *20*, 145-151.
[46] Yang, N. S., Sun, W. H. Gene gun and other non-viral approaches for cancer gene therapy. *Nature Med.,* **1995**, *1*, 481-483.
[47] Kawano, T.; Yamagata, M.; Takahashi, H.; Niidome, Y.; Yamada, S.; Katayama, Y.; Niidome, T. Stabilizing of plasmid DNA in vivo by PEG-modified cationic gold nanoparticles and the gene expression assisted with electrical pulses. *J. Control. Release,* **2006**, *111*, 382-389.
[48] Sandhu, K. K.; McIntosh, C. M.; Simard, J. M.; Smith, S. W.; Rotello, V. M. Gold nanoparticle-mediated transfection of mammalian cells. *Bioconjugate Chem.,* **2002**, *13*, 3-6.
[49] Thomas, M.; Klibanov, A. M. Conjugation to gold nanoparticles enhances polyethylenimine's transfer of plasmid DNA into mammalian cells. *Proc. Natl. Acad. Sci. USA,* **2003**, *100*, 9138-9143.
[50] Salem, A. K.; Searson, P. C.; Leong, K. W. Multifunctional nanorods for gene delivery. *Nature Mater.,* **2003**, *2*, 668-671.
[51] Paciotti, G. F.; Myer, L.; Weinreich, D.; Goia, D.; Pavel, N.; McLaughlin, R. E.; Tamarkin, L. Colloidal gold: a novel nanoparticle vector for tumor directed drug delivery. *Drug Deliv.,* **2004**, *11*, 169-183.
[52] Visaria, R. K.; Griffin, R. J.; Williams, B. W.; Ebbini, E. S.; Paciotti, G. F.; Song, C. W.; Bischof, J. C. Enhancement of tumor thermal therapy using gold nanoparticle-assisted tumor necrosis factor-alpha delivery. *Mol. Cancer Ther.,* **2006**, *5*, 1014-1020.
[53] Joshi, H. M.; Bhumkar, D. R.; Joshi, K.; Pokharkar, V.; Sastry, M. Gold nanoparticles as carriers for efficient transmucosal insulin delivery. *Langmuir,* **2006**, *22*, 300-305.
[54] Gu, H.; Ho, P. L.; Tong, E.; Wang, L.; Xu, B. Presenting vancomycin on nanoparticles to emhance antimicrobial activities. *Nano Lett.,* **2003**, *3*, 1261-1263.
[55] Ren, L.; Chow, G. M. Synthesis of nir-sensitive Au–Au_2S nanocolloids for drug delivery. *Mater. Sci. Eng. C,* **2003**, *23*, 113-116.
[56] Oldenburg, S. J.; Averitt, R. D.; Wetcott, S. L.; Halas, N. J. Nanoengineering of optical resonances. *Chem. Phys. Lett.,* **1998**, *288*, 243-247.
[57] Hirsch, L. R.; Stafford, R. J.; Bankson, J. A.; Sershen, S. R.; Rivera, B.; Price, R. E.; Hazle, J. D.; Halas, N. J.; West, J. L. Nanoshell-mediated near-infrared thermal therapy of tumors under magnetic resonance guidance. *Proc. Natl. Acad. Sci. USA,* **2003**, *100*, 13549-13554.
[58] O'Neal, D. P.; Hirsch, L. R.; Halas, N. J.; Payne, J. D.; West, J. L. Photo-thermal tumor ablation in mice using near infrared-absorbing nanoparticles. *Cancer Lett.,* **2004**, *209*, 171-176.
[59] Sershen, S. R.; Westcott, S. L.; Halas, N. J.; West, J. L. Temperature-sensitive polymer-nanoshell composites for photothermally modulated drug delivery. *J. Biomed. Mater. Res.,* **2000**, *51*, 293-298.
[60] Chen, J.; Saeki, F.; Wiley, B. J.; Cang, H.; Cobb, M. J.; Li, Z. Y.; Au, L.; Zhang, H.; Kimmey, M. B.; Li, X.; Xia, Y. Gold nanocages: bioconjugation and their potential use as optical imaging contrast agents. *Nano Lett.,* **2005**, *5*, 473-477.
[61] Mashino, T.; Shimotohno, K.; Ikegami, N.; Nishikawa, D.; Okuda, K.; Takahashi, K.; Nakamura, S.; Mochizuki, M. Human immunodeficiency virus-reverse transcriptase

inhibition and hepatitis C virus RNA-dependent RNA polymerase inhibition activities of fullerene derivatives. *Bioorg. Med. Chem. Lett.,* **2005**, *15*, 1107-1109.
[62] Rajagopalan, P.; Wudl, F.; Schinazi, R. F.; Boudinot, F. D. Pharmacokinetics of a water-soluble fullerene in rats. *Antimicrob. Agents. Chemother.,* **1996**, *40*, 2262-2265.
[63] Bolskar, R. D.; Benedetto, A. F.; Husebo, L. O.; Price, R. E.; Jackson, E. F.; Wallace, S.; Wilson, L. J.; Alford, J. M. First soluble M@C60 derivatives provide enhanced access to metallofullerenes and permit in vivo evaluation of Gd@C60[C(COOH)2]10 as a MRI contrast agent. *J. Am. Chem. Soc.,* **2003**, *125*, 5471-5478.
[64] Okumura, M.; Mikawa, M.; Yokawa, T.; Kanazawa, Y.; Kato, H.; Shinohara, H. Evaluation of water-soluble metallofullerenes as MRI contrast agents. *Acad. Radiol.,* **2002**, *Suppl 2*, S495-497.
[65] Toth, E.; Bolskar, R. D.; Borel, A.; Gonzalez, G.; Helm, L.; Merbach, A. E.; Sitharaman, B.; Wilson, J. Water-soluble gadofullerenes: toward high-relaxivity, pH-responsive MRI contrast agents. *J. Am. Chem. Soc.,* **2005**, *127*, 799-805.
[66] Gharbi, N.; Pressac, M.; Hadchouel, M.; Szwarc, H.; Wilson, S. R.; Moussa, F. [60]fullerene is a powerful antioxidant in vivo with no acute or subacute toxicity. *Nano Lett.,* **2005**, *5*, 2578-2585.
[67] Wang, J.; Chen, C.; Li, B.; Yu, H.; Zhao, Y.; Sun, J.; Li, Y.; Xing, G.; Yuan, H.; Tang, J.; Chen, Z.; Meng, H.; Gao, Y.; Ye, C.; Chai, Z.; Zhu, C.; Ma, B.; Fang, X.; Wan, L. Antioxidative function and biodistribution of [Gd@C82(OH)22]n nanoparticles in tumor-bearing mice. *Biochem. Pharmacol.,* **2006**, *71*, 872-881.
[68] Mashino, T.; Nishikawa, D.; Takahashi, K.; Usui, N.; Yamori, T.; Seki, M.; Endo, T.; Mochizuki, M. Antibacterial and antiproliferative activity of cationic fullerene derivatives. *Bioorg. Med. Chem. Lett.,* **2003**, *13*, 4395-4397.
[69] Tsao, N.; Kanakamma, P. P.; Luh, T. Y.; Chou, C. K.; Lei, H. Y. Inhibition of Escherichia coli-induced meningitis by carboxyfullerence. *Antimicrob. Agents. Chemother.,* **1999**, *43*, 2273-2277.
[70] Gonzalez, K. A.; Wilson, L. J.; Wu, W.; Nancollas, G. H. Synthesis and in vitro characterization of a tissue-selective fullerene: vectoring C(60)(OH)(16)AMBP to mineralized bone. *Bioorg. Med. Chem.,* **2002**, *10*, 1991-1997.
[71] Mirakyan, A. L.; Wilson, L. J. Functionalization of C_{60} with diphosphonate groups: a route to bone-vectored fullerenes. *J. Chem. Soc. Perkin Trans. 2,* **2002**, 1173-1176.
[72] Rancan, F.; Helmreich, M.; Molich, A.; Jux, N.; Hirsch, A.; Roder, B.; Witt, C.; Bohm, F. Fullerene-pyropheophorbide a complexes as sensitizer for photodynamic therapy: uptake and photo-induced cytotoxicity on Jurkat cells. *J. Photochem. Photobiol. B,* **2005**, *80*, 1-7.
[73] Tabata, Y.; Murakami, Y.; Ikada, Y. Photodynamic effect of polyethylene glycol-modified fullerene on tumor. *Jpn. J. Cancer Res.,* **1997**, *88*, 1108-1116.
[74] Isobe, H.; Nakanishi, W.; Tomita, N.; Jinno, S.; Okayama, H.; Nakamura, E. Nonviral gene delivery by tetraamino fullerene. *Mol. Pharm.,* **2006**, *3*, 124-134.
[75] Isobe. H.; Nakanishi, W.; Tomita, N.; Jinno, S.; Okayama, H.; Nakamura, E. Gene delivery by aminofullerenes: structural requirements for efficient transfection. *Chem. Asian J.,* **2006**, *1-2*, 167-175.
[76] Yamago, S.; Tokuyama, H.; Nakamura, E.; Kikuchi, K.; Kananishi, S.; Sueki, K.; Nakahara, H.; Enomoto, S.; Ambe, F. In vivo biological behavior of a water-miscible fullerene: 14C labeling, absorption, distribution, excretion and acute toxicity. *Chem. Biol.,* **1995**, *2*, 385-389.

[77] Zakharian, T. Y.; Seryshev, A.; Sitharaman, B.; Gilbert, B. E.; Knight, V.; Wilson, L. J. A fullerene-paclitaxel chemotherapeutic: synthesis, characterization, and study of biological activity in tissue culture. *J. Am. Chem. Soc.*, **2005**, *127*, 12508-12509.

[78] Montellano, A.; Da Ros, T.; Bianco, A.; Prato, M. Fullerene C_{60} as a multifunctional system for drug and gene delivery. *Nanoscale,* **2011**, *3*, 4035-4041.

[79] Iijima, S.; Yudasaka, M.; Yamada, R.; Bandow, S.; Suenaga, K.; Kokai, F.; Takahashi, K. Nano-aggregates of single-walled graphitic carbon nano-horns. *Chem. Phys. Lett.,* **1999**, *309*, 165-170.

[80] Ajima, K.; Yudasaka, M.; Suenaga, K.; Kasuya, D.; Azami, T.; Iijima, S. Material storage mechanism in porous nanocarbon. *Adv. Mater.,* **2004**, *16*, 397-401.

[81] Murata, K.; Kaneko, K.; Steele, W. A.; Kokai, F.; Takahashi, K.; Kasuya, D.; Hirahara, K.; Yudasaka, M.; Iijima, S. Molecular potential structures of heat-treated single-wall carbon nanohorn assemblies. *J. Phys. Chem. B,* **2001**, *105*, 10210-10216.

[82] Isobe, H.; Tanaka, H.; Maeda, R.; Noiri, E.; Solin, N.; Yudasaka, M.; Iijima, S.; Nakamura, E. Preparation, purification, characterization, and cytotoxicity assessment of water-soluble, transition-metal-free carbon nanotube aggregates. *Angew. Chem. Int. Ed. Engl.,* **2006**, *45*, 6676-6680.

[83] Murata, K.; Kaneko, K.; Kanoh, H.; Kasuya, D.; Takahashi, K.; Kokai, F.; Yudasaka, M.; Iijima, S. Adsorption mechanism of supercritical hydrogen in internal and interstitial nanospaces of single-wall carbon nanohorn assembly. *J. Phys. Chem. B,* **2002**, *106*, 11132-11138.

[84] Shvedova, A. A.; Castranova, V.; Kisin, E. R.; Schwegler-Berry, D.; Murray, A. R.; Gandelsman, V. Z.; Maynard, A.; Baron, P. Exposure to carbon nanotube material: assessment of nanotube cytotoxicity using human keratinocyte cells. *J. Toxicol. Environ. Health. A,* **2003**, *66*, 1909-1926.

[85] Yudasaka, M.; Ajima, K.; Suenaga, K.; Ichihashi, T.; Hashimoto, A.; Iijima, S. Nano-extraction and nano-condensation for C_{60} incorporation into single-wall carbon nanotubes in liquid phases. *Chem. Phys. Lett.,* **2003**, *380*, 42-46.

[86] Murakami, T.; Ajima, K.; Miyawaki, J.; Yudasaka, M.; Iijima, S.; Shiba, K. Drug-loaded carbon nanohorns: adsorption and release of dexamethasone in vitro. *Mol. Pharm.,* **2004**, *1*, 399-405.

[87] Ajima, K.; Murakami, T.; Mizoguchi, K.; Tsuchida, K.; Ichihashi, S.; Iijima, S.; Yudasaka, M. Enhancement of in vivo anticancer effects of cisplatin by incorporation inside single-wall carbon nanohorns. *ACS Nano,* **2008**, *2*, 2057-2064.

[88] Murakami, T.; Sawada, H.; Tamura, G.; Yudasaka, M.; Iijima, S.; Tsuchida, K. Water-dispersed single-wall carbon nanohorns as drug carriers for local cancer chemotherapy. *Nanomed,* **2008**, *3*, 453-463.

[89] Yamamoto, K. R. Steroid receptor regulated transcription of specific genes and gene networks. *Annu. Rev. Genet.,* **1985**, *19*, 209-252.

[90] Nakamura, M.; Tahara, Y.; Ikehara, Y.; Murakami, T.; Tsuchida, K.; Iijima, S.; Waga, I.; Yudasaka, M. Single-walled carbon nanohorns as drug carriers: adsorption of prednisolone and anti-inflammatory effects on arthritis. *Nanotechnol,* **2011**, *22*, 465102.

[91] Kerr, J. B. *Atlas of Functional Histology,* Mosby International Ltd.: London, **1999**.

[92] Murakami, T.; Fan, J.; Yudasaka, M.; Iijima, S.; Shiba, K. Solubilization of single-wall carbon nanohorns using a PEG-doxorubicin conjugate. *Mol. Pharm.,* **2006**, *3*, 407-414.

[93] Brigger, I.; Dubernet, C.; Couvreur, P. Nanoparticles in cancer therapy and diagnosis. *Adv. Drug Deliv. Rev.,* **2002**, *54*, 631-651.

[94] Zhang, M.; Murakami, T.; Ajima, K.; Tsuchida, K.; Sandanayaka, A. S. D.; Ito, O.; Iijima, S.; Yudasaka, M. Fabrication of ZnPc/protein nanohorns for double photodynamic and hyperthermic cancer phototherapy. *Proc. Natl. Acad. Sci. USA,* **2008**, *105*, 14773-14778.

[95] Bergen, J. M.; von Recum, H. A.; Goodman, T. T.; Massey, A. P.; Pun, S. H. Gold nanoparticles as a versatile platform for optimizing physicochemical parameters for targeted drug delivery. *Macromol. Biosci.,* **2006**, *6*, 506-516.

[96] Fiorito, S.; Serafino, A.; Andreola, F.; Togna, A.; Togna, G. Toxicity and biocompatibility of carbon nanoparticles. *J. Nanosci. Nanotech.,* **2006**, *6*, 591-599.

[97] Sayes, C. M.; Gobin, A. M.; Ausman, K. D.; Mendez, J.; West, J. L.; Colvin, V. L. Nano-C60 cytotoxicity is due to lipid peroxidation. *Biomaterials,* **2005**, *26*, 7587-7595.

[98] Miyawaki, J.; Yudasaka, M.; Imai, H.; Yorimitsu, H.; Isobe, H.; Nakamura, E.; Iijima, S. In vivo magnetic resonance imaging of single-walled carbon nanohorns by labeling with magnetite nanoparticles. *Adv. Mater.,* **2006**, *18*, 1010-1014.

[99] Mastrobattista, E.; Koning, G. A.; Storm, G. Immunoliposomes for the targeted delivery of antitumor drugs. *Adv. Drug Deliv. Rev.,* **1999**, *40*, 103-127.

[100] Noble, C. O.; Kirpotin, D. B.; Hayes, M. E.; Mamot, C.; Hong, K.; Park, J. W.; Benz, C. C.; Marks, J. D.; Drummond, D. C. Development of ligand-targeted liposomes for cancer therapy. *Expert Opin. Ther. Targets,* **2004**, *8*, 335-353.

[101] Miyawaki, J.; Yudasaka, M.; Imai, H.; Yorimitsu, H.; Isobe, H.; Nakamura, E.; Iijima, S. Synthesis of ultrafine Gd2O3 nanoparticles inside single-wall carbon nanohorns. *J. Phys. Chem. B,* **2006**, *110*, 5179-5181.

[102] Murata, K.; Hashimoto, A.; Yudasaka, M.; Kasuya, D.; Kaneko, K.; Iijima, S. The use of charge transfer to enhance the methane-storage capacity of single-walled, nanostructured carbon. *Adv. Mater.,* **2004**, *16*, 1520-1522.

[103] Yuge, R.; Ichihashi, T.; Shimakawa, Y.; Kubo, Y.; Yudasaka, M.; Iijima, S. Preferential deposition of Pt nanoparticles inside single-walled carbon nanohorns. *Adv. Mater.,* **2004**, *16*, 1420-1423.

[104] Ajima, K.; Yudasaka, M.; Maigné, A.; Miyawaki, J.; Iijima, S. Effect of functional groups at hole edges on cisplatin release from inside single-wall carbon nanohorns. *J. Phys. Chem. B,* **2006**, *110*, 5773-5778.

[105] Hashimoto, A.; Yorimitsu, H.; Ajima, K.; Suenaga, K.; Isobe, H.; Miyawaki, J.; Yudasaka, M.; Iijima, S.; Nakamura, E. Selective deposition of a gadolinium(III) cluster in a hole opening of single-wall carbon nanohorn. *Proc. Natl. Acad. Sci. USA,* **2004**, *101*, 8527-8530.

Send Orders for Reprints to reprints@benthamscience.net
Recent Advances in Medicinal Chemistry, Vol. 1, **2014, 99-130**

CHAPTER 4

Aromatase Inhibitors: A New Reality for the Adjuvant Endocrine Treatment of Early-Stage Breast Cancer in Postmenopausal Women

Colozza Mariantonietta[1,*], Minenza Elisa[2], Nunzi Martina[1], Sabatini Silvia[1], Dinh Phuong[3], Califano Raffaele[4] and De Azambuja Evandro[3]

1Medical Oncology Department, Terni Hospital, Via Tristano di Joannuccio 1, 05100 Terni, Italy; 2Medical Oncology Department, Macerata Hospital, Via Santa Lucia 2, 62100 Macerata, Italy; [3]Medical Oncology Clinic, Jules Bordet Institute, Boulevard de Waterloo, 125, 1000 Brussels, Belgium and [4]Department of Medical Oncology, The Christie NHS Foundation Trust, Wilmslow Road, Manchester, M20 4BX, United Kingdom

Abstract: Tamoxifen, a selective estrogen receptor modulator (SERM), has been used for many decades as the "gold standard" adjuvant treatment for patients with hormone-receptor-positive early breast cancer. This drug, when administered for 5 years, reduces the risk for recurrence, contralateral breast cancer (BC) and death. These benefits have been observed up to 15 years and are independent of the patient's age, menopausal status, nodal status, hormonal receptor status, and the use of adjuvant chemotherapy. The optimal duration of tamoxifen in the adjuvant setting has not been established yet, but it has been demonstrated that 5 years are better than shorter treatment while it is still unclear if a prolongation of the treatment for more than 5 years is worthwhile. Tamoxifen is usually well-tolerated, but important adverse events such as endometrial cancer, cerebrovascular accidents and thromboembolic events can occur, and the increase in absolute risk of these adverse events appears to be age-correlated.

In the last decade, third generation aromatase inhibitors (AIs), either steroidal (exemestane) or non-steroidal (anastrozole, letrozole), have shown to be an effective alternative to tamoxifen in postmenopausal patients with BC regardless of its stage. These agents act by blocking the aromatase enzyme which converts androgens into estrogens. Trials comparing AIs to tamoxifen in postmenopausal women with metastatic disease have shown a superiority of AIs over tamoxifen and a more favourable safety profile. In the adjuvant setting, AIs have been shown to be more effective than tamoxifen given for 5 years either in the up-front administration or after 2-3 years (early switch). Two randomised trials which have evaluated the two strategies of AIs

***Address correspondence to Mariantonietta Colozza:** S.C. Oncologia Medica, Azienda Ospedaliera, Via Tristano di Joannuccio 1, 05100 Terni, Italy; Tel: 0039 0744 205576; Fax: 0039 0744 205632; E-mail: mariantonietta.colozza@tin.it administration have shown superimposable results in terms of efficacy with AIs given up-front or in sequence to tamoxifen. AIs seem to give benefits in comparison to placebo if given after 5 years (late switch) of tamoxifen. At the moment, therefore, the treatment decision should be based on individual factors such as risk of relapse, absolute benefit, and comorbidities.

Keywords: Adjuvant endocrine therapy, aromatase inhibitors, breast cancer, postmenopausal women.

INTRODUCTION

Breast cancer is the leading cause of cancer-related death in women despite the fact that, in the last years, a reduction in mortality has been observed in different countries due to the extensive implementation of screening programs and the use of more effective adjuvant therapies. In postmenopausal women, hormone-responsive cancers *i.e.* tumours that express estrogen receptors (ER) and/or progesterone receptors (PgR) represent about two-thirds of all breast cancers. In these patients, estrogens are a potent stimulus for the proliferation and progression of tumour cells. Two different strategies have been developed to reduce the effect of estrogens on tumour growth: 1) blockade of estrogen binding with its receptor or 2) reduction of estrogen circulating levels. The anti-estrogenic drugs compete with endogenous estrogens for the binding to their receptor and tamoxifen is the most commonly used drug in the adjuvant setting both for pre-and post-menopausal patients. Tamoxifen, when administered for five years, produces a reduction of the risk of relapse and death with improvement in 10-year survival of 12.6% in node-positive patients and 5.3% in node-negative patients. Furthermore, the clinical benefits of tamoxifen benefits persist after the completion of therapy (carry over effect) for about 15 years and are independent of patients' age, menopausal status, PgR status and the use of adjuvant chemotherapy [1]. The optimal duration of tamoxifen administration is still undefined although it is well established that 5 years of treatment are superior to shorter periods. A prolongation of the treatment for 10 years or indefinitely did not provide further benefit and was associated with a worse outcome, even if not statistically significant, in two relatively small trials [2, 3]. However, a small study that enrolled patients with node-positive disease treated also with adjuvant chemotherapy has shown a longer time to relapse (TTR) and longer time to development of contralateral tumour in patients receiving tamoxifen for 10 years [4]. Two large randomised trials, that have completed the accrual, are evaluating different durations of tamoxifen treatment: the ATLAS (Adjuvant Tamoxifen Longer Against Shorter) and, the aTTOm (adjuvant Tamoxifen Treatment offers more?) but, since their results are not yet available, the standard duration of tamoxifen treatment remains 5 years.

Tamoxifen is usually well tolerated but serious adverse events such as thromboembolic events, endometrial cancer and cerebrovascular accidents can occur because of its partial-agonist activity. The incidence of these side effects is higher in older women [5]. Furthermore, it should be noted that about 23-40% of patients discontinue tamoxifen prematurely due to side-effects.

The third generation aromatase inhibitors (AIs) are a new class of drugs that were initially used in postmenopausal women with hormone-receptor positive metastatic breast cancer as second-line therapy after tamoxifen failure [6-11]. Third generation AIs have been evaluated in the adjuvant setting utilizing different strategies: as direct comparison with tamoxifen (up-front), after 2-3 years of tamoxifen (early switch) or after 5 years of tamoxifen (late switch). AIs have significantly improved DFS or relapse-free survival (RFS) in all trials and, in some, also distant disease-free survival (DDFS). A statistically significant improvement in OS has been observed in four trials [12-15] but in one trial only when ER negative patients were excluded [14], in another when adjusting for treatment crossover [15]. Tables **1** and **2** summarise the patient and tumour characteristics of these trials.

Table 1: Up-front and "early switch" trials: patient and tumour characteristics

	ATAC[20]	BIG 1-98[13]	IES[32]	ITA [38]	ARNO95/ ABCSG8[40]	TEAM [31]	N-SAS BC03[43]
No. randomised patients	6241	8010	4724	448	3224	9779	706
Median age (years)	64.1	61	64	63	62.1	64	59.8
Node-negative % patients	60.7	57.3	51.6	0	74	47	59.8
T size < 2 cm % patients	63.4	62	47.7	46.5	70	58	79.3 (<3 cm)
Grade 3 % patients	23.5	NR	18.6	NR	5.5	25	NR

Table 1: contd…

H-R positive % patients	83.5	99.7	88.1	88.5	98	98	92.9 (ER+)
Adjuvant CT % patients	21.5	25.3	32.6	67	0	36	53.6

Abbreviations: NR=not reported; T=tumour; H-R=hormone –receptor; CT=chemotherapy.

Table 2: "Late switch" trials: patient and tumour characteristics

	MA.17 [48]	**NSABP B-33** [53]	**ABCSG 6a** [52]
No. randomised patients	5187	1598	856
Median age (years)	62	*50%<60*	61.8
Node-negative % patients	50	52	67.4
T size ≤ 2 cm % patients	NR	61	62.7
Grade % patients	NR	NR	20
Hormone-receptor positive % patients	98	96	~94
Adjuvant CT % patients	45.3	55.5	NR

Abbreviations: NR=not reported; T=tumour; CT=chemotherapy.

Mechanism of Action

Estradiol, which is the most potent endogenous estrogen, is biosynthesized from androgens by the cytochrome P450 enzyme complex called aromatase. The ovaries in premenopausal women, the placenta in pregnant women, and the peripheral adipose tissues in postmenopausal women and in men produce the highest levels of this enzyme [16]. In addition, the expression of aromatase is highest in or near breast tumour sites [17].

In postmenopausal women, AIs act by inhibiting the cytochrome P-450-enzyme aromatase that promotes the conversion of androstenedione and testosterone to estrone and estradiol respectively, mostly in adipose tissue, liver, muscle, brain and breast cancer tissue thereby reducing estrogen circulating levels.

The enzyme complex is bound in the endoplasmic reticulum of the cell and is comprised of two major proteins [16]. One protein is cytochrome $P450_{arom}$, a hemoprotein that converts C_{19} steroids (androgens) into C_{18} steroids (estrogens) containing a phenolic A ring, whereas the second protein is the NADPH-cytochrome P450 reductase, which is a flavoprotein and is responsible for transferring reducing equivalents from NADPH to any microsomal form of cytochrome $P450_{arom}$ [16]. For the conversion of one mole of substrate into one mole of estrogen product, three moles of NADPH and three moles of oxygen are necessary. Aromatization of androstenedione, which is the preferred substrate, proceeds *via* three successive oxidation steps, with the first two being hydroxylations of the angular C-19 methyl group. The final oxidation step proceeds with the aromatization of the A ring of the steroid and loss of the C-19 carbon atom as formic acid. This last final step cleaves the C_{10}-C_{19} bond.

Aromatization is a unique reaction in steroid biosynthesis and is the last step so that its blockade should not affect production of other steroids. Therefore, aromatase is a suitable target for inhibition by selective compounds.

AIs may be divided into two subtypes, steroidal and nonsteroidal. Steroidal inhibitors (*e.g.* exemestane) or enzyme inactivators are analogues of androstenedione that bind irreversibly to the substrate binding site on the aromatase molecule. The nonsteroidal compounds (*e.g.* anastrozole and letrozole) bind reversibly to the haem group of the enzyme. These differences in molecular structure and mechanism of action, may, indeed, underline the differences in reducing estrogen circulating levels and in modifying lipid profiles [18] and, therefore, determine a different cardiovascular toxicity despite similar clinical activity. Trials comparing these drugs are ongoing, and their results will hopefully clarify this issue.

Up-Front Strategy

The main results of up-front aromatase inhibitors are summarised in Table **3**. Two large randomised, double-blinded studies have compared tamoxifen with an AI. The ATAC trial randomised 9366 postmenopausal women with ER/PgR+ or unknown receptor status early-stage breast cancer to receive tamoxifen alone, anastrozole alone or tamoxifen plus anastrozole [19]. The primary endpoint was DFS, defined as the time to the earliest occurrence of local or distant recurrence, new primary breast cancer, or death from any cause. Secondary endpoints were time to recurrence (TTR) (including new contralateral tumours, but not patients who had died from non-breast-cancer causes before recurrence), incidence of new contralateral primary breast tumours, time to distant recurrence (TTDR) and OS. The tamoxifen plus anastrozole arm was prematurely stopped because the first analysis showed no superiority over tamoxifen alone. In the first analysis of the trial [20], at a median follow-up of 33 months, no differences were seen in DFS between the two arms neither for patients with node-positive disease nor for those treated with chemotherapy. After a median follow-up of 120 months, anastrozole, in comparison to tamoxifen, significantly improved DFS (HR 0.91, 95% CI 0.83-0.99, p=.04), and TTR (HR 0.84, 95% CI 0.75-0.93, p=.001) and these effects were greater in hormone receptor-positive patients [21]. No effect was seen for any endpoints in patients with negative or unknown hormone receptor status. In this trial, it has been shown that the reduction of the relapses with anastrozole increases over time even after 5 years of treatment and remains significant for the entire 10-year follow-up period, although the benefit is smaller after 8 years. Anastrozole seems to provide the same carry-over effect as tamoxifen [22] but a longer follow-up is needed to confirm these data. A significant overall benefit in TTDR in favour of anastrozole was demonstrated (481 *vs.* 544 events, HR 0.87, 95% CI 0.77-0.99, p=.03), as well as in the subset of hormone receptor-positive patients (HR 0.85, 95% CI 0.73-0.98, p=.02) with an absolute difference in the distance recurrence rates of 2.6% at 10 years. The incidence of contralateral breast cancer was significantly reduced by anastrozole in all patients (73 *vs.* 105, 32% reduction, 95% CI 9-50, p=.01) and in hormone receptor positive-patients (38% reduction, 95% CI 15-55, p=.003). No statistically significant difference in OS between the two treatment arms (HR 0.95, 95% CI 0·84-1·06, p=0.4) was observed even if there was a 12% reduction in deaths from breast cancer in the anastrozole group (HR 0.88, 95% CI 0.74-1.05, p=0.2). However, since the patients enrolled had a relatively good prognosis, a longer follow-up is needed to potentially observe a difference in survival.

Table 3: "upfront and "early switch" trials: results

	ATAC[20]	BIG 1-98 [13]*	IES [14]	ITA[39]	ARNO 95/ ABCSG 8 [40] **	TEAM [31]	N-SAS BC03 [43]
Median follow-up (months)	120	97,2	91	64	28	61,2	42
Patients on treatment at the moment of the analysis (%)	0	0	0	0	45	60	NR
Primary end-point HR (95%CI)	DFS 0.91 (0.83-0.99)	DFS 0.82 (0.74-0.92)	DFS 0.81 (0.72-0.91)	RFS 0.56 (0.35-0.89) & EFS 0.57 (0.38-0.85)	EFS/ 0.60 (0.44-0.81)	DFS 0,97 (0,88-1,08)	DFS 0.69 (0.42-1.14)
Absolute benefit in DFS (%)	4.3 at 120 months	4.4 at 103 months	4.5 at 91 months	NR	EFS 3.1 at 36 months	NR	3.6 at 36 months
DDFS HR (95%CI)	0.87 (0.77-0.99)	0.79 (0.68-0.92)	0.84 (0.73-0.97)	0.49 (0.22-1.059) at 36 months	0.54 (0.37-0.80)	0.93 (0,81-1.07)	NR
Overall survival HR (95%CI)	0.95 (0.84-1.06)	0.79 (0.69-0.90)	0.86 (0.75-0.99)	0.56 (0.28-1.15)	NR	1,00 (0,89-1,14)	0.59 NR

Abbreviations: NR=not reported; HR=hazard ratio; DFS=disease-free survival; DDFS=distant disease-free survival; RFS=relapse-free survival; EFS=event-free survival.

*The reported results for BIG 1-98 refer to the comparison of letrozole monotherapy to tamoxifen monotherapy.

** Only the results of the combined analysis of the two trials is reported.

A retrospective, unplanned subgroup analysis showed that ER+/PgR-patients had a higher clinical benefit with anastrozole [23]. These data are interesting because they raise the possibility that ER+/PgR-could be a surrogate marker of an extreme activation of growth factor receptors resulting in worse clinical benefit for tamoxifen [24]. However, a recent centralized analysis of ER, PgR and HER-2 status was undertaken in a subgroup of patients enrolled in the ATAC trial and no difference in survival among ER+/PgR-patients or HER-2-positive patients was shown between the two arms but HER-2-positive tumours tended to be less sensitive to endocrine therapy [25].

Treatment-related adverse events on treatment or within 14 days of discontinuation occurred significantly less often with anastrozole than with tamoxifen (61% *vs.* 68%, p<.0001), as well as serious treatment-related adverse events (5% *vs.* 9%; p<.0001) and adverse events leading to withdrawal (11% *vs.* 14%; p=.0002). There were no significant differences in the incidence of ischemic cardiovascular events (p=0.1), the most common of which was angina (p=.07). Patients on tamoxifen had a higher incidence of endometrial cancer (p=.02), cerebrovascular events (p=.03), deep venous thromboembolic events (p=.0004), hot flushes (p<.0001), vaginal bleeding (p<.0001) and vaginal discharge (p<.0001), while patients receiving anastrozole presented a higher incidence of fractures (p<.0001) but, interestingly, the rate of hip fractures was low in both groups (1% in both arms) and was similar in the post-treatment follow-up period. Hypercholesterolemia was more common in the anastrozole arm (p=.0001). Other adverse events including muscle cramps, anaemia, nail disorders, fungal infections and urinary tract infection were less common in the anastrozole arm while a higher incidence of carpal-tunnel syndrome, paresthaesia, mouth dryness, decrease of libido and dispareunia was reported with anastrozole [21, 26]. The higher incidence of osteoporosis and osteopenia was observed mostly in the first two years of treatment with a reduction in the next three years and it is noteworthy that no patients with a baseline normal bone mineral density developed osteoporosis during the treatment [27].

No differences in non-breast cancer causes of death were apparent and the incidence of other cancers was similar between groups (425 *vs.* 431) and continue to be higher with anastrozole for colorectal (66 *vs.* 44) and lung cancer (51 *vs* 44) and with tamoxifene for endometrial cancer (24 *vs.* 6), melanoma (19 *vs.* 8), and ovarian cancer (28 *vs.* 17). No new safety concerns were reported [21]. Quality of life evaluation in 1091 patients enrolled in the three initial arms of the study did not show any significant difference [28].

The Breast International Group study (BIG 1-98) randomised 8010 postmenopausal patients with ER+ and/or PgR+ early breast cancer in two arms of monotherapy option to tamoxifen for 5 years (911) or to letrozole for 5 years (917) or in the four-arm option, to letrozole for 5 years (1546), to tamoxifen for 5 years (1548), to letrozole for 2 years followed by tamoxifen for 3 years (1540), and to tamoxifen for 2 years followed by letrozole for 3 years (1548). The first analysis, conducted at a median follow up of 25.8 months, evaluated tamoxifen and letrozole excluding events and follow up after 2 years of treatment in the two sequential arms [29]. The primary endpoint of the study was DFS, defined as the time from randomization to the first of one of the following events: recurrence at local, regional, or distant sites; a new invasive cancer in the contralateral breast; any second non-breast cancer; or death without a prior cancer event. Secondary endpoints included OS (defined as the time from randomization to death from any cause), DDFS. Patients in the letrozole arm had a significantly better DFS (HR 0.81, 95% CI 0.70-0.93, p=.003) with a significant reduction of distant relapses (HR 0.73, 95% CI 0.60-0.88, p=.001). Prospectively planned subgroup analyses showed greater benefits of letrozole in patients treated with chemotherapy, node-positive patients and patients who did not receive radiotherapy. DDFS was significantly greater in women treated with letrozole (HR 0.83, 95% CI 0.72-0.97, p=.02) and, although fewer women died in the letrozole group, no statistically significant difference in OS was observed. In 2005, based on these results, a protocol amendment facilitated the crossover to letrozole of patients who were still receiving tamoxifen alone. A centralized analysis of ER, PgR and HER2 status of patients enrolled in this study was performed and no significant difference between the two drugs was shown in different subgroups according to hormone receptor status and HER-2 status [30]. Life-threatening or fatal protocol-specified adverse events were similar in the two arms (1.7% in both arms). In the letrozole arm, fractures were significantly more frequent than in the tamoxifen arm (P<.001), while a lower incidence of thromboembolic events (P<.001), vaginal bleeding (p<.001) and invasive endometrial cancers (p=0.18) was observed. The overall incidence of grade 3-5 adverse cardiovascular events was similar in the two groups (3.7% in the letrozole group and 4.2% in the tamoxifen group), but more women in the letrozole group had grade 3-5 cardiac events (2.1% *vs.* 1.1%, p<.001) and grade 1 hypercholesterolemia.

At a median follow-up of 8.1 years all 8010 patients were analysed [13]. Cox models and Kaplan-Meier estimates with inverse probability of censoring weighting (IPCW) were used to account for selective crossover to letrozole of patients (n=619) in the tamoxifen arm.

4922 patients were included in the monotherapy analysis and 6182 in the sequential treatment analysis. At a median follow-up of 8.7 years from randomization letrozole monotherapy significantly improved DFS (HR 0.82, 95% CI 0.74-0.92, p=.0002), distant recurrence-free interval (DRFI) (HR 0.79, 95% CI 0.68-0.92, p=.003), OS (HR 0.79, 95% CI 0.69-0.90, p=.0006) and breast cancer-free interval (BCFI) (HR 0.80, 95% CI 0.70-0.92, p=.002) in comparison to tamoxifen by IPCW. The relative treatment effects expressed as HRs were homogeneous across node-positive and node-negative subgroups. Statistically significant improvements, even if of smaller magnitude, were obtained by intention-to-treat analysis.

In the sequential treatment analysis at a median follow-up of 8.0 years there were no statistically significant differences in any of the four endpoints between the two sequential arms and the letrozole monotherapy.

Up-Front AI *vs.* Sequential Therapy After Tamoxifen

The results of the BIG 1-98 have been reported above. The other trial that has evaluated this strategy is the TEAM (The Tamoxifen Exemestane Adjuvant Multinational) trial [31] where 9779 postmenopausal women with hormone-receptor-positive breast cancer were randomly assigned to open-label exemestane alone (4904) or tamoxifen for 2.5-3 years followed by exemestane for a total of 5 years (4875). Originally this trial was designed to compare exemestane with tamoxifen but it was modified after the publication of the results of the IES trial [32]. The primary endpoint was DFS at 2.75 years and 5.0 years, defined as the time from randomization to the earliest documentation of disease relapse or death from any cause. Disease relapse was defined as tumor recurrence (locoregional or distant), or ipsilateral or contralateral breast cancer. Secondary endpoints included OS, RFS, and safety. TTDR was included in an exploratory analysis. At a median follow-up of 5.1 years 60% of patients had completed at least 5 years of follow-up. No significant difference in DFS at five years between the two arms (HR 0.97, 95% CI 0.88-1.08, p=0.60) was reported in the intention to treat analysis. The effect of treatment was consistent in all subgroups. Also in the first analysis, at 2.75 years, no significant difference was noted in DFS between the two groups [33]. The OS at five years was similar (HR 1.00, 95% CI 0.89-1.14, p>0.99) in the two arms. Cardiac-related deaths were numerically higher with exemestane but the difference was not significant (p=0.11). No difference was observed in the cumulative incidence of distant metastases (HR 0.93, 95% CI 0.81-1.07, p=0.30). In the safety analysis a higher incidence of gynaecological symptoms (20% *vs.* 11%), venous thrombosis (2% *vs.* 1%), and endometrial abnormalities (4% *vs.* <1%) was noted in the sequential arm. Musculoskeletal adverse events in general (50% *vs.* 44%), and osteoporosis (10% *vs.* 6%) and fractures (5% *vs.* 35%) in particular were reported more frequently with exemestane alone.

Hypertension (6% *vs.* 5%), and hyperlipidaemia (6% *vs.* 5%) were also more frequently observed in the exemestane alone arm.

In a prospectively planned pathology substudy the predictive value of PgR expression for outcome has been tested based on the hypothesis that low PgR expression predicts benefit for patients receiving initial AIs treatment rather than tamoxifene in the first 2.75 years of therapy [34]. Five of nine countries that participated in the TEAM trial (United Kingdom/Ireland, The Netherlands, Belgium, Germany, and Greece) provided tumor samples for this substudy. Of 4,325 eligible ER-positive patients, 23% were PgR-poor (Allred score, 0 to 4) and 77% PgR-rich (Allred score, 5 to 8). Overall, the pathology substudy population was well matched to the patient population from which the samples were collected, but the patients were of higher risk than those in the United States. Multivariate Cox regression analysis, including treatment-by-marker analysis, showed no evidence that PgR expression was predictive of a differential DFS benefit for exemestane compared to tamoxifen (PgR-poor HR 0.85, 95% CI 0.61-1.19, PgR-rich HR 0.83, 95% CI 0.65-1.05, test for interaction p=.88). Both ER and PgR expression levels were quantitatively associated with DFS and in an unplanned analysis patients with breast cancer expressing high ER levels were shown to potentially benefit preferentially from initial treatment with exemestane over tamoxifen.

Sequential Therapy After 2-3 Years of Tamoxifen (Early Switch)

The main results of the trials evaluating this strategy are summarised in Table **3**. In the IES (Intergroup Exemestane Study) study, a randomised, double-blinded, multi-centre trial, 4724 postmenopausal women with unknown or hormone receptor-positive breast cancer who remained free of disease after receiving adjuvant tamoxifen therapy for 2-3 years were randomised to receive tamoxifen or exemestane for other 3-2 years [32]. The primary endpoint was DFS, defined as the time from randomization to recurrence of breast cancer at any site, diagnosis of a second primary breast cancer, or death from any cause. Secondary endpoints included breast cancer-free survival (BCFS), defined as DFS in which deaths without recurrence are censored, OS, time to contralateral breast cancer (CLBC), TTDR, and long-term tolerability [14].

The first analysis [32], at a median follow-up of 30.6 months, with 90% of patients who had completed treatment, showed a longer DFS in the exemestane arm (HR 0.68%, 95% CI 0.56-0.82, p=.00005) which corresponded to an absolute benefit of 4.7% at three years, longer DDFS (HR 0.66, 95% CI 0.52-0.83, p=.0004) and a reduction of the risk of contralateral breast cancer (HR 0.44, 95% CI 0.20-0.98, p=.04). There was no statistically significant difference in OS.

The last analysis focused on 4599 patients resulted ER-positive/ER-unknown in a subsequent hormone receptor status re-assessment [14]. At a median follow-up of 91 months, with 90.7% of patients followed for ≥ 6 years or died, a statistically significant benefit in DFS continued to be observed in the exemestane arm which translated to an absolute benefit of 4.5%. Multivariate analysis gave an HR of 0.81%, (95% CI 0.72-0.91, p <.001). Considering rates of each DFS event type, local recurrence and second breast cancer each occurred at a rate of approximately 0.5% per year, distant recurrence remained the most common event occurring in 2% to 2.5% per year, and deaths without recurrence were initially rare but increased with longer follow-up and corresponding increase in patients' age (approximately 1.5%). A significant benefit of switching to exemestane was also observed in BCFS (unadjusted HR 0.81%, 95% CI 0.71-0.92, p=.001), and in TTDR (HR 0.84%, 95% CI 0.73-0.97, p=.01) while fewer patients had a new primary contralateral breast cancer but the difference was not statistically significant. Risk of non-breast cancer second primary cancers also appeared to be reduced by switching to exemestane and with the exception of endometrial cancer remains unexplained. Notably, an improvement in OS was shown in the exemestane arm (HR 0.86, 95% CI 0.75-0.99, p=.04), with an absolute difference in survival outcome at 8 years of 2.4%. No significant difference in the incidence of grade 3-4 adverse events was observed between the two arms, as reported in the second analysis at 55.7 months [35]. Thromboembolic events and muscle cramps were more frequent in the tamoxifen arm while patients in the sequential arm experienced more diarrhoea, gastric ulcer, musculoskeletal pain, arthritis, arthralgia, carpal tunnel syndrome, paresthesia, joint stiffness, osteoporosis and fractures. The incidence of cardiovascular events was similar in the two arms, but a higher percentage of myocardial infarction was seen in the exemestane arm. No statistically significant difference was observed between the two arms for endometrial cancer and necessity of hysterectomy, but more vaginal bleeding, uterine polyps/fibroids and endometrial hyperplasia occurred in the tamoxifen group. The updated analysis focused also on adverse events after treatment cessation; in particular, musculoskeletal pain, arthritis and osteoporosis became less frequent and a reduction in carpal tunnel syndrome incidence was noted. Also serious gynaecologic events reported in the tamoxifen arm did not persist into the follow-up period.

In an ancillary study, the effects of exemestane on bone mineral density (BMD) were evaluated in 206 patients enrolled in the IES study [36]. The primary endpoint of this study was mean annual changes from baseline in lumbar spine and total hip BMD assessed by dual-energy X-ray absorptiometry (DXA) in both arms. Secondary endpoints included changes in biochemical markers of bone turnover in the two treatment groups at each time point, within-group changes in BMD and bone markers, and assessment of the links between changes in biochemical markers of bone metabolism and changes in BMD. The incidences of fractures were also assessed. A statistically significant reduction in BMD compared with baseline was seen in the exemestane group within the first six months of treatment both at the lumbar spine and at the total hip. Thereafter, the reduction in BMD progressively slowed in months 6-12 and 12-24, but continued to decline. The decline in BMD after the switch to exemestane resulted in a higher incidence of osteopenia at 24 months, and five patients, who were osteopenic at baseline, developed osteoporosis. The changes in bone markers from baseline were significantly different in patients in the exemestane arm at all time-points. Generally, there were significant negative correlations between bone-marker and BMD changes at 24 months. A significantly higher incidence of fractures occurred in the exemestane group (7% *vs.* 5%, p=.003) but no patients with a normal baseline BMD developed a fracture. The rate of fractures observed in 24 months was so low that it was not possible to correlate it with changes in BMD or changes in biochemical markers of bone metabolism. The quality of life assessment on 582 enrolled patients did not show any differences between the two groups [37].

Three European studies investigated five years of tamoxifen *vs.* tamoxifen for 2-3 years followed by anastrozole for 3-2 years. The ITA (Italian Tamoxifen Anastrozole) trial randomised 448 node-positive ER+ postmenopausal women who received 2-3 years of tamoxifen to complete 5 years of treatment with tamoxifen or to receive 2-3 years of anastrozole [38]. The primary endpoint was RFS including both locoregional and distant recurrences except contralateral breast cancer. Event-free survival (EFS) included as events any of the following: locoregional recurrences, distant metastases, second primary tumours including contralateral breast cancer and breast cancer-unrelated deaths. Secondary endpoints were incidence of deaths, whatever the cause, and adverse events. All second primary tumours (except contralateral breast cancer) were included among serious adverse events. At a median follow up of 36 months, women who switched to anastrozole had a significantly longer EFS (HR 0.35, 95% CI 0.20-0.63, p=.0002) and RFS (HR 0.35, 95% CI 0.18-0.68, p=.001). Women in the anastrozole group also had significantly longer loco-regional recurrence-free survival (HR 0.15, 95% CI 0.03-0.65, p=.003), whereas the difference in DDFS did not reach statistical significance (HR 0.49, 95% CI 0.22-1.05, p=.06). Overall, more patients in the anastrozole arm presented adverse events but more patients in the tamoxifen arm presented serious adverse events even if the difference was not statistically significant. These results were confirmed at a median follow-up of 64 months [39]. RFS (HR 0.56, 95% CI 0.35-0.89, p=.01) and EFS (HR 0.57, 95% CI 0.38-0.85, P=.005) were longer in the anastrozole group while the difference in OS was not statistically significant (P=0.1). Overall more patients in the anastrozole group experienced at least one adverse event (209 *vs.* 151; p=.000) but the number of patients experiencing serious adverse events was comparable (p=0.7). However, gynaecological problems, including endometrial cancer, were significantly more frequent in the tamoxifen group (p=.006).

Two multi-centre studies with similar design, the ABCSG 8 (Austrian Breast and Colorectal Cancer Study Group) and the ARNO 95 (ARIMIDEX/NOLVADEX), randomised postmenopausal hormone receptor-positive patients, not treated with adjuvant chemotherapy, to receive anastrozole after 2 years of tamoxifen or to continue tamoxifen for overall 5 years. It is noteworthy that, in the first trial, patients were randomised before starting the adjuvant therapy while in the second trial, after completing 2 years of tamoxifen which was given at the dose of 20 mg/day in the ABCSG 8 trial and 30 mg/day in the ARNO 95 trial. A combined analysis of these two trials was performed, at a median follow-up of 28 months, with 3224 patients enrolled of whom 55% had completed the treatment [40]. Anastrozole reduced the risk of new events of 40% (HR=0.60, 95% CI 0.44-0.81, p=.0009) with an absolute benefit at 3 years of 3.1%. This advantage was independent of nodal status, age or hormone receptor status although there was a suggestion that the benefit of anastrozole in ER+/PgR+ patients was higher. Patients in the anastrozole arm presented a decreased risk of distant metastases as first event only (HR 0.54, 95% CI 0.37-0.80, p=.0016) while there was no statistically significant difference in OS. Sequential therapy was associated with more fractures, less thrombotic events and a non-significant reduction of embolic events and endometrial cancers. Patients in the anastrozole group experienced more nausea and arthralgia.

Data from the ARNO 95 trial were analysed separately at a median follow-up of 30.1 months with 42.5% of patients who had completed the planned 5 years of treatment [12]. The primary endpoint was DFS (the time from random assignment to the occurrence of local or distant recurrence, new primary breast cancer, or death); secondary endpoints included OS, safety, and tolerability. Patients who switched to anastrozole had a statistically significant improvement in DFS (HR 0.66, 95% CI 0.44-1.00, p=.049) with an absolute difference of 4.2% at 3 years and in OS (HR=0.53, 95% CI 0.28-0.99, p=.045). After adjustment for potential prognostic factors (age, tumour size and grade, lymph node status, and type of primary surgery) these results were confirmed. The overall safety profile for anastrozole was consistent with previous reports. The incidence of serious adverse events was lower with anastrozole. The small number of any recurrence (36 *vs.* 47) and deaths (15 *vs.* 28) in the sequential arm and in the tamoxifen arm, respectively, is noteworthy.

Data from the ABCSG 8 trial were also analysed separately at a median follow-up of 60 months [41]. This trial included 3714 post-menopausal patients with endocrine receptor-positive, G1 or G2 tumors. Seventy-five percent of the patients had node-negative disease and 75% had T1 tumors. The primary end-point was RFS, defined as the time from random assignment to the earliest occurrence of local or distant recurrence or death of any cause. Exploratory end points were DFS, DRFS and OS. In this population with breast cancer at limited risk of relapse, the sequence strategy showed a trend for a longer RFS, but not statistically significant (ITT: HR 0.80, 95% CI 0.63-1.01, p=.06) with an absolute difference in 60-month survival rates of 1.6%. ITT analysis did not indicate a benefit for DFS and OS. The difference in recurrence did not seem to come from a reduction in local recurrence because DRFS revealed a 22% reduction in risk in favour of anastrozole (ITT: HR 0.78, 95% CI 0.60-0.99, p <.05).

There were no significant differences detected in serious adverse events grouped as cardiovascular or thromboembolic. Patients in the tamoxifen arm experienced more uterine disorders (polyps and endometrial hyperplasia) and less musculoskeletal disorder. Hot flushes and vaginal bleeding or discharge were equally distributed.

A meta-analysis of the ITA, ABCSG 8 and ARNO 95 trials confirmed the results of the single trials and showed a reduction of relapses and deaths in the anastrozole arm with a significant longer DFS (HR 0.59, 95% CI 0.48-0.74, p<.0001), EFS (HR 0.55; 95% CI 0.42-0.71; p<.0001), DDFS (HR 0.61, 95% CI 0.45-0.83, p=.002), and OS (HR 0.71, 95% CI 0.52-0.98, p=.04). Although these data are interesting, it is worthwhile to point out that the three trials differ in time of randomisation, primary endpoints, patient characteristics and follow-up duration [42]. The most important difference regards the time of randomisation so that the ARNO 95 and ITA trials having randomised patients after 2-3 years of tamoxifen, excluded those with early relapses and therefore less hormone-responsive disease.

In the N-SAS BC03 study, a randomised, open-label, multi-institutional, 706 Japanese postmenopausal women with hormone-responsive breast cancer who remained disease-free after receiving adjuvant tamoxifen therapy for 1-4 years were randomised to receive tamoxifen or anastrozole for total treatment duration of 5 years [43].The primary endpoints were DFS (defined as the time from randomization to recurrence of breast cancer at any site, diagnosis of a secondary cancer, or death from any cause) and adverse events. The secondary endpoints were RFS defined as time from randomization to locoregional relapse and distant metastasis, OS and health-related quality of life (HRQOL).

The result of switching adjuvant therapy from tamoxifen to anastrozole is effective in Asian women as in Western women irrespective of ethnic differences in polymorphisms of the CYP2D6 or CYP19 genes. At a median follow-up of 42 months, 696 patients were evaluable. The unadjusted HR was 0.69 (95% CI 0.42-1.14, p=.014) for DFS and 0.54 (95% CI 0.29-1.02, p=.06) for RFS, both in favour of anastrozole. The range of the 95% CI was large because of the relatively small number of events but the HR for DFS in this study was in line with the results of the other trials of early switch. There was no statistically significant difference in OS. Patients in the anastrozole group experienced more arthralgia but less hot flashes and vaginal discharge. There were no thromboembolic events in the tamoxifen group and one grade 4 in the anastrozole group. The incidence of fractures was low in both groups but lower in the anastrozole group (5 *vs.* 9) but the difference was not statistically significant. No differences between the two groups in the incidence of contralateral breast cancer and second malignancies. A planned subgroup analysis suggested that PgR status did not influence the efficacy of anastrozole.

Sequential Therapy After 5 Years of Tamoxifen (Late Switch)

The main results of the trials evaluating this strategy are summarised in Table **4**. Data from the Early Breast Cancer Trialists' Collaborative Group (EBCTCG) meta-analysis and some retrospective studies suggest that the risk of relapse for patients with early stage breast cancer is high and persists for many years after diagnosis [44-47]. In particular, patients with ER positive tumours who received adjuvant tamoxifen for 5 years have a risk of relapse at 10 and 15 years of 24.7% and 33.2%, respectively, and a risk of death at 10 and 15 years of 17.8% and 25.6%, respectively [1]. It appears, therefore, appropriate to prolong the hormonotherapy beyond 5 years in the attempt to reduce the risk or relapse and death. So far, tamoxifen given for more than 5 years, has not been shown to further improve survival [2, 3] and it is associated with the risk of serious adverse events and the chance of acquired resistance. Therefore, a few years ago, AIs started to be evaluated in this setting in postmenopausal women.

Table 4: "late switch" trials: results

	MA.17 [15*]	**NSABP B-33** [53]	**ABCSG 6a** [52]
Median follow-up (months)	64	30	60
Patients on treatment at the moment of the analysis (%)	88,6	NR	0
Primary end-point **HR** **(95%CI)**	DFS 0.68 (0.56-0.83)	DFS RR=0.68 NS	RFS 0.64 0.41-0.99
Absolute benefit in DFS (%)	at 4 years 4.6	at 4 years 2	NR
DDFS HR **(95%CI)**	0.81 (0.63-1.04)	RR=0.69 NS	NR
Overall survival HR **(95% CI)**	0.99 (0.79-1.24)	RR=1.20	0.90 (0.59-1.34)

Abbreviations: NR=not reported; DFS=disease-free survival; RFS=relapse-free survival; DDFS=distant disease-free survival; NS=not significant.
*The results of MA.17 trial refer to ITT analysis.

The NCIC CTG MA.17 study, a randomised double blinded placebo-controlled trial, evaluated the efficacy of letrozole in 5187 postmenopausal women with positive or unknown hormone receptor breast cancer who were disease-free after 4-6 years of therapy with tamoxifen [48]. The primary endpoint was DFS, defined as time from randomization to loco-regional recurrence, distant metastasis or a contralateral new primary breast cancer. The first interim analysis was conducted at a median follow-up of 2.4 years and based on the positive results obtained with letrozole, this trial was prematurely closed with only 14 patients having completed 5 years of treatment with letrozole. A statistically significantly longer DFS (HR 0.57, 95% CI 0.43-0.75, p=.00008) was obtained with letrozole and this advantage was evident both for node-positive (HR 0.60, p=.003) and for node-negative patients (HR 0.47; p=.005). OS was the same in both arms. In a recent update, at a median follow-up of 30 months, this advantage in DFS was confirmed (HR 0.58, 95% CI 0.45-0.76, p<.001) with an absolute reduction of the risk of relapse of

4.6% at 4 years and a longer DDFS (HR=0.60; 95% CI 0.43-0.84, p=.002) that was independent of nodal status, previous use of adjuvant chemotherapy and duration of tamoxifen treatment (≤ 5 years or ≥5 years). The incidence of contralateral breast cancer was also reduced in the letrozole arm but the difference was not statistically significant. No difference in OS between the two arms was observed but a pre-planned subgroup analysis showed a longer OS for node-positive patients (HR=0.61; 95% CI 0.38-0.98; p=.04). With the limits of a subgroup analysis, this was the first trial to demonstrate an advantage in OS with AIs in the adjuvant setting [49].

A retrospective analysis, based on ER and PgR status, suggested that the clinical benefits obtained with letrozole were greater in patients with ER+/PgR+ tumours which represented the largest subgroup (73%). The DFS hazard ratio for letrozole *vs.* placebo in women with ER+/PgR+ tumours was 0.49 (95% CI 0.36-0.67) *vs.* 1.21 (95% CI 0.63 to 2.34) in women with ER+/PgR-tumours. Similar results were observed for DDFS (HR 0.53, 95%CI 0.35-0.80) and OS (HR 0.58, 95% CI 0.37-0.90) in the ER+/PgR+ subgroup. A statistically significant difference in treatment effect between ER+/PgR+ and ER+/PgR-subgroups was reported for DFS (p=.02), but not for DDFS (p=.06) or OS (p=.09). These results, which are discordant from those reported in other trials comparing AIs and tamoxifen in which subgroup analysis based on hormone receptor status was retrospectively done, should be interpreted with caution because this was a retrospective unplanned analysis. Moreover, the determination of the hormone-receptor status was not measured centrally and all the groups except for the ER+/PgR+ subgroup had a relatively small number of patients [50].

Analyses were conducted to examine the relationships between duration of treatment and outcomes [51]. The hazard rates for DFS, DDFS and OS at 6, 12, 24, 36 and 48 months of follow-up and the hazard ratios for letrozole to placebo and associated 95% confidence intervals for DFS, DDFS and OS at these time points were estimated. Patients in the placebo arm had a continuous increase of the risk of relapse over time while in the letrozole arm this risk peaked at 2 years and then slowly decreased. These analyses suggest that, at least up to 48 months, longer duration of letrozole treatment is associated with greater benefit in the extended adjuvant therapy setting.

Hot flushes, arthralgia, muscle pain, anorexia and alopecia were significantly more frequent in the letrozole group. The higher incidence of vaginal bleeding that occurred in the placebo arm can be explained with the inhibition of endometrial proliferation due to the AIs mechanism of action. Furthermore, patients who received letrozole experienced more fractures, cardiovascular accidents and osteoporosis, although only osteoporosis reached a statistically significant difference.

A recent publication has shown results of analyses based on longer follow-up and adjusting for treatment cross-over [15]. When the trial was closed, after the first interim analysis, women in the placebo arm were offered to switch to letrozole. Indeed, among 2587 patients originally assigned to placebo, 204 (7.9%) experienced recurrence or death before the date of unblinding, 804 (31.1%) remained on placebo and 1579 (61%) crossed over to letrozole. The median time from random assignment to treatment crossover was 2.7 years (range 1.1 to 7.0 years). A higher proportion of these women were young, were white, had more often an Eastern Cooperative Oncology Group (ECOG) performance status of 0, a longer period between initial diagnosis and random assignment, positive axillary nodes, positive tumor hormone receptor status at diagnosis, axillary dissection, prior adjuvant chemotherapy. Two statistical approaches have been used to adjust for the potential effects of treatment cross-over: one was based on the IPCW Cox model and the other on a Cox model with time-dependent covariates (SCC). At a median follow-up of 64 months, both IPCW and SCC analyses showed significant improvement for letrozole *vs.* placebo for DFS, DDFS and OS. In the IPCW analyses, the HRs for letrozole and placebo were 0.52 (95% CI 0.45-0.61, $p<.001$) for DFS, 0.51 (95% CI 0.42-0.61, $p<.001$) for DDFS and 0.61 (95% CI 0.52-0.71, $p<.001$) for OS. In the SCC analyses, the HRs for letrozole and placebo were 0.58 (95% CI 0.47-0.72, $p<.001$) for DFS, 0.68 (95% CI 0.52-0.88, $p=.004$) for DDFS and 0.77 (95% CI 0.61-0.97, $p=.03$) for OS. At the same follow-up the ITT analyses confirmed advantage only in DFS for the letrozole group (HR 0.68, 95%CI 0.56-0.83, $p<.001$)

A new randomised trial (MA.17R) is ongoing to evaluate a prolongation of letrozole treatment for ten years in patients who are disease-free after completing 5 years of letrozole (MA.17 trial or routine clinical practice) or 5 years of any other adjuvant AI.

An Austrian trial (ABCSG 6) randomised 1986 postmenopausal hormone receptor-positive early breast cancer patients to receive tamoxifen for 5 years or tamoxifen plus aminoglutethimide for 2 years followed by 3 years of tamoxifen. Eight hundred and fifty-six women were then enrolled in another trial (ABCSG 6a) in which they were randomised to continue hormonotherapy with anastrozole for 3 years or to receive no treatment [52]. Primary endpoint was RFS and at a median follow-up of 5 years, patients in the letrozole arm had a longer RFS (HR=0.64, 95%CI 0.41-0.99, p=.047) but no difference in OS was seen. So far, no toxicity data have been reported.

Another study, the NSABP B-33 randomised postmenopausal, hormone-receptor positive women with stage I-II breast cancer to receive 2 years of exemestane or placebo after the standard 5 years of adjuvant tamoxifen [53]. This trial had planned to enrol 3000 patients but, was prematurely closed after the publication of the MA.17 trial's results with only 1598 patients enrolled and 1577 patients eligible. In the placebo group, 44% of patients accepted to start exemestane and 72% of patients in the exemestane group completed the treatment. The primary endpoint of the study was DFS defined as time to earliest occurrence of local or distant recurrence, new primary breast cancer, second tumour, or death without recurrence. At a median follow-up of 30 months, there were no statistically significant differences in DFS, DDFS and OS even if the number of events was reduced in the exemestane arm, but, exemestane significantly prolonged RFS (p=.004) and reduced the incidence of contralateral cancer (p=.05). A statistically significant benefit in DFS was seen in patients with node-positive disease, tumour size >2cm and who had received adjuvant chemotherapy. There were no treatment-related deaths. In the exemestane arm occurred significantly more grade 3-4 fatigue, arthralgia and muscle pain but there was no difference in the incidence of fractures.

Cost-Benefit of Aromatase Inhibitors in Early Breast Cancer

The ATAC trial was used as the model for analysing cost-effectiveness in the UK. Modelled for 25 years, anastrozole was shown to be a cost-effective alternative to generic tamoxifen in the adjuvant treatment of postmenopausal HR positive early BC, resulting in an estimated incremental cost-effectiveness of 17 656 pounds per quality-adjusted life year (QALY) gained [54]. Also, in a cohort of 1000 postmenopausal women with HR positive early BC, another model from the US healthcare system, showed that anastrozole *vs.* tamoxifen would lead to 257 QALYs gained (0.26 QALYs gained per patient), at an additional cost of $5.21 million over 25 years ($5,212 per patient) [55]. Based on the ARNO and ATAC studies, sequential tamoxifen-AI is the preferred cost-effective adjuvant strategy compared to upfront AI in postmenopausal women with HR positive BC, whereas AI upfront appears to be the preferred cost-effective strategy for patients with ER-positive and PgR-negative BC [56]. Using the two AIs upfront trials, the BIG 1-98 and ATAC trials, incremental cost per QALY gained for letrozole *vs.* tamoxifen is $33,536 (95% CI $20,409 to $70,566) and for anastrozole *vs.* tamoxifen is $38,967 (95% CI $23,826 to $81,904). Compared with anastrozole, letrozole is less costly ($9,647 *vs.* $10,190) and gains more QALYs (0.29 *vs.* 0.26), although differences in costs (95% CI $1,669 to $671) and QALYs (95% CI 0.16 to 0.22) are not statistically significant [57].

Comparing the ATAC, IES, and MA.17 studies, tamoxifen for 2-3 years followed by an aromatase inhibitor for 3-2 years provided the lowest cost/QALY estimates, while administration of an AI subsequent to 5 years on tamoxifen provided the highest values. The difference between strategies increased with patient age. Cost/QALY estimates were sensitive to an increase in hip fracture risk and to cost reductions due to relapse prevention [58]. A similar study performed in the UK showed that compared to 5 years of tamoxifen, adjuvant treatment of postmenopausal HR positive women with letrozole or anastrozole for 5 years, or 2 years tamoxifen followed by exemestane for 3 years, is a cost-effective therapy. The mean results indicate that upfront use of an AI is a more cost-effective therapy than switching to an AI after 2-3 years of tamoxifen, although the difference is not significant [59]. The sequential exemestane treatment (IES study) in early BC showed that switching to an AI after 2-3 years of tamoxifen is a cost-effective option compared with tamoxifen alone in Sweden [60].

In base-case analyses using data from the MA.17 trial, extended adjuvant letrozole *vs.* no extended adjuvant therapy results in an expected gain of 0.34 QALYs per patient (13.62 *vs.* 13.28 QALYs), at an additional lifetime cost of $9,699 per patient ($55,254 *vs.* $45,555). The incremental cost per QALY gained with letrozole is $28,728, which is within the range of other generally accepted medical interventions in the United States. Cost-effectiveness is sensitive to the assumed reduction in risk of breast cancer events with letrozole but is insensitive to the risks, costs, and quality-of-life effects of osteoporosis and hip fracture [61].

CONCLUSIONS

The results from the above-mentioned studies suggest that AIs are superior to tamoxifen given for 5 years both as up-front therapy (anastrozole and letrozole) and as sequential therapy (exemestane and anastrozole) after 2-3 years of treatment with tamoxifen. Mathematical models, based on the data from all these studies, have been utilized to evaluate the best strategy but, conflicting results have been obtained [62, 63]. At the present time, there does not seem to be a significant difference between the two strategies (up-front and early switch) based on the results of two trials, BIG 1-98 and TEAM, that have compared them. However, we do not know exactly if there are subgroups of patients that could benefit more from one or the other.

A prolongation of adjuvant hormonotherapy with an AI for 5 years after tamoxifen given for 5 years has significantly improved DFS, DDFS and OS after adjustment for treatment crossover in MA.17 trial, DFS in ABCSG 6a trial and RFS in NSABP-B33 trial. Furthermore the benefit of letrozole seems to increase over time. The short-term toxicity of AIs is acceptable although vaginal dryness and joint pain/stiffness are not insignificant problems as well as bone loss that requires monitoring. The long-term toxicity is still largely unknown.

All these remarkable results leave, however, several unanswered questions. Among these: the necessity to identify subgroups of patients that should be approached differently according to tumour characteristics or host factors that may be predictors of efficacy and toxicity.

Pharmacogenomics and translational research could be important to better understand these differences. It has been reported in a retrospective study [64] that the levels of endoxifen, one of the most potent tamoxifen metabolites, vary with the number of mutant alleles of the cytochrome P450 CYP2D6 enzyme and therefore, could identify good and poor tamoxifen metabolizers. A variety of subsequent articles both supported and refused these observations [65]. However, two recent papers appear to refuse the utility of CYP2D6 genotyping for predicting either tamoxifen benefit in the adjuvant setting or side effects [66, 67]. Furthermore, it has been observed that polymorphisms of CYP 19, the aromatase gene, are differently distributed across ethnic groups and could be associated with a different risk-benefit profile of AIs [68-70]. These observations need, however, to be evaluated in prospective trials.

It is also important to identify the subgroup of patients who should receive a prolonged adjuvant endocrine treatment, since in the decision-making process it is important to balance the absolute benefits with the risk of side effects and the costs of treatment.

The optimal duration of adjuvant hormonotherapy is still unknown. A treatment with AIs for 5 years has been chosen based on the results obtained with tamoxifen. Ongoing trials are evaluating longer durations of AIs administration and hopefully will help to clarify this issue.

A better knowledge of the long-term adverse effects on the cardiovascular system, cognitive functions and bone metabolism of AIs is also crucial. A recent meta-analysis by Amir *et al.* focused on toxicity of adjuvant AIs [71] in the attempt to explain the lack of OS benefit with AIs in comparison to tamoxifen in all trials. The authors evaluated seven randomized trials that enrolled 30023 patients. The meta-analysis was carried out in three different cohorts: 1) 5 years of AI *vs.* 5 years of tamoxifen (up-front strategy), 2) tamoxifen for 2-3 years followed by AI for 2-3 years *vs.* tamoxifen for 5 years, 3) tamoxifen for 2-3 years followed by AI for 2-3 years *vs.* AI for 5 years. The odds of cardiovascular disease and bone fractures were increased with longer duration of AIs use compared with tamoxifen use while the odds of venous thrombosis and endometrial carcinoma were decreased with longer AIs use. There were no differences in the odds of cerebrovascular disease and other second cancers. The up-front strategy with AIs was associated with a non-statistically significant increased odds of death without breast cancer recurrence compared to the use of tamoxifen alone or to the early switch strategy. The results of this meta-analysis suggest that switching strategies are also rational and effective.

Last but not least the different molecular structures and power of AIs could translate into different clinical efficacy and toxicity but only a direct comparison between them can give a definitive answer and there are currently studies underway.

For the moment AIs should be part of endocrine therapy for postmenopausal women with early breast cancer. The choice of the best strategy for the individual patient should be the result of an extensive discussion with the patient, taking into account the risk of disease relapse, the toxicities of the treatment and the patient's wishes.

ACKNOWLEDGEMENTS

Declared none.

CONFLICT OF INTEREST

The authors confirm that this chapter contents have no conflict of interest.

DISCLOSURE

The chapter submitted for eBook Series entitled: "**Recent Advances in Medicinal Chemistry, Volume 1**" is an update of our article published in **Mini-Reviews in Medicinal Chemistry, Volume 8, Number 6, pp. 564 to 574**, with additional text and references.

REFERENCES

[1] Early Breast Cancer Trialists' Collaborative Group. Effects of chemotherapy and hormonal therapy for early breast cancer on recurrence and 15-year survival: an overview of the randomised trials. *Lancet,* **2005,** 365 (9472), 1687-1717.

[2] Fisher, B.; Dignam, J.; Bryant, J.; Wolmark, N. Five versus more than five years of tamoxifen for lymph node-negative breast cancer: updated findings from the National Surgical Adjuvant Breast and Bowel Project B-14 randomized trial. *J. Natl .Cancer Inst.* **2001,** 93 (9), 684-690.

[3] Stewart, H. J.; Prescott, R. J.; Forrest, A. P. Scottish adjuvant tamoxifen trial: a randomized study updated to 15 years. *J. Natl. Cancer Inst,* **2001,** 93 (6), 456-462.

[4] Bryant, J.; Fisher, B.; Dignam, J. Duration of adjuvant tamoxifen therapy. *J. Natl.Cancer Inst. Monogr.* **2001**, 56-61.

[5] Passage, K. J.; McCarthy, N. J. Critical review of the management of early-stage breast cancer in elderly women. *Intern. Med. J.* **2007,** 37 (3), 181-189.
[6] Buzdar, A.; Douma, J.; Davidson, N.; Elledge, R.; Morgan, M.; Smith, R.; Porter, L.; Nabholtz, J.; Xiang, X.; Brady, C. Phase III, multicenter, double-blind, randomized study of letrozole, an aromatase inhibitor, for advanced breast cancer versus megestrol acetate. *J. Clin.Oncol.* **2001,** 19 (14), 3357-3366.
[7] Buzdar, A. U.; Jonat, W.; Howell, A.; Jones, S. E.; Blomqvist, C. P.; Vogel, C. L.; Eiermann, W.; Wolter, J. M.; Steinberg, M.; Webster, A.; Lee, D. Anastrozole versus megestrol acetate in the treatment of postmenopausal women with advanced breast carcinoma: results of a survival update based on a combined analysis of data from two mature phase III trials. Arimidex Study Group. *Cancer* **1998,** 83 (6), 1142-1152.
[8] Buzdar, A. U.; Jones, S. E.; Vogel, C. L.; Wolter, J.; Plourde, P.; Webster, A., A. Phase III trial comparing anastrozole (1 and 10 milligrams), a potent and selective aromatase inhibitor, with megestrol acetate in postmenopausal women with advanced breast carcinoma. Arimidex Study Group. *Cancer* **1997,** 79 (4), 730-739.
[9] Jonat, W.; Howell, A.; Blomqvist, C.; Eiermann, W.; Winblad, G.; Tyrrell, C.; Mauriac, L.; Roche, H.; Lundgren, S.; Hellmund, R.; Azab, M. A. Randomised trial comparing two doses of the new selective aromatase inhibitor anastrozole (Arimidex) with megestrol acetate in postmenopausal patients with advanced breast cancer. *Eur. J. Cancer* **1996,** 32A (3), 404-412.
[10] Kaufmann, M.; Bajetta, E.; Dirix, L. Y.; Fein, L. E.; Jones, S. E.; Zilembo, N.; Dugardyn, J. L.; Nasurdi, C.; Mennel, R. G.; Cervek, J.; Fowst, C.; Polli, A.; di Salle, E.; Arkhipov, A.; Piscitelli, G.; Miller, L. L.; Massimini, G. Exemestane is superior to megestrol acetate after tamoxifen failure in postmenopausal women with advanced breast cancer: results of a phase III randomized double-blind trial. The Exemestane Study Group. *J. Clin.Oncol.* **2000,** 18 (7), 1399-1411.
[11] Dombernowsky, P.; Smith, I.; Falkson, G.; Leonard, R.; Panasci, L.; Bellmunt, J.; Bezwoda, W.; Gardin, G.; Gudgeon, A.; Morgan, M.; Fornasiero, A.; Hoffmann, W.; Michel, J.; Hatschek, T.; Tjabbes, T.; Chaudri, H. A.; Hornberger, U.; Trunet, P. F. Letrozole, a new oral aromatase inhibitor for advanced breast cancer: double-blind randomized trial showing a dose effect and improved efficacy and tolerability compared with megestrol acetate. *J. Clin. Oncol.* **1998,** 16 (2), 453-461.
[12] Kaufmann, M.; Jonat, W.; Hilfrich, J.; Eidtmann, H.; Gademann, G.; Zuna, I.; von Minckwitz, G. Improved overall survival in postmenopausal women with early breast cancer after anastrozole initiated after treatment with tamoxifen compared with continued tamoxifen: the ARNO 95 Study. *J. Clin. Oncol.* **2007,** 25 (19), 2664-2670.
[13] Regan, M.M.M; Neven, P.; Giobbie-Hurder, A.; Goldhrirsch, A.; Ejlertsen, B.; Mauriac, L.; Forbes, J.F.; Smith, I.; Lang, I.; Wardley, A.; Rabaglio, M.; Price, K.N.; Gelber, R.D.; Coates A.S.; Thurlimann, B. Assessment of letrozole and tamoxifen alone in sequence for postmenopausal women with steroid hormone receptor-positive breast cancer: The BIG 1-98 randomized clinical trial at 8.1 years median follow-up. *Lancet Oncol.* **2011,** 12 (12), 1101-1108.
[14] Bliss, J. M.; Kilburn L. S.; Coleman, R. E.; Forbes J.F.; Coates, A. S.; Jones, S. E.; Jassem, J.; Delozier, T.; Andersen, J.; van de Velde, C.; Lonning, P. E.; Morden J.; Reise J.; Cisar L.; Menschik T.; Coombes, R. Disease –related outcome with long-term follow-up: an update analysis of the Intergroup Exemestane Study. *J. Clin. Oncol.* **2012,** 30 (7) 709-717.

[15] Jin, H.; Tu, D.; Zhao, N;, Shepherd, L.E.; Goss, P.E. Longer-term outcomes of letrozole versus placebo after 5 years of tamoxifen in the NCIC CTG MA.17 trial: analyses adjusting for treatment crossover. *J. Clin. Oncol.* **2012**, 30 (7) 718-721.

[16] Simpson, E. R.; Mahendroo, M. S.; Means, G. D.; Kilgore, M. W.; Hinshelwood, M. M.; Graham-Lorence, S.; Amarneh, B.; Ito, Y.; Fisher, C. R.; Michael, M. D.; et al. Aromatase cytochrome P450, the enzyme responsible for estrogen biosynthesis. *Endocr. Rev.* **1994,** 15 (3), 342-355.

[17] Miller, W. R.; O'Neill, J. The importance of local synthesis of estrogen within the breast. *Steroids* **1987,** 50 (4-6), 537-548.

[18] McCloskey, E.; Hannon, R.; Lakner, G.; Clack, G.; Miyamoto, A.; Finkelman, R.; Gastell, R. Complete data from the letrozole (L), exemestane (E), and anastrozole (A) pharmacodynamic (PD) 'LEAP' trial: direct comparison of safety parameters between aromatase inhibitors (AIs) in healthy postmenopausal women. *Breast Cancer Res. Treat.* **2006,** 100 (Suppl 1), abstr. 2092, S111.

[19] Howell, A.; Cuzick, J.; Baum, M.; Buzdar, A.; Dowsett, M.; Forbes, J. F.; Hoctin-Boes, G.; Houghton, J.; Locker, G. Y.; Tobias, J. S. Results of the ATAC (Arimidex, Tamoxifen, Alone or in Combination) trial after completion of 5 years' adjuvant treatment for breast cancer. *Lancet* **2005,** 365 (9453), 60-62.

[20] Baum, M.; Budzar, A. U.; Cuzick, J.; Forbes, J.; Houghton, J. H.; Klijn, J. G.; Sahmoud, T. Anastrozole alone or in combination with tamoxifen versus tamoxifen alone for adjuvant treatment of postmenopausal women with early breast cancer: first results of the ATAC randomised trial. *Lancet* **2002,** 359 (9324), 2131-2139.

[21] Cuzick, J.; Sestak, .;I Baum, M.; Howell, A.; Dowsett, M.; Forbes, J.F.; on behalf of the ATAC/LATTE investigators. Effect of anastrozole and tamoxifen as adjuvant treatment for early-stage breast cance: 10-year analysis of the ATAC trial. *Lancet Oncol.* **2011**, 11, 1135-1141

[22] Early Breast Cancer Trialists' Collaborative Group. Tamoxifen for early breast cancer: an overview of the randomised trials. *Lancet* **1998,** 351 (9114), 1451-1467.

[23] Dowsett, M.; Cuzick, J.; Wale, C.; Howell, T.; Houghton, J.; Baum, M. Retrospective analysis of time to recurrence in the ATAC trial according to hormone receptor status: an hypothesis-generating study. *J. Clin. Oncol.* **2005,** 23 (30), 7512-7517.

[24] Osborne, C. K.; Schiff, R.; Arpino, G.; Lee, A. S.; Hilsenbeck, V. G. Endocrine responsiveness: understanding how progesterone receptor can be used to select endocrine therapy. *Breast* **2005,** 14 (6), 458-465.

[25] Dowsett, M.; Allred, D. C. Relationship between quantitative estrogen and progesteron receptor expression and human epidermal growth factor receptor 2 (HER2) status with recurrence in the Arimidex, Tamoxifen, Alone or in Combination trial. *J. Clin. Oncol.* **2008,** 26 (7), 1059-1065.

[26] Buzdar, A.; Howell, A.; Cuzick, J.; Wale, C.; Distler, W.; Hoctin-Boes, G.; Houghton, J.; Locker, G. Y.; Nabholtz, J. M. Comprehensive side-effect profile of anastrozole and tamoxifen as adjuvant treatment for early-stage breast cancer: long-term safety analysis of the ATAC trial. *Lancet Oncol.* **2006,** 7 (8), 633-643.

[27] Eastell, R.; Adams, J.E.; Coleman, R. E.; Howell, A.; Hannon, R.A.; Cuzick, J.; Mackey, J.R.; Beckmann, M.W.; and Clack, G., Effect of anastrozole on bone mineral density: 5-year results from the Arimidex, Tamoxifen, Alone or in combination trial 18233230. *J. Clin. Oncol.* **2008,** 26 (7), 1051-1058.

[28] Fallowfield, L.; Cella, D.; Cuzick, J.; Francis, S.; Locker, G.; Howell, A. Quality of life of postmenopausal women in the Arimidex, Tamoxifen, Alone or in Combination (ATAC) Adjuvant Breast Cancer Trial. *J. Clin. Oncol.* **2004,** 22 (21), 4261-4271.

[29] Thurlimann, B.; Keshaviah, A.; Coates, A. S.; Mouridsen, H.; Mauriac, L.; Forbes, J. F.; Paridaens, R.; Castiglione-Gertsch, M.; Gelber, R. D.; Rabaglio, M.; Smith, I.; Wardley, A.; Price, K. N.; Goldhirsch, A. A comparison of letrozole and tamoxifen in postmenopausal women with early breast cancer. *N. Engl. J. Med.* **2005,** 353 (26), 2747-2757.

[30] Viale, G.; Regarm, M.; Dell'Orto, P.; Del Curto, B.; Braye, S.; Orosz, Z.; Brown, R.; Olszewski, W. P.; Knox, F.; Oehlschlegel; Thurlimann, B., Central review of ER, PgR and HER-2 in BIG 1-98 evaluating letrozole vs. tamoxifen as adjuvant endocrine therapy for postmenopausal women with receptor-positive breast cancer. *Breast Cancer Res. Treat.* **2005,** 94 (Suppl 1), abstr. 44, S13-14 .

[31] van de Valde, C.J.H.; Rea, D.; Seynaeve, C.; Putter, H.; Hasemburg, A.; Vannetzel, J-M.; Paridaens, R.; Markopoulos, C.; Hozumi, Y.; Hille, E.T.M.; Kieback, D.G.; Asmar, L.; Smeets, J.; Nortier, J.W.R.; Hadji, P.; Bartlett, J.M.S.; Jones, S.E. Adjuvant tamoxifen and exemestane in early breast cancer (TEAM) a randomised phase III trial. *Lancet* **2011,** *377,* 321-331.

[32] Coombes, R. C.; Hall, E.; Gibson, L. J.; Paridaens, R.; Jassem, J.; Delozier, T.; Jones, S. E.; Alvarez, I.; Bertelli, G.; Ortmann, O.; Coates, A. S.; Bajetta, E.; Dodwell, D.; Coleman, R. E.; Fallowfield, L. J.; Mickiewicz, E.; Andersen, J.; Lonning, P. E.; Cocconi, G.; Stewart, A.; Stuart, N.; Snowdon, C. F.; Carpentieri, M.; Massimini, G.; Bliss, J. M.; van de Velde, C. A randomized trial of exemestane after two to three years of tamoxifen therapy in postmenopausal women with primary breast cancer. *N. Engl. J. Med.* **2004,** 350 (11), 1081-1092.

[33] Jones, S.E.; Seynaeve, C.; Hasenburg, A.; Rae,D.; Vannetzel, J-M.; Paridaens, R.; Markopoulos, C.; Hozumi, Y.; Putter, H.; Hille, E.; Kieback, D.; Asmar, L.; Smeets, J.; Urbanski, R.; Bartlett, J.M.S.; van de Valde, C.J.H. Results of the first planned analysis of the TEAM (tamoxifen exemestane adjuvant multinational) prospective randomized phase III trial in hormone sensitive postmenopausal early breast cancer. *Cancer Res.* **2009**, 69 (suppl), abstr.15, 67s.

[34] Bartlett, J.M.S.; Brookes, C.L.; Robson, T.; van de Velde, C.J.H.; Billingham, L.J.; Campbell, F.M.; Grant, M.; Hasenburg, A.; Hille, E.T.M.; Kay, C.; Kieback, D.G.; Putter, H.; Markopoulos, C;, Kranenbarg, E. M-K.; Mallon, E.A.; Dirix, L.; Seynaeve, C.; and Rea, D. Estrogen receptor and progesterone receptor as predictive biomarkers of response to endocrine therapy: a prospectively powered pathology study in the tamoxifen and exemestane adjuvant multinational trial. *J. Clin. Oncol.* **2011**, 29 (12), 1531-1538.

[35] Coombes, R. C.; Kilburn, L. S.; Snowdon, C. F.; Paridaens, R.; Coleman, R. E.; Jones, S. E.; Jassem, J.; Van de Velde, C. J.; Delozier, T.; Alvarez, I.; Del Mastro, L.; Ortmann, O.; Diedrich, K.; Coates, A. S.; Bajetta, E.; Holmberg, S. B.; Dodwell, D.; Mickiewicz, E.; Andersen, J.; Lonning, P. E.; Cocconi, G.; Forbes, J.; Castiglione, M.; Stuart, N.; Stewart, A.; Fallowfield, L. J.; Bertelli, G.; Hall, E.; Bogle, R. G.; Carpentieri, M.; Colajori, E.; Subar, M.; Ireland, E.; Bliss, J. M. Survival and safety of exemestane versus tamoxifen after 2-3 years' tamoxifen treatment (Intergroup Exemestane Study): a randomised controlled trial. *Lancet* **2007,** 369 (9561), 559-570

[36] Coleman, R. E.; Banks, L. M.; Girgis, S. I.; Kilburn, L. S.; Vrdoljak, E.; Fox, J.; Cawthorn, S. J.; Patel, A.; Snowdon, C. F.; Hall, E.; Bliss, J. M.; Coombes, R. C. Skeletal effects of exemestane on bone-mineral density, bone biomarkers, and fracture incidence in postmenopausal women with early breast cancer participating in the Intergroup Exemestane Study (IES): a randomised controlled study. *Lancet Oncol.* **2007,** 8 (2), 119-127.

[37] Fallowfield, L. J.; Bliss, J. M.; Porter, L. S.; Price, M. H.; Snowdon, C. F.; Jones, S. E.; Coombes, R. C.; Hall, E. Quality of life in the intergroup exemestane study: a randomized trial of exemestane versus continued tamoxifen after 2 to 3 years of tamoxifen in postmenopausal women with primary breast cancer. *J. Clin .Oncol.* **2006,** 24 (6), 910-917.

[38] Boccardo, F.; Rubagotti, A.; Puntoni, M.; Guglielmini, P.; Amoroso, D.; Fini, A.; Paladini, G.; Mesiti, M.; Romeo, D.; Rinaldini, M.; Scali, S.; Porpiglia, M.; Benedetto, C.; Restuccia, N.; Buzzi, F.; Franchi, R.; Massidda, B.; Distante, V.; Amadori, D.; Sismondi, P., Switching to anastrozole versus continued tamoxifen treatment of early breast cancer: preliminary results of the Italian Tamoxifen Anastrozole Trial. *J. Clin. Oncol.* **2005,** 23 (22), 5138-5147.

[39] Boccardo, F.; Rubagotti, A.; Guglielmini, P.; Fini, A.; Paladini, G.; Mesiti, M.; Rinaldini, M.; Scali, S.; Porpiglia, M.; Benedetto, C.; Restuccia, N.; Buzzi, F.; Franchi, R.; Massidda, B.; Distante, V.; Amadori, D.; Sismondi, P. Switching to anastrozole verus continued tamoxifen treatment for early breast cancer. Updated results of the Italian Tamoxifen Anastrozole (ITA) trial. *Ann. Oncol.* **2006,** 17 (Suppl 7), abstr. vii10-vii14.

[40] Jakesz, R.; Jonat, W.; Gnant, M.; Mittlboeck, M.; Greil, R.; Tausch, C.; Hilfrich, J.; Kwasny, W.; Menzel, C.; Samonigg, H.; Seifert, M.; Gademann, G.; Kaufmann, M.; Wolfgang, J. Switching of postmenopausal women with endocrine-responsive early breast cancer to anastrozole after 2 years' adjuvant tamoxifen: combined results of ABCSG trial 8 and ARNO 95 trial. *Lancet* **2005,** 366 (9484), 455-462.

[41] Dubsky, P.C.; Mlineritsch, B.; Postlberger, S.; Samonigg, H.; Kwasny, W.; Tausch, C.; Stoger, H.; Haider, K.; Fitzal, F.; Singer, C.F.; Stiere, M.; Sevelda, P.; Luschin-Ebengreuth, G.; Taucher, S.; Rudas, M.; Bartsch, R.; Steger, G.G.; Greil, R.; Filipcic, L. and Gnant M., Tamoxifene and anastrozole as a sequencing strategy: a randomized controlled trial in postmenopausal patients with endocrine-responsive early breast cancer from the Austrian Breast and Colorectal Cancer Study Group. *J. Clin. Oncol.* **2012**, 30 (7) 722-728.

[42] Jonat, W.; Gnant, M.; Boccardo, F.; Kaufmann, M.; Rubagotti, A.; Zuna, I.; Greenwood, M.; Jakesz, R. Effectiveness of switching from adjuvant tamoxifen to anastrozole in postmenopausal women with hormone-sensitive early-stage breast cancer: a meta-analysis. *Lancet Oncol.* **2006,** 7 (12), 991-996.

[43] Aihara, T.; Takatsuka, Y.; Ohsumi, S.; Aogi, K.; Hozumi, Y.; Imoto, S.; Mukai, H.; Iwata, H.; Watanabe, T.; Shimizu , C.; Nakagami, K.; Tamura, M.; Ito, T.; Masuda, N.; Ogino, N.; Hisamatsu, K.; Mitsuyama, S.; Abe, H.; Tanaka, S.; Yamaguchi, T.; Ohashi, Y. Phase III randomized adjuvant study of tamoxifen alone versus sequential tamoxifen and anastrozole in Japanese postmenopausal women with hormone-responsive breast cancer: N-SAS BC03 study. *Breast Cancer Res. Treat.* **2010**, 121 (2), 379-387

[44] Saphner, T.; Tormey, D. C.; Gray, R. Annual hazard rates of recurrence for breast cancer after primary therapy. *J. Clin. Oncol.* **1996,** 14 (10), 2738-2746.

[45] Kennecke, H.; Speers, C.; Chia, S.; Norris, B.; Gelmon, K.; Bryce, C.; Barnett, J.; Olivotto, I. 10 year event free survival in postmenopausal women with early breast cancer during the

second five years of tamoxifen *Breast Cancer Res. Treat.* **2004,** 88 (Suppl 1), abstr. 1049, S57.

[46] Hortobagyi, G. N.; Kau, S.-W.; Budzar, A. U.; Theriault, R. L.; Booser, D. J.; Gwyn, K.; Valero, V. What is the prognosis of patients with operable breast cancer (BC) five years after diagnosis? *J. Clin. Oncol.* **2004,** 22 (14S), abstr. 585, 23s.

[47] Chia, S. K.; Speers, C. H.; Bryce, C. J.; Hayes, M. M.; Olivotto, I. A. Ten-year outcomes in a population-based cohort of node-negative, lymphatic, and vascular invasion-negative early breast cancers without adjuvant systemic therapies. *J. Clin. Oncol.* **2004,** 22 (9), 1630-1637.

[48] Goss, P. E.; Ingle, J. N.; Martino, S.; Robert, N. J.; Muss, H. B.; Piccart, M. J.; Castiglione, M.; Tu, D.; Shepherd, L. E.; Pritchard, K. I.; Livingston, R. B.; Davidson, N. E.; Norton, L.; Perez, E. A.; Abrams, J. S.; Therasse, P.; Palmer, M. J.; Pater, J. L. A randomized trial of letrozole in postmenopausal women after five years of tamoxifen therapy for early-stage breast cancer. *N. Engl. J. Med.* **2003,** 349 (19), 1793-1802.

[49] Goss, P. E.; Ingle, J. N.; Martino, S.; Robert, N. J.; Muss, H. B.; Piccart, M. J.; Castiglione, M.; Tu, D.; Shepherd, L. E.; Pritchard, K. I.; Livingston, R. B.; Davidson, N. E.; Norton, L.; Perez, E. A.; Abrams, J. S.; Cameron, D. A.; Palmer, M. J.; Pater, J. L. Randomized trial of letrozole following tamoxifen as extended adjuvant therapy in receptor-positive breast cancer: updated findings from NCIC CTG MA.17. *J. Natl. Cancer Inst.* **2005,** 97 (17), 1262-1271.

[50] Goss, P. E.; Ingle, J. N.; Martino, S.; Robert, N. J.; Muss, H. B.; Piccart, M. J.; Castiglione, M.; Tu, D.; Shepherd, L. E.; Pritchard, K. I.; Livingston, R. B.; Davidson, N. E.; Norton, L.; Perez, E. A.; Abrams, J. S.; Cameron, D. A.; Palmer, M. J.; Pater, J. L. Efficacy of letrozole extended adjuvant therapy according to estrogen receptor and progesterone receptor status of the primary tumor: National Cancer Institute of Canada Clinical Trials Group MA.17. *J. Clin. Oncol.* **2007,** 25 (15), 2006-2011.

[51] Ingle, J. N.; Tu, D.; Pater, J. L.; Martino, S.; Robert, N. J.; Muss, H. B.; Piccart, M. J.; Castiglione, M.; Shepherd, L. E.; Pritchard, K. I.; Livingston, R. B.; Davidson, N. E.; Norton, L.; Perez, E. A.; Abrams, J. S.; Cameron, D. A.; Palmer, M. J.; Goss, P. E. Duration of letrozole treatment and outcomes in the placebo-controlled NCIC CTG MA.17 extended adjuvant therapy trial. *Breast Cancer Res. Treat.* **2006,** 99 (3), 295-300.

[52] Jakesz, R.; Somonigg, H.; Greil, R.; Gnant, M.; Schmid, M.; Kwasny, W.; Kubista, E.; Millneritsch, B.; Tausch, C.; Stierer, M. Extended adjuvant treatment with anatrozole: results from the Austrian Breast and Colorectal Cancer Study Group trial 6a (ABCSG-6a). *J. Clin. Oncol.* **2005,** 20 (Suppl 1), abstr. 527,10s.

[53] Mamounas, E.; Jeong, J.-H.; Wickerham, L.; Smith, R.; Geyer, C.; Ganz, P.; Land, S.; Hupchins, L.; Eisen, A.; Ingle, J. N.; Constantino, J.; Wolmark, N., Benefit from exemestane (EXE) as extended adjuvant therapy after 5 years of tamoxifen (TAM): intent-to-treat analysis of NSABP B-33. *Breast Cancer Res. Treat.* **2006,** 100 (Suppl 1), abstr. 49, S22.

[54] Mansel, R.; Locker, G.; Fallowfield, L.; Benedict, A.; Jones, D. Cost-effectiveness analysis of anastrozole vs tamoxifen in adjuvant therapy for early stage breast cancer in the United Kingdom: the 5-year completed treatment analysis of the ATAC ('Arimidex', Tamoxifen alone or in combination) trial. *Br. J. Cancer* **2007,** 97 (2), 152-161.

[55] Locker, G. Y.; Mansel, R.; Cella, D.; Dobrez, D.; Sorensen, S.; Gandhi, S. K. Cost-effectiveness analysis of anastrozole versus tamoxifen as primary adjuvant therapy for

postmenopausal women with early breast cancer: a US healthcare system perspective. The 5-year completed treatment analysis of the ATAC ('Arimidex', Tamoxifen Alone or in Combination) trial. *Breast Cancer Res. Treat.* **2007** 106(2):229-238.

[56] Younis, T.; Rayson, D.; Dewar, R.; & Skedgel, C. Modelling for cost-effective- adjuvant aromatase inhibitor strategies for postmenopausal women with breast cancer. *Ann. Oncol.* **2007,** 18 (2), 293-298.

[57] Delea, T. E.; Karnon, J.; Barghout, V.; Thomas, S. K.; Papo, N. L. Cost-effectiveness of letrozole and anastrozole as adjuvant therapy for hormone receptor positive early breast cancer in postmenopausal women. *J. Clin. Oncol.* **2006,** 24 (18S), abstract 10577,574s.

[58] Lonning, P. E., Comparing cost/utility of giving an aromatase inhibitor as monotherapy for 5 years versus sequential administration following 2-3 or 5 years of tamoxifen as adjuvant treatment for postmenopausal breast cancer. *Ann. Oncol.* **2006,** 17 (2), 217-225.

[59] Karnon, J.; Delea, T. E.; Kaura Meng, S.; di Trapani, F. Comparison of the cost-effectiveness of upfront letrozole or anastrozole, or switched exemestane versus tamoxifen for early breast cancer in hormone receptor positive (HR+) postmenopausal women: the UK perspective. *Breast Cancer Res. Treat.* **2006,** 100 (Suppl 1), abstr. 5049, S229.

[60] Lundkvist, J.; Wilking, N.; Holmberg, S.; Jonsson, L., Cost-effectiveness of exemestane versus tamoxifen as adjuvant therapy for early-stage breast cancer after 2-3 years treatment with tamoxifen in Sweden. *Breast Cancer Res. Treat.* **2007,** 102 (3), 289-299.

[61] Delea, T. E.; Karnon, J.; Smith, R. E.; Johnston, S. R.; Brandman, J.; Sung, J. C.; Gross, P. E. Cost-effectiveness of extended adjuvant letrozole therapy after 5 years of adjuvant tamoxifen therapy in postmenopausal women with early-stage breast cancer. *Am. J. Manag. Care* **2006,** 12 (7), 374-386.

[62] Punglia, R. S.; Kuntz, K. M.; Winer, E. P.; Weeks, J. C.; Burstein, H. J. Optimizing adjuvant endocrine therapy in postmenopausal women with early-stage breast cancer: a decision analysis. *J. Clin. Oncol.* **2005,** 23 (22), 5178-5187.

[63] Cuzick, J.; Sasieni, P.; Howell, A. Should aromatase inhibitors be used as initial adjuvant treatment or sequenced after tamoxifen? *Br. J. Cancer* **2006,** 94 (4), 460-464.

[64] Goetz, M. P.; Rae, J. M.; Suman, V. J.; Safgren, S. L.; Ames, M. M.; Visscher, D. W.; Reynolds, C.; Couch, F. J.; Lingle, W. L.; Flockhart, D. A.; Desta, Z.; Perez, E. A.; Ingle, J. N. Pharmacogenetics of tamoxifen biotransformation is associated with clinical outcomes of efficacy and hot flashes. *J. Clin .Oncol.* **2005,** 23 (36), 9312-9318.

[65] Higgins, M.J.; Rae, J.M.; Flockhart, D.A.; Hayes, D.F.; Stearns, V. Pharmacogenetics of tamoxifen: who should undergo CYP2D6 genetic testing? *J. Natl. Compr. Cancer Netw.* **2009,** 7 (2), 203-213

[66] Regan, M.; Leyland-Jones, B.; Bouzyk, M.; Pagani, O.; Tang, W.; Kammler, R.; Dell'Orto, P.; Biasi, M.O.; Thurlimann, B.; Lyng, M.B.; Ditzel, H.; Neven, P.; Debled, M.; Maibach, R.; Price, K.; Gelber, R.; Coates, A.; Goldhirsch, A.; Rae, J.; Viale, G. CYP2D6 Genotype and Tamoxifen Response in Postmenopausal Women with Endocrine-Responsive Breast Cancer: The Breast International Group 1-98 Trial. *J. Natl. Cancer Inst.* ***2012,*** 104 (6), 441-451

[67] Rae, J.M.; Drury, S; Hayes, D.F.; Stearns, W.; Thibert, J; Haynes, B.P.; Salter, J; Sestak, I; Cuzick, J.; Dowsett, M., on behalf of the ATAC trialists; CYP2D6 and UGT2B7 Genotype and Risk of Recurrence in Tamoxifen-Treated Breast Cancer Patients *J. Natl. Cancer Inst.* **2012,** 104 (6), 452-460

[68] Ma, C. X.; Adjei, A. A.; Salavaggione, O. E.; Coronel, J.; Pelleymounter, L.; Wang, L.; Eckloff, B. W.; Schaid, D.; Wieben, E. D.; Adjei, A. A.; Weinshilboum, R. M. Human aromatase: gene resequencing and functional genomics. *Cancer. Res.* **2005,** 65 (23), 11071-11082.

[69] Moy, B.; Tu, D.; Pater, J. L.; Ingle, J. N.; Shepherd, L. E.; Whelan, T. J.; Goss, P. E., Clinical outcomes of ethnic minority women in MA.17: a trial of letrozole after 5 years of tamoxifen in postmenopausal women with early stage breast cancer. *Ann. Oncol.* **2006,** 17 (11), 1637-1643.

[70] Wang,L.;, Ellsworth, K.A.; Moon, I.; Pelleymounter, L.L.; Eckloff, B.W.; Martin, Y.N.; Fridley, B.L.; Jenkins, G.D.; Batzler, A.; Suman, V.J.; Ravi, S.; Dixon, J.M.; Miller, W.R.; Wieben, E.D.; Buzdar, A.; Weinshilboum, R.M.; and Ingle, J.N. Functional genetic polymorphisms in the aromatase gene CYP 19 vary the response of breast cancer patients to neoadjuvant therapy with aromatase inhibitors. *Cancer Res.* **2010,** 70 (1), 319-328.

[71] Amir, E.; Seruga, B.; Niraula, S.; Carlsson, L.; Ocaña, A. Toxicity of adjuvant endocrine therapy in postmenopausal breast cancer patients: a systematic review and meta-analysis. *J.Natl. Cancer Inst.***2011** 103 (17), 1299-1309

Send Orders for Reprints to reprints@benthamscience.net
Recent Advances in Medicinal Chemistry, Vol. 1, 2014, 131-158

CHAPTER 5

Bacterial FabH: Towards the Discovery of New Broad-Spectrum Antibiotics

Yunierkis Pérez-Castillo[1,2,3], Matheus Froeyen[2], Ann Nowé[3] and Miguel Ángel Cabrera-Pérez[1,4,*]

[1]Molecular Simulation & Drug Design Group. Centro de BioactivosQuímicos. Universidad Central de Las Villas, Santa Clara, 54830, Villa Clara, Cuba; [2]Laboratory for Medicinal Chemistry, Rega Institute for Medical Research, Katholieke Universiteit Leuven, Minderbroedersstraat 10, B-3000 Leuven, Belgium; [3]Computational Modeling Lab (CoMo), Department of Computer Sciences, Faculty of Sciences, Vrije Universiteit Brussel, Pleinlaan 2, B-1050 Brussel, Belgium and [4]Engineering Department, Pharmacy and Pharmaceutical Technology Area, Faculty of Pharmacy, University Miguel Hernandez, Alicante 03550, Spain

Abstract: The emergence of drug resistant strains of important human pathogens has made urgent the necessity of finding new targets and novel antimicrobial agents. One of the most promising targets is FabH. Here we summarize the progress made in the design of FabH inhibitors and the role played by the 3D-structure of the enzyme as well as by the modeling studies in the design of new FabH inhibitors.

Keywords: Drug design, FabH, FAS, FabH inhibitors.

INTRODUCTION

Bacterial diseases have not yet been overcome. More than one-third of the world population is likely infected by bacterial pathogens and two million fatalities occur per year from bacterial infections [1]. Antibiotic resistance has increased over the past two decades with almost every human pathogen and every class of antimicrobials in clinical use acquiring resistance [2-4]. Contributing to the

***Address correspondence to Miguel Ángel Cabrera-Pérez:** Molecular Simulation & Drug Design Group. Centro de BioactivosQuímicos. Universidad Central de Las Villas, Santa Clara, 54830, Villa Clara, Cuba; Tel: (53)-42-281473; (53)-42-281192; Fax: (53)-42-281130; E-mails: macabrera@uclv.edu.cu; migue@gammu.com; macabreraster@gmail.com

Atta-ur-Rahman, Muhammad Iqbal Choudhary and George Perry (Eds)

dilemma, only two new classes of antibiotics have been introduced over the past 30 years: the oxazolidinones, represented by linezolid, and the lipopeptide daptomycin [5-9]. The relatively small number of antibiotic classes currently in use is directed against asmall subset of essential bacterial targets. Clearly, new approaches to the discovery of novel anti bacterial are required and innovative strategies will be necessary to identify novel and effective candidates.

During the last decade, with the beginning of the genomics era, it was assumed that genomics would provide a plethora of novel targets and hence a flood of new therapeutic agents [10]. Nevertheless, not all proteins in the bacteria proteome can serve as targets for broad spectrum antimicrobial agents because they do not possess the desirable properties for an ideal bacterial target such as being essential for cell survival, present in multiple bacterial species and being selective or not having a close homologue in the human genome.

Historically, different targets in key areas of the bacterial cell cycle have been studied such as those related with the transcription and translation of genetic code, cell wall biosynthesis, metabolic pathways, cell division, virulence factors, resistance mechanism and so on [11]. Among them, one of the most attractive biochemical pathways to be used as the target for new antibacterial agents is fatty acid biosynthesis (FAS). This pathway has been demonstrated to be essential for the bacteria cell survival [12], and differs considerably from the human FAS pathway. While in humans fatty acid synthesis occurs *via* a homodimeric multifunctional enzyme [13, 14], in bacteria the pathway is composed of various discrete enzymes and each one can be considered a putative molecular target. These features make the type II FAS pathway a potential target for new antimicrobial agents.

A key enzyme in this pathway is the β-ketoacyl-acyl carrier protein synthase III (FabH), which is the enzyme responsible for the first pathway reaction and plays an important regulatory role. FabH has also been demonstrated to be essential for organism survival and it is present in a wide number of important human pathogens. Furthermore, some chemical compounds have been shown to inhibit FabH from diverse microorganisms, including multi-drug resistant strains. These facts support the idea that FabH can be used as an effective molecular target for the development of new antimicrobial agents.

Taking into consideration the recent availability of three-dimensional (3D) structures of FabH from different microorganisms, in this review we attempt to summarize the recent progress made in the field of FabH inhibitors and the implications of those crystal structures for the structure-based drug design of new broad-spectrum antibacterial agents.

BACTERIAL TYPE II FAS THE BIOLOGICAL PROCESS

In bacteria, each enzyme catalyzes a particular reaction and some steps in the pathway can be catalyzed by more than one enzyme, while the acyl carrier protein (ACP) is the responsible for the transfer of substrates along the whole pathway. The full pathway have been extensively reviewed elsewhere [15, 16] and highly characterized in *E. coli* [17, 18]. A schematic view of the entire pathway is depicted in Fig. (**1**).

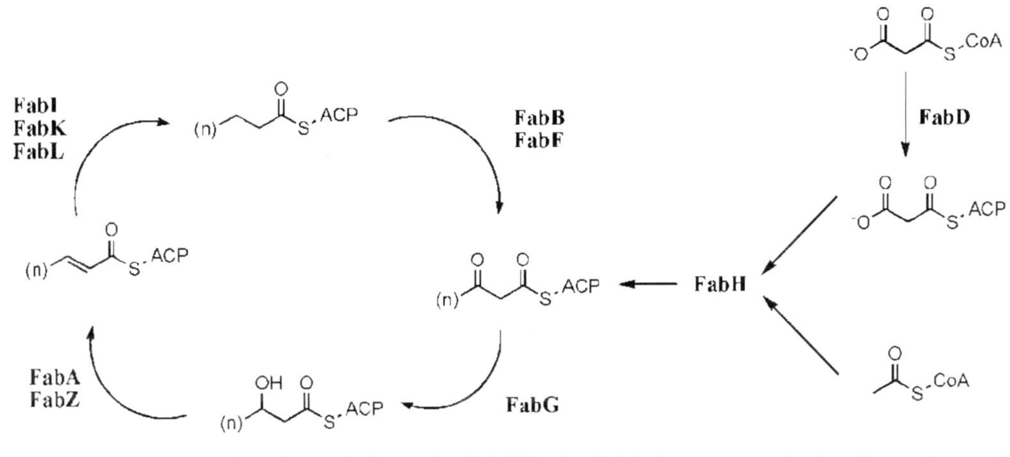

Figure 1: Type II FAS pathway in bacteria. FabH is the initiating enzyme in the elongation cycle and plays a key regulatory role in the whole pathway. The same reaction step can be catalyzed by more than one enzyme. In some microorganisms more than one enzyme catalyzing this step can be found. On the other hand, FabH is present in a wide range of pathogens and no other enzyme has this function.

The first step in the fatty acid biosynthetic pathway is the synthesis of malonyl-ACP from ACP and malonyl-CoA. This reaction is catalyzed by malonyl-CoA: ACP transacyclase (FabD) [19]. Malonyl-ACP is then condensed with acetyl-CoA by β-ketoacyl-ACP synthase III (FabH) yielding acetoacetyl-ACP [20, 21], which initiates the elongation cycle and is reduced by β-ketoacyl-ACP reductase (FabG)

[22]. The β-hydroxybutyryl-ACP is dehydrated later by either FabA or FabZ to form trans-2-enoyl-ACP [23], which is further reduced by enoyl-ACP reductase (FabI) to form acyl-ACP [24]. In subsequent steps malonyl-ACP is condensed with acyl-ACP either by FabB or FabF [20, 25, 26]. Mature acyl-ACPs are obtained through further cycles while the cycle is regulated by the feedback inhibition of FabH and FabI [23, 26].

All the enzymes involved in the fatty acid biosynthetic pathway have been successfully used as drug targets for the development of new potent antibacterial agents and have been previously reviewed [16, 27-29]. Inhibitors have been developed targeting FabD [30], the condensing enzymes FabB and FabF [31-37], the reductase FabG [38], dehydratases FabA [39] and FabZ [40], enoyl-ACP reductases I (FabI) [41-46] and II (FabK) [45] and β-ketoacyl-acyl carrier protein synthase III (FabH) [47-62].

FabH AS A DRUG TARGET DIFFERENCES AND ADVANTAGES

As shown in Fig. (**1**), FabH initiates the elongation cycles catalyzing the condensation of acyl-CoA and malonyl-ACP to form acetoacetyl-ACP [20] and regulates the entire pathway *via* a feedback inhibition mechanism by long chain acyl-ACP [63]. FabH is also an essential enzyme [64, 65]. This enzyme differs, in sequence and structure, from the rest of the condensing enzymes (FabB and FabF) and its catalytic triad (Cys, His, Asn) is different compared with the FabB and FabF enzymes (Cys, His, His) [66-69]. FabH has no close homologue in humans and its selectivity for acetyl-CoA over acyl-ACP diverges from the other condensing enzymes [20, 70].

On the other hand, FabH is present in many important human pathogens, most of them having multiresistant strains, such as *Escherichia coli* (*ec*FabH), *Staphylococcus aureus* (*sa*FabH), *Mycobacterium tuberculosis* (*mt*FabH), *Enteroccoccus faecium* (*ef*FabH), *Streptococcus pneumonia* (*sp*FabH), *Pseudomonas aeruginosa* (*pa*FabH), *Neisseria meningitidis* (*nm*FabH) and *Haemophilus influenzae* (*hi*FabH) [20, 50, 71-74]. In Table **1** are listed the most important residues involved in ligand binding in *E. coli* and the corresponding aminoacids in various microorganisms. The numbering of residues corresponds to *E. coli* and is the same numbering scheme used in the rest of the paper.

Table 1: Conservation of the residues in the binding tunnel of FabH across different species. Catalytic residues are in italics and mutated ones are in bold. Numbering corresponds to that of *E. coli* and amino acids are represented through the standard one letter code. Remarkable is that most residues are strictly conserved or are conservatively substituted. The major substitution influencing substrate specificity and differences in inhibitors potency across species is that which occurs at F87

Organism	**R36**	**T37**	**W32**	**R151**	**I155**	**I156**	**F157**	***N274***	**M207**	**L189**	**G209**	**N210**	**C112**
E. Coli	R	T	W	R	I	I	F	*N*	M	L	G	N	*C*
H. influenzae	R	T	W	R	**V**	**L**	F	*N*	M	L	G	N	*C*
M. tuberculosis	R	T	W	R	**F**	I	F	*N*	**L**	**I**	G	**P**	*C*
N. meningitidis	R	T	W	R	**V**	**L**	F	*N*	M	L	G	**P**	*C*
P. aeruginosa	R	T	W	R	**A**	**L**	F	*N*	M	L	G	**R**	*C*
S. aureus	**M**	T	W	R	**V**	**L**	F	*N*	M	L	G	**R**	*C*
S. pneumoniae	R	T	W	R	**V**	**L**	F	*N*	M	L	G	**R**	*C*
E. faecium	R	T	W	R	**V**	**L**	F	*N*	M	L	G	**R**	*C*
Organism	**G305**	**G306**	**F304**	**R249**	**N247**	**A246**	**F213**	**V212**	**A216**	**H244**	**I250**	**F87**	
E. Coli	G	G	F	R	N	A	F	V	A	*H*	I	F	
H. influenzae	G	G	F	R	N	A	F	**T**	A	*H*	I	**Y**	
M. tuberculosis	G	**A**	**Y**	R	N	A	F	V	A	*H*	I	**T**	
N. meningitidis	G	G	F	R	N	A	F	V	A	*H*	I	F	
P. aeruginosa	G	**A**	F	R	N	A	F	V	A	*H*	I	**C**	
S. aureus	G	G	F	R	N	A	F	V	A	*H*	I	F	
S. pneumoniae	G	G	**I**	R	N	A	F	V	A	*H*	I	**M**	
E. faecium	G	G	**Y**	R	N	A	F	**I**	A	*H*	I	**T**	

From Table **1** it is clear that most residues involved in ligand binding are conserved across species or are conservatively substituted. The most important difference is the substitution of Phe87 which determines the active site tunnel size and the substrate specificity across species. These differences and their implications for the design of new FabH inhibitors will be widely discussed in this review.

All the FabH attributes described above strongly suggest that this enzyme is a potential target for the development of new potent antibacterial agents having broad-spectrum activity.

THREE DIMENSIONAL STRUCTURE OF FabH

FabH in *E. coli* is a 35kDa protein which is active, in solution, as a 70kDa homodimer [75, 76]. The crystal structure of FabH has been solved for various species. Microorganisms for which the three dimensional structures are available in the Protein Data Bank (PDB) are *E. coli* [47, 66-68, 77, 78], *M. tuberculosis*

[69, 79-82], *S. aureus* [78, 83], *H. influenzae* [78], *E. faecalis* [78], *P. aeruginosa* [84] and *T. thermophilus*. The overall structure of the enzyme monomer is the same for these seven micro organisms, as shown in Fig. (**2a**).

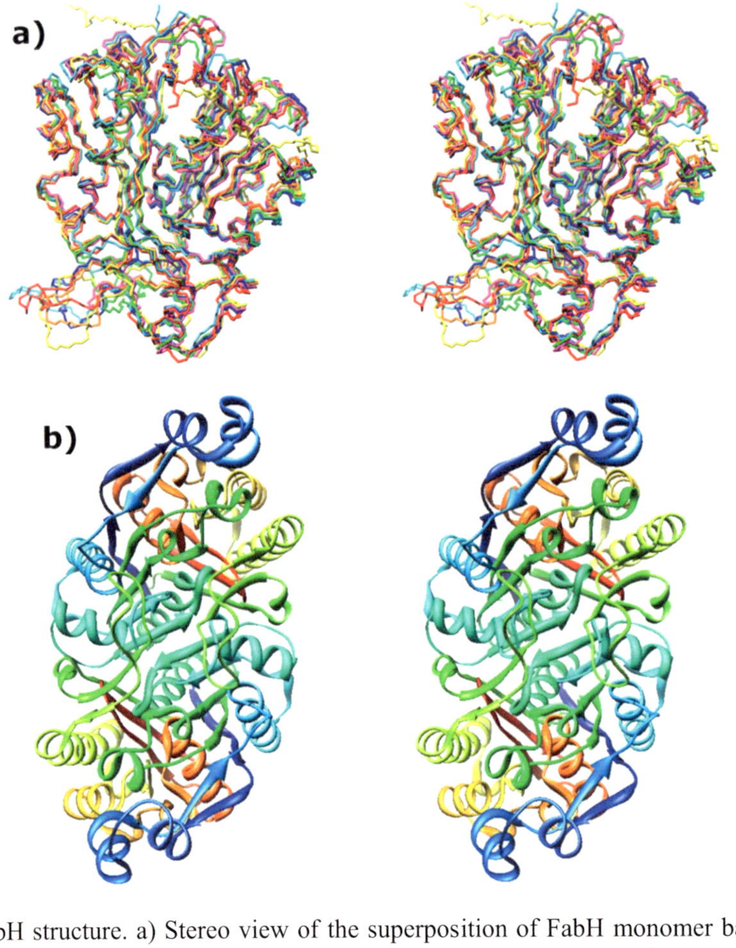

Figure 2: FabH structure. a) Stereo view of the superposition of FabH monomer backbone from *E. coli* (PDB code 1EBL, blue), *S. aureus* (1ZOW, magenta), *T. termophylus* (1UB7, green), *M. tuberculosis* (1U6E, yellow), *P. aeruginosa* (2X3E, red), E. *faecalis* (3IL4, cyan) and *H. influenza* (3IL3, orange) showing the identical overall fold for the seven species for which crystal structures are available. b) Ribbon diagram of *E. coli* FabH dimmer, the central β sheet corresponds to the continuous ten stranded beta sheet formed upon dimer formation. Both images were generated using the USCF CHIMERA package [85].

The FabH monomer structure can be reduced to the duplication of two almost identical parts (N and C terminals), despite the only 11% sequence identity of both. In *E. coli* each part corresponds to residues 1-171 and 172-317, respectively.

The overall structure of *E. coli* FabH is a mixed α/β structure made up of five β-strands and three α-helices, where the five β-strands form a mixed β-sheet. In the dimer, the third β-strand of the N-terminal part of one monomer interacts with the same strand from the other monomer to form a ten stranded β-sheet (see Fig. (**2b**)). On the other hand, the fold of this enzyme has been determined to be very similar to that in the thiolases [86, 87].

The crystal structure of *E. coli* FabHis available in the *apo* form [67, 68, 78], as well as in a complex with the acetylated enzyme and the enzyme with CoA and malonyl-CoA [66-68]. Also the crystal structure of the *E. coli* FabH in complex with a bound inhibitor has been determined [47]. These crystal structures have revealed that the active site of the enzyme is a deep and hydrophobic tunnel where the catalytic residues (Cys112, His244 and Asn274) are positioned at the end.

In the dimer form, two active sites are located on opposite sides of the molecule and each one is mainly formed by residues of its own subunit. In the case of *E. coli* FabH the size of the hydrophobic tunnel is limited by the presence of a phenylalanine residue in position 87 of the other chain and consequently, the size of the acyl chain that can be bound is limited to four carbons. This is different for the *M. tuberculosis* FabH where acyl chains of 16 carbons can bind due to the presence of threonine in place of Phe87 (see Table **1**) [69]. This substitution opens a new channel for this microorganism that in *E. coli* is blocked by the presence of the phenylalanine which can bind long chain acyl substrates. In Fig. (**3**) are represented the binding channels of the primer in *E. coli* and *M. tuberculosis*. As will be discussed in the next section, the presence of Phe87 is one of the main factors influencing the differences in substrate specificity among different species.

Despite the overall structure of FabH remains the same across all the solved X-ray structures regardless the microorganism the enzyme comes from, there are some findings provided by these crystallographic studies that indicate a highly flexible enzyme. The first prove of this high flexibility in the FabH enzyme was provided by the Qiu *et al.* [68] when they solved the X-ray structure of the ligand-free FabH in the tetragonal form. This structure showed large-scale conformational changes in several important regions of FabH, mainly involving four loops that participate in the formation of the dimer interface. Besides, some of these disordered residues participate in the recognition and binding of the ligands to the enzyme. From the analysis of the data that they obtained, these authors concluded that FabH should exist in an open and a closed conformations and that the presence of a hydroxyl anion in the oxyanion hole of the enzyme can be critical in the stabilization of the enzyme in the closed conformation. Furthermore, they suggested that this apo-protein structure in the tetragonal form suggested a catalytically impaired conformation for the enzyme.

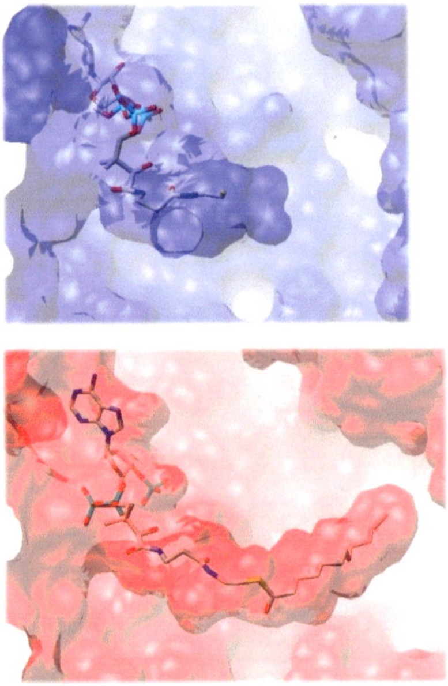

Figure 3: Comparison of the size of the FabH primer binding pocket in *E. coli* and *M. tuberculosis*. Cross section of FabH from *E. coli* (upper image, PDB code 1EBL) and *M. tuberculosis* (lower image, PDB code 1U6S) showing the acyl primer binding pocket in both microorganisms. Before surface generation the two enzymes were superposed, so the proteins are represented at the same scale. Stick lines inside the binding tunnels represent acetyl-CoA in ecFabH and lauroyl-CoA in mtFabH. As shown in the picture the primer binding pocket in mtFabH is much larger than in ecFabH. The picture was produced using the UCSF CHIMERA package [85].

In order to study the catalytic mechanism of FabH, Alhamadsheh *et al.* carried out crystallographic and kinetic studies of ecFabH using Alkyl-CoA disulfides inhibitors [77]. Their results supported and expanded the previous proposal by Qiu *et al.* [68] that apo-ecFabH exists in a partially disordered form that becomes ordered on binding of substrate. They also demonstrated that this structurally disordered apo-ecFabH dimer orders on binding of the substrate or inhibitor to one monomer and afterwards the same binding process takes place in the second subunit at a much slower rate. Based on their results, they proposed that the highly disordered loops found by Qiu *et al.* [68] can have a critical role in the ligand induced stabilization of the dimer.

SUBSTRATE BINDING AND SPECIFICITY

As discussed before, the length of acyl-CoA primers that can be bound to FabH range from two to eighteen carbons, across species [50, 75, 83, 88, 89]. In the same way, in some microorganism FabH only bind straight chain acyl-CoA substrates and in others, such as *S. aureus*, where branched chain fatty acids are needed for survival, FabH is capable of using branched chain acyl-CoA primers as substrate, although the aminoacids in direct contact with the acyl primer are the same as those for the *ec*FabH [88].

Determination of the three dimensional structure of the *S. aureus* FabH was crucial for the explanation of this fact. As discussed by Qiu *et al.* [83], the acyl primer binding pocket in *S. aureus* FabH is larger than for the *E. coli* enzyme, as a consequence of a shift in residues in contact with the substrate with respect to their positions in the *E. coli* enzyme. As a consequence of this shift, the side chain of Phe87 moves 3.0 Å away from the position of this residue in the *E. coli* structure. This shift can be explained on the basis of the differences in sequences between the two species at the dimer interface residues that allows an increased volume of the primer binding pocket.

Additional information provided by the crystal structures is related to the binding mode of the acyl-CoA primer. In all cases the pantethiene moiety of CoA fits in the active site tunnel and the phosphate groups of CoA form salt bridges with the conserved Arg36 and possibly with Arg151. Furthermore the adenine group of CoA stacks between Trp32 and Arg151, while the acyl group side chain is accommodated in the bottom of the tunnel.

CATALYTIC MECHANISM AND SUBSTRATE BINDING

In all organisms the FabH enzyme catalyzes the condensation of malonyl-ACP with acyl-CoA by a unique mechanism that consistsof a two steps (ping-pong) Claisen condensation reaction [20, 21]. The first step is the transfer of the acyl group to the cysteine active site and the release of CoA, while the second one is the condensation of the cysteine bound acyl group with malonyl-ACP yielding acetoacyl-ACP.

The general mechanism of the reaction has been extensively studied and is the same for all bacteria despite the differences in substrate specificity that exist among species [66, 68, 80, 83]. The first half of the whole reaction includes the binding of the acyl-CoA primer to the enzyme, the deprotonation of the active site cysteine where the essential thiol group for all condensing enzymes is present [90], and the nucleophilic attack on the acyl-CoA by the thiolate anion of the deprotonated cysteine, the subsequent acylation of that cysteine and the release of CoA-SH. This first half reaction is represented in Fig (**4**).

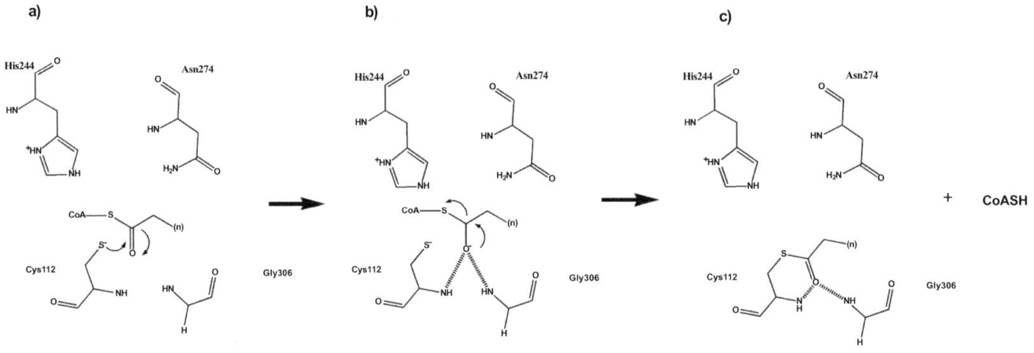

Figure 4: Acetylation half reaction mechanism.
a) The first step is the binding of the acyl-CoA primer to the enzyme, deprotonation of the cysteine active site and nucleophilic attack on the acyl-CoA by the thiolate anion of the cysteine active site.
b) The developing negative charge of the transition state becomes stabilized by the oxyanion hole formed by the backbone amides of Cys112 and Gly306.
c) Finally, the enzyme becomes acetylated and CoA-SH is released from the enzyme that is then ready to enter the second half of the condensation reaction. The oxyanion hole serves to stabilize the carbonyl group of the acetylated cysteine.

The mechanism responsible for the deprotonation of the active site cysteine differs considerably in the enzymes that form acetylated intermediates, where a hydrogen bond between this cysteine and a hydrogen bond acceptor residue promotes the deprotonation of the active cysteine [91, 92]. In the case of FabH, both the active site histidine (His244) and asparagine (Asn274) are too far from Cys112 to promote its deprotonation. Furthermore, site directed mutagenesis studies of the *E. coli* enzyme have confirmed the non essential role of His244 and Asn274 for the transacylation half reaction [66]. In this study, H244A and N274A mutants showed increased transacylation activities when compared with the wild type enzyme. On the basis of these results and the crystallographic structure of the enzyme that shows that Cys112 is located at the N-terminus of a long α helix, Davies *et al.* [38] proposed a mechanism by which the nucleophilicity of the cysteine active site is due to the effect of the α helix dipole moment [93] and in this position Cys112 can receive the benefit of a half unit of positive charge generated by the helix dipole.

More recently, site directed mutagenesis experiments conducted on *M. tuberculosis* FabH has shown some influence of the active site histidine (His258) on the transacylation reaction [80]. Unlike the above mentioned results in *E. coli*, the *M. tuberculosis* H258A FabH mutant retains only 22.6% of the activity of the wild type enzyme. Using this result and considering the presence of a water molecule in the crystallographic structure located between His258 and the active site cysteine (Cys122), a new role for His258 was proposed in which the epsilon nitrogen (Nε) of His258 is capable of extracting a proton from the water molecule and the formed hydroxyl anion can deprotonate Cys122. The fact that the H258A mutant still retains some transacylation activity shows that this residue helps in the deprotonation of Cys122, but transacylation appears to be partially promoted *via* the α helix dipole moment as occurs in *E. coli*.

Another important feature of FabH is the so-called oxyanion hole, a region of the active site formed by the backbone amide groups of Cys112 and Gly306. The function of this region is to stabilize the developing negative charge of the tetrahedral transition state during the formation of the acetyl enzyme intermediate (Fig. (**4b**)). In various FabH crystal structures, an electronic density is observed in the oxyanion hole and in one of them this density is identified as a water molecule or an hydroxyl anion that makes hydrogen bond interactions with the nitrogen backbone atoms of Cys112 and Gly306 and that can share a proton with the thiol group of Cys112 [68]. Thus, this water molecule has also been proposed to be relevant for the deprotonation of Cys112.

Once the acyl enzyme intermediate is formed, the carbonyl oxygen of the acetyl-thioester intermediate displaces the water molecule of the oxyanion hole and forms strong hydrogen bonds with the amides of Cys112 and Gly306. Then CoA is released and malony-ACP, that serves as the other substrate for the condensation reaction and has been shown to bind preferentially to the acetylated form of FabH [70], binds to FabH. This last step of the acetylation half reaction is shown in Fig. (**4c**)

In both *E. coli* and *M. tuberculosis,* His244 and Asn274 have been demonstrated to be essential for the second half of the reaction, specifically for the decarboxylation reaction [66, 83]. The crystal structures of FabH show how, as a consequence of a network of hydrogen bonds, the Nε atom of His244 and the Nδ Asn274 point towards the same location in the active site. This configuration of the catalytic residues Asn274 and His244 "mimics" the oxyanion hole and helps to stabilize the developing negative charge on the malonyl-ACP thioester carbonyl to promote the formation of the enol intermediate and the formation of a carbanion on the C2 of malonate. This carbanion then attacks the acyl group bound to the active site cysteine and the resulting tetrahedral transition state is again stabilized by the oxyanion hole. As a final result of the whole reaction acetoacyl-ACP and CO_2 are produced. This second half of the FabH whole reaction is shown in Fig. (**5**).

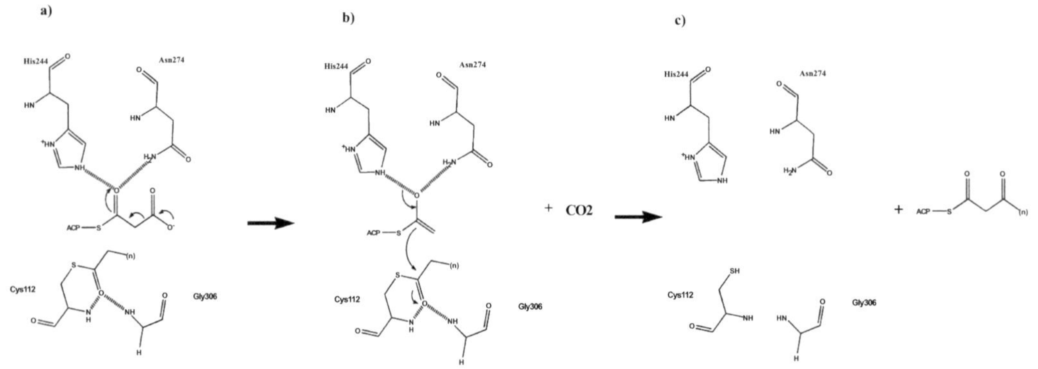

Figure 5: Second half of the FabH condensation reaction.
a) A network of hydrogen bonds orients Asn274 and His244 to pointto the same position in the active site.
b) The configuration of His244 and Asn274 "mimics" the oxyanion hole and helps to stabilize the developing negative charge on the malonyl-ACP thioester carbonyl to promote the formation of the enol intermediate and the formation of a carbanion on the C2 of malonate that attacks the acyl group bound to the active site cysteine; the resulting tetrahedral transition state is again stabilized by the oxyanion hole and a CO_2 molecule is released.
c) Finally, acetoacyl-ACP is released.

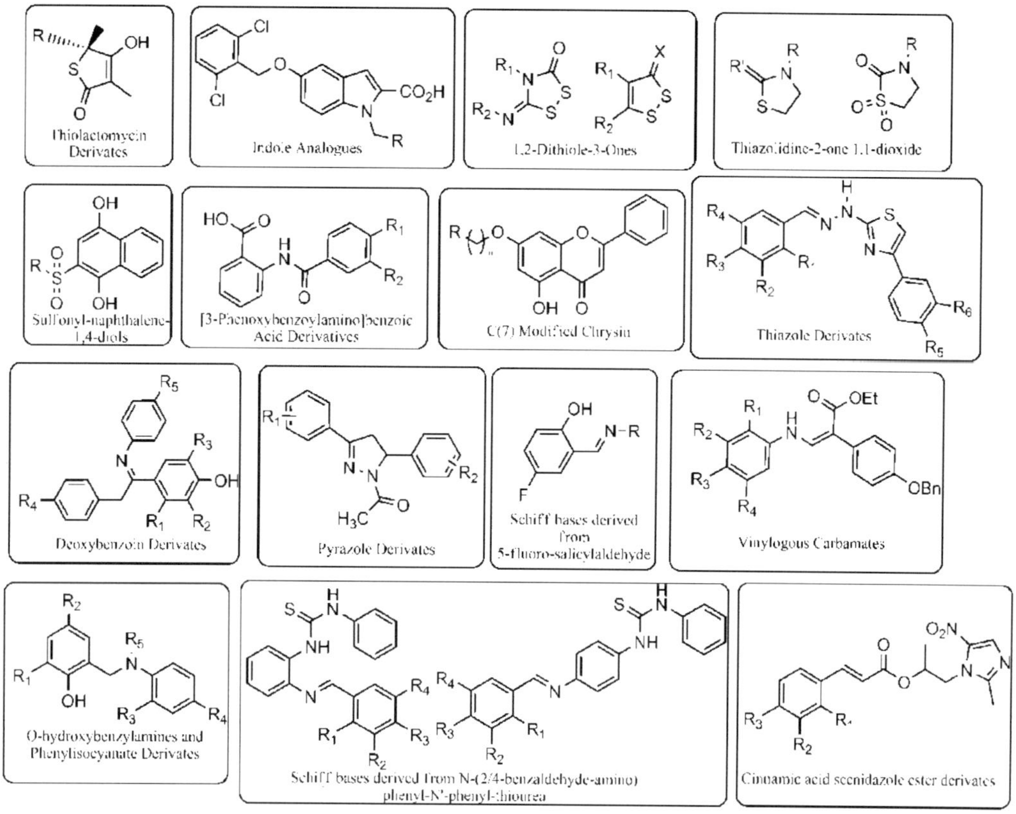

Figure 6: Fifteen known types of FabH inhibitors.

Many research groups have focused on the search for new chemical compounds that can selectivity inhibit FabH from various microorganisms. Some inhibitors have been found that inhibit the enzyme of various species, including important human pathogens such as *E. coli, H. influenzae, S. pneumoniae, M. tuberculosis, E. faecalis, S. pyogenes, P. falciparum*and *S. aureus*. On the other hand, the determination of the crystal structure of this enzyme from various species has aided in the structure-based design of new potent inhibitors. Fifteen basic types of chemical compounds have been demonstrated to inhibit FabH (Fig. (**6**)).

One of the compounds that weakly inhibit FabH is thiolactomycin (TLM, see Fig (**6**)), a natural product produced by actinomycetes [94]. TLM inhibits the three condensing enzymes, although FabH has been shown to be less sensitiveto this compound. The IC_{50} of TLM for *E. coli* FabH has been determined to be 110 μM

[95] and recent works have demonstrated that the potency of TLM differs from one species to another [50, 89]. Thus, *M. tuberculosis* FabH has been shown to be 3 times more sensitive to TLM than the *E. coli* enzyme [89], while the enzymes from *S. pneumoniae* and *H. influenzae* were found to be 4-5 fold more susceptible to TLM [50] than *E. coli* FabH.

The results obtained with TLM were used as the starting point for the design of TLM analogues with improved activity against *M. tuberculosis* FabH [96, 97]. Biphenyl and acetylene based analogues of TLM have been synthesized and assayed against *mt*FabH. The inhibition values obtained for both acetylene and biphenyl based analogues were improved compared with those obtained with TLM. Kim *et al.* have also investigated the influence of different substitutions at position 5 of TLM [98]. Only three out of 31 substitutions tested showed modest improved activity against *M. tuberculosis* FabH, while none of the TLM variants showed activity against *E. coli* FabH. The only variants able to improve the IC_{50} value of TLM were those that included long aliphatic chains (12-14 carbon atoms) at the 5 position. In this work the essentiality of each of the two conjugated double bonds adjacent to the thiolactone ring was also demonstrated.

Although there is no experimental or theoretical evidence to support this fact, it can be surmised that a bulky substituent as the biphenyl group at the 5 position of TLM will only fit the active site of an enzyme with a large volume in the cavity, which is the case of *M. tuberculosis* FabH. In the same way, a long aliphatic chain at the position 5 of TLM could perfectly fit the channel that extends from the active site cysteine in *M. tuberculosis* and accommodates the long chain acyl primers. This is consistent with the fact that no inhibition is reported for the *E. coli* FabH when these variants of TLM are assayed.

The first crystallographic structure of FabH in complex with an inhibitor molecule was solved by Daines *et al.* [47]. Indole inhibitors of FabH (Fig. (**6**)) were first discovered in a high throughput screening effort against *S. pneumonia* FabH. Attempts were made for the co-crystallization of these compounds with either *E. coli* or *S. pneumoniae* FabH, but no crystal could be obtained, mainly due to its poor solubility. Then a structure-based drug design study was carried out and eight compounds with improved solubility and unmodified key substituents were assayed against *E. coli* and *S. pneumoniae* FabH. Finally, a co-crystal structure of FabH and an inhibitor was solved and the predicted binding mode of indole inhibitors was experimentally corroborated. Interestingly, this type of inhibitor does not interact directly with the catalytic triad but with the rest of the active site tunnel.

The results of a virtual screening effort for the identification of FabH inhibitors have been reported [48, 52, 53]. This work led to the identification of 1,2-dithiole-3-ones [48], thiazolidine-2-one 1,1-dioxides [53] and sulfonyl-naphthalene-1,4-diols [52] as FabH inhibitors (see Fig. (**6**)).

Several analogues of 1,2-dithiole-3-ones were identified and assayed for FabH inhibition. Despite the results showed low inhibitory concentrations of these compounds against *E. coli* and *S. aureus* FabH, the reactivity of 1,2-dithiole-3-ones renders them capable of targeting multiple cellular processes, thus limiting the specificity of this kind of inhibitors. Furthermore, the mechanism by which 1,2-dithiole-3-ones inhibit FabH is not clear and it can involves the covalent linkage of the compound to the enzyme through the opening of the ring at the S-2 position because of a nucleophilic attack by Cys112.

In the case of the thiazolidine-2-one 1,1-dioxides FabH inhibitors [53], 17 compounds were tested against ecFabH, mtFabH and *P. falciparum* KASIII. Nine of these compounds were shown to be ecFabH inhibitors while no inhibition was detected at 100 μM for the *M. tuberculosis* and *P. falciparum* enzymes. Despite FabH is well conserved among many human pathogens, the observed lack of activity against *M. tuberculosis* and *P. falciparum* could be explained based on the structural differences that exist at the mouth of the CoA binding channel that can limit the access of these compounds to it. Based on the results of the above mentioned virtual screening effort, Alhamadsheh *et al.* also studied a series of 14 sulfonyl-naphthalene-1,4-diols FabH inhibitors [52]. Most of the compounds assayed were active against ecFabH, pfFabH and mtFabH and they were shown to be non-covalent FabH inhibitors.

Nie *et al.* found a novel series of benzoylaminobenzoic acids as potent inhibitors of *E. faecalis* FabH (Fig. (**6**)) [51]. The authors proposed an approach that involved filtering three million chemicals from various commercial sources using computational filters, including substructure, 2D similarity, pharmacophore with size and shape constraints, and docking with FlexX. After applying all filters about 2500 compounds were selected for the primary FabH inhibition screening and 27 exhibited *in vitro* activities below 10 μM against *E. faecalis* and *H. influenzae* FabH. These 27 compounds were grouped in four structurally diverse series and benzoylaminobenzoic acids were selected for further optimization.

Co-crystallization studies were carried out with the most potent compound and the *E. faecalis* FabH [78]. The crystallographic analysis showed that two fundamental interactions guide the binding process: an ionic interaction with the protonated nitrogen of the active site histidine and a hydrogen bond with amine group of the side chain of the catalytic asparagine. On the basis of these results, a new campaign of structure based drug design was carried out and yielded as a final result a set of potent *in vitro* inhibitors of FabH, many of them having good antibacterial activity when evaluated against various Gram-positive and Gram-negative pathogens. More interestingly, the inhibitors proposed by Nie *et al.* showed good inhibitory potency when tested *in vivo* against *E. coli, S. pneumoniae, S. pyogenes, E. faecalis, N. meningitides* and a Methicillin resistant *S. aureus* strain (MRSA).

Recent research conducted at the State Key Laboratory of Pharmaceutical Biotechnology of the Nanjing University has lead to the discovery of nine new types of FabH inhibitors. These inhibitor types (see Fig. (**6**)) are: C (7) modified chrysin derivates [62], thiazole derivates [61], deoxybenzoin derivates [60], pyrazole derivates [59], Schiff bases derived from 5-fluoro-salicylaldehyde [58], vinylogous carbamates [57], o-hydroxybenzylamines and phenylisocyanate derivates [56], schiff bases derived from N-(2/4-benzaldehyde-amino) phenyl-N'-phenyl-thiourea [55] and cinnamic acid secnidazole ester derivates [54]. All the chemical scaffolds proposed by this research group were shown to be inhibitors of ecFabH in the μM range. Besides, the inhibitors they proposed showed activity against both Gram positive and Gram negative microorganisms such as *E. coli, E. cloacae*, *P. aeruginosa*, *P. fluorescence, B. subtilis*, *S. aureus* and *E. faecalis*.

All these results in the search of FabH inhibitors show that this protein remains an attractive target for the development of new broad spectrum antibacterial agents. Furthermore, they demonstrate that the inhibition of FabH can be used as a novel mechanism of action of new antibiotics to overcome the developed resistance to the drugs currently in clinical use.

MOLECULAR MODELING STUDIES OF FABH

Although there is many experimental data available regarding FabH and its inhibitors, hitherto only a few structure and ligand-based modeling studies have been carried out. Regarding the ligand-based drug design approaches, the set of 46 efFabH reported by Nie *et al.* [51], have been used by Ashek *et al.* and by Singh *et al.* for molecular docking and QSAR studies [99-101]. In one of these studies [99], the authors found a correlation between the calculated docking scores and the inhibitory activities They were also able to describe in detail and quantify the influence of each substitution in the activity of benzoylaminobenzoic acids and to find a good correspondence between the CoMSIA field and the receptor three dimensional structures when they were overlapped. The same authors also reported a HQSAR study for this data set where they could explain the relationship between the individual atomic contributions and the overall activity [101].

In the study of Singh *et al.* [100], the authors developed several QSAR models using the same data set as above and concluded that the hydrophobicity, aromaticity, molar refractivity, and presence of a hydroxyl group in a fixed position are the main factors that influence the inhibitory activity of this compounds. They were also able to propose, based on the information provided by the developed QSAR models, which kind of substituents should be placed on each position of the inhibitor scaffold in order to positively contribute to their inhibitory activity. They also studied the inhibitory activity of the 10 most active compounds against *S. aureus, S. pneumoniae, S. pyogenes, E. faecalis, N. meningitides* and *E. coli*, and concluded that hydrophobocity, aromaticity and presence of OH group on a specific position are conducive for the inhibitory activity of this class of inhibitors.

As we discussed before, there is strong experimental evidences that support that FabH is a highly flexible enzyme. This flexibility is one of the key factors to take into account for the structure-based design of new FabH inhibitors. In this sense, such catalytic or inhibition processes where large conformational changes take place, can't be modeled based on a single structure and often the study of the time dependent evolution of the system at the atomic level is essential. To explore the conformational space of FabH, get more insights into the conformational changes that take place on both ligand-bound and ligand-free enzymes and demonstrate the implications of the enzyme flexibility for the structure-based design of new FabH inhibitors; we have carried out Molecular Dynamics (MD) simulations of ecFabH [102].

We ran a 10 ns MD simulation for two systems: an unliganded enzyme and a complex of it with a dichlorobenzyloxy-indole-carboxylic acid inhibitor using the same starting protein conformation on both systems. When setting the complex system, we used only one ligand bound to one of the active sites of the dimer in order to study the ligand-induced stabilization of the free subunit due to the presence of a bound ligand in the other one. After analyzing the temperature factors derived from the MD simulations of both systems, we found that the four loop regions previously reported as highly flexible were also the most flexible regions on both systems during the simulations. Significant peaks of the temperature factor in the ligand-free protein were observed for residues: W32, R36, R151, G152, G209, F304, G305 and G306. The first four residues are located in the entrance of the binding pocket and have a high influence on its topology as can be seen when comparing, for instance, the structures of the ecFabH solved in complex with a dichlorobenzyloxy-indole-carboxylic acid inhibitor by Daines *et al.* and the one solved by Gajiwala *et al.* [47, 78]. On the other hand, residues F304, G305 and G306 are located at the bottom of the binding pocket and form the "oxyanion hole" reported to stabilize the acetyl group of the acetylated intermediate formed after the first halve reaction. We also observed stabilization of the ligand free binding pocket as well as of the highly flexible loops in the complexed system. This last observation was supported by the higher temperature factors of the residues in these regions on any of the dimer subunits of the unliganded system when compared to the complex one.

Taking this into account and to make a deeper exploration of the conformational space of ecFabH, we extended the MD simulation of the ligand free system up to 100 ns. Through a Principal Component Analysis (PCA) of this MD trajectory, we found that 26.97% of the total system conformational variance (first eigenvector) was mainly distributed among the large 185-217 loops while 39% of it (first two eigenvectors) was mainly distributed on the experimentally highly flexible regions. This PCA was also useful to identify concerted motions in the enzyme that were mainly concentrated on these disordered loops. A very interesting result derived from the PCA and the analysis of the solvent occupancy on the oxyanion hole, was that as the occupancy of the solvent in this region increased, the observed flexibility along the first PC decreased. This finding served as a demonstration of the importance of the presence of a solvent molecule in the oxyanion hole of ecFabH for its conformational stabilization, mainly of loops 84-86, 146-152 and 185-217 that have a high importance in dimer stability and ligand binding.

To study how the information provided by the MD simulations could aid in the structure-based design of FabH inhibitors, we used cluster analysis to select representative structures of the binding pocket; then we used them to model ten benzoylaminobenzoic acid inhibitors published by Nie *et al.* which were shown to be broad spectrum FabH inhibitors and were assayed against *E. coli* [51]. Since no crystal structure of these inhibitors in complex with ecFabH is available, we evaluated the value of a set of representative conformations of the binding pocket extracted from the MD simulation of the free enzyme and the available structures of ecFabH to reproduce the right orientation of the ten benzoylaminobenzoic acid inhibitors inside the cavity. The optimal number of representative binding pocket conformations was found to be 15. These representative binding pocket conformations were shown to be a diverse set that included conformations where the mouth of the pocket was completely blocked to others where it was completely open.

The results of the molecular docking of these inhibitors to the available X-ray structures of ecFabH, showed that poses of the compounds inside the binding pocket can be obtained for only three out of the seven available x-ray receptor conformations. The best conformation per compound obtained using the crystallographic structures of ecFabH didn't agreed with the experimental pose determined in complex with efFabH as shown by the RMSD values between 2.6-5.3 Å. This high RMSD values demonstrated the inability to find correctly oriented poses of the compounds under study on this X-ray receptor conformations. On the other hand, the results of the docking to the MD-derived conformations showed RMSD values between 1-2.5 Å for 9 out of the 10 compounds under study when the global lowest scored conformation was selected. Furthermore, the scoring values of the poses obtained using the MD provided receptor conformations were considerably lower than those obtained for the ecFabH crystallographic structures. More important, the ligand-enzyme interactions observed in the inhibitor-efFabH X-ray structure were only reproduced when the compounds were docked into the representative ecFabH conformations derived from the MD simulation.

CONCLUDING REMARKS

As discussed here, many advances have been reported in the development of new FabH inhibitors in the last few years and various research groups are searching for novel inhibitors for this enzyme. The results achieved in the design of new FabH inhibitors shows that, despite differences in the three dimensional structure and substrate specificity of the enzyme among different species, the same compounds can inhibit FabH from more than one microorganism.

The availability of crystal structures of FabH from various species in complex with natural substrates and, more important, with inhibitors, have aided for a better understanding of the mechanism of this enzyme. Although, this structural data doesn't cover the whole conformational space of FabH, even for the most studied ecFabH. In this sense, we have shown that the use of the structure-based drug design techniques such as MD simulations can provide important and unique information that can be used to complement the information provided by the 3D structures of the enzyme; and to increase the accuracy in the structure-based design of FabH inhibitors.

ACKNOWLEDGEMENTS

Pérez-Castillo Y. thanks the Flemish Interuniversity Council (VLIR) for financial support through the project: "Strengthening research and PhD formation in Computer Sciences and its applications" in the framework of the VLIR-UCLV collaborative program. Cabrera-Pérez M. A. thanks the Spanish Agency of International Cooperation for the Development (AECID) for financial support through the project "Montaje de un laboratorio de química computacional, con fines académicos y científicos, para el diseño racional de nuevos candidatos a fármacos en enfermedades de alto impacto social" (A1/036687/11).

CONFLICT OF INTEREST

The authors confirm that this chapter contents have no conflict of interest.

DISCLOSURE

The chapter submitted for eBook Series entitled: "**Recent Advances in Medicinal Chemistry, Volume 1**" is an update of our article published in **Mini-Reviews in Medicinal Chemistry, Volume 8, Number 1, pp. 36 to 45**, with additional text and references.

REFERENCES

[1] Monaghan, R.L.; Barrett, J.F. Antibacterial drug discovery--then, now and the genomics future. *Biochem Pharmacol,* 2006, *71* (7), 901-909.

[2] Levy, S.B.; Marshall, B. Antibacterial resistance worldwide: causes, challenges and responses. *Nat Med,* 2004, *10* (12 Suppl), S122-129.

[3] Nikaido, H. Multidrug Resistance in Bacteria. *Annual Review of Biochemistry,* 2009, *78* (1), 119-146.

[4] Livermore, D.M. The need for new antibiotics. *Clinical Microbiology & Infection,* 2004, *10* (s4), 1-9.

[5] Pucci, M.J. Use of genomics to select antibacterial targets. *Biochem Pharmacol,* 2006, *71* (7), 1066-1072.

[6] Conly, J.; Johnston, B. Where are all the new antibiotics? The new antibiotic paradox. *Can J Infect Dis Med Microbiol,* 2005, *16* (3), 159-160.

[7] Barbachyn, M.R.; Ford, C.W. Oxazolidinone structure-activity relationships leading to linezolid. *Angew Chem Int Ed Engl,* 2003, *42* (18), 2010-2023.

[8] Fischbach, M.A.; Walsh, C.T. Antibiotics for emerging pathogens. *Science,* 2009, *325* (5944), 1089-1093.

[9] Kern, W.V. Daptomycin: first in a new class of antibiotics for complicated skin and soft-tissue infections. *Int J Clin Pract,* 2006, *60* (3), 370-378.

[10] Black, M.T.; Hodgson, J. Novel target sites in bacteria for overcoming antibiotic resistance. *Adv Drug Deliv Rev,* 2005, *57* (10), 1528-1538.

[11] Yoneda, H.; Katsumata, R. Antibiotic Resistance in BActeria and Its Future for Novel Antibiotic Development. *Biosci Biotechnol Biochem,* 2006, *70* (5), 1060-1075.

[12] Heath, R.J.; Rock, C.O. Fatty acid biosynthesis as a target for novel antibacterials. *Curr Opin Investig Drugs,* 2004, *5* (2), 146-153.

[13] Jayakumar, A.; Tai, M.H.; Huang, W.Y.; al-Feel, W.; Hsu, M.; Abu-Elheiga, L.; Chirala, S.S.; Wakil, S.J. Human fatty acid synthase: properties and molecular cloning. *Proc Natl Acad Sci U S A,* 1995, *92* (19), 8695-8699.

[14] Jayakumar, A.; Huang, W.Y.; Raetz, B.; Chirala, S.S.; Wakil, S.J. Cloning and expression of the multifunctional human fatty acid synthase and its subdomains in Escherichia coli. *Proc Natl Acad Sci U S A,* 1996, *93* (25), 14509-14514.

[15] Campbell, J.W.; Cronan, J.E., Jr. Bacterial fatty acid biosynthesis: targets for antibacterial drug discovery. *Annu Rev Microbiol,* 2001, *55*, 305-332.

[16] White, S.W.; Zheng, J.; Zhang, Y.M.; Rock The structural biology of type II fatty acid biosynthesis. *Annu Rev Biochem,* 2005, *74*, 791-831.

[17] Heath, R.J.; White, S.W.; Rock, C.O. Lipid biosynthesis as a target for antibacterial agents. *Prog Lipid Res,* 2001, *40* (6), 467-497.

[18] Magnuson, K.; Jackowski, S.; Rock, C.O.; Cronan, J.E., Jr. Regulation of fatty acid biosynthesis in Escherichia coli. *Microbiol Rev,* 1993, *57* (3), 522-542.

[19] Serre, L.; Verbree, E.C.; Dauter, Z.; Stuitje, A.R.; Derewenda, Z.S. The Escherichia coli malonyl-CoA:acyl carrier protein transacylase at 1.5-A resolution. Crystal structure of a fatty acid synthase component. *J Biol Chem,* 1995, *270* (22), 12961-12964.

[20] Tsay, J.T.; Oh, W.; Larson, T.J.; Jackowski, S.; Rock, C.O. Isolation and characterization of the beta-ketoacyl-acyl carrier protein synthase III gene (fabH) from Escherichia coli K-12. *J Biol Chem,* 1992, *267* (10), 6807-6814.

[21] Jackowski, S.; Rock, C.O. Acetoacetyl-acyl carrier protein synthase, a potential regulator of fatty acid biosynthesis in bacteria. *J Biol Chem,* 1987, *262* (16), 7927-7931.

[22] Heath, R.J.; Rock, C.O. Enoyl-acyl carrier protein reductase (fabI) plays a determinant role in completing cycles of fatty acid elongation in Escherichia coli. *J Biol Chem,* 1995, *270* (44), 26538-26542.

[23] Heath, R.J.; Rock, C.O. Roles of the FabA and FabZ beta-hydroxyacyl-acyl carrier protein dehydratases in Escherichia coli fatty acid biosynthesis. *J Biol Chem,* 1996, *271* (44), 27795-27801.

[24] Bergler, H.; Wallner, P.; Ebeling, A.; Leitinger, B.; Fuchsbichler, S.; Aschauer, H.; Kollenz, G.; Hogenauer, G.; Turnowsky, F. Protein EnvM is the NADH-dependent enoyl-ACP reductase (FabI) of Escherichia coli. *J Biol Chem,* 1994, *269* (8), 5493-5496.

[25] Garwin, J.L.; Klages, A.L.; Cronan, J.E., Jr. Structural, enzymatic, and genetic studies of beta-ketoacyl-acyl carrier protein synthases I and II of Escherichia coli. *J Biol Chem,* 1980, *255* (24), 11949-11956.

[26] Subrahmanyam, S.; Cronan, J.E., Jr. Overproduction of a functional fatty acid biosynthetic enzyme blocks fatty acid synthesis in Escherichia coli. *J Bacteriol,* 1998, *180* (17), 4596-4602.

[27] Heath, R.J.; White, S.W.; Rock, C.O. Inhibitors of fatty acid synthesis as antimicrobial chemotherapeutics. *Appl Microbiol Biotechnol,* 2002, *58* (6), 695-703.

[28] Khandekar, S.S.; Daines, R.A.; Lonsdale, J.T. Bacterial beta-ketoacyl-acyl carrier protein synthases as targets for antibacterial agents. *Curr Protein Pept Sci,* 2003, *4* (1), 21-29.

[29] Zhang, Y.M.; White, S.W.; Rock, C.O. Inhibiting bacterial fatty acid synthesis. *J Biol Chem,* 2006, *281* (26), 17541-17544.

[30] Liu, W.; Han, C.; Hu, L.; Chen, K.; Shen, X.; Jiang, H. Characterization and inhibitor discovery of one novel malonyl-CoA: acyl carrier protein transacylase (MCAT) from Helicobacter pylori. *FEBS Lett,* 2006, *580* (2), 697-702.

[31] Young, K.; Jayasuriya, H.; Ondeyka, J.G.; Herath, K.; Zhang, C.; Kodali, S.; Galgoci, A.; Painter, R.; Brown-Driver, V.; Yamamoto, R.; Silver, L.L.; Zheng, Y.; Ventura, J.I.; Sigmund, J.; Ha, S.; Basilio, A.; Vicente, F.; Tormo, J.R.; Pelaez, F.; Youngman, P.; Cully, D.; Barrett, J.F.; Schmatz, D.; Singh, S.B.; Wang, J. Discovery of FabH/FabF inhibitors from natural products. *Antimicrob Agents Chemother,* 2006, *50* (2), 519-526.

[32] Ondeyka, J.G.; Zink, D.L.; Young, K.; Painter, R.; Kodali, S.; Galgoci, A.; Collado, J.; Tormo, J.R.; Basilio, A.; Vicente, F.; Wang, J.; Singh, S.B. Discovery of bacterial fatty acid synthase inhibitors from a Phoma species as antimicrobial agents using a new antisense-based strategy. *J Nat Prod,* 2006, *69* (3), 377-380.

[33] Kauppinen, S.; Siggaard-Andersen, M.; von Wettstein-Knowles, P. beta-Ketoacyl-ACP synthase I of Escherichia coli: nucleotide sequence of the fabB gene and identification of the cerulenin binding residue. *Carlsberg Res Commun,* 1988, *53* (6), 357-370.

[34] Hayashi, T.; Yamamoto, O.; Sasaki, H.; Okazaki, H.; Kawaguchi, A. Inhibition of fatty acid synthesis by the antibiotic thiolactomycin. *J Antibiot (Tokyo),* 1984, *37* (11), 1456-1461.

[35] Nishida, I.; Kawaguchi, A.; Yamada, M. Effect of thiolactomycin on the individual enzymes of the fatty acid synthase system in Escherichia coli. *J Biochem (Tokyo),* 1986, *99* (5), 1447-1454.

[36] Herath, K.B.; Jayasuriya, H.; Guan, Z.; Schulman, M.; Ruby, C.; Sharma, N.; MacNaul, K.; Menke, J.G.; Kodali, S.; Galgoci, A.; Wang, J.; Singh, S.B. Anthrabenzoxocinones from Streptomyces sp. as liver X receptor ligands and antibacterial agents. *J Nat Prod,* 2005, *68* (9), 1437-1440.

[37] Kodali, S.; Galgoci, A.; Young, K.; Painter, R.; Silver, L.L.; Herath, K.B.; Singh, S.B.; Cully, D.; Barrett, J.F.; Schmatz, D.; Wang, J. Determination of selectivity and efficacy of fatty acid synthesis inhibitors. *J Biol Chem,* 2005, *280* (2), 1669-1677.

[38] Zhang, Y.M.; Rock, C.O. Evaluation of epigallocatechin gallate and related plant polyphenols as inhibitors of the FabG and FabI reductases of bacterial type II fatty-acid synthase. *J Biol Chem,* 2004, *279* (30), 30994-31001.

[39] Kass, L.R. The antibacterial activity of 3-decynoyl-n-acetylcysteamine. Inhibition *in vivo* of beta-hydroxydecanoyl thioester dehydrase. *J Biol Chem,* 1968, *243* (12), 3223-3228.

[40] Sharma, S.K.; Kapoor, M.; Ramya, T.N.; Kumar, S.; Kumar, G.; Modak, R.; Sharma, S.; Surolia, N.; Surolia, A. Identification, characterization, and inhibition of Plasmodium falciparum beta-hydroxyacyl-acyl carrier protein dehydratase (FabZ). *J Biol Chem,* 2003, *278* (46), 45661-45671.

[41] Levy, C.W.; Roujeinikova, A.; Sedelnikova, S.; Baker, P.J.; Stuitje, A.R.; Slabas, A.R.; Rice, D.W.; Rafferty, J.B. Molecular basis of triclosan activity. *Nature,* 1999, *398* (6726), 383-384.

[42] Banerjee, A.; Dubnau, E.; Quemard, A.; Balasubramanian, V.; Um, K.S.; Wilson, T.; Collins, D.; de Lisle, G.; Jacobs, W.R., Jr. inhA, a gene encoding a target for isoniazid and ethionamide in Mycobacterium tuberculosis. *Science,* 1994, *263* (5144), 227-230.

[43] Heath, R.J.; Yu, Y.T.; Shapiro, M.A.; Olson, E.; Rock, C.O. Broad spectrum antimicrobial biocides target the FabI component of fatty acid synthesis. *J Biol Chem,* 1998, *273* (46), 30316-30320.

[44] Stewart, M.J.; Parikh, S.; Xiao, G.; Tonge, P.J.; Kisker, C. Structural basis and mechanism of enoyl reductase inhibition by triclosan. *J Mol Biol,* 1999, *290* (4), 859-865.

[45] Payne, D.J.; Miller, W.H.; Berry, V.; Brosky, J.; Burgess, W.J.; Chen, E.; DeWolf Jr, W.E., Jr.; Fosberry, A.P.; Greenwood, R.; Head, M.S.; Heerding, D.A.; Janson, C.A.; Jaworski, D.D.; Keller, P.M.; Manley, P.J.; Moore, T.D.; Newlander, K.A.; Pearson, S.; Polizzi, B.J.; Qiu, X.; Rittenhouse, S.F.; Slater-Radosti, C.; Salyers, K.L.; Seefeld, M.A.; Smyth, M.G.; Takata, D.T.; Uzinskas, I.N.; Vaidya, K.; Wallis, N.G.; Winram, S.B.; Yuan, C.C.; Huffman, W.F. Discovery of a novel and potent class of FabI-directed antibacterial agents. *Antimicrob Agents Chemother,* 2002, *46* (10), 3118-3124.

[46] Seefeld, M.A.; Miller, W.H.; Newlander, K.A.; Burgess, W.J.; DeWolf, W.E., Jr.; Elkins, P.A.; Head, M.S.; Jakas, D.R.; Janson, C.A.; Keller, P.M.; Manley, P.J.; Moore, T.D.; Payne, D.J.; Pearson, S.; Polizzi, B.J.; Qiu, X.; Rittenhouse, S.F.; Uzinskas, I.N.; Wallis, N.G.; Huffman, W.F. Indole naphthyridinones as inhibitors of bacterial enoyl-ACP reductases FabI and FabK. *J Med Chem,* 2003, *46* (9), 1627-1635.

[47] Daines, R.A.; Pendrak, I.; Sham, K.; Van Aller, G.S.; Konstantinidis, A.K.; Lonsdale, J.T.; Janson, C.A.; Qiu, X.; Brandt, M.; Khandekar, S.S.; Silverman, C.; Head, M.S. First X-ray cocrystal structure of a bacterial FabH condensing enzyme and a small molecule inhibitor achieved using rational design and homology modeling. *J Med Chem,* 2003, *46* (1), 5-8.

[48] He, X.; Reeve, A.M.; Desai, U.R.; Kellogg, G.E.; Reynolds, K.A. 1,2-dithiole-3-ones as potent inhibitors of the bacterial 3-ketoacyl acyl carrier protein synthase III (FabH). *Antimicrob Agents Chemother,* 2004, *48* (8), 3093-3102.

[49] He, X.; Reynolds, K.A. Purification, characterization, and identification of novel inhibitors of the beta-ketoacyl-acyl carrier protein synthase III (FabH) from Staphylococcus aureus. *Antimicrob Agents Chemother,* 2002, *46* (5), 1310-1318.

[50] Khandekar, S.S.; Gentry, D.R.; Van Aller, G.S.; Warren, P.; Xiang, H.; Silverman, C.; Doyle, M.L.; Chambers, P.A.; Konstantinidis, A.K.; Brandt, M.; Daines, R.A.; Lonsdale, J.T. Identification, substrate specificity, and inhibition of the Streptococcus pneumoniae beta-ketoacyl-acyl carrier protein synthase III (FabH). *J Biol Chem,* 2001, *276* (32), 30024-30030.

[51] Nie, Z.; Perretta, C.; Lu, J.; Su, Y.; Margosiak, S.; Gajiwala, K.S.; Cortez, J.; Nikulin, V.; Yager, K.M.; Appelt, K.; Chu, S. Structure-based design, synthesis, and study of potent inhibitors of beta-ketoacyl-acyl carrier protein synthase III as potential antimicrobial agents. *J Med Chem,* 2005, *48* (5), 1596-1609.

[52] Alhamadsheh, M.M.; Waters, N.C.; Sachdeva, S.; Lee, P.; Reynolds, K.A. Synthesis and biological evaluation of novel sulfonyl-naphthalene-1,4-diols as FabH inhibitors. *Bioorg Med Chem Lett,* 2008, *18* (24), 6402-6405.

[53] Alhamadsheh, M.M.; Waters, N.C.; Huddler, D.P.; Kreishman-Deitrick, M.; Florova, G.; Reynolds, K.A. Synthesis and biological evaluation of thiazolidine-2-one 1,1-dioxide as inhibitors of Escherichia coli beta-ketoacyl-ACP-synthase III (FabH). *Bioorg Med Chem Lett,* 2007, *17* (4), 879-883.

[54] Zhang, H.J.; Zhu, D.D.; Li, Z.L.; Sun, J.; Zhu, H.L. Synthesis, molecular modeling and biological evaluation of beta-ketoacyl-acyl carrier protein synthase III (FabH) as novel antibacterial agents. *Bioorg Med Chem,* 2011, *19* (15), 4513-4519.

[55] Zhang, H.J.; Qin, X.; Liu, K.; Zhu, D.D.; Wang, X.M.; Zhu, H.L. Synthesis, antibacterial activities and molecular docking studies of Schiff bases derived from N-(2/4-benzaldehyde-amino) phenyl-N'-phenyl-thiourea. *Bioorg Med Chem,* 2011, *19* (18), 5708-5715.

[56] Li, Z.L.; Li, Q.S.; Zhang, H.J.; Hu, Y.; Zhu, D.D.; Zhu, H.L. Design, synthesis and biological evaluation of urea derivatives from o-hydroxybenzylamines and phenylisocyanate as potential FabH inhibitors. *Bioorg Med Chem,* 2011, *19* (15), 4413-4420.

[57] Li, H.Q.; Luo, Y.; Zhu, H.L. Discovery of vinylogous carbamates as a novel class of beta-ketoacyl-acyl carrier protein synthase III (FabH) inhibitors. *Bioorg Med Chem,* 2011, *19* (15), 4454-4459.

[58] Shi, L.; Fang, R.Q.; Zhu, Z.W.; Yang, Y.; Cheng, K.; Zhong, W.Q.; Zhu, H.L. Design and synthesis of potent inhibitors of beta-ketoacyl-acyl carrier protein synthase III (FabH) as potential antibacterial agents. *Eur J Med Chem,* 2010, *45* (9), 4358-4364.

[59] Lv, P.C.; Sun, J.; Luo, Y.; Yang, Y.; Zhu, H.L. Design, synthesis, and structure-activity relationships of pyrazole derivatives as potential FabH inhibitors. *Bioorg Med Chem Lett,* 2010, *20* (15), 4657-4660.

[60] Li, H.Q.; Luo, Y.; Lv, P.C.; Shi, L.; Liu, C.H.; Zhu, H.L. Design and synthesis of novel deoxybenzoin derivatives as FabH inhibitors and anti-inflammatory agents. *Bioorg Med Chem Lett,* 2010, *20* (6), 2025-2028.

[61] Lv, P.C.; Wang, K.R.; Yang, Y.; Mao, W.J.; Chen, J.; Xiong, J.; Zhu, H.L. Design, synthesis and biological evaluation of novel thiazole derivatives as potent FabH inhibitors. *Bioorg Med Chem Lett,* 2009, *19* (23), 6750-6754.

[62] Li, H.Q.; Shi, L.; Li, Q.S.; Liu, P.G.; Luo, Y.; Zhao, J.; Zhu, H.L. Synthesis of C(7) modified chrysin derivatives designing to inhibit beta-ketoacyl-acyl carrier protein synthase III (FabH) as antibiotics. *Bioorg Med Chem,* 2009, *17* (17), 6264-6269.

[63] Heath, R.J.; Rock, C.O. Regulation of fatty acid elongation and initiation by acyl-acyl carrier protein in Escherichia coli. *J Biol Chem,* 1996, *271* (4), 1833-1836.

[64] Lai, C.Y.; Cronan, J.E. Beta-ketoacyl-acyl carrier protein synthase III (FabH) is essential for bacterial fatty acid synthesis. *J Biol Chem,* 2003, *278* (51), 51494-51503.

[65] Revill, W.P.; Bibb, M.J.; Scheu, A.K.; Kieser, H.J.; Hopwood, D.A. Beta-ketoacyl acyl carrier protein synthase III (FabH) is essential for fatty acid biosynthesis in Streptomyces coelicolor A3(2). *J Bacteriol,* 2001, *183* (11), 3526-3530.

[66] Davies, C.; Heath, R.J.; White, S.W.; Rock, C.O. The 1.8 A crystal structure and active-site architecture of beta-ketoacyl-acyl carrier protein synthase III (FabH) from escherichia coli. *Structure,* 2000, *8* (2), 185-195.

[67] Qiu, X.; Janson, C.A.; Konstantinidis, A.K.; Nwagwu, S.; Silverman, C.; Smith, W.W.; Khandekar, S.; Lonsdale, J.; Abdel-Meguid, S.S. Crystal structure of beta-ketoacyl-acyl carrier protein synthase III. A key condensing enzyme in bacterial fatty acid biosynthesis. *J Biol Chem,* 1999, *274* (51), 36465-36471.

[68] Qiu, X.; Janson, C.A.; Smith, W.W.; Head, M.; Lonsdale, J.; Konstantinidis, A.K. Refined structures of beta-ketoacyl-acyl carrier protein synthase III. *J Mol Biol,* 2001, *307* (1), 341-356.

[69] Scarsdale, J.N.; Kazanina, G.; He, X.; Reynolds, K.A.; Wright, H.T. Crystal structure of the Mycobacterium tuberculosis beta-ketoacyl-acyl carrier protein synthase III. *J Biol Chem,* 2001, *276* (23), 20516-20522.

[70] Heath, R.J.; Rock, C.O. Inhibition of beta-ketoacyl-acyl carrier protein synthase III (FabH) by acyl-acyl carrier protein in Escherichia coli. *J Biol Chem,* 1996, *271* (18), 10996-11000.

[71] Fleischmann, R.D.; Adams, M.D.; White, O.; Clayton, R.A.; Kirkness, E.F.; Kerlavage, A.R.; Bult, C.J.; Tomb, J.F.; Dougherty, B.A.; Merrick, J.M.; *et al.* Whole-genome random

sequencing and assembly of Haemophilus influenzae Rd. *Science,* 1995, *269* (5223), 496-512.

[72] Cole, S.T.; Brosch, R.; Parkhill, J.; Garnier, T.; Churcher, C.; Harris, D.; Gordon, S.V.; Eiglmeier, K.; Gas, S.; Barry, C.E., 3rd; Tekaia, F.; Badcock, K.; Basham, D.; Brown, D.; Chillingworth, T.; Connor, R.; Davies, R.; Devlin, K.; Feltwell, T.; Gentles, S.; Hamlin, N.; Holroyd, S.; Hornsby, T.; Jagels, K.; Krogh, A.; McLean, J.; Moule, S.; Murphy, L.; Oliver, K.; Osborne, J.; Quail, M.A.; Rajandream, M.A.; Rogers, J.; Rutter, S.; Seeger, K.; Skelton, J.; Squares, R.; Squares, S.; Sulston, J.E.; Taylor, K.; Whitehead, S.; Barrell, B.G. Deciphering the biology of Mycobacterium tuberculosis from the complete genome sequence. *Nature,* 1998, *393* (6685), 537-544.

[73] Kuroda, M.; Ohta, T.; Uchiyama, I.; Baba, T.; Yuzawa, H.; Kobayashi, I.; Cui, L.; Oguchi, A.; Aoki, K.; Nagai, Y.; Lian, J.; Ito, T.; Kanamori, M.; Matsumaru, H.; Maruyama, A.; Murakami, H.; Hosoyama, A.; Mizutani-Ui, Y.; Takahashi, N.K.; Sawano, T.; Inoue, R.; Kaito, C.; Sekimizu, K.; Hirakawa, H.; Kuhara, S.; Goto, S.; Yabuzaki, J.; Kanehisa, M.; Yamashita, A.; Oshima, K.; Furuya, K.; Yoshino, C.; Shiba, T.; Hattori, M.; Ogasawara, N.; Hayashi, H.; Hiramatsu, K. Whole genome sequencing of meticillin-resistant Staphylococcus aureus. *Lancet,* 2001, *357* (9264), 1225-1240.

[74] Stover, C.K.; Pham, X.Q.; Erwin, A.L.; Mizoguchi, S.D.; Warrener, P.; Hickey, M.J.; Brinkman, F.S.; Hufnagle, W.O.; Kowalik, D.J.; Lagrou, M.; Garber, R.L.; Goltry, L.; Tolentino, E.; Westbrock-Wadman, S.; Yuan, Y.; Brody, L.L.; Coulter, S.N.; Folger, K.R.; Kas, A.; Larbig, K.; Lim, R.; Smith, K.; Spencer, D.; Wong, G.K.; Wu, Z.; Paulsen, I.T.; Reizer, J.; Saier, M.H.; Hancock, R.E.; Lory, S.; Olson, M.V. Complete genome sequence of Pseudomonas aeruginosa PA01, an opportunistic pathogen. *Nature,* 2000, *406* (6799), 959-964.

[75] Han, L.; Lobo, S.; Reynolds, K.A. Characterization of beta-ketoacyl-acyl carrier protein synthase III from Streptomyces glaucescens and its role in initiation of fatty acid biosynthesis. *J Bacteriol,* 1998, *180* (17), 4481-4486.

[76] Khandekar, S.S.; Konstantinidis, A.K.; Silverman, C.; Janson, C.A.; McNulty, D.E.; Nwagwu, S.; Van Aller, G.S.; Doyle, M.L.; Kane, J.F.; Qiu, X.; Lonsdale, J. Expression, purification, and crystallization of the Escherichia coli selenomethionyl beta-ketoacyl-acyl carrier protein synthase III. *Biochem Biophys Res Commun,* 2000, *270* (1), 100-107.

[77] Alhamadsheh, M.M.; Musayev, F.; Komissarov, A.A.; Sachdeva, S.; Wright, H.T.; Scarsdale, N.; Florova, G.; Reynolds, K.A. Alkyl-CoA disulfides as inhibitors and mechanistic probes for FabH enzymes. *Chem Biol,* 2007, *14* (5), 513-524.

[78] Gajiwala, K.S.; Margosiak, S.; Lu, J.; Cortez, J.; Su, Y.; Nie, Z.; Appelt, K. Crystal structures of bacterial FabH suggest a molecular basis for the substrate specificity of the enzyme. *FEBS Lett,* 2009, *583* (17), 2939-2946.

[79] Musayev, F.; Sachdeva, S.; Scarsdale, J.N.; Reynolds, K.A.; Wright, H.T. Crystal structure of a substrate complex of Mycobacterium tuberculosis beta-ketoacyl-acyl carrier protein synthase III (FabH) with lauroyl-coenzyme A. *J Mol Biol,* 2005, *346* (5), 1313-1321.

[80] Brown, A.K.; Sridharan, S.; Kremer, L.; Lindenberg, S.; Dover, L.G.; Sacchettini, J.C.; Besra, G.S. Probing the mechanism of the Mycobacterium tuberculosis beta-ketoacyl-acyl carrier protein synthase III mtFabH: factors influencing catalysis and substrate specificity. *J Biol Chem,* 2005, *280* (37), 32539-32547.

[81] Sachdeva, S.; Musayev, F.N.; Alhamadsheh, M.M.; Scarsdale, J.N.; Wright, H.T.; Reynolds, K.A. Separate entrance and exit portals for ligand traffic in Mycobacterium tuberculosis FabH. *Chem Biol,* 2008, *15* (4), 402-412.

[82] Sachdeva, S.; Musayev, F.; Alhamadsheh, M.M.; Neel Scarsdale, J.; Tonie Wright, H.; Reynolds, K.A. Probing reactivity and substrate specificity of both subunits of the dimeric Mycobacterium tuberculosis FabH using alkyl-CoA disulfide inhibitors and acyl-CoA substrates. *Bioorg Chem,* 2008, *36* (2), 85-90.

[83] Qiu, X.; Choudhry, A.E.; Janson, C.A.; Grooms, M.; Daines, R.A.; Lonsdale, J.T.; Khandekar, S.S. Crystal structure and substrate specificity of the beta-ketoacyl-acyl carrier protein synthase III (FabH) from Staphylococcus aureus. *Protein Sci,* 2005, *14* (8), 2087-2094.

[84] Oke, M.; Carter, L.G.; Johnson, K.A.; Liu, H.; McMahon, S.A.; Yan, X.; Kerou, M.; Weikart, N.D.; Kadi, N.; Sheikh, M.A.; Schmelz, S.; Dorward, M.; Zawadzki, M.; Cozens, C.; Falconer, H.; Powers, H.; Overton, I.M.; van Niekerk, C.A.; Peng, X.; Patel, P.; Garrett, R.A.; Prangishvili, D.; Botting, C.H.; Coote, P.J.; Dryden, D.T.; Barton, G.J.; Schwarz-Linek, U.; Challis, G.L.; Taylor, G.L.; White, M.F.; Naismith, J.H. The Scottish Structural Proteomics Facility: targets, methods and outputs. *J Struct Funct Genomics,* 2010, *11* (2), 167-180.

[85] Pettersen, E.F.; Goddard, T.D.; Huang, C.C.; Couch, G.S.; Greenblatt, D.M.; Meng, E.C.; Ferrin, T.E. UCSF Chimera--a visualization system for exploratory research and analysis. *J Comput Chem,* 2004, *25* (13), 1605-1612.

[86] Mathieu, M.; Modis, Y.; Zeelen, J.P.; Engel, C.K.; Abagyan, R.A.; Ahlberg, A.; Rasmussen, B.; Lamzin, V.S.; Kunau, W.H.; Wierenga, R.K. The 1.8 A crystal structure of the dimeric peroxisomal 3-ketoacyl-CoA thiolase of Saccharomyces cerevisiae: implications for substrate binding and reaction mechanism. *J Mol Biol,* 1997, *273* (3), 714-728.

[87] Modis, Y.; Wierenga, R.K. Crystallographic analysis of the reaction pathway of Zoogloea ramigera biosynthetic thiolase. *J Mol Biol,* 2000, *297* (5), 1171-1182.

[88] Choi, K.H.; Heath, R.J.; Rock, C.O. beta-ketoacyl-acyl carrier protein synthase III (FabH) is a determining factor in branched-chain fatty acid biosynthesis. *J Bacteriol,* 2000, *182* (2), 365-370.

[89] Choi, K.H.; Kremer, L.; Besra, G.S.; Rock, C.O. Identification and substrate specificity of beta -ketoacyl (acyl carrier protein) synthase III (mtFabH) from Mycobacterium tuberculosis. *J Biol Chem,* 2000, *275* (36), 28201-28207.

[90] Huang, W.; Jia, J.; Edwards, P.; Dehesh, K.; Schneider, G.; Lindqvist, Y. Crystal structure of beta-ketoacyl-acyl carrier protein synthase II from E.coli reveals the molecular architecture of condensing enzymes. *Embo J,* 1998, *17* (5), 1183-1191.

[91] Kamphuis, I.G.; Kalk, K.H.; Swarte, M.B.; Drenth, J. Structure of papain refined at 1.65 A resolution. *J Mol Biol,* 1984, *179* (2), 233-256.

[92] Jez, J.M.; Noel, J.P. Mechanism of chalcone synthase. pKa of the catalytic cysteine and the role of the conserved histidine in a plant polyketide synthase. *J Biol Chem,* 2000, *275* (50), 39640-39646.

[93] Hol, W.G.; van Duijnen, P.T.; Berendsen, H.J. The alpha-helix dipole and the properties of proteins. *Nature,* 1978, *273* (5662), 443-446.

[94] Sasaki, H.; Oishi, H.; Hayashi, T.; Matsuura, I.; Ando, K.; Sawada, M. Thiolactomycin, a new antibiotic. II. Structure elucidation. *J Antibiot (Tokyo),* 1982, *35* (4), 396-400.

[95] Price, A.C.; Choi, K.H.; Heath, R.J.; Li, Z.; White, S.W.; Rock, C.O. Inhibition of beta-ketoacyl-acyl carrier protein synthases by thiolactomycin and cerulenin. Structure and mechanism. *J Biol Chem,* 2001, *276* (9), 6551-6559.

[96] Senior, S.J.; Illarionov, P.A.; Gurcha, S.S.; Campbell, I.B.; Schaeffer, M.L.; Minnikin, D.E.; Besra, G.S. Biphenyl-based analogues of thiolactomycin, active against Mycobacterium tuberculosis mtFabH fatty acid condensing enzyme. *Bioorg Med Chem Lett,* 2003, *13* (21), 3685-3688.

[97] Senior, S.J.; Illarionov, P.A.; Gurcha, S.S.; Campbell, I.B.; Schaeffer, M.L.; Minnikin, D.E.; Besra, G.S. Acetylene-based analogues of thiolactomycin, active against Mycobacterium tuberculosis mtFabH fatty acid condensing enzyme. *Bioorg Med Chem Lett,* 2004, *14* (2), 373-376.

[98] Kim, P.; Zhang, Y.M.; Shenoy, G.; Nguyen, Q.A.; Boshoff, H.I.; Manjunatha, U.H.; Goodwin, M.B.; Lonsdale, J.; Price, A.C.; Miller, D.J.; Duncan, K.; White, S.W.; Rock, C.O.; Barry, C.E., 3rd; Dowd, C.S. Structure-activity relationships at the 5-position of thiolactomycin: an intact (5R)-isoprene unit is required for activity against the condensing enzymes from Mycobacterium tuberculosis and Escherichia coli. *J Med Chem,* 2006, *49* (1), 159-171.

[99] Ashek, A.; Cho, S.J. A combined approach of docking and 3D QSAR study of beta-ketoacyl-acyl carrier protein synthase III (FabH) inhibitors. *Bioorg Med Chem,* 2006, *14* (5), 1474-1482.

[100] Singh, S.; Soni, L.K.; Gupta, M.K.; Prabhakar, Y.S.; Kaskhedikar, S.G. QSAR studies on benzoylaminobenzoic acid derivatives as inhibitors of beta-ketoacyl-acyl carrier protein synthase III. *Eur J Med Chem,* 2008, *43* (5), 1071-1080.

[101] Ashek, A.; San Juan, A.A.; Cho, S.J. HQSAR study of beta-ketoacyl-acyl carrier protein synthase III (FabH) inhibitors. *J Enzyme Inhib Med Chem,* 2007, *22* (1), 7-14.

[102] Perez-Castillo, Y.; Froeyen, M.; Cabrera-Perez, M.A.; Nowe, A. Molecular dynamics and docking simulations as a proof of high flexibility in *E. coli* FabH and its relevance for accurate inhibitor modeling. *J Comput Aided Mol Des,* 2011, *25* (4), 371-393.

Send Orders for Reprints to reprints@benthamscience.net
Recent Advances in Medicinal Chemistry, Vol. 1, **2014, 159-208**

CHAPTER 6

Tannins and their Influence on Health

Kateřina Macáková[1], Vít Kolečkář[2], Lucie Cahlíková[1], Jakub Chlebek[1], Anna Hošťálková[1], Kamil Kuča[2], Daniel Jun[2] and Lubomír Opletal[1,*]

[1]Department of Pharmaceutical Botany and Ecology, ADINACO Research Group, Faculty of Pharmacy, Charles University, Hradec Kralove, Czech Republic and [2]Centre of Advanced Studies, Faculty of Military Health Sciences, University of Defence, Hradec Kralove, Czech Republic

Abstract: Natural polyphenols are a wide class of secondary plant metabolites and represent an abundant antioxidant component of human diet. An important, but often neglected, group of natural polyphenols are tannins. This review offers a general description of chemistry of hydrolysable and condensed tannins (proanthocyanidins), and phlorotannins, the mechanisms of their antioxidation action, like free radical scavenging activity, chelation of transition metals, inhibition of prooxidative enzymes and lipid peroxidation. The mechanisms of action of inhibition of various enzyme systems, antibacterial, antiviral, antiprotozoal, anticarcinogenic, antidiabetic, hepatoprotective, cardiovascular system preventing, immunomodulation, antiallergic and anti-inflammatory effects as well as the absorption, metabolic fate and positive *in vivo* effects of tannins are enclosed.

Keywords: Adverse effects, biological activity, metabolism, phlorotannins, perspectives, tannins.

1. INTRODUCTION

There are many natural compounds with wide scale of biological activities. Especially, there is a great interest in polyphenols, which have been studied for many years and are still very active domain of research because of their prospective use in health protection. Polyphenols are a wide class of substances, which contain over 8000 compounds, from those with simple structure (*e.g.* phenolic acids) to the polymeric substances like some condensed tannins [1, 2].

***Address correspondence to Lubomír Opletal:** Department of Pharmaceutical Botany and Ecology, ADINACO Research Group, Faculty of Pharmacy, Charles University, Hradec Kralove, Czech Republic; Tel: +420495067111; Fax: +420495518002; E-mail: opletal@faf.cuni.cz

Atta-ur-Rahman, Muhammad Iqbal Choudhary and George Perry (Eds)

Flavonoids are very important and well known group of compounds with various pharmacological effects [3-7]. An important, but often neglected group of polyphenols are also tannins with lower molar weight. Since 1950, about 25 thousand papers, thereof approximately 1700 reviews, were dedicated to tannins in wider scale.

In 1957, Bate-Smith and Swain defined plant tannins as water-soluble phenolic compounds having a molecular weight between 500 and 3000 Dalton [8]. Their characteristic properties include forming of insoluble complexes with proteins, polysaccharides, nucleic acids, or alkaloids. Within this general character, tannins exhibit number of various bioactivities, which are often related to their antioxidant activity. Tannins are classified into two major groups on the basis of their structure: the hydrolysable and the condensed tannins. Another group of tannins is formed by complex tannins [9]. Currently, phlorotannins, condensates of phloroglucinol which are present in sea algae, are also classified as tannins [9]. This review is focused on all groups of tannins, nevertheless more attention is given to cendensed tannins, because they are represented in nature more widely and are important components of human food, and on relatively new group of substances from seaweed - phlorotannins. This review continues the review study of the authors group published some time ago and extends the view of the very important group of natural substances that are commonly utilized in the diet [10, 11].

Recently, series of review papers focused on these substances, especially hydrolysable tannins - compounds of medicinal and food plants [12], chemical reactivity of C-glycosylated ellagitannins in relation to wine chemistry and biological activity [13], synthesis of ellagitannins [14] and biological activities [15], chemical-ecological aspects [16], and pharmacological activity, or antinutritive factors in plants [17]. Some species of plants are interesting concerning their possible use: *e.g. Emblica offcinalis* [18, 19], *Phyllanthus amarus* [20] and *Terminalia arjuna* [21].

2. CLASSIFICATION OF TANNINS

2.1. Hydrolysable Tannins

Hydrolysable tannins are compounds containing a central core of glucose or another polyol esterified with gallic acid, also called gallotannins, or with hexahydroxydiphenic acid, also called ellagitannins. Pentagalloylglucose (PGG) is a basic unit of the metabolism of hydrolysable tannins, from which other molecules are derived. Gallotannins consist of a central molecule, such as glucose, surrounded by gallic acid (**1**) units. Ellagitannins contain hexahydroxydiphenic acid, or its dilactone form, ellagic acid (**3**) (Fig. **1**). The great variety in the structure of these compounds is due to the many possibilities in formation of oxidative linkages. Intermolecular oxidation reactions give rise to many oligomeric compounds having a molecular weight between 2000 and 5000 Dalton [22].

Plants are able to biosynthesize gallotannins, ellagitannins, or form the mixture of both types of hydrolysable tannins. While condensed tannins are presented in many species of higher plants, presence of hydrolysable tannins is limited to Angiospermae, Dicotyledons. Gallic acid derivatives are presented in several families, *e.g.* Ericaceae, Geraniaceae, or *Fagaceae*. Ellagitannins are presented in subclasses *Hamamelidae*, *Dilleniidae* and *Rosidae* species [23-25].

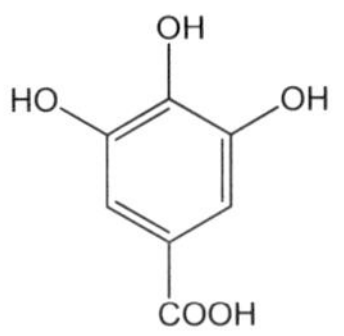

1

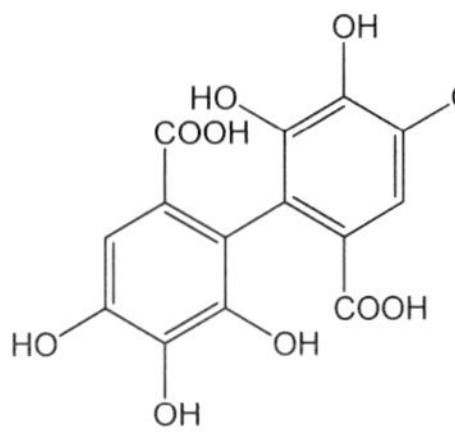

2

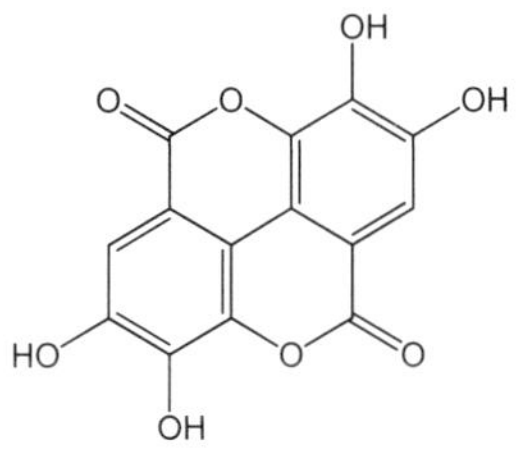

3

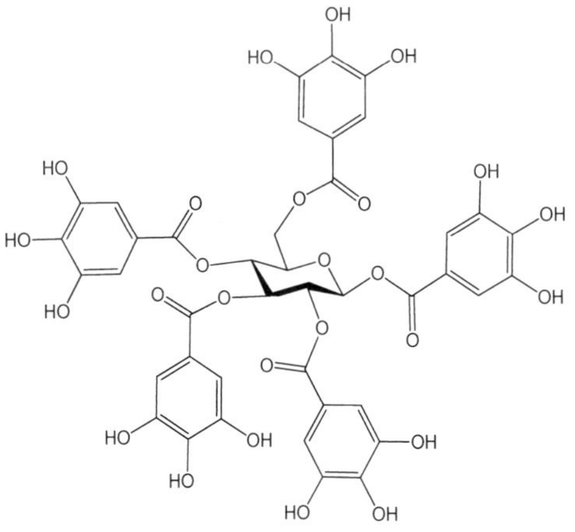

4

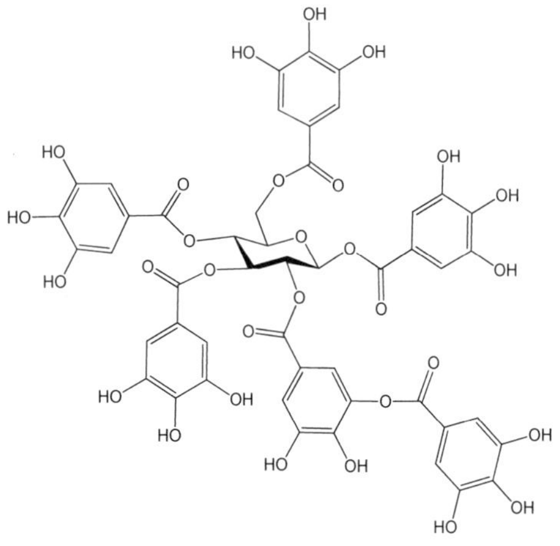

5

Fig. 1: contd….

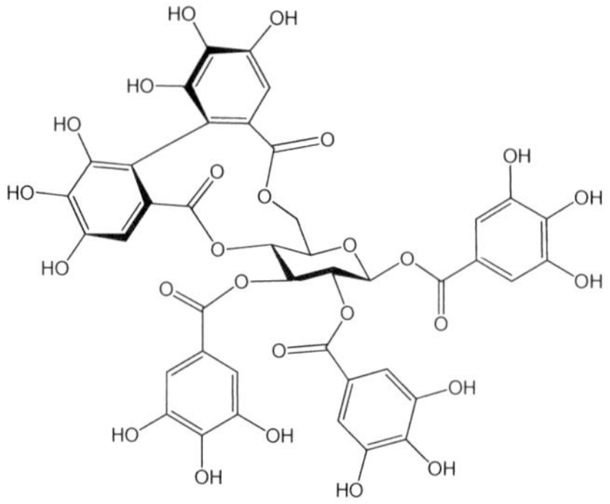

6

Figure 1: Characteristic structures of hydrolysable tannins: gallic acid (1); hexahydroxydiphenic acid (2); ellagic acid (3); pentagalloylglucose (4), the basic unit of hydrolysable tannins; 2-*O*-digalloyl-1,3,4,6-tetra-*O*-galoyl-β-D-glucopyranose (5), the example of gallotannin; tellimagradin II (6), the typical ellagitannin.

2.2. Condensed Tannins

Condensed tannins are oligomers or polymers composed of flavan-3-ol nuclei. They are also called proanthocyanidins, because they are decomposed to anthocyanidins in heated ethanol solutions (under acidic conditions). The most frequent basic units of condensed tannins are derivatives of flavan-3-ols: (+)-catechin (**7**), (-)-epicatechin (**8**), (+)-gallocatechin (**9**) and major polyphenols of green tea: (-)-epigallocatechin (EGC, **10**) and (-)-epigallocatechin gallate (EGCG, **11**) (Fig. **2**). The structural diversity is caused by variation in hydroxylation pattern, stereochemistry at the three chiral centers, and the location and type of interflavan linkage. Furthermore, derivatisations as *O*-methylation, *C*- and *O*-glycosylation, and *O*-galloylation are frequently reported. Proanthocyanidins are classified according to their hydroxylation pattern into several subgroups, *e.g.* procyanidins (3,5,7,3',4'-OH), prodelphinidins (3,5,7,3',4',5'-OH), propelargonidins (3,5,7,4'-OH), profisetinidins (3,7,3',4'-OH), prorobinetinidins (3,7,3',4',5'-OH), proguibourtinidins (3,7,4'-OH), proteracacinidins (3,7,8,4'-OH) and promelacacinidins (3,7,8, 3',4'-OH) [26, 27].

Procyanidins of the B-type (dimeric) and C-type (trimeric) are characterized by single linked flavanyl units, usually between C-4 of the flavan-3-ol of the upper unit and C-6 or C-8 of the lower unit. Proanthocyanidins of the A-type possess an additional ether linkage between C-2 of the upper unit and a 7 or 5-OH of the lower unit (Fig. **3**) [26, 27]. Polymeres, composed of up to fifty monomers, are formed by the addition of more flavans. Especially polyepicatechins and copolymers of procyanidins and prodelphinidins are common.

Proanthocyanidins have been isolated from many species of plants, and they are also important components of human food. The largest group of proanthocyanidins is formed by procyanidins. Procyanidin B-1 (**14**) is presented in grapefruit, sorghum, and cranberries, B-2 (**15**) in apples, cocoa beans, and cherries, B-3 (**16**) in strawberries and hops and B-4 in raspberries and blackberries [28, 29]. Well-known source of protoanthocyanidins is also red wine, green tea, cocoa and chocolate [30].

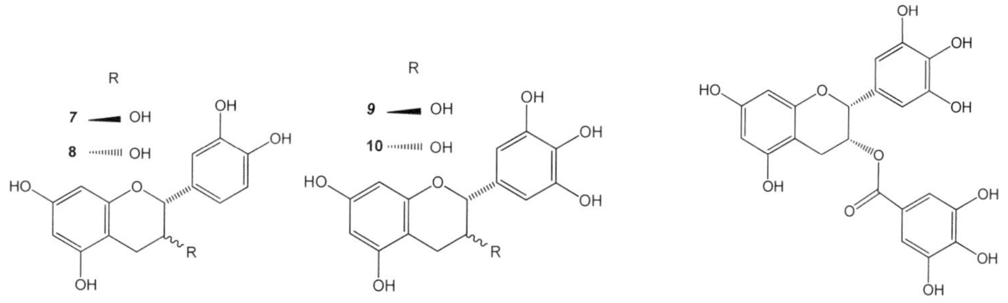

11

Figure 2: The most frequent structure units of condensed tannins: (+)-catechin (7); (-)-epicatechin (8); (+)-gallocatechin (9); (-)-epigallocatechin (10); (-)-epigallocatechin gallate (11).

procyanidins A

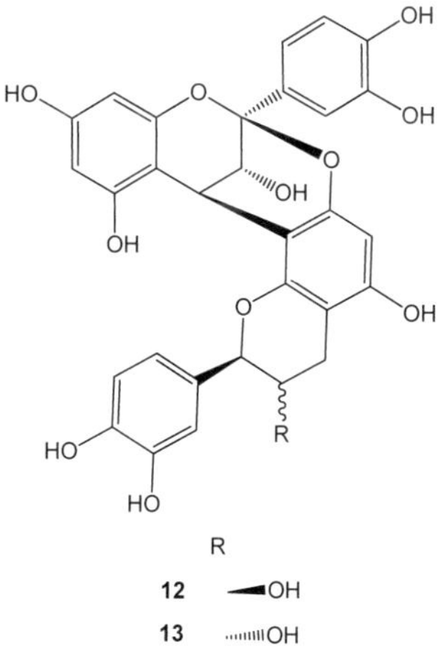

procyanidins B

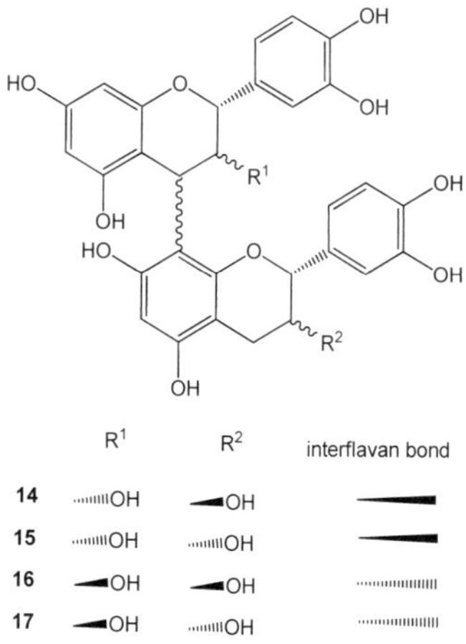

Fig. 3: contd….

procyanidins C

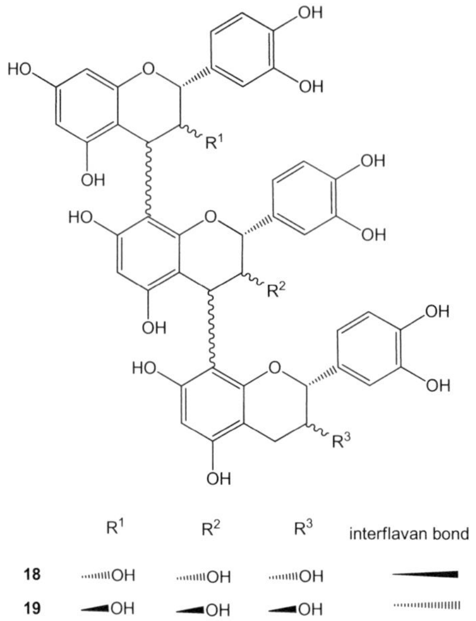

Figure 3: Oligomeric procyanidins: procyanidin A-1 (epicatechin-(4β→8,2 β→7)-catechin) (12); procyanidin A-2 (epicatechin-(4β→8,2β→7)-epicatechin) (13); procyanidin B-1 (epicatechin-(4β→8)-catechin) (14); procyanidin B-2 (epicatechin-(4β→8)-epicatechin) (15); procyanidin B-3 (catechin-(4β→8)-catechin) (16); procyanidin B-4 (catechin-(4β→8)-epicatechin) (17); procyanidin C-1 (epicatechin-(4β→8)-epicatechin-(4β→8)-epicatechin) (18); procyanidin C-2 (epicatechin-(4β→8)-catechin-(4β→8)-catechin) (19).

2.3. Phlorotannins

Widely presented group of polyphenolic substances based on phloroglucinol (**20**), phlorotannins, was found in marine brown algae (Phaeophyta). The phloroglucinol units are differently bound in these substances; they are considerably hydrophilic substances, with the relative molecular weight between 126 and 650 kDa, containing both phenyl- and phenoxy-groups. On the basis of the condensation of the units, they can be classified into four subgroups: fuhalols, phlorethols (phlorotannins with ether bond), fucophlorethols (with ether bond and phenyl connection) and eckols (with dibenzodioxin structural elements). The basic structures are depicted on Fig. **4**. These substances were found especially in *Ecklonia cava, Ecklonia stolonifera, Ecklonia kurome, Eisenia bicyclis, Ishige okamurae, Sargassum thunbergii, Hizikia fusiformis, Undaria pinnatifida* and *Laminaria japonica* [9].

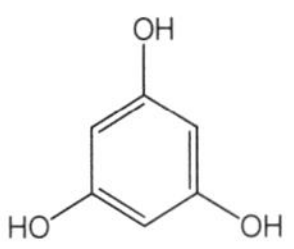

20

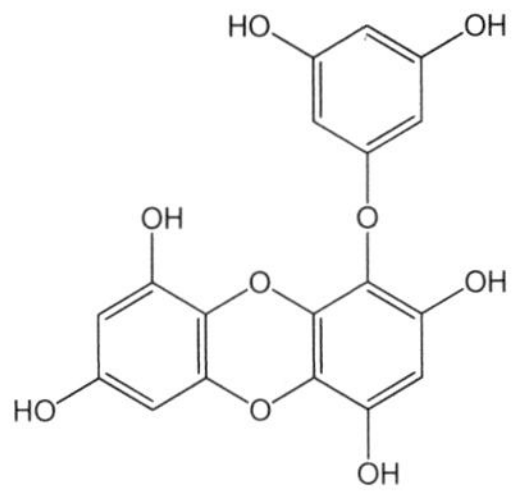

21

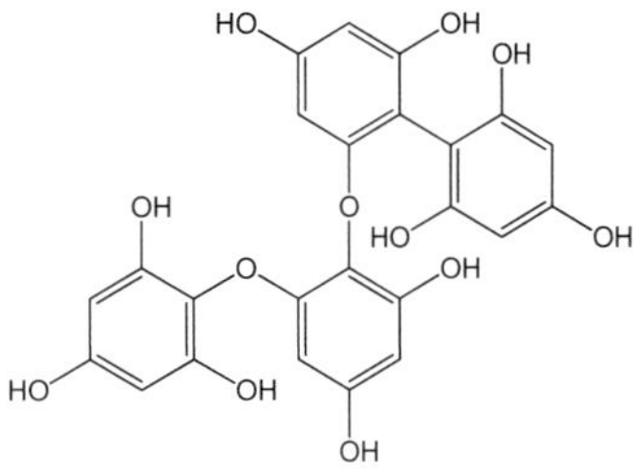

22

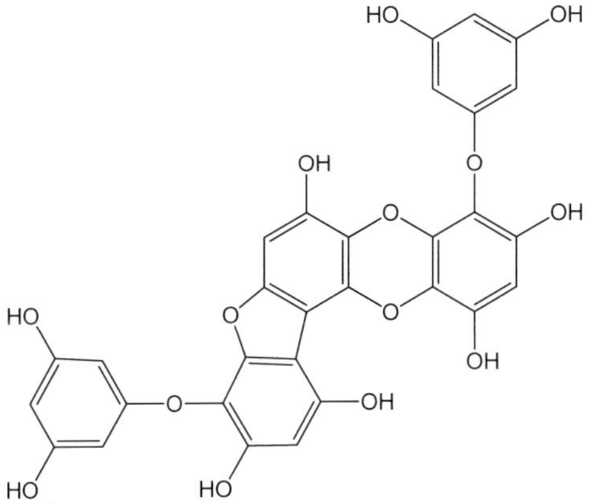

23

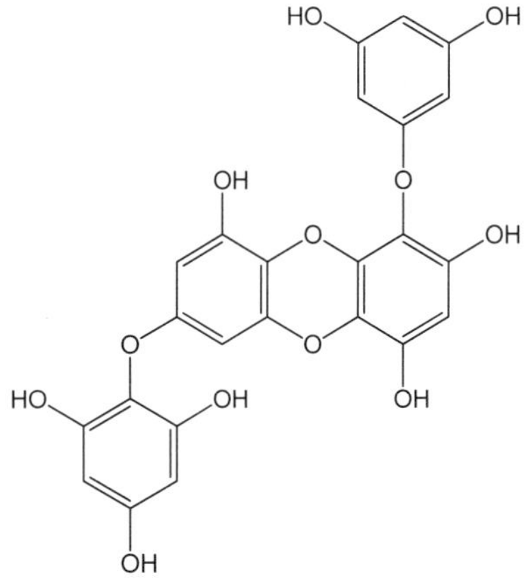

24

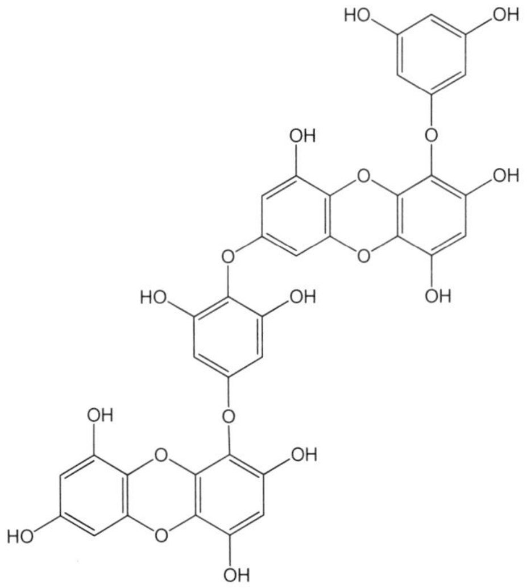

25

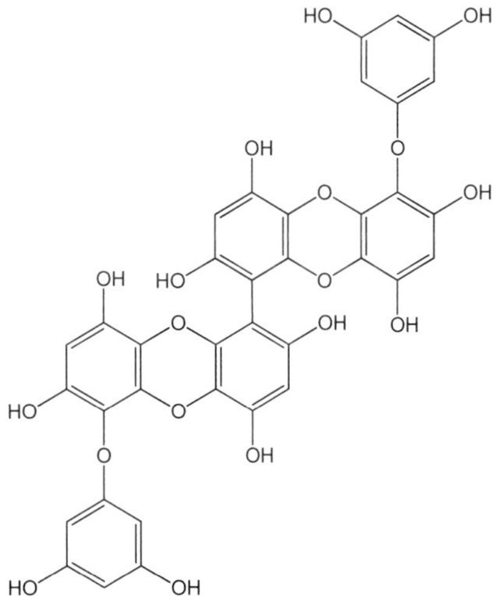

26

Figure 4: Basic structures of phlorotannins isolated from seaweeds: phloroglucinol (20), eckol (21), fucodiphloroethol, G (22), phlorofucofuroeckol A (23), 7-phloroeckol (24), dieckol (25), and 6,6′-bieckol (26).

It is very interesting and recently developing group of polyphenolics, which biological activities are just revealed; since the beginning of their study, about 300 papers were published, thereof 30 reviews, since 2006 about 170 papers (thereof about 20 reviews). They posses anti-cancerogenic effect [31], antioxidant [9], inhibit some enzyme systems [9], show antibacterial and anti-HIV activity [9], are radioprotective [9] and with antiallergic effects [9]. Many effects are similar to the effects of tannins isolated from higher plants, especially in case of polyphenolic substances of constant chemical composition [12]. It appears, that these substances will become an important component in the production of food supplements and medical foods [31-35]. They can be used also in cosmetics [36]. Phlorotannins are still waiting for their overall evaluation.

3. ABSORPTION AND METABOLISM OF TANNINS

It was assumed for a long time that tannins are not absorbed due to their high molecular weight and their ability to form insoluble complexes with components of food, such as amino acids and proteins. Though there are some studies which confirm that the absorption of tannins is higher than it was assumed there are still many questions about their biological availability [37]. Generally, absorption of tannins decreases with the increasing polymerization degree.

It was found that dimeric and trimeric procyanidins are absorbed by intestinal epithelium without any considerable limitations. Caco-2 cell line was used as an *in vitro* model of intestinal epithelium. The absorption decrease with further polymerization, and the transport of hexamers did not proceed [38].

The *in vitro* study which dealt with the stability of procyanidins in acidic environment, as found in the gastric milieu, stated that high-molecular substances are fragmented to absorbable monomers and dimers [39]. Another study claims that higher procyanidins are not degraded in stomach environment of six healthy volunteers, and only original monomers and dimers are absorbed [40]. This discrepance could be interpreted in view of various characters of tested proanthocyanidins: some compounds can be more disposed to degradation than others.

According to one study epicatechin is the primary bioavailable form of procyanidin dimers B-2 (**15**) and B-5 after perfusion of isolated small intestine, while other study in rats arrived at a conclusion that procyanidin B-2 (**15**) is absorbed and excreted predominantly unmodified and is metabolized only partly [41, 42]. These results show that quantitative composition of a tested mixture and also the type of used biological model affect the process of absorption.

Human intestinal microflora plays an important role in metabolization of tannins. It is known that the part of non-absorbed tannins is degraded *in vitro* by colonic microflora to various phenolic products, which can be absorbed and participate on various pharmacological effects [43]. Aura presents an instructive view on microbial metabolism of dietary tannins in the colon: proanthocyanidins, which enter to the colon, are not much metabolized to phenolic acids. They are degraded by colon bacteria to various derivatives of phenylvaleric, phenylpropionic, phenylacetic and benzoic acid with different rate of hydroxylation. The rate of polymeration is very Important: The higher polymeration the lower is the rate of metabolization in the colon. Ellagitannins are substances with big molecule and relatively limited biological availability. They are degraded to ellagic acid (**3**) which is excreted by urine and feces: in the colon, two lactones (urolithin A and B), which are conjugated with glucuronic acid, are formed [44]. The detailed data about biological availability, metabolism and biological activity of ellagic acid (**3**) and similar substances presented in diet was presented recently [45]. Non-absorbed high-molecular tannins and tannin-protein complexes have also an important role for protection of intestinal tract because they keep their antioxidant activity [46]. Although tannins are able to act against various pathophysiological processes, mechanism of the effect of many of these interactions is not known. However, the last studies show that these substances bond with cell membranes and biochemical reactions are subsequently set off. Solid-State NMR Spectroscopy begins to be used for the study of these interactions and clarification of possible activity [47].

No details about the bioavailability of phlorotannins have been presented yet.

4. BIOLOGICAL ACTIVITY

4.1. ROS and Antioxidants

Antioxidants are defined like substances that, when present at low concentrations compared to those of an oxidizable substrate, significantly delays or prevent oxidation of that substrate. Recently, great attention has been given to antioxidant agents by reason of their medical use. It is given by association of many human diseases with oxidative stress. Free radicals play an important role in pathogenesis of ageing, various cardiovascular diseases, type 2 diabetes mellitus or cancer. For example, in the radiation-induced carcinogenesis, highly reactive hydroxyl radicals are recognized as primary cause of the disease, whereas in diseases like atherosclerosis or rheumatoid arthritis, an oxidative stress is not the inciting agent, but supports their pathology [26, 48-50].

Primarily, free radicals play an important physiological role, for example, they are mediators of energy transfer, immunity defense factors, or signal molecules of cell regulation. In special conditions they may become undesirable and injure the organism. For maintenance of the redox balance and prevention of increased formation of free radicals, organism uses some mechanisms, such as scavenging of free radicals, prevention of new formation through regulation of enzymes, which form them, support of antioxidant enzymes, or inactivation of transition metals which support formation of free radicals [51].

Secondary plant metabolites, such as flavonoids or tannins, can be also involved in complex system of antioxidant defense. The basic mechanisms of antioxidant activity of tannins are free radical scavenging activity, chelation of transition metals and inhibition of prooxidative enzymes.

4.2. Free Radical Scavenging Activity

The basic concept of free radical scavenging activity of polyphenols, including tannins, is the ability of antioxidant to donate electron to a free radical and produce a more stable and therefore less harmful radical structure. DPPH (1,1-diphenyl-2-(2,4,6-trinitrophenyl)hydrazyl) or $ABTS^+$ ((2,2-azinobis)3-ethyl-2,3-dihydrobenzothiazol-6-sulphonic acid) radicals are often used for *in vitro* determination of free radical scavenging activity [52, 53]. A number of tested condensed and hydrolysable tannins scavenged these radicals [54-58]. It is possible to found it beneficial that this biological effect was found in plants used as a component of the diet, *e.g.* green lentil (*Lens culinaris*, seeds) [59], mangosteen fruit (*Garcinia mangostana*, rind) [60], peanut (*Arachis hypogea*, skin) [61], pomegranate (*Punica granatum*, peel) [62], chilean blackberry

(*Aristotelia chilensis*, fruit) [63], pecan (*Carya illinoinensis*, shell) [64] and acai (*Euterpe oleracea,* fruit) [65]. Out of other radicals, free radical scavenging activity against superoxide radical [66], hydroxyl radical [67, 68], and peroxyl or nitric oxide (NO) [69, 70] was determinated [71, 72].

Generally, it applies to proanthocyanidins that scavenging activity increases with the number of hydroxyls, especially if they are at ortho position on benzen nucleus and if gallic acid is introduced. The activity is also influenced by the size of molecule; it increases from monomers to trimers, afterwards it decreases [55, 56].

The comparison of antioxidant activities of B-type procyanidins showed ambiguous results. One study found that procyanidin B-2 (**15**) is more active than B-3 and B-5 in DPPH assay, while other study did not prove any differences in $ABTS^+$ scavenging activity of six different B-type procyanidins [54, 56]. In superoxide radical scavenging assay, procyanidins B-1 (**14**) and B-3 (**16**) showed similar scavenging activity [66]. These differences are not surprising because one compound can act differently against different radicals.

Tannins, due to their higher molecular weight and high degree of hydroxylation of aromatic rings, show high antioxidant potential. In comparison to *in vitro* antioxidant activity of various types of polyphenols, dimeric procyanidins are the most active on the basis of scavenging of $ABTS^+$ radical, hypochlorous acid, or in FRAP test for the evaluation of reducing power. Dimeric procyanidines were followed by flavanols, flavonols, hydroxycinnamic acids and simple phenolic acids [73].

It was found that, unlike other phenolic antioxidants, procyanidins might not show prooxidative activity. Quercetin and other flavon(ol)s form *o*-quinone structures in quenching reactions that may act as prooxidants in further redox reactions. While the study, which used electron spin resonance technique (ESR) which dealt with mechanism of antioxidant activity of hydrolysable and condensed tannins, showed the contrary. *o*-Quinones of proanthocyanidins formed after reaction with radical are subjected to subsequent nucleophilic addition reactions and as a result more complicated structures are formed, which keep great number of hydroxyl groups and thereby also their antioxidant activity (Fig. **5**). From this point of view proanthocyanidins are better antioxidants than monomeric flavon(ol)s. The similar experiments with hydrolysable tannins were not so unambiguous [74]. Another study which evaluated anti- and pooxidation properties of gallic and ellagic acid (**3**) and tannin (tannic acid) according to the DNA injury in mussels *Unio tumidus* cells found that high concentrations of these substances injured DNA and showed prooxidant activity [75].

Oligomeric procyanidins (B-1 (**14**), B-3 (**16**), and others) and phenolic monomers (catechin, epicatechin, and flavonoid taxifolin) are the main bioactive compounds of standardized *Pinus maritima* bark extract (Pycnogenol®). This pine originated from south part of France and in many countries is used as a material for production of the mixture known as Pycnogenol®, which is the principal component of dietetic preparations for cardiovascular system protection. It causes vasodilation effect, inhibition of angiotensin converting enzyme (ACE), or increase in capillary permeability. Pine bark extract also contains phenolic acids (such as caffeic, ferulic, and *p*-hydroxybenzoic acids) as minor constituents and glycosylation products, *e.g.* glucopyranosyl derivatives of either flavonoids or phenolic acids as minute constituents. Recently, the great scientific attention regards the medicinal use of this extract. Studies indicate that Pycnogenol® components are highly bioavailable. The complex extract exhibits higher biological activity than isolated substances, which means that the constituents act synergistically. The extract has strong free radical scavenging activity against RNOS. The procyanidins contribute significantly to the ESR free radical signal. Pycnogenol® modulates NO metabolism in activated macrophages by quenching the NO radical and inhibiting both iNOS mRNA experession and iNOS activity. Oligomeric procyanidins highly participate on this high radical scavenging activity. The ability to regenerate ascorbyl radical and protect endogenous vitamin E and glutathione (GSH) against oxidative damage also support the complex antioxidant properties of Pycnogenol® [76].

The antioxidant (free radical scavenging) activity is also demonstrated by phlorotannins, *e.g.* from *Ascophyllum nodosum* [77], some species of genus *Eisenia* [78, 79], *Ecklonia* [78] and many others.

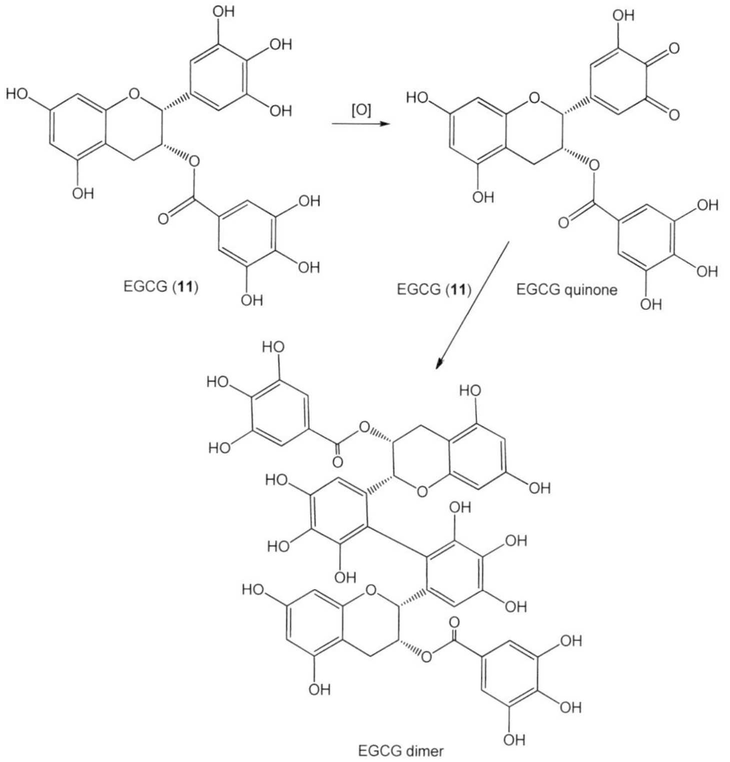

Figure 5: Polyphenolic reaction of epigallocatechin gallate (EGCG, 11).

Recently, it is clear that *sensu lato* antioxidant activity synergizes with some other biological effects, as it was proved in many plant extracts containing tannins, *e.g.* they participate favorably in chemoprotective (hepatoprotective) effects (*Boerhaavia diffusa*) [80], antimicrobial activity (*Camellia sinensis*) [81], decrease of genotoxicity (*Quercus resinosa*) [82] and abrogates oxidative stress-induced functional alterations in murine macrophages (*Quercus infectoria*) [83].

The attention is also paid to oligomers of procyanidins concerning the synthetic point of view - especially structure activity relationships (DPPH radical scavenging activity and Maillard reaction inhibitory activity) [84].

4.3. Chelation of Transition Metals

It is known for a long time, that polyphenolics [85],including tannins [86] are able to bond cations of transition metals and act as protective agents against progression of some diseases, *e.g.* Alzheimer's or Parkinson´s disease [87].

The transition metals, especially iron, mangan and copper perform many functions in human organism. For example, they play an important physiological role like cofactors of antioxidant enzymes such as superoxide dismutase, catalase or glutathione peroxidase. They are usually bound to proteins, such as ferritin or caeuroplasmin [88]. When they occur separately, they can catalyze the radical reactions. The typical example of these reactions is well known Fenton reaction [89]:

$$Fe^{2+} + H_2O_2 \rightarrow Fe^{3+} + OH^- + HO\bullet$$

There are many studies, which examined chelation activity of polyphenols, especially flavonoids [90-93]. In the study which dealt with antioxidant activity of tannin, it was found that tannic acid was more efficient in protecting against 2-deoxyribose degradation then classical hadroxyl radical scavengers. The *in vitro* study concluded that chelation of iron ions by this substance may be more important for its antioxidant activity than the mechanism of quenching of free radicals themself [94]. Other study determined that chelation of iron and resulting inhibition of Fenton reaction participates in high antioxidant activity of *Vitis vinifera* procyanidins. Antioxidant potency of procyanidins was studied in phosphatidylcholine liposomes, using iron-promoted lipid peroxidation [95].

The ability to chelate metals was also proved by the study which dealt with stability of aluminium-proanthocyanidin complexes. This study concludes that existence of phenolic groups, especially in B-ring *o*-position, is very important for chelating activity. The stability of complexes increased with increasing polymerization degree [96].

The recent study which dealt with beneficial influence of oligomeric proanthocyanidins on lead-induced neurotoxicity in rats showed ambiguous results. It was confirmed that proanthocyanidins exhibited antioxidant and chelating activity in *in vitro* studies, but the lead-induced toxicity did not decrease in *in vivo* experiment and Pb^{2+} was cumulated even in some organs [97].

The polyphenolics of tea (*Camelia sinensis*) containing especially substances of flavan-3-ol type (catechins and proanthocyanidins) play an important role in this field. It is known about these substances that they act as non-toxic iron chelators in brain [98], or more precisely they can act widely protectively on human organism [99]. The tea polyphenolics decrease also the toxicity of vanadium on rat kidney [100].

However, some proanthocyanidins (delphinidin) can catalyze the degradation of cellular DNA by their prooxidant activity in the presence of transition metals (copper) and thus show the anticancerogenic effect [101]. These opinions are recently discussed frequently [102, 103].

4.4. Inhibition of Enzymes

4.4.1. Enzymes Interfering Redox Processes

Antioxidant activity can also be exhibited through inhibition of prooxidative enzymes.

Tannins decrease formation of NO through inhibition of nitric oxide synthases (NOS). For example, hydrolysable tannins isolated from East Asian plant *Melastoma dodecandrum* inhibited the induction of the iNOS in the course of macrophage activation with LPS and recombinant mouse interferon-γ [104]. Next *in vitro* study which tested monomeric catechins up to octameric procyanidins isolated from hops (*Humulus lupulus*), showed that procyanidin B-2 (**15**) was the most active against brain NOS, while procyanidin B-3 (**16**), catechin (**7**), and epicatechin (**8**) did not inhibit this enzyme [105].

Xanthin oxidase is also ranked among enzymes with prooxidative activity. For example, ellagitannins isolated from New Caledonian plant *Cunonia macrophylla* inhibited *in vitro* this enzyme. The enzyme activity was measured spectrophotometrically following the conversion of xanthine to uric acid. Ellagic acid-4-*O*-β-D-xylopyranoside (Fig. **6**) was the most active from the tested compounds [106].

Several studies dealt with effect of tannins on lipoxygenases (LOX), which can damage membrane lipids. It was found that tannins called phlorotannins isolated from brown seaweed *Eisenia bicyclis* exhibited high *in vitro* inhibition activity of soybean LOX and 5-LOX. The phlorotannins are structurally different from hydrolysable and condensed tannins which are produced by terrestrial plants.

Eckol (**21**, trimer), phlorofucofuroeckol A (**23**, pentamer), dieckol (**25**, hexamer), and 8,8′-bieckol (hexamer) were even more active then known inhibitor of LOX epigallocatechin gallate (EGCG, **11**). These compounds had also pronounced *in vitro* inhibitory effects on secretory phospholipase $sPLA_2$ [107].

The influence of polymerization on inhibition of 5-LOX was showed in cocoa procyanidins (*Theobroma cacao*). Recombinant human 5-LOX was significantly inhibited by (-)-epicatechin (**8**) in a dose-dependent manner. Among the procyanidin fractions, only the dimer fraction and, to a lesser extent, the trimer through pentamer fractions exhibited comparable effects, whereas the larger procyanidins (hexamer through nonamer) were almost inactive [108].

The study with tannins rich Pycnogenol® showed the *in vitro* inhibition of horseradish peroxidase, LOX, NOS, and xanthine oxidase. Authors concluded that inhibition of these enzymes is probably non-specific, and it is given by high affinity of polyphenols to proteins [76, 109]. Non-competitive enzyme inhibition was also described in other plant extracts. For example, *Vitis vinifera* procyanidins inhibited xanthine oxidase, proteolytic enzymes elastase and collagenase, as well as β-glucuronidase and hyaluronidase [95].

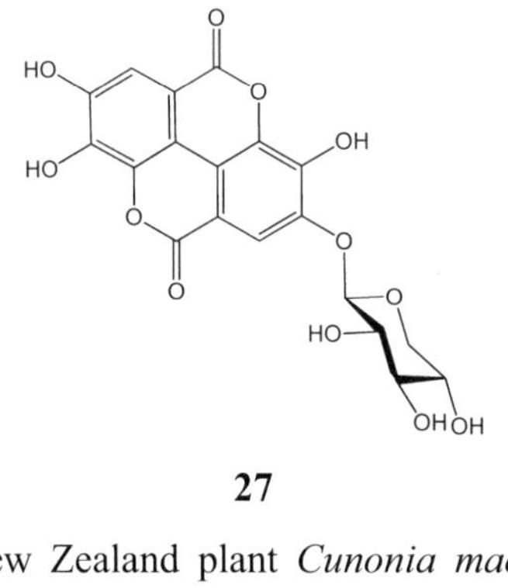

27

Figure 6: Ellagitannin from New Zealand plant *Cunonia macrophylla*: Ellagic acid-4-*O*-β-D-xylopyranoside (27).

The mixture of catechins from green tea protects against the decrease of glutathione peroxidase activity and age related rat brain damage and decreases levels of carbonyl proteins [110]. Although series of studies with the important medicinal plant *Terminalia arjuna* bark extract containing tannins on various animal models discuss the increased levels of catalase, further studies will have to be performed, because the results are questionable: it was also proved that the ethanol bark extract did not improve the activity of catalase *in vitro* [111].

4.4.2. Other Enzyme Systems

The extract from the mango seed kernel (*Mangifera indica*) containing phenolic substances with high antioxidant activity inhibited the activity of tyrosinase [112]; hydrolysable tannins isolated from leaves of eucalypt (*Eucalyptus globulus*) (the mixture contained 5 ellagitannins and 4 gallotannins) acted similarly, and the inhibition of hyaluronidase was found simultaneously [113]. Hyaluronidase is also inhibited by phlorotannins from brown alga *Eisenia arborea* [79], and other seaweed species [78]. Therefore these substances can be used as anti-inflammatory and antiallergic components of diet and also in cosmetics.

Hydrolysable tannins from 6 species of genus *Terminalia* inhibited the activity of α-glucosidase. The highest activity was exhibited by the mixture of these substances from leaves of *T. kaerbachii* (IC_{50} 0.27 ± 0.17 μg/ml) which can be perspective in influencing of diabetes mellitus [114]. The postprandial hyperglycemia can be also favorably affected by administration of polar extract of cinnamon (*Cinnamomum verum*) bark containing inhibitors of α-glucosidase as was proved on diabetic rats [115]. The high α-amylase activity was found in ellagitannins from strawberry (*Fragaria* x *ananassa*) fruit *in vitro* [116].

Pomegranate (*Punica granatum*) polyphenolics (punicalagin, **28**) (Fig. **7**) from the juice can be applied in protection of cardiovascular system: they increase the expression of hepatic PON1 *via* the intercellular signaling cascade PPARγ-PAK-cAMP and thus act anti-atherogenically [117]. 5 Phlorotannins were isolated from the thallus of seaweed *Ecklonia cava*; dieckol showed the highest inhibitory activity against ACE (IC_{50} 0.96 mg/ml). This material is a potential source for production of nutraceuticals [118].

The mixture of condensed tannins (70% acetone extract) from some woody plants showed significant activity against steroid 5α-reductase. This fact would be used in the BHP and prostate cancer therapy [119].

4.5. Lipid Peroxidation and Impact on Cardiovascular System

Lipid peroxidation belongs to important pathological processes. This process is involved in oxidative modification of low-density lipoproteins (LDL) which ultimately leads to the formation of atherosclerotic lesions [120], they assert significantly in liver impairments [121]. It was found that proanthocyanidins can protect LDL against oxidation. For example, catechins and their oligomers of coffee-beans (*Coffea arabica*) inhibited *in vitro* human LDL oxidation in the following order: procyanidin C-1 (**18**) > procyanidin B-2 (**15**) > (+)-catechin (**7**) > (-)-epicatechin (**8**). It was also confirmed that the number of hydroxyl groups is related to the antioxidant activity [122].

Generally, it is possible to say that the substances which inhibit lipid peroxidation act through the mechanism of quenching of initiatory radicals (hydroxyls), or already formed oxidative products (peroxyl, alcoxyl). The mechanism of chelation of transition metals can be also involved. Many studies proved *in vitro* inhibition of lipid peroxidation by hydrolysable as well as condensed tannins [57, 66, 123-125]. For example hydrolysable tannin punicalagin (**28**), contained in pomegranate, (*Punica granatum*) participates in high antioxidant potential of prepared extracts and drinks. Punicalagin inhibits lipid peroxidation induced by Fe^{2+} in a liposomal mode [123]. Also tannins isolated from cranberry (*Rhodococcum vitis-idaea*) exhibited high *in vitro* inhibition of lipid peroxidation. Out of six tested substances, cinnamtannin B-1 (**29**) (Fig. **7**) was the most active [66].

Above mentioned studies used various methods for evaluation of inhibition of lipid peroxidation and they generally showed that the ability to inhibit lipid peroxidation is high, comparable with vitamin E.

Lipid peroxidation asserts significantly in damage of cardiovascular system; the most widely studied sources are compounds of common stimulating foods (tea, wine) and some fruits. In case of tea (*Camellia sinensis*), the impact of catechin derivatives on eNOS stimulation and phosphorylation in aorta cells was definitely proved. The monomeric units are considered to be more effective than proanthocyanidin oligomers, the esters with gallic acid act very favorably. However, it appeared that fermented teas containing theaflavins and thearubigins can protect cardiovascular system equally like polyphenolics from green tea

[126]. The gallotannins from green tea and red wine are strong inhibitors of CaCC (on the contrary to catechin (**7**) and epicatechin (**8**)), they inhibit contractions of endothelial cells of smooth muscles and intestinal secretion of Cl^- ions and act cardioprotectively [127].

The question of the possible antihyperlipidemic effect of dietary tannins is still not definitely answered; *e.g.* water extract from leaves of mango (*Mangifera indica*) decreased significantly the levels of cholesterol, TAG, LDL, VLDL, and increased HDL [128]. However, it is early to assume that it is caused by the tannins, because this extract contained also saponins and simple phenol glycosides like it was in other cases.

The opinion that tannins possess antiobesity effect is often promoted (especially by the commercial subjects producing nutraceuticals). It is unquestionable that these substances can be contributive to the solution of difficult problem like overweight. They have specific antinutritive properties (inhibition of digestive enzymes and nutrients absorption from GIT), but they do not feature thermogenic effect. Their main effect consists rather in wider antioxidant activity, which assert favorably in affection of obesity and its consequences, as it was proved in ellagitannins from pomegranate (*Punica granatum*) [129].

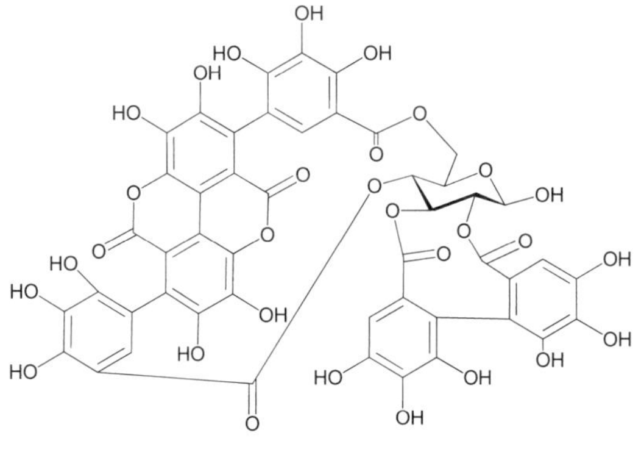

28

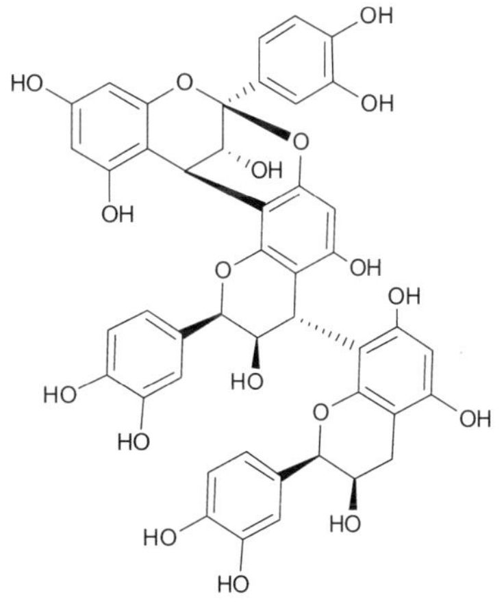

29

Figure 7: Some tannins with antioxidant activity: pomegranate ellagitannin from *Punica granatum*, punicalagin (28) and cinnamtannin B-1 (29), cranberry trimeric proanthocyanidin (*Rhodococcum vitis-idaea*).

Proanthocyanidins have cardiovascular protection effect due to their antioxidant activity, inhibition of LDL oxidation, ability of vasodilation, antiplatelet activity and protection against ischemia-reperfusion injury. Antioxidant activity of proanthocyanidins and the ability to protect lipid peroxidation was described above.

The ability to relax blood vessels is due to NO released from endothelial cells by the activity of proanthocyanidins and subsequently increase in cyclic GMP levels in the vascular smooth muscle cells. The NO/cyclic GMP pathway is known to be involved in many cardiovascular protective roles. *In vitro* vasodilation activity of isolated grape seeds proanthocyanidins tended to increase with degree of polymerization, epicatechin (**8**) content and with galloylation [130].

Proanthocyanidins can inhibit the renin-angiotensin-aldosteron system through affecting angiotensin-I converting enzyme (ACE) [131], or through antagonism on angiotensin receptor [132]. Rennin-angitensin-algosteron system is an endocrine system which influences vascular tone, fluid and electrolyte balance and the sympathetic nervous system. The *in vitro* analysis of inhibition of ACE activity and consequently assessement of ACE activity using angiotensin-I as substrate in both cultured HUVEC and purified enzyme indicated the close relevance in structure/activity relationship of tested epicatechin (**8**), procyanidin dimer, tetramer and hexamer. Procyanidin tetramer significantly inhibited the ACE activity by cultured HUVEC, whereas dimer and hexamer caused a nonsignificant inhibition. When ACE activity was assayed using the isolated rabbit enzyme, maximal ACE inhibition was exerted by tetramer and hexamer. The influence of the presence of plasma protein albumin on the activity of tested compounds was also *in vitro* determined. The presence of albumin did not reverse the ACE inhibition by dimer and tetramer, but decreased hexamer inhibition by 65%. The study concluded that although the number of epicatechin (**8**) units in procyanidin is one determinant of the specificity and extent of inhibition, the way that epicatechin (**8**) units are bond should also be considered [131]. It was demonstrated *in vitro* that proanthocyanidins could inhibit angiotensin-II binding to the angiotensin-I receptor. Inhibitory activity increased with the degree of polymerization to a maximal activity for pentamers and hexamers [132].

The increased platelet aggregation contributes to pathology of cardiovascular diseases. Several *in vitro* studies described inhibition of platelet aggregation by procyanidin rich cocoa. Consumption of flavonol-rich cocoa inhibited several measures of platelet activity including, epinephrine- and ADP-induced glycoprotein IIb/IIIa and P-selectin expression, platelet microparticle formation, and epinephrine-collagen and ADP-collagen induced primary hemostasis [133, 134]. High inhibiton of thrombine-induced platelet aggregation was exhibited by tannins isolated from leaves of *Arbutus unedo*, so called "Strawberry Tree", growing in Mediterranean region, which is used traditionally as a remedy for decrease of high blood pressure [135].

Other positive effects of tannins on cardiovascular system include their ability to decrease tissue injury induced by ischemia and reperfusion. For example, the antiarrhithmic and cytoprotective effect of an oral three-week pretreatment with oligomeric procyanidins of *Vitis vinifera* was investigated on the isolated perfused heart after global no-flow ischemia [136].

4.6. Impact on Diabetes Mellitus

In the process of influencing of this pathophysiological state, a prudent diet is recommended, including especially consumption of vegetable in which the polyphenolics are also present and interfere in the progression of the disease - inhibition of α-glucosidase (chapter 4.4.2). However, the safe natural substances with antidiabetic effect, utilizable as nutraceuticals, are noticeable object of research and polyphenolics, or more precisely tannins, count among them. The attractiveness of these substances is increased by the fact that additionally they have significant antioxidant and free radical scavenging effects. The interesting effects were found in gallotannins: they act preventively against activation of PPAR and inhibit also PARG in case of nephropathy in streptozotocin diabetes [137]. The insulinomimetic effect was found in two gallotannins from fruits of *Capparis moonii* [138]. There are many papers which present antihyperglycemic activity of extracts from various plant parts, *e.g.* [139, 140]; but it is necessary to approach them differentially and discover the involvement of tannins on this effect.

No references on the antidiabetic effect of phlorotanins have been presented yet.

4.7. Hepatoprotective Effects

During last 10 years, about 15 papers concerning hepatoprotective effect of tannins were published. However, it is still difficult to evaluate if these data are valid. They were mainly examined the summary polar extracts of various parts of plants which usually contain also flavonoids and terpenoids thus it is hard to specify the effect of tannins in these cases. There is an opinion that they contribute on decrease of the effect of toxic metabolites (ROS) and signal molecules (NF-κB, NO, FPT) on liver tissue by their antioxidant and free radical scavenging activity. There are unequivocal reports on hepatoprotective effect of some ellagitannins from *Alnus* sp. [141]. However, it is possible to assume that tannins contribute largely on antihepatotoxic activity of extracts from aerial part of *Phyllanthus atropurpureus*, because they comprise the main part of the secondary metabolites profile [142]. Similarly, it is possible to assume it in the extract from leaves of *Terminalia arjuna*, which acts antihepatotoxically against paracetamol induced damage of rat liver [143]. Both species are important medicinal plants of eastern medicine with high content of tannins [144].

4.8. Influence on Inflammation, Allergy and Immunity

These three pathophysiological processes are closely related by metabolism of various tissue factors and small signal molecules, thus their mutual separation is frequently problematic concerning biologically active compounds which can influence them.

Anti-inflammatory activity is one of the important effects of tannins. The mechanisms of action have not been definitely solved yet due to the complex character of inflammatory processes and thereby many possibilities of their interference. Most of the studies, which dealt with anti-inflammatory activity of tannins, targeted their antioxidant activity and interference of nuclear factor - κB (NF-κB), which is an important regulator of gene expression and promotes transcription of many inflammatory cytokinins including *e.g.* IL-8, TNF-α, or RANKL. The effects on the activation of extracellular signal-regulated protein kinase (ERK), c-Jun N-terminal kinase (JNK), and p38 mitogen-activated protein kinase (p38MAPK), which are upstream enzymes known to regulate COX-2 expression in many cell types, were also examined [145-148].

Topical application of pomegranate fruit acetone extract, containing anthocyanins and hydrolysable tannins, to mice inhibited TPA (12-*O*-tetradecanoylphorbol-13-acetate) - mediated increase in skin edema and hyperplasia, epidermal ODC activity and protein expression on ODC, and COX-2. The study also found, that topical application of extract resulted in inhibition of TPA-induced fosforylation of ERK and JNK, as well as activation of NF-κB. The study provided clear evidence that pomegranate extract possesses antiskin-tumor-promoting effects in CD-1 mouse [149].

Gallotannin pentagalloylglucose PGG inhibit *in vitro* NF-κB activation and IL-8 production in human monocytic cells (U937) stimulated with phorbol myristate acetate or TNF-α. Furthermore, PGG prevented degradation of the NF-κB inhibitory protein 1-κBα. The paper concluded that (PGG) can inhibit IL-8 gene expression in a mechanism involving its inhibition of NF-κB activation, which is dependent on 1-κBα degradation [146].

Ellagitannin furosin (**30**) (Fig. **8**) isolated from *Euphorbia helioscopia*, was examined for the effects on bone metabolism. Furosin *in vitro* decreased the differentiation of both murine bone marrow mononuclear cells and Raw 264.7 cells into osteoclasts. Furosin targeted at the early stage osteoclastic differentiation and had no cytotoxic effect on osteoclast precursors. The mechanism of effect was due to the inhibition of the receptor activator of NF-κB ligand (RANKL)-induced activation of p38 mitogen-activated protein kinase (p38MAPK) and c-Jun N-terminal kinase (JNK)/activating protein-1 (AP-1). Furthermore, furosin reduced resorption pit formation in osteoclasts, which was accompanied by disruption of the actin rings. The study concluded that the furosin would be the potential candidate for treatment of bone diseases [147].

The study which compared the effect of flavonoid monomers, dimers, trimer and Pycnogenol® on NO production, TNF-α secretion and NF-κB activity demonstrated that procyanidins act as modulators of the immune response in macrophages. Monomers and dimers repressed NO production, TNF-α secretion and NF-κB dependent gene expression induced by interferon-γ, whereas the trimeric procyanidin C-2 (**19**) and Pycnogenol® enhanced these parameters. In addition, in unstimulated Raw 264.7 macrophages, both procyanidin C-2 (**19**) and

Pycnogenol® increased TNF-α secretion in a concentration- and time-dependent manner [148].These results show the importance of the influence of the structure and polymerization on the anti-inflammatory activity of these substances.

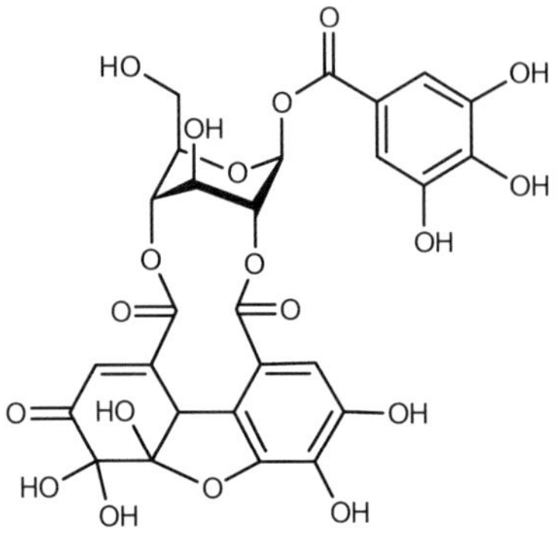

Figure 8: Ellagitannin from *Euphorbia helioscopia*, furosin (30).

The anti-inflammatory activity of polyphenolic substances is already known for many years, especially in flavonoids, but also the tannins contribute on it [150]. In many cases, as was mentioned earlier, it is not possible to evaluate if it is *via* tannin effect or the effect of other substances like flavonoids and triterpenoids [151]. Up to now, there is an insufficient number of papers reporting only the effect of tannins in order to be possible to define the impact of individual types of substances and guarantee the relation between the activity and structure. Seven gallotannins were isolated from several species of *Euphorbia* sp. and their impact on the metabolism of NO as inflammation mediator (LPS macrophage stimulation) was observed. Two of these substances (gallotannins 15 and 23) showed the inhibitory effect on LPS-induced inflammation reaction in dose-dependent manner (0.1-10 μg/ml) [152]. Similar effect on NO, or more preciously macrophage mediated immunity, was performed by 1,2,3,6-tetra-*O*-galloyl-β-D-allopyranose (gallotannin 24) from *Euphorbia jolkini* [153]. Anthocyans and hydrolysable tannins from the juice of pomegranate (*Punica granatum*) inhibit process of inflammation by activation of human mast cells and basophils. This can also have interesting therapeutic benefit for affection of inflammatory diseases by the inhibition of activation of MAP kinases and NF-κB [154].

Anti-inflammatory and anti-allergic effects were found in some algal phlorotannins (42 species of seaweeds); the substances from *Eisenia arborea* exhibited the highest activity; 6 isolated substances (eckol (**21**), 6,6′-bieckol (**26**), 6,8′-bieckol, 8,8′-bieckol, phlorofucofuroeckol A (**23**) and B) showed different level of inhibition against COX-2, LOX, PLA_2, and HA [155]. The antiallergic effect of dry extract from the leaves of clove tree (*Syzygium cumini*) is caused by hydrolysable tannins together with flavonoids [156]. The antiallergic effects were found also in phlotannins from *Ecklonia cava* [34].

4.9. Antiinvasive Effect

4.9.1. Anti-Cancer Activity (Cytotoxicity)

Though many studies indicated an interesting anti-tumor activity of hydrolysable as well as condensed tannins, mechanisms of action have not been clearly determined yet. The antioxidant activity can play a positive role. Tannins suppress the oxidative stress, which is important for pathogenesis of cancer and influence apoptosis of cells. The study, which investigated whether the anti-cancer effects of oligomeric proanthocyanidins are induced by apoptosis on human colorectal cancer cell line (SNU-C4) pronounced that cytotoxic effect of proanthocyanidins on SNU-C4 cells appeared in a dose-dependent manner. Proanthocyanidin treatment revealed typical morphological apoptic features, increased level of Bax and caspase-3, and decreased level of Bcl-2 mRNA expression. Bax and Bcl-2 belong to the group of genes which promote (Bax) or inhibit (Bcl-2) apoptosis. Caspase-3 enzyme activity was also significantly increased by treatment of proanthocyanidins. The study concluded that proanthocyanidins caused cell death by apoptosis through caspase pathway [157].

Several studies examined anti-tumor activity of punicalagin (**28**), tannin extract and juice from pomegranate [123, 145]. Tested samples exhibit antiproliferative activity on humanoral (KB, CAL27), colon (HT-29, HCT116, SW480, SW620) and prostate (RWPE-1, 22Rv1) tumor cells and induced apoptosis in HT-29 and HCT116 cancer colon cells [123]. Continuous study examined the effects of pomegranate on inflammatory cell signals (TNF-α, NF-κB, Akt) in the HT-29 human colon cancer cell line. TNF-α, cytokine mainly secreted by macrophages, is involved in the regulation of a wide spectrum of biological processes including cell proliferation, apoptosis, lipid metabolism, and coagulation. The most active suppressor of TNF-α induced COX-2 protein expression was pomegranate juice followed by tannin extract and punicalagin (**28**). Additionally, pomegranate juice, tannin extract and punicalagin, but no ellagic acid (**3**) reduced phosphorylation of the p65 subunit and binding to the NF-κB response element, which is an important regulator of gene expression and promotes transcription of many inflammatory mediators. Pomegranate juice also abolished TNF-α-induced Akt activation. Akt, genes implicated in cellular signaling, are needed for NF-κB activity. The study concluded that polyphenolic phytochemicals in pomegranate play an important role in the modulation of inflammatory cell signaling in colon cancer cells [145].

Except ellagitanins isolated from pomegranate juice, also their microbial metabolites, urolithins, contribute on decrease of proliferation in mentioned cell lines [158]. Hamamelitannin (**31**) (Fig. **9**) from Witch Hazel (*Hamamelis virginiana*) is cytotoxic against HT-29 cells [159].

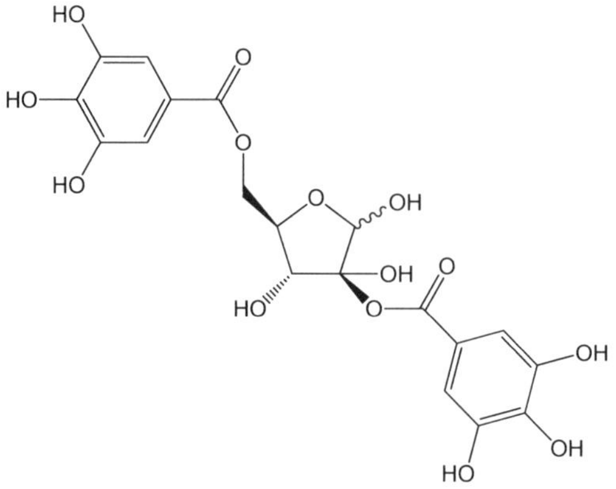

Figure 9: Tannins from *Hamamelis virginiana*, hamamellitannin (31).

Tannins are considered to be non-mutagenic substances; some of them even exhibit anti-mutagenic activity. The initial step in the formation of cancer is damage to the genome of a somatic cell producing a mutation in an oncogene or a tumor-suppressor gene. For example, inhibition of mutagenmethyl methansulphonate and metabolically activated carcinogen benzo(a)pyrene was described in juices and organic solvent extract from cranberries, raspberries, and blueberries. Of prepared solvent extract, the hydrolysable tannin containing fraction from strawberries (Sweet Charlie cultivar) was the most effective at inhibiting mutations [160]. Another study described the effect of cocoa liquor proanthocyanidins on pyridine derivate induced mutagenesis *in vitro* and on *in vivo* carcinogenesis in female rats. Study concluded that cocoa proanthocyanidins inhibit *in vitro* mutagenicity of pyridine derivate, as well as rat pancreatic carcinogenesis in the initiation stage, but not mammary carcinogenesis induced by pyridine derivate [161].

The effect of hydrolysable tannins from green tea (*Camelia sinensis*) on formation of benzo(a)pyrene induced DNA adducts was described many times in the past, recently the relationship between structure and activity of these substances has been elaborated [162].

1,2,3,4,6-Penta-*O*-galloyl-β-D-glucose, which occurs in many medicinal plants, *e.g.* in Witch Hazel, or roots of *Paeonia* sp. [163], possess anticancerogenic effects including anti-angiogenesis, anti-proliferative actions through inhibition of DNA replicative synthesis, S-phase arrest, and G(1) arrest, induction of apoptosis [164]. The induction of apoptosis of tumor cells (B16 mouse melanoma cells and BALB-MC.E12 mouse mammary tumor cells) *via* activation of caspase-3 was found in procyanidins from apples [165]. Proanthocyanidins esterified by gallic acid, or more precisely procyanidin B-2-3,3′-di-*O*-gallate is the main compound of the tannin mixture from grape seeds (*Vitis vinifera*) causing the inhibition of growth and apoptotic death of DU145 human prostate carcinoma cells [166]. This fact is not surprising to a certain extent, because the semisynthetically prepared esters of gallic acid act against TNF-α induced activation of NF-κB (239/NF-κB-Luc human embryonic kidney cells) and thus protect the expression of some genes which start cancerogenesis [167].

The protective effect of tannins can be also linked to inhibition of ornithine decarboxylase (ODC), an enzyme, which participates in biosynthesis of polyamines; increased polyamines expression is a marker of tumor development. Inhibition of ODC by plant metabolites was determined in several studies [168]. Proanthocyanidin-rich fraction of Amarican cranberries (*Vaccinium macrocarpon*) exhibited significant *in vitro* chemoprotective actvitiy indicated by an ornithine-dacarboxylase assay [168].

4.9.2. Antiviral, Antibacterial and Antiprotozoal Activity

The ability of tannins to antagonize various pathogens has been recognized for a long time. Mechanisms, which facilitate inhibition of bacteria or fungi growth include: *e.g.* non-specific ability of tannins to bound bacterial enzymes, direct action on metabolism of pathogens through inhibition of oxidative phosphorylation, or ability to complex transition metals ions, which are important for pathogens growth [26]. *In vitro* studies which dealt with this topic proved inhibition of many strains of bacteria including genus *Aeromanas, Bacillus, Clostridium, Enterobacter, Helicobacter, Klebsiella, Proteus, Pseudomonas, Shigella, Escherichia, Staphylococcus,* or *Streptococcus*. Out of fungi, it is possible to mention *Aspergillus, Coniophora*, or *Penicillium* [169, 170].

The mechanism of antiviral activity can be due to the linkage of tannins on protein surface of viruses or on cell membrane of host cells. Through this adsorption, penetration and eventually virus uncoating is restricted. Inhibition of enzymes, *e.g.* reverse transcriptase, can be also included. There are several studies, which determined inhibition of *Herpes simplex* virus (HSV), or Human immunodeficiency virus (HIV) by various condensed and hydrolysable tannins [171, 172]. 1,2,3,4,6-Penta-*O*-galloyl-β-D-glucose showed the inhibition of influenza A virus replication and release of the virus from infected cells (chicken erythrocytes) [173]. However, only the future shows if this effect can be exploitable. Other tannins: geraniin and 1,3,4,6-tetra-*O*-galloyl-β-D-glucose (*Phyllanthus urinaria*) were effective against *Herpes simplex* viruses *in vitro*: 1,3,4,6- tetra-*O*-galloyl-β-D-glucose IC_{50} 19.2±4.0 μM and geraniin IC_{50} 18.4±2.0 μM for HSV-2 [174]. Geraniin is also effective against human enterovirus 71, effectively inhibits the replication of the virus in the rhabdomyosarcoma cells (IC_{50} 10 μg/ml) [175]. It appears that this substance possesses relatively wide spectrum of biological activities [176].

From the practical point of view, the greatest attention was paid on the effect of tannins on bacteria. The whole systematic units are subjected to the screening tests for inhibition of common pathogens [177]. Results of these tests should be finding of the species suitable for further studies. In the last 10 years, there is quite large number of these papers (*e.g. Punica granatum* [178], *Mangifera indica*

[179], *Momordica dioica* and *Moringa oleifera* [180], *Polygonum capitatum* [181], *Abutilon indicum* [182] and others). Until now, no small-molecule secondary metabolite which could be used for moderation of the growth of obligatory bacterial infections was found, although various sophisticated methods are suggested (study of antimicrobial activity of medicinal plants against various multiple drug resistance pathogens and their molecular characterization and it's bioinformatics analysis of antibiotic gene from genomic databases with degenerate primer prediction) [183].

American cranberries (*Vaccinium macrocarpon*) have been used for treatment of urinary infections caused by bacteria *Escherichia coli* for a long time. It was found that just proanthocyanidins adhere the urinary tract epithelium and protect *E. coli* adhesion. The activity of various tannins isolated from cranberries was compared, and it was found that the A-type proanthocyanidin trimers are the most active. A-type dimers possessed lower activity then trimers and substances with B-type connection and monomers didn't show any antibacterial activity [184, 185]. It appears that procyanidins from cranberries have the greatest practical importance [186-188]. The interest in the study of natural compounds which would influence the growth of cariogenic *Streptococcus mutans* is increasing in last 5 years. Tannins belong to the group of substances to which the attention is paid. The polar extracts from *Terminalia glaucescens* possess the promising results [189]. Some species of genus *Potentilla*, especially *P. fruticosa*, containing both hydrolysable and condensed tannins, form highly effective surface anti-biofilm (up to 100 μg/ml) preventing formation of dental plaque, appear to be perspective [190]. It is positive from the practical point of view that the polyphenolics in wine (including tannins) do not influence the biological activity of bacteriocines nisin and pediocin PA-1, which are used as biopreservatives in wine [191].

Polymeric proanthocyanidin purified from the fruit of *Zanthoxylum piperitum* decreased the minimum inihibitory concentrations of β-lactam antibiotics for methicillin-resistant *Staphylococcus aureus*. *In vitro* study of the effects of the compound indicated that it suppressed the activity of β-lactamase and largely decreased the stability of the bacterial cell membrane [171].

Tannins are also examined as antiprotozoal agents. For example, recent study which dealt with antileishmanial activity of 67 tannins examined macrophage activation for release of NO, TNF-α and interferon-like activities. The effect of tannis on macrophage functions were further assessed by expression analysis (iNOS, IFN-α, IFN-γ, TNF-α, IL-1, IL-10, IL-12, IL-18). The most of tannins revealed little direct toxicity for extracellular promastigote *Leishmania* strains. In contrast, many polyphenols appreciably reduced the survival of the intracellular, amastigote parasite form *in vitro*. Data from functional bioassays suggested that the effects of polyphenols on intracellular *Leishmania* parasites were due to macrophage activation rather than direct antiparasitic activity. Gene expression analyses confirmed functional data, however, *in vitro* experiments are essential to prove the therapeutic benefits of polyphenolic immunomodulators [192]. The ethanol extract from date palm (*Phoenix dactylifera*) exhibited the activity against extracellular forms of promastigotes *Leishmania tropica*; the hydrolysable tannins were identified in the extract [193], but it is not possible to define if they participate on the antiprotozoal activity.

5. HUMAN INTERVENTION STUDIES

The most of above mentioned studies described only *in vitro* activity of tannins. These studies dealt with biological activity of tannins, but do not consider absorption, bioavailability and metabolism, and that is why the results of *in vitro* activities do not always correspond with *in vivo* efficiency. The studies on enzymes, tissue cultures or animals and especially human long-lasting intervention studies which evaluate modifications of many biomarkers can bring confirmation of gained results. This part of review examines the effect of proanthocyanidins demonstrated in some of human intervention studies.

Recently, review from Williamson and Manach on clinical data of polyphenols including proanthocyanidins was published [194]. Williamson and Manach mentioned nearly fourty human intervention studies in which the health effects of proanthocyanidins rich food were demonstrated. It's necessary to point out that the human intervention studies did not use pure compounds, but only proanthocyanidins rich food (extracts and juices from apples, grape seeds, pomegranates, cranberries, blueberries, black currant, furthermore various types of chocolate, cocoa, red wine, or Pycnogenol®), which is due to high difficulty of preparation sufficient amount of pure compounds. Subsequently, also other active substances, except tested proanthocyanidins and their metabolic products, can be included in the activity; *e.g.* catechins, which occur with proanthocyanidins very often. However, it is necessary to put a question if it is meaningful to perform these studies with pure compounds. They have only an importance from the toxicological point of view, but this viewpoint is out of the question in human. Additionally, tannins will not be most likely used as drugs, and their use in diet is verified historically.

The Williamson and Manach stated that predominant health effects of proanthocyanidin rich food are on cardiovascular system. Biomarkers affected are: increased general plasmatic antioxidant activity, decreased platelet aggregation, decreased lipid peroxides plasmatic concentration, decreased LDL concentration and increased HDL concentration, decreased disposition of LDL to oxidation, endothelium induced vasodilatation and decreased blood pressure, positive effect on capillary permeability and fragility, increased ascorbic acid plasmatic concentration, decreased expression of P-selectin, decreased thromboxane serum concentration, increased microvessels diameter, increased homocysteine and vitamin B6 plasmatic concentration, endothelium functions support, increased platelet subjected production of NO, inhibition of superoxide, increased α-tocopherol concentration, and decreased concentration of self antibody against oxidized LDL [194]. Increased general immunity, decreased UV sensitivity, menstrual and abdominal pain control, and decreased urinary infections relapse number are other effects caused by proanthocyanidins rich food and mentioned by the clinical review of Williamson and Manach [194].

The number of health effects determined in *in vivo* studies is often superior to results from pure substances in *in vitro* tests. It can be due to the effect of potentiation of activities of individual compounds, due to the action of various metabolites formed by colon microflora, or due to the effects of previously mentioned ballast substances (*e.g.* catechins). Therefore in the process of evaluation of *in vivo* effects of tannins we can find several studies providing only partial view which evaluate these substances in the complex with other substances and deal more with the influence on human health. Since 2006, only several wider studies, which tried to evaluate the effect of these substances in wider health context, were published. They involve raspberries [195], strawberries [196], raisins [197] and relationship of natural phenolic substances to bone metabolism [198]. The review study evaluating comprehensive antioxidant tannins from tea was also published [199].

6. ADVERSE AND UNDESIRABLE EFFECTS

Tannins form nonabsorbable complexes with proteins, sugars, digestive enzymes, or metal ions and decrease nutritive value of food. Therefore, it is not advisable to ingest high extent of tannins. It was proved that animals, which were fed with tannin free feed, had higher weight gain compared with those fed with tannin rich feed. Tannins influenced also vitamins utilization; tannin included in rats' diet decrease vitamin A content in liver and utilization of vitamin B_{12}. Also the absorption of iron was decreased by formation of insoluble complexes in people, who ate tannin rich sorghum [200]. A series of review studies concerning interactions between proteins and tannins was published [201].

However, the interaction with enzymes, which decreases utilization of food, remains an undesirable effect: tannins form the non-functional complexes with trypsin [202], on the contrary the reaction of proanthocyanidins from grape seeds (*Vitis vinifera*) with α-amyloid [203] can be advantageous for decrease of postprandial hyperglycemia.

Gallotannin has been shown to produce hepatic necrosis in humans and grazing animals. The breakdown of polyribosomes in mouse liver and inhibition of the incorporation of aminoacids into hepatic protein was found after subcutaneous injections of gallotannin to mice [204]. The reports on carcinogenic effect of tannins has been published [205], but it is necessary to check them out precisely, because the majority of polyphenols of this type possesses anticarcinogenic effect.

In the past, several studies have informed about possible mutagenic and carcinogenic activities of tannins [206]. For example, in Caribbean region great consumption of tannin rich food was associated with higher occurrence of esophageal cancer [207]. Nevertheless, direct association between the incidence of cancer and tannins has never been documented and many other studies presented that tannins are nonmutagenic and noncarcinogenic [25, 57, 204].

As it was mentioned above, one study, which evaluated anti- and pooxidation properties of gallic and ellagic acid (**3**) and gallotannin according to the DNA injury in mussels *Unio tumidus* cells, found that high concentrations of these substances injure DNA and showed prooxidant activity. The study concluded that the bioactivity of gallic acid, ellagic acid (**3**) and gallotannin would be more complicated [75].

7. CONCLUSIONS

Tannins show various health benefit activities; especially antioxidant, antitumor, cardioprotective, anti-inflammatory and antimicrobial activity. Mechanisms of activity, as well as their bioavailability have not been satisfactorily clarified yet, even though they are consumed daily and their structure has been known for 100 years. It is unquestionable that the research of these polyphenolics will be extended because of the mentioned reasons, it will result in determination of recommended daily allowance and clarification of interactions with other secondary metabolites with which the population come into contact. It is not adviseable to ingest large amount of tannins, until other studies aimed at finding average daily intake and possible adverse effects in higher doses of proanthocyanidins are done. However, small doses of tannin rich food can be beneficial to human health.

Recently it is evident that the attention will be paid on phlorotannins from brown seaweeds within the extending phytochemical research. These substances will be important for production of food products [208], but also as new cosmeceuticals [32].

The determination of tannins utilization in decrease of methanogenesis in beef breeding is the second important area of their study. *E.g.* it appeared that almost one half of the greenhouse gases emissions on New Zealand originated from animal production and enteral production of methane form about 30 % of total emissions of carbone dioxide. The intestinal emissions of methane increased by 9 % since 1990 [209]. Just tannins are the substances which can improve digestive processes and decrease the production of these undesirable gases [210-214].

The study of production and utilization of especially procyanidins from tea dust, wastes from production of wine, cocoa and other food sources, which will become valuable material for industrial use of these substances in larger scale then now, will play an important role in next few years.

All of these facts will place demands (in the area of plant polyphenolics) on the research oriented to the area of plant physiology, or more precisely to influence on metabolism of these substances from both quantitative and qualitative points of view.

ACKNOWLEDGEMENTS

This work has been supported by the European Social Fund and the state budget of the Czech Republic. TEAB, project no. CZ.1.07/2.3.00/20.0235.

CONFLICT OF INTEREST

The authors confirm that this chapter contents have no conflict of interest.

DISCLOSURE

The chapter submitted for eBook Series entitled: "**Recent Advances in Medicinal Chemistry, Volume 1**" is an update of our article published in **Mini-Reviews in Medicinal Chemistry, Volume 8, Number 5, pp. 436 to 447**, with additional text and references.

ABBREVIATIONS

$ABTS^+$ radical	=	(2,2-azinobis)3-ethyl-2,3-dihydrobenzothiazol-6-sulphonic acid)
ACE	=	angiotensin converting enzyme
ADP	=	adenosine monophosphate
Akt	=	protein kinase B
AP-1	=	activator protein 1
Bax	=	pro-apoptic Bcl-2 protein

Bcl-2	=	B-cell lymphoma 2
BHP	=	benign hyperplasia prostate
CaCC	=	calcium activated Cl channels
cAMP	=	cyclic adenosine monosphoshate
COX-2	=	cyclooxygenase 2
DPPH	=	1,1-diphenyl-2-(2,4,6-trinitrophenyl)hydrazyl) radical
EGC	=	(-)-epigallocatechin
EGCG	=	(-)-epigallocatechin gallate
eNOS	=	endothelial nitric oxide synthase
ERK	=	extracellular signal-regulated kinase
ESR	=	electron spin resonance
FPT	=	farnesylprotein transferase
FRAP	=	fluorescence recovery after photobleaching
GIT	=	gastrointestinal tract
GMP	=	guanosine monophosphate
GSH	=	glutathion
HA	=	hyaluronic acid
HDL	=	high-density lipoprotein
HIV	=	human immunodeficiency virus
HSV	=	Herpes simplex viruses

HUVEC	=	human umbilical vein endothelial cells
IFN	=	interferon
IL	=	interleukin
iNOS	=	inducible nitric oxide synthase
JNK	=	c-Jun N-terminal kinases
LDL	=	low-density lipoprotein
LOX	=	lipoxygenase
5-LOX	=	5-lipoxygenase
LPS	=	lipopolysaccharide
MAP	=	mitogen activated protein kinases
NF-κB	=	nuclear factor κB
NO	=	nitric oxide
ODC	=	ornithine decarboxylase
P38MAPK	=	p38 mitogen-activated protein kinases
PARG	=	poly(ADP-ribose) glycohydrolase
PARP	=	poly(ADP-ribose) polymerase
PGG	=	pentagalloylglucose
PLA_2	=	phospholipase A_2
PON1	=	paraoxonase 1
PPARγ	=	peroxisome proliferator-activated receptor γ

RANKL = soluble receptor activator of NK-κB ligand

RNOS = reactive nitrogen and oxygen species

RNS = reactive nitrogen species

ROS = reactive oxygen species

TAG = triacylglycerols

TNF-α = tumor necrosis factor-α

TPA = 12-*O*-tetradecanoylphorbol-13-acetate

VLDL = very low-density lipoproteins

REFERENCES

[1] Bravo, L. Polyphenols: chemistry, dietary sources, metabolism, and nutritional significance. *Nutr. Rev.*, **1998**, *56*(11), 317-333.

[2] Balasundram, N.; Sundram, K.; Samman, S. Phenolic compounds in plants and agri-industrial by-products: antioxidation activity, occurrence, and potential uses. *Food Chem.*, **2006**, *99*(1), 191-203.

[3] Bors, W.; Michel, C.; Stettmaier, K. Antioxidant effects of flavonoids. *BioFactors,* **1997**, *6*(4), 399-402.

[4] Lopez-Revuelta, A.; Sanchez-Gallego, J.I.; Hernandez-Hernandez, A.; Sanchez-Yague, J.; Llanillo, M. Membrane cholesterol contents influence the protective effects of quercetin and rutin in erythrocytes damaged by oxidative stress. *Chem. Biol. Interact.,* **2006**, *161*(1), 79-91.

[5] Youdim, K.A.; Shukitt-Hale, B.; Joseph, J.A. Flavonoids and the brain: interactions at the blood-brain barrier and their physiological effects on the central nervous system. *Free Radic. Biol. Med.,* **2004**, *37*(11), 1683-1693.

[6] Russoa, A.; Cardileb, V.; Lombardob, L.; Vanellaa, L.; Acquavivaa, R. Genistin inhibits UV light-induced plasmid DNA damage and cell growth in human melanoma cells. *J. Nutr. Biochem.,* **2006**, *17*(2), 103-108.

[7] Cimino, F.; Saija, A. Flavonoids in skin cancer chemoprevention. *Curr. Top. Nutr. Res.,* **2005**, *3*(4), 243-258.

[8] Bate-Smith, E.C.; Swain, T. In *Comparative Biochemistry*; Mason, H.S., Florkin, A.M., Eds; Academic Press: New York, **1962**; vol. *3*, p. 764.

[9] Li, Y.-X.; Wijesekara, I.; Li, Y.; Kim, S.-K. Phlorotannins as bioactive agents from brown algae. *Process Biochem.*, **2011**, *46*(12), 2219-2224.

[10] Koleckar, V.; Kubikova, K.; Rehakova, Z.; Kuca, K.; Jun, D.; Jahodar, L.; Opletal, L. Condensed and hydrolysable tannins as antioxidants influencing the health. *Mini-Rev. Med. Chem.,* **2008**, *8*(5), 436-447.

[11] Kolečkář, V.; Řeháková, Z.; Brojerová, E.; Kuča, K.; Jun, D.; Macáková, K.; Opletal, L.; Drašar, P.; Jahodář, L.; Chlebek, J.; Cahlíková, L. Proanthocyanidins and their antioxidation activity. *Chem. Listy,* **2012**, *106*(2), 113-121.

[12] Okuda, T.; Ito, H. Tannins of constant structure in medicinal and food plant – hydrolyzable tannins and polyphenols related to tannins. *Molecules*, **2011**, *16*, 2191-2217.

[13] Quideau, S.; Jourdes, M.; Lefeuvre, D.; Pardon, P.; Saucier, C.; Teissedre, P.-L.; Glories, Y. Ellagitannins – an underestimated class of plant polyphenols: chemical reactivity of C-glucosidic ellagitannins in relation to wine chemistry and biological activity. *Rec. Adv. Polyphenol Res.,* **2010**, *2*, 81-137.

[14] Pouysegu, L.; Deffieux, D.; Malik, G.; Natangelo, A.; Quideau, S. Synthesis of ellagitannin natural products. *Nat. Prod. Rep.*, **2011**, *28*(5), 853-874.

[15] Pereira, D.D.; Valentao, P.; Pereira, J.A.; Andrade, P.B. Phenolics: from chemistry to biology. *Molecules*, **2009**, *14*(6), 2202-2211.

[16] Salminen, J.-P.; Karonen, M.; Sinkkonen, J. Chemical ecology of tannins: recent developments in tannin chemistry reveal new structures and structure-activity patterns. *Chemistry - Eur. J.*, **2011**, *17*(10), 2806-2816.

[17] Soetan, K.O. Pharmacological and other beneficial effects of anti-nutritional factors in plants – A rewiev. *Afr. J. Biotechnol.*, **2008**, *7*(25), 4713-4721.

[18] Chaudhuri, R.K.; Lascu, Z.; Puccetti, G. Inhibitory effects of *Phyllanthus emblica* tannins on melanin synthesis. *Cosmet. Toilet.*, **2007**, *122*(2), 73-4, 76, 78, 80.

[19] Farooqi, A.H.A.; Kumar, V.S.; Abdul-Khaliq. Rewiev of biological and therapeutic effects of *Emblica officinalis* and its active constituents. *J. Med. Aromat. Plant Sci.*, **2011**, *33*(1), 3-12.

[20] Patel, J.R.; Tripathi, P.; Sharma, V.; Chauhan, N.S.; Dixit, V.K. *Phyllanthus amarus*: ethnomedical uses, phytochemistry and pharmacology: A rewiev. *J. Ethnopharmacol.*, **2011**, *138*(2), 286-313.

[21] Choudhari, A.B.; Nazim, S.; Gomase, P.V.; Khairnar, A.S.; Shaikh, A.; Choudhari, P. Phytopharmacological review of Arjuna bark. *J. Pharm. Res.*, **2011**, *4*(3), 580-581.

[22] Khanbabee, K.; van Ree, T. Tannins: classification and definition. *Nat. Prod. Rep.,* **2001**, *18*(6), 641-649.

[23] Okuda, T.; Yoshida, T.; Hatano, T. Oligomeric hydrolyzable tannins, a new class of plant polyphenols. *Heterocycles,* **1990**, *30*(2), 1195–1218.

[24] Haslam, E. In: *The biochemistry of plants;* Conn, E.E., Ed.; Academic Press: London, **1981**, Vol. *7*, pp. 527-556.

[25] De Bruyne, T.; Pieters, L.; Deelstra, H.; Vlietinck, A. Condensed vegetable tannins: biodiversity in structure and biological activities. *Biochem. Syst. Ecol.,* **1999**, *27*(4), 445-459.

[26] Cos, P.; De Bruyne, T.; Hermans,N.; Apers, S.; Berghe, D.V.; Vlietinck, A.J. Proanthocyanidins in health care: current and new trends. *Curr. Med. Chem.,* **2004**, *11*(10), 1345-1359.

[27] Ferreira, D.; Slade, D. Oligomeric proanthocyanidins: naturally occuring O-heterocycles. *Nat. Prod. Rep.,* **2002**, *19*(5), 517-541.

[28] Haslam, E. Symmetry and promiscuity in procyanidin biochemistry. *Phytochemisty,* **1997**, *16*(11), 1625-1640.

[29] Xie, D.Y.; Dixon, R.A. Proanthocyanidin biosynthesis – still more questions than answers? *Phytochemistry,* **2005**, *66*(18), 2127-2144.

[30] Scalbert, A.; Williamson, G. Dietary intake and bioavailability of polyphenols. *J. Nutr.*, **2000**, *130*(8), 2073-2085.

[31] Yuan, Y.V. Antioxidants from edible seaweeds., *ACS Symp. Ser.*, **2007**, *956*, 268-301.

[32] Wijesekara, I.; Yoon, N.Y.; Kim, S.-K. Phlorotannins from *Ecklonia cava* (Phaeophyceae): biological activities and potential health benefits. *BioFactors*, **2010**, *36*(6), 408-414.

[33] Ngo, D.-H.; Wijesekara, I.; Vo, T.-S.; Van Ta, Q.; Kim, S.-K. Marine food-derived functional ingredients as potential antioxidants in the food industry: An overview. *Food Res. Int.*, **2011**, *44*(2), 523-529.

[34] Gupta, S.; Abu-Ghannam, N. Bioactive potential and possible health effects of edible brown seaweeds. *Trends Food Sci. Technol.*, **2011**, *22*(6), 315-326.

[35] Kim, S.-K.; Thomas, N.V.; Li, X. Anticancer compounds from marine macroalgae and their applications as medicinal foods. *Adv. Food Nutr. Res.*, **2011**, *64*, 213-224.

[36] Wijesinghe, W.A.J.P.; Jeon, Y.-J Biological activities and potential cosmeceutical applications of bioactive components from brown seaweeds: a review. *Phytochem. Rev.*, **2011**, *10*(3), 431-443.

[37] Ross, J.A.; Kasum, Ch.M. Dietary flavonoids: bioavailability, metabolic effects, and safety. *Annu. Rev. Nutr.,* **2002**, *22*, 19-34.

[38] Deprez, S.; Mila, I.; Huneau, J.F.; Tome, D.; Scalbert, A. Transport of proanthocyanidin dimer, trimer, and polymer across monolayers of human intestinal epithelial Caco-2 cells. *Antioxid. Redox. Sign.,* **2001**, *3*(6), 957-967.

[39] Spencer, J.P.E.; Chaudry, F.; Pannala, A.S.; Srai, S.K.; Debnam, E.; Rice-Evans, C. Decomposition of cocoa procyanidins in the gastric milieu. *Biochem. Biophys. Res. Commun.*, **2000**, *272*(1), 236-241.

[40] Rios, L.Y.; Bennett, R.N.; Lazarus, S.A.; Remesy, C.; Scalbert, A.; Williamson, G. Cocoa procyanidins are stable during gastric transit in humans. *Am. J. Clin. Nutr.,* **2002**, *76*(5), 1106-1110.

[41] Spencer, J.P.E.; Schroeter, H.; Shenoy, B.; Srai, S.K.S.; Debnam, E.S.; Rice-Evans, C. Epicatechin is the primary bioavailable form of the procyanidin dimers B2 and B5 after transfer across the small intestine. *Biochem. Biophys. Res. Commun.,* **2001**, *285*(3), 588-593.

[42] Baba, S.; Osakabe, N.; Natsume, M.; Terao, J. Absorption and urinary excretion of procyanidin B2 [epicatechin-(4β-8)epicatechin] in rats. *Free Radic. Biol. Med.,* **2002**, *33*(1), 142-148.

[43] Deprez, S.; Brezillon, C.; Rabot, S.; Philippe, C.; Mila, I.; Lapierre, C.; Scalbert, A. Polymeric proanthocyanidins are catabolized by human colonic microflora into low-molecular-weight phenolic acids. *J. Nutr.,* **2000**, *130*(11), 2733-2738.

[44] Aura, A.-M. Microbial metabolism of dietary phenolic compounds in the colon. *Phytochem. Rev.*, **2008**, *7*(3), 407-429.

[45] Tomas-Barberan, F.A.; Garcia-Conesca, M.T.; Larrosa, M.; Cerda, B.; Gonzalez-Barrio, R.; Bermudez-Soto, M.J.; Gonzalez-Sarrias, A.; Espin, J.C. Bioavailability, metabolism, and bioactivity of food ellagic acid and related polyphenols. *Rec. Adv. Polyphenol Res.,* **2008**, *1*, 263-277.

[46] Riedl, K.M.; Hagerman, A.E. Tannin-protein complexes as radical scavengers and radical sinks. *J. Agric. Food Chem.,* **2001**, *49*(10), 4917–4923.

[47] Yu, X.-T.; Chu, S.-D.; Hagerman, A.E.; Lorigan, G.A. Probing the interaction of polyphenols with lipid bilayers by solid-state NMR spectroscopy. *J. Agric. Food Chem.,* **2011**, *59*(12), 6783-6789.

[48] Gutteridge, J.M. Free radicals in disease processes: a compilation of cause and consequence. *Free Radic. Res. Commun.,* **1993**, *19*(3), 141–158.
[49] Kehrer, J.P. Free radicals as mediators of tissue injury and disease. *Crit. Rev. Toxicol.,* **1993**, *23*(1), 21–48.
[50] Storz, P. Reactive oxygen species in tumor progression. *Front. Biosci.,* **2005**, *10*(2), 1881-1896.
[51] Aruoma, O.I. Nutrition and health aspects of free radicals and antioxidants. *Food Chem. Toxicol.,* **1994**, *32*(7), 671–683.
[52] Bondet, V.; Brand-Williams, W., Berset, C. Kinetics and mechanism of antioxidant activity using the DPPH free radical method. *Lebensm. Wiss. Technol.,* **1997**, *30*(6), 609-615.
[53] Miller, N.J.; Rice-Evans, C.A. Spectrophotometric determination of antioxidant activity. *Redox Rep.,* **1996**, *2*(3), 161-171.
[54] Hatano, T.; Miyatake, H.; Natsume, M.; Osakabe, N.; Takizawa, T.; Ito, H.; Yoshida, T. Proanthocyanidin glycosides and related polyphenols from cacao liquor and their antioxidant effects. *Phytochemistry,* **2002**, *59*(7), 749-758.
[55] Cai, Y.Z.; Sun, M.; Xing, J.; Luo, Q.; Corke, H. Structure-radical scavenging activity relationship of phenolic compounds from traditional Chinese medicinal plants. *Life Sci.,* **2006**, *78*(25), 2872-2888.
[56] Plumb, G.W.; De Pascual-Teresa, S.; Santos-Buelga, C.; Cheynier, V.; Williamson, G. Antioxidant properties of catechins and proanthocyanidins: effect of polymerization, galloylation and glycosylation. *Free Radic. Res.*, **1998**, *29*(4), 351-358.
[57] Okuda, T. Systematics and health effects of chemically distinct tannins in medicinal plants. *Phytochemistry,* **2005**, *66*(17), 2012-2031.
[58] Hagerman, A.E.; Riedl, K.M.; Jones, G.A.; Sovik, K.N.; Ritchard, N.T.; Hartzfeld, P.W.; Riechel, T.L. High molecular weight plant polyphenolics (tannins) as biological antioxidants. *J. Agric. Food Chem.,* **1998**, *46*(5), 1887-1892.
[59] Amarowicz, R.; Estrella, I.; Hernandez, T.; Robredo, S.; Troszynska, A.; Kosinska, A.; Pegg, R.B. Free radical-scavenging capacity, antioxidant activity, and phenolic composition of green lentil (Lens culinaris). *Food Chem.*, **2010**, *121*(3), 705-711.
[60] Pothitirat, W.; Chomnawang, M.T.; Supabphol, R.; Gritsanapan, W. Free radical scavenging and anti-acne activities of mangosteen fruit extracts prepared by different extraction methods. *Pharm. Biol.*, **2010**, *48*(2), 182-186.
[61] Craft, B.D.; Hargrove, J.L.; Greenspan, P.; Hartle, D.K.; Amarowicz, R.; Pegg, R.B. LC characterisation of peanut skin phytonutrients: antioxidant, radical-scavenging, and biological activities. *Spec. Publ. Royal Soc. Chem.*, **2010**, *326*, 283-296.
[62] Soni, H.; Nayak, G.; Mishra, K.; Singhai, A.K.; Pathak, A.K. Evaluation of phytopharmaceutical and antioxidant potential of methanolic extract of peel of *Punica granatum*. *Res. J. Pharm. Technol.*, **2010**, *3*(4), 1170-1174.
[63] Cespedes, C.L.; El-Hafidi, M.; Pavon, N.; Alarcon, J. Antioxidant and cardioprotective activities of phenolic extracts from fruits of Chilean blackberry *Aristotelia chilensis* (Elaeocarpaceae), Maqui. *Food Chem.*, **2008**, *107*(2), 820-829.
[64] Pinheiro do Prado, A,C.; Aragao, A.; Fett, R.; Block, J.M. Phenolic compounds and antioxidant activity of pecan [Carya illinoinensis (Wangenh.) C. Koch] shell extracts. *Brazil. J. Food Technol.*, **2009**, *12*(4), 323-332.
[65] de Souza, M.C.; Figueiredo, R.W.; Maia, G.A.; Alves, R.E.; Brito, E.S.; Moura, C.F.H.; Rufino, M.S. M. Bioactive compounds and antioxidation activity on fruits from different acai (*Euterpene oleracea* Mart.) progenies. *Acta Hort.*, **2009**, *841*, 455-458.

[66] Ho, K.Y.; Huang, J.S.; Tsai, C.C.; Lin, T.C.; Hsu, Y.F.; Lin, C.C. Antioxidant activity of tannin components from *Vaccinium vitis-idaea* L. *J. Pharm. Pharmacol.,* **1999**, *51*(9), 1075-1078.

[67] Metodiewa, D.A.; Jaiswal, A.K.C.; Cenas, N.B.; Dickancaite, E.B.; Segura-Aguilar, J.A. Quercetin may act as a cytotoxic prooxidant after its metabolic activation to semiquinone and quinoidal product. *Free Radic. Biol. Med.,* **1999**, *26*(1-2), 107-116.

[68] Bors, W.; Michel, C. Antioxidant capacity of flavanols and gallate esters: pulse radioanalysis studies. *Free Radic. Biol. Med.,* **1999**, *27*(11-12), 1413-1426.

[69] Goncalves, C.; Dinis, T.; Batista, M.T. Antioxidant properties of proanthocyanidins of *Uncaria tomentosa* bark decoction: a mechanism for anti-inflammatory activity. *Phytochemistry,* **2005**, *66*(1), 89-98.

[70] Yoshimura, Y.; Nakazawa, H.; Yamaguchi, F. Evaluation of the NO scavenging activity of procyanidin in grape seed by use of the TMA-PTIO/NOC 7 ESR system. *J. Agric. Food Chem.,* **2003**, *51*(22), 6409-6412.

[71] Singhai, A.; Nayak, G.; Budhwani, A.; Singhai, A. Nitric oxide radical scavenging activity of aqueous extract of *Terminalia belerica* bark. *Int. J. Chem. Sci.*, **2009**, *7*(4), 2617-2623.

[72] Ramos-Escudero, F.; Munoz, A.M.; Ureta, C.A.-O.; Yanez, J.A. Anthocyanins, polyphenols and antioxidant activity of purple Sachapapa (Dioscorea trifida L.) and evaluation of lipid peroxidation in human serum. *Revista Soc. Quim. Peru,* **2010**, *76*(1), 61-72.

[73] Soobrattee, M.A.; Neergheen, V.S.; Luximon-Ramma, A.; Aruoma, O.I.; Bahorun, T. Phenolics as potential antioxidant therapeutic agents: mechanism and actions. *Mutat. Res-Fund. Mol. M.,* **2005**, *579*(1-2), 200-213.

[74] Bors, W.; Michel, C.; Stettmaier, K. Electron paramagnetic resonance studies of radical species of proanthocyanidins and gallate esters. *Arch. Biochem. Biophys.,* **2000**, *374*(2), 347-355.

[75] Labieniec, M.; Gabryelak, T.; Falcioni, G. Antioxidant and pro-oxidant effects of tannins in digestive cells of the freshwater mussel *Unio tumidis*. *Mutat. Res.-Gen. Tox. En.,* **2003**, *539*(1-2), 19-28.

[76] Packer, L.; Rimbach, G.; Virgili, F. Antioxidant activity and biologic properties of a procyanidin-rich extract from pine (Pinus maritima) bark, pycnogenol. *Free Radic. Biol. Med.,* **1999**, *27*(5-6), 704-724.

[77] Breton, F.; Cerantola, S.; Ar Gall, E. Distribution and radical scavenging activity of phenols in *Ascophyllum nodosum* (Phaeophyceae). *J. Exp. Marine Biol. Ecol.*, **2011**, *399*(2), 167-172.

[78] Thomas, N.V.; Kim, S.-K. Potential pharmacological applications of polyphenolic derivatives from marine brown algae. *Environ. Toxicol. Pharmacol.*, **2011**, *32*(3), 325-335.

[79] Sugiura, Y.; Matsuda, K.; Yamada, Y.; Imai, K.; Kakinuma, M.; Amano, H. Radical scavenging and hyaluronidase inhibitory activities of phlorotannins from the edible brown alga *Eisenia arborea. Food Sci. Technol. Res.*, **2008**, *14*(6), 595-598.

[80] Olaleye, M. T.; Akinmoladun, A.C.; Ogunboye, A. A.; Akindahunsi, A.A. Antioxidant activity and hepatoprotective property of leaf extracts of *Boerhaavia diffusa* Linn against acetaminophen-inducer liver damage in rats. *Food Chem. Toxicol.*, **2010**, *48*(8-9), 2200-2205.

[81] Almajano, M. P.; Carbo, R.; Jimenez, J. A.L.; Gordon, M. H. Antioxidant and antimicrobial activities of tea infusion. *Food Chem.*, **2008**, *108*(1), 55-63.

[82] Rocha-Guzman, N.E.; Gallegos-Infante, J.A.; Gonzalez-Laredo, R.F.; Reynoso-Camacho, R.; Ramos-Gomez, M.; Garcia-Gasca, T.; Rodriguez-Munoz, M.E.; Guzman-Maldonado, S.H.; Medina-Torres, L.; Lujan-Garcia, B.A. Antioxidant activity and genotoxic effect on HeLa cells of phenolic compounds from infusions of *Quercus resinosa* leaves. *Food Chem.*, **2009**, *115*(4), 1320-1325.

[83] Kaur, G.; Athar, M.; Alam, M. S. Quercus infectoria galls posses antioxidant activity and abrogates oxidative stress-induced functional alterations in murine macrophages. *Chem. Biol. Interact.*, **2008**, *171*(3), 272-282.

[84] Saito, A.; Nakajima, N. Structure-activity relationship of synthesized procyanidin oligomers: DPPH radical scavenging activity and Maillard reaction inhibitory activity. *Heterocycles*, **2010**, *80*(2), 1081-1090.

[85] Kostalova, D.; Peciar, B. Natural polyphenols – their occurence, clasiffication and pharmaceutical activities. *Farm. Obzor*, **1998**, *67*(1), 7-10.

[86] Haslam, E. Natural polyphenols (vegetable tannins) as drugs: possible modes of action. *J. Nat. Prod.*, **1996**, *59*(2), 205-15.

[87] Weinreb, O.; Mandel, S.; Amit, T.; Youdim, M.B.H. Neurological mechanism of green tea polyphenols in Alzheimer's and Parkinson's disease. *J. Nutr. Biochem.*, **2004**, *15*(9), 506-516.

[88] Vaya, J.; Aviram, M. Nutritional antioxidants: mechanism of action, analyses of activities and medical applications. *Curr. Med. Chem. Immunol. Endocr. Metab. Agents,* **2001**, *1*(1), 99-118.

[89] Piterkova, J.; Tomankova, K.; Luhova, L.; Petrivalsky, M.; Pec, P. Oxidative stress: localization of reactive oxygen species formation and degradation in plant tissues. *Chem. Listy,* **2005**, *99*(7), 455-466.

[90] Moran, J.F.; Klucas, R.V.; Grayer, R.J.; Abian, J.; Becana, M. Complexes of iron with phenolic compounds from soybean nodules and other legume tissues: prooxidant and antioxidant properties. *Free Radic. Biol. Med.,* **1997**, *22*(5), 861-870.

[91] Morel, I.; Lescoat, G.; Cogrel, P.; Sergent, O.; Pasdeloup, N.; Brissot, P.; Cillard, P.; Cillard, J. Antioxidant and iron-chelating activities of the flavonoids catechin, quercetin and diosmetin on iron-loaded rat hepatocyte cultures. *Biochem. Pharmacol.,* **1993**, *45*(1), 13-19.

[92] Guo, Q.; Zhao, B.; Li, M.; Shen, S.; Xin, W. Studies on protective mechanism of four components of green tea polyphenols against lipid peroxidation in synaptosomes. *Biochim. Biophys. Acta,* **1996**, *1304*(3), 210-222.

[93] Mladěnka, P.; Macáková, K.; Filipský, T.; Zatloukalová, L.; Jahodář, L.; Bovicelli, P.; Silvestry, I.P.; Hrdina, R.; Saso, L. In vitro analysis of iron chelating activity of flavonoids. *J. Inorg. Biochem.* **2011**, *105*, 693-701.

[94] Lopes, G.K.; Schulman, H.M.; Hermes-Lima M. Polyphenol tannic acid inhibits hydroxyl radical formation from Fenton reaction by complexing ferrous ions. *Biochim. Biophys. Acta,* **1999**, *1472*(1-2), 142-152.

[95] Maffei-Facino, R.; Carini, M.; Aldini, G.; Bombardelli, E.; Morazzoni, P.; Morelli, R. Free radicals scavenging action and anti-enzyme activities of procyanidines from *Vitis vinifera*: a mechanism for their capillary protective action. *Arzneim-Forsch.,* **1994**, *44*(5), 592-601.

[96] Yoneda, S.; Nakatsubo, F. Effects of the hydroxylation patterns and degrees of polymerization of condensed tannins on their metal-chelating capacity. *J. Wood Chem. Technol.,* **1998**, *18*(2), 193-205.

[97] Zhang, J.; Wang, X.F.; Lu, Z.B.; Liu, N.Q.; Zhao, B.-L. The effects of meso-2,3-dimercaptosuccinic acid and oligomeric procyanidins on acute lead neurotoxicity in rat hippocampus. *Free Radic. Biol. Med.,* **2004**, *37*(7), 1037-1050.

[98] Mandel, S.; Weinreb, O.; Reznichenko, L.; Kalfon, L.; Amit, T. Green tea catechins as brain-permeable, non toxic, iron chelators to iron out iron from the brain. *J. Neural Transmis.*, **2006**, Suppl., *71*(Oxidative Stress and Neuroprotection), 249-257.

[99] Lambert, J.D.; Elias, R.J. The antioxidant and pro-oxidant activities of green tea polyphenols: a role in cancer prevention. *Arch. Biochem. Biophys.*, **2010**, *501*(1), 65-72.

[100] Soussi, A.; Murat, J.C.; Gaubin, Y.; Croute, F.; Soleilhavoup, J.P.; El Feki, A. Green tea drinking reduces the effects of vanadium poisoning in rat kidney. *Food Sci. Technol. Res.*, **2009**, *15*(4), 413-422.

[101] Hanif, S.; Shamim, U.; Ullah, M.F.; Azmi, A. S.; Bhat, S.H.; Hadi, S.M. The anthocyanidin delphinidin mobilizes endogenous copper ions from human lymphocytes leading to oxidative degradation of cellular DNA. *Toxicology*, **2008**, *249*(1), 19-25.

[102] Forester, S.C.; Lambert, J.D. The role of antioxidant versus pro-oxidant effects of green tea polyphenols in cancer prevention. *Mol. Nutr. Food Res.*, **2011**, *55*(6), 844-854.

[103] Goodman, B.A.; Severino, J. Ferreira; Pirker, K.F. Reactions of green and black teas with Cu(ii). *Food & Function*, **2012**, *3*(4), 399-409.

[104] Ishii, R.; Saito, K.; Horie, M.; Shibano, T.; Kitanaka, S.; Amano, F. Inhibitory effects of hydrolyzable tannins from *Melastoma dodecandrum* Lour. on nitric oxide production by a murine macrophage-like cell line, RAW264.7, activated with lipopolysaccharide and interferon-γ. *Biol. Pharm. Bull.,* **1999**, *22*(6), 647-653.

[105] Stevens, J.F.; Miranda, C.L.; Wolthers, K.R.; Schimerlik, M.; Deinzer, M.L.; Buhler, D.R. Identification and in vitro biological activities of hop proanthocyanidins: inhibition of nNOS activity and scavenging of reactive nitrogen species. *J. Agric. Food Chem.,* **2002,** *50*(12), 3435-3443.

[106] Fogliani, B.; Raharivelomanana, P.; Bianchini, J.P.; Bouraima-Madjebi, S.; Hnawia, E. Bioactive ellagitannins from *Cunonia macrophylla*, an endemic Cunoniaceae from New Caledonia. *Phytochemistry,* **2005**, *66*(2), 241-247.

[107] Shibata, T.; Nagayama, K.; Tanaka, R.; Yamaguchi, K.; Nakamura, T. Inhibitory effects of brown algal phlorotannins on secretory phospholipase A2s, lipoxygenases and cyclooxygenases. *J. Appl. Phycol.,* **2003**, *15*(1), 61-66.

[108] Schewe, T.; Kuhn, H.; Sies, H. Flavonoids of cocoa inhibit recombinant human 5-lipoxygenase. *J. Nutr.,* **2002**, *132*(7), 1825-1829.

[109] Moini, H.; Guo, Q.; Packer, L. Enzyme inhibition and protein-binding action of the procyanidin-rich French maritime pine bark extract, pycnogenol: effect on xanthine oxidase. *J. Agric. Food Chem.,* **2000**, *48*(11), 5630-5639.

[110] Kishido, T.; Unno, K.; Yoshida, H.; Choba, D.; Fukutomi, R.; Asahina, S.; Iguchi, K.; Oku, N.; Hoshino, M. Decline in glutathione peroxidase activity is a reason for brain senescence: consumption of green tea catechin prevents the decline in its activity and protein oxidative damage in ageing mouse brain. *Biogerontology*, **2007**, *8*(4), 423-430.

[111] Sree, T.N.P.; Kumar, S.K.; Senthilkumar, A.; Aradhyam, G.K.; Gummadi, S.N. In vitro effect of Terminalia arjuna bark extract on aditional enzyme catalase. *J. Pharmacol. Toxicol.*, **2007**, *2*(8), 698-708.

[112] Maisuthisakul, P.; Gordon, M.H. Antioxidant and tyrosinase inhibitory activity of mango seed kernel by product. *Food Chem.*, **2009**, *117*(2), 332-341.

[113] Sugimoto, K.; Nakagawa, K.; Hayashi, S.; Amakura, Y.; Yoshimura, M.; Yoshida, T.; Yamaji, R.; Nakano, Y.; Inui, H. Hydrolizable tannins as antioxidants in the leaf extract of *Eucalyptus globulus* possessing tyrosinase and hyaluronidase inhibitory activities. *Food Sci. Technol. Res.*, **2009**, *15*(3), 331-336.
[114] Anam, K.; Widharna, R.M.; Kusrini, D. α-glucosidase inhibitor activity of Terminalia species. *Int. J. Pharmacol.*, **2009**, *5*(4), 277-280.
[115] Shihabudeen, H.; Mohamed, S.; Priscilla, D.H.; Thirumurugan, K. Cinnamon extract inhibits α-glucosidase activity and dampens postprandial glucose excursion in diabetic rats. *Nutr. Metabol.*, **2011**, *8*, 46-56.
[116] da Silva Pinto, M.; Ernesto de Carvalho, J.; Lajolo, F.M.; Genovese, M.I.; Shetty, K. Evaluation of antiproliferative, anti-type 2 diabetes, and antihypertension potentials of ellaginatannins from strawberries (*Fragaria* x *ananassa* Duch.) using in vitro models. *J. Med. Food*, **2010**, *13*(5), 1027-1035.
[117] Khateeb, J.; Gantman, A.; Kreitenberg, A.J.; Aviram, M.; Fuhrman, B. Paraoxonase 1 (PON1) expression in hepatocytes is upregulated by pomegranate polyphenols: a role for PPAR-γ-pathway. *Atherosclerosis*, **2010**, *208*(1), 119-25.
[118] Wijesinghe, W.A.J.P.; Ko, S.-Ch.; Jeon, Y.-J. Effects of phlorotannins isolated from *Ecklonia cava* on angiotensin I-converting enzyme (ACE) inhibitory activity. *Nutr. Res. Pract.*, **2011**, *5*(2), 93-100.
[119] Liu, J.; Ando, R.; Shimizu, K.; Hashida, K.; Makino, R.; Ohara, S.; Kondo, R. Steroid 5α-reductase inhibitory activity of condensed tannins from woody plants. *J. Wood Sci.*, **2008**, *54*(1), 68-75.
[120] Witztum, J.L. The oxidation hypothesis of atherosclerosis. *Lancet,* **1994**, *344*(8925), 793-795.
[121] Kang J.O.; Kim S.-J.; Kim H. Effect of astaxanthin on the hepatotoxicity, lipid peroxidation and antioxidative enzymes in the liver of CCl_4-treated rats. *Methods Findings Exp. Clin. Pharmacol.*, **2001**, *23*(2), 79-84.
[122] Osakabe, N.; Yasuda, A.; Natsume, M.; Takizawa, T.; Terao, J.; Kondo, K. Catechins and their oligomers linked by C4→C8 bonds are major cacao polyphenols and protect low-density lipoprotein from oxidation in vitro. *Exp. Biol. Med.,* **2002**, *227*(1), 51-56.
[123] Seeram, N.P.; Adams, L.S.; Henning, S.M.; Niu, Y.; Zhang, Y.; Nair, M.G.; Heber, D. In vitro antiproliferative, apoptotic and antioxidant activities of punicalagin, ellagic acid and a total pomegranate tannin extract are enhanced in combination with other polyphenols as found in pomegranate juice. *J. Nutr. Biochem.,* **2005**, *16*(6), 360-367.
[124] Sanchez-Moreno, C.; Jimenez-Escrig, A.; Saura-Calixto, F. Study of low-density lipoprotein oxidizability indexes to measure the antioxidant activity of dietary polyphenols. *Nutr. Res.,* **2000**, *20*(7), 941-953.
[125] Hashimoto, F.; Ono, M.; Masuoka, C.; Ito, Y.; Sakata, Y.; Shimizu, K.; Nonaka, G.I.; Nishioka, I.; Nohara, T. Evaluation of the anti-oxidative effect (in vitro) of tea polyphenols. *Biosci. Biotechnol. Biochem.,* **2003**, *67*(2), 396-401.
[126] Lorenz, M.; Urban, J.; Engelhardt, U.; Baumann, G.; Stangl, K.; Stangl, V. Green and black tea are equally potent stimuli of NO production and vasodilation: new insights into tea ingredients involved. *Basic Res. Cardiol.*, **2009**, *104*(1), 100-110.
[127] Namkung, W.; Thiagarajah, J.R.; Phuan, P.-W.; Verkman, A.S. Inhibition of Ca^{2+} activated Cl^- channels by gallotannins as a possible molecular basis for health benefits of red wine and green tea. *Offic. Publ. Feder. Am. Soc. Exp. Biol.*, **2010**, *24*(11), 4178-86.

[128] Shah, K.A.; Patel, M. B.; Shah, S.S.; Chauhan, K.N.; Parmar, P.K.; Patel, N.M. Antihyperlipidemic activity of Mangifera indica I. Leaf extract on rats fed with high cholesterol diet. *Pharm. Sin.*, **2010**, *1*(2), 156-161.

[129] Heber, D.; Seeram, N.P.; Wyatt, H.; Henning, S.M.; Zhang, Y.; Ogden, L.G.; Dreher, M.; Hill, J.O. Safety and antioxidant activity of a pomegranate ellagitannin-enriched polyphenol dietary supplement in overweight individuals with increased waist size. *J. Agric. Food Chem.*, **2007**, *55*(24), 10050-10054.

[130] Fitzpatrick, D.F.; Bing, B.; Maggi, D.A.; Fleming, R.C.; O'Malley, R.M. Vasodilating procyanidins derived from grape seeds. *Ann. N. Y. Acad. Sci.,* **2002**, *957*, 78-89.

[131] Caballero-George, C.; Vanderheyden, P.M.L.; De Bruyne, T.; Shahat, A.A.; Van Den Heuvel, H.; Solis, P.N.; Gupta, M.P.; Claeys, M.; Pieters, L.; Vauquelin, G.; Vlietinck, A.J. In vitro inhibition of [3H]-angiotensin II binding on the human AT1 receptor by proanthocyanidins from *Guazuma ulmifolia* bark. *Planta Med.,* **2002**, *68*(12), 1066-1071.

[132] Ottaviani, J.I.; Actis-Goretta, L.; Villordo, J.J.; Fraga, C.G. Procyanidin structure defines the extent and specificity of angiotensin I converting enzyme inhibition. *Biochimie,* **2006**, *88*(3-4), 359-365.

[133] Pearson, D.A.; Paglieroni, T.G.; Rein, D.; Wun, T.; Schramm, D.D.; Wang, J.F.; Holt, R.R.; Gosselin, R.; Schmitz, H.H.; Keen, C.L. The effects of flavanol-rich cocoa and aspirin on ex vivo platelet function. *Thromb. Res.,* **2002**, *106*(4-5), 191-197.

[134] Holt, R.R.; Schramm, D.D.; Keen, C.L.; Lazarus, S.A.; Schmitz, H.H. Chocolate consumption and platelet function. *JAMA,* **2002**, *287*(17), 2212-2213.

[135] Mekhfi, H.; El Haouari, M.; Bnouham, M.; Aziz, M.; Ziyyat, A.; Legssyer, A. Effects of extracts and tannins from Arbutus unedo leaves on rat platelet aggregation. *Phytother. Res.,* **2006**, *20*(2), 135-139.

[136] Makdessi, S.A.; Sweidan, H.; Jacob, R. Effect of oligomer procyanidins on reperfusion arrhythmias and lactate dehydrogenase release in the isolated rat heart. *Arzneim-Forsch.*, **2006**, *56*(5), 317-321.

[137] Gandrak, P.G.; Gaikwad, A.B.; Tikoo, K. Gallotannin ameliorates the development of streptozotocin-induced diabetic nephropathy by preventing the activation of PARP. *Phytotherapy Res.,* **2009**, *23*(1), 72-77.

[138] Kanaujia, A.; Duggar, R.; Pannakal, S.T.; Yadav, S.S.; Katiyar, Ch.K.; Bansal, V.; Anand, S.; Sujatha, S.; Lakshmi, B.S. Insulinomimetic activity of two new gallotannins from the fruits of *Capparis moonii. Bioorg. Med. Chem.*, **2010**, *18*(11), 3940-3945.

[139] Paramesha, M.; Ramesh, C.K.; Krishna, V.; Parvathi, K.M.M.; Kuppast, I.J. Antihyperglycemic activity of methanolic extract of *Carthamus tinctorius* L. annigere-2. *Asian J. Exp. Sci.*, **2009**, *23*(3), 497-502.

[140] Ghosh, G.; Kar, D.M.; Subudhi, B.B.; Mishra, S.K. Anti-hyperglycemic and antioxidant activity of stem bark of *Polyathia longifolia* var. *angustifolia. Pharm. Lettre*, **2010**, *2*(2), 206-216.

[141] Sati, S.Ch.; Sati, N.; Sati, O.P. Bioactive constituents and medicinal importance of genus Alnus. *Pharmacognosy Rev.*, **2011**, *5*(10), 174-183.

[142] Sarg, T.; Abdel Ghani, A.; Zayed, R.; El-Sayed, M. Antihepatotoxic activity of *Phyllanthus atropurpureus* cultivated in Egypt. *Z. Naturforsch. C*, **2011**, *66*(9/10), 447-452.

[143] Biswas, M.; Karan, T.K.; Kar, B.; Bhattacharya, S.; Ghosh, A.K.; Kumar, R.B. Suresh; H.P.K. Hepatoprotective activity of *Terminalia arjuna* leaf against paracetamol-induced liver damage in rats. *Asian J. Chem.*, **2011**, *23*(4), 1739-1742.

[144] Blaschek, W.; Hilgenfeldt, U.; Holzgrabe, U.; Reichling, F.; Ruth, P.; Schulz, V. *Hager's Enzyklopädie der Arzneistoffe und Drogen*; HagerROM 2010: Springer, Stuttgart **2011**.

[145] Adams, L.S.; Seeram, N.P.; Aggarwal, B.B.; Takada, Y.; Sand, D.; Heber, D. Pomegranate juice, total pomegranate ellagitannins and punicalagin suppress inflammatory cell signaling in colon cancer cells. *J. Agric. Food Chem.,* **2006**, *54*(3), 980-985.

[146] Oh, G.S.; Pae, H.O.; Choi, B.M.; Lee, H.S.; Kim, I.K.; Yun, Y.G.; Kim, J.D.; Chung, H.T. Penta-O-galloyl-β-D-glucose inhibits phorbol myristate acetate-induced interleukin-8 gene expression in human monocytic U937 cells through its innactivation of nuclear factor-kB. *Int. Immunopharmacol.,* **2004**, *4*(3), 377-386.

[147] Park, E.K.; Kim, M.S.; Lee, S.H.; Kim, K.H.; Park, J.Y.; Kim, T.H.; Lee, I.S.; Woo, J.T.; Jung, J.C.; Shin, H.I.; Choi, J.Y.; Kim, S.Y. Furosin, an ellagitannin, suppresses RANKL-induced osteoclast differentiation and function through inhibition of MAP kinase activation and actin ring formation. *Biochem. Biophys. Res. Commun.,* **2004**, *325*(4), 1472-1480.

[148] Park, Y.C.; Rimbach, G.; Saliou, C.; Valacchi, G.; Packer, L. Activity of monomeric, dimeric, and trimeric flavonoids on NO production, TNF-α secretion, and NF-κB-dependent gene expression in RAW 264.7 macrophages. *FEBS Lett.,* **2000**, *465*(2-3), 93-97.

[149] Afaq, F.; Saleem, M.; Krueger, C.G.; Reed, J.D.; Mukhtar, H. Anthocyanin- and hydrolyzable tannin-rich pomegranate fruit extract modulates MAPK and NF-κB pathways and inhibits skin tumorigenesis in CD-1 mice. *Int. J. Cancer.,* **2005**, *113*(3), 423-433.

[150] Lee, Ch.S.; Jang, E.R.; Kim, Y.-J.; Seo, S.J.; Choi, S.E.; Lee, M.W. Polyphenol acertannin prevents TRAIL-induced apoptosis in human keratinocytes by suppressing apoptosis-related protein activation. *Chem.-Biol. Interact.*, **2011**, *189*(1-2), 52-59.

[151] Hegde, K.; Koshy, S; Joshi, A.B. Anti-nociceptive, anti-inflammatory and antiarthritic activity of *Carissa carandas* root extract. *Int. J. Pharmacol. Biol. Sci.*, **2010**, *4*(2), 25-34.

[152] Kim, M.-S.; Park, S.-B.; Suk, K.; Kim, I.K.; Kim, S.-Y.; Kim, J.-A.; Lee, S.H.; Kim, S.-H. Gallotannin isolated from Euphorbia species, 1,2,6-tri-O-galloyl-β-D-allose, decreases nitric oxide production through inhibition of nuclear factor-κB and downstream inducible nitric oxide synthase. *Biol. Pharm. Bull.*, **2009**, *32*(6), 1053-1056.

[153] Park, S.-B.; Kim, M.-S.; Lee, H.S.; Lee, S.H.; Kim, S.-H. 1,2,3,6-tetra-O-galloyl-ß-D-allopyranose gallotannin isolated from *Euphorbia jolkini*, attenuates LPS-induced nitric oxide production in macrophages. *Phytother. Res.*, **2010**, *24*(9), 1329-1333.

[154] Rasheed, Z.; Akhtar, N.; Anbazhagan, A.N.; Ramamurthy, S.; Shukla, M.; Haqqi, T.M. Polyphenol-rich pomegranate fruit extract (POMx) suppressed PMACI-induced expression of pro-inflemmatory cytokines by inhibiting the activation of MAP kinases and NF-κB in human KU812 cells. *J. Inflamm.*, **2009**, (6), No pp. given; Chem. Abstr. **2009**:922362.

[155] Sugiura, Y.; Imai, K.; Amano, H. The anti-allergic and anti-inflammatory effects of seaweed polyphenol (phlorotannin). *Foods & Food Ingred. J. Japan*, **2011**, *216*(1), 46-55.

[156] Brito, F.A.; Lima, L.A.; Ramos, M.F.S.; Nakamura, M.J.; Cavalher-Machado, S.C.; Siani, A.C.; Henriques, M.G.M.O.; Sampaio, A.L.F. Pharmacological study of anti-allergic activity of Syzygium cumini (L.) skeels. *Brazil. J. Med. Biol. Res.*, **2007**, *40*(1), 105-115.

[157] Kim, Y.J.; Park, H.J.; Yoon, S.H.; Kim, M.J.; Leem, K.H.; Chung, J.H.; Kim, H.-K. Anticancer effects of oligomeric proanthocyanidins on human colorectal cancer cell line, SNU-C4. *World J. Gastroenterol.,* **2005**, *11*(30), 4674-4678.

[158] Kasimsetty, S.G.; Bialonska, D.; Reddy, M.K.; Ma, G.; Khan, S.I.; Ferreira, D. Colon cancer chemopreventive activities of pomegranate ellagitannins and urolithins. *J. Agric. Food Chem.*, **2010**, *58*(4), 2180-2187.

[159] Sanchez-Tena, S.; Fernandez-Cachon, M.L.; Carreras, A.; Mateos-Martin, M.L.; Costoya, N.; Moyer, M.P.; Nunez, M.J.; Torres, J.L.; Cascante, M. Hamamelitannin from witch hazel (*Hamamelis virginiana*) displays specific cytotoxic activity against colon cancer cells. *J. Nat. Prod.*, **2012**, *75*(1), 26-33.

[160] Smith, S.H.; Tate, P.L.; Huang, G.; Magee, J.B.; Meepagala, K.M.; Wedge, D.E.; Larcom, L.L. Antimutagenic activity of berry extracts. *J. Med. Food,* **2004**, *7*(4), 450-455.

[161] Yamagishi, M.; Natsume, M.; Osakabe, N.; Nakamura, H.; Furukawa, F.; Imazawa, T.; Nishikawa, A.; Hirose, M. Effects of cacao liquor proanthocyanidins on PhIP-induced mutagenesis in vitro, and in vivo mammary and pancreatic tumori genesis in female Sprague-Dawley rats. *Cancer Lett.,* **2002**, *185*(2), 123-130.

[162] Cao, P.; Cai, J.; Gupta, R.C. Effect of green tea catechins and hydrolyzable tannins on benzo[a]pyrene-induced DNA adducts and structure-activity relationship. *Chem. Res. Toxicol.*, **2010**, *23*(4), 771-777.

[163] Li, W.; Zhao, Y.; Yang, Y.; Zhang, Z.; Lai, J.; Zhuang, L. RP-HPLC with UV switch determination of 9 components in white peony root pieces. *Yaowu Fenxi Zazhi*, **2011**, *31*(12), 2208-2212; Chem. Abstr. **2012**:314476.

[164] Zhang, J.; Li, L.; Kim, S.-H.; Hagerman, A.E.; Lu, J. Anti-cancer, anti-diabetic and other pharmacologic and biological activities of penta-galloyl-glucose. *Pharm. Res.*, **2009**, *26*(9), 2066-2080, (2009).

[165] Miura, T.; Chiba, M.; Kasai, K.; Nozaka, H.; Nakamura, T.; Shoji, T.; Kanda, T.; Ohtake, Y.; Sato, T. Apple procyanidins induce tumor cell apoptosis though mitochondrial pathway activation of caspase-3. *Carcinogenesis,* **2008**, *29*(3), 585-593.

[166] Ksouri, R.; Falleh, H.; Megdiche, W.; Trabelsi, N.; Mhamdi, B.; Chaieb, K.; Bakrouf, A.; Magne, C.; Abdelly, Ch. Antioxidant and antimicrobial activities of the edible medicinal halophyte Tamarix gallica L. and related polyphenolic constituents. *Food Chem. Toxicol.*, **2009**, *47*(8), 2083-2091.

[167] Morais, M.C.C.; Luqman, S.; Kondratyuk, T. P.; Petronio, M.S.; Regasini, S.O.; Silva, D.H.S.; Bolzani, V.S.; Sares, Ch.P.; Pezzuto, J. Suppresion of THF-α induced NFκB activity by gallic acid and its semi-synthetic esters: possible role in cancer chemoprevention. *Nat. Prod. Res.*, **2010**, *24*(18), 1758-1765.

[168] Kandil, F.E.; Smith, M.A.; Rogers, R.B.; Pepin, M.F.; Song, L.L.; Pezzuto, J.M.; Seigler, D.S. Composition of a chemoprotective proanthocyanidin-rich fraction from cranberry fruits responsible for the inhibition of 12-O-tetradecanoyl phorbol-13-acetate (TPA)-induced ornithine decarboxylase (ODC) activity. *J. Agric. Food Chem.,* **2002,** *50*(5), 1063-1069.

[169] Chung, K.T.; Wei, C.I.; Johnson, M.G. Are tannins a double-edged sword in biology and health? *Trends Food Sci. Tech.,* **1998**, *9*(4), 168-175.

[170] Jayaprakasha, G.K.; Selvi, T.; Sakariah, K.K. Antibacterial and antioxidant activities of grape (*Vitis vinifera*) seed extract. *Food Res. Int.,* **2003**, *36*(2), 117-122.

[171] Fukuchi, K.; Sakagami, H.; Okuda, T.; Hatano, T.; Tanuma, S.; Kitajima, K.; Inoue, Y.; Inoue, S.; Ichikawa, S.; Nonomiya, M.; Konno, K. Inhibition of herpex simpex virus infection by tannins and related compounds. *Antiviral Res.,* **1989**, *11*(5-6), 285-97.

[172] Sakagami, H.; Satoh, K.; Ida, Y.; Koyama, N.; Premanathan, M.; Arakaki, R.; Nakashima, H.; Hatano, T.; Okuda, T.; Yoshida, T. Induction of apoptosis and anti-HIV activity by tannin- and lignin-related substances. *Basic Life Sci.,* **1999***, 66*, 595-611.

[173] Liu G.; Xiong, S.; Xiang, Y.-F.; Guo, Ch.-W.; Ge, F.; Yang, Ch.-R.; Zhang, Y.-J.; Wang, Y.-F.; Kitazato, K. Antiviral activity and possible mechanism of action of

pentagylloylglucose (PGG) against influenza A virus. *Arch. Virol.*, **2011**, *156*(8), 1359-1369.

[174] Yang, Ch.-M., Cheng, H-Y., Lin, T.-Ch., Chiang, L.-Ch.; Li, Ch.-Ch. The in vitro activity of geraniin and 1,3,4,6-tetra-O-galloyl-ß-D-glucose isolated from Phyllanthus urinaria against herpes simpex virus type 1 and type 2 infection. *J. Ethnopharmacol.*, **2007**, *110*(3), 555-558.

[175] Yang ,Y.; Zhang, L.; Fan, X.; Qin, Ch.; Liu, J. Antiviral effects of geraniin on human enterovirus 71 in vitro and in vivo. *Bioorg. Med. Chem. Lett.*, **2012**, *22*(6), 2209-2211.

[176] Yoshida, T.; Hatano, T.; Ito, H. Geraniin-related ellagitannins of the Euphorbiaceae. *Int. Congr. Ser.*, **1998**, *1157*, 549-560; Chem. Abstr. **1999**:30535.

[177] Khan, Z.S.; Shinde, V.N.; Bhosle, N.P.; Nasreen, S. Chemical composition and antimicrobial activity of Angiospermic plants. *Middle East J. Sci. Res.*, **2010**, *6*(1), 56-61.

[178] Al-Zoreky, N.S. Antimicrobial activity of pomegranate (*Punica granatum* L.) fruit peels. *Int. J. Food Microbiol.*, **2009**, *134*(3), 244-248.

[179] Engels, Ch.; Knodler, M.; Zhao, Y.-Y.; Carle, R.; Ganzle, M.G.; Schieber, A. Antimicrobial activity of gallotannins isolated from mango (*Mangifera indica* L.) kernels. *J. Agric. Food Chem.*, **2009**, *57*(17), 7712-7718.

[180] Katariya, P.K.; Mathur, M.; Yadav, S.; Kamal, R. Phytochemical and antimicrobial screening of *Momordica dioica* Roxb. and *Moringa oleifera* Lam. *Asian J. Exp. Sci.*, **2010**, *24*(2), 263-267.

[181] Liao, S.-G.; Zhang, L.-J.; Sun, F.; Zhang, J.-J.; Chen, A.-Y.; Lan, Y.-Y.; Li, Y.-J.; Wang, A.-M.; He, X.; Xiong, Y.; Dong, L.; Chen, X.-J., Li, Y.-T., Zuo, L., Wang, Y.-L. Antibacterial and anti-inflammatory effects of extracts and fractions from *Polygonum capitatum*. *J. Ethnopharmacol.*, **2011**, *134*(3), 1006-1009.

[182] Ramasubraramaniaraja, R.; Babu, M.N. Pharmacognostical phytochemical and antibacterial (Gram-positive and Gram-negative pathogens) evaluation of ethanolic leaf extract of *Abutilon indicum* (Linn). *J. Pharm. Res.*, **2011**, *4*(8), 2500-2503.

[183] Singh, K.; Tiwari, V.; Prajapat, R. Study of antimicrobial activity of medicinal plants against various multiple drug resistance pathogens and their molecular characterization and it´s bioinformatics analysis of antibiotic gene from genomic database with degenerate primer prediction. *Int. J. Biol. Technol.*, **2010**, *1*(2), 15-19.

[184] Hutchinson, J. Do cranberries help prevent urinary tract infections? *Nurs. Times,* **2005**, *101*(47), 38-40.

[185] Foo, L.Y.; Lu, Y.; Howell, A.B.; Vorsa, N. A-type proanthocyanidin trimers from cranberry that inhibit adherence of uropathogenic P-fimbriated Escherichia coli. *Nat. Prod.,* **2000**, *63*(9), 1225-1228.

[186] Meyer, A.L. Health-promoting properties of the American cranberry (*Vaccinium macrocarpon*) and products thereof: urinary tract infections and other aspects. *Ernaehrung*, **2011**, *35*(3), 111-114.

[187] Rossi, R.; Porta, S.; Canovi, B. Overview on cranberry and urinary tract infections in females. *J. Clin. Gastroenterol.*, **2010**, *44*(Suppl. 1), S61-S62.

[188] Cote, J.; Caillet, S.; Doyon, G.; Sylvain, J.-F.; Lacroix, M. Bioactive compounds in cranberries and their biological properties. *Crit. Rev. Food Sci. Nutr.*, **2010**, *50*(7), 666-679.

[189] Ogundiya, M.O.; Kolapo, A L.; Okunade M.B.; Adejumobi, J.A. Assessment of phytochemical composition and antimicrobial activity of *Terminalia glaucescens* against some oral pathogens. *EJEAFChe, Electron. J. Environ., Agric. Food Chem.,* **2009**, *8*(7), 466-471.

[190] Tomczyk, M.; Pleszczynska, M.; Wiater, A. Variation in total polyphenolics content of aerial parts of Potentilla species and their anticariogenic activity. *Molecules*, **2010**, *15*, 4639-4651.

[191] Knoll, C.; Divol, B.; du Toit, M. Influence of phenolic compounds on activity of nisin and pediocin PA-1. *Am. J. Enol. Viticult.*, **2008**, *59*(4), 418-421.

[192] Shahat, A.A.; Cos, P.; De Bruyne, T.; Apers, S.; Hammouda, F.M.; Ismail, S.I.; Azzam, S.; Claeys, M.; Goovaerts, E.; Pieters, L.; Vanden Berghe, D.; Vlietinck, A.J. Antiviral and antioxidant activity of flavonoids and proanthocyanidins from Crataegus sinaica. *Planta Med.,* **2002**, *68*(6), 539-541.

[193] Harsha, N.S.; Kumar, G.S.; Al-wesali, M.S.; Rasool, S.T.; Ibrahim, F.A.S. The chronotropic and inotropic effects of aqueous-ethanolic extract of *Achillea millefolium* on rat´s isolated heart. *Pharmacologyonline*, **2009**, (3), 791-799.

[194] Chung, K.T.; Wong, T.Y.; Wei, C.I.; Huang, Y.W.; Lin, Y. Tannins and human health: A review. *Crit. Rev. Food Sci. Nutr.,* **1998**, *38*(6), 421-464.

[195] Rao, A.V.; Snyder, D.M. Raspberries and human health: A review. *J. Agric. Food Chem.*, **2010**, *58*(7), 3871-3883.

[196] Giampieri, F.; Tulipani, S.; Alvarez-Suarez, J. M; Quiles, J.L; Mezzetti, B.; Battino, M. The strawberry: composition, nutritional quality, and impact on human health. *Nutrition*, **2012**, *28*(1), 9-19.

[197] Williamson, G.; Carughi, A. Polyphenol content and health benefits of raisins. *Nutr. Res.*, **2010**, *30*(8), 511-519.

[198] Habauzit, V.; Horcajada, M.-N. Phenolic phytochemicals and bone. *Phytochemistry Rev.*, **2008**, *7*(2), 313-344.

[199] Frei, B.; Higdon, J.V. Antioxidant activity of tea polyphenols in vivo: evidence from animal studies *J. Nutr.*, **2003**, *133*(10), 3275S-3284S.

[200] Kolodziej, H.; Kiderlen, A.F. Antileishmanial activity and immune modulatory effects of tannins and related compounds on Leshmania parasitized RAW 264.7 cells. *Phytochemistry,* **2005**, *66*(17), 2056-2071.

[201] Haslam, E. In: *Handbook of Plant Science*; Keith, R., Ed.; John Wiley & Sons Ltd.: Chichester, **2007**; Vol. *2*, pp. 984-987.

[202] Goncalves, R.; Mateus, N.; Pianet, I.; Laguerre, M.; de Freitas, V. Mechanism of tannin-induced trypsin inhibition: a molecular approach. *Langmuir*, **2011**, *27*(21), 13122-13129.

[203] Goncalves, R.; Mateus, N.; de Freitas, V. Inhibition of α-amylase activity by condensed tannins. *Food Chem.*, **2010**, *125*(2), 665-672.

[204] Williamson, G.; Manach, C. Bioavailability and bioefficacy of polyphenols in humans. II. Review of 93 intervention studies. *Am. J. Clin. Nutr.,* **2005**, *81*(1), 243-255.

[205] Dixit, A.K.; Shekhawat, M.S. A review on carcinogenic potential hazards of herbal medications. *Nat. Environ. Pollution Technol.*, **2007**, *6*(2), 343-350.

[206] Oterdoom, H.J. Tannin, sorghum, and oesophageal cancer. *Lancet,* **1985**, *2*(8450), 330.

[207] Chung, K.T.; Wei, C.I.; Johnson, M.G. Are tannins a double-edged sword in biology and health? *Trends Food Sci. Technol.,* **1998**, *9*(4), 168-175.

[208] Li, Y.X.; Wijesekara, I.L.Y.; Kim, S-K. Phlorotannins as bioactive agents from brown algae. *Process Biochem.*, **2011**, *46*(12), 2219-2224.

[209] Clark, H.; Kelliher, F.; Pinares-Patino, C. Reducing CH_4 emissions from ruminants in New Zealand: challenges and opportunities. *Asian-Australasian J. Anim. Sci.*, **2011**, *24*(2), 295-302.

[210] Kamra, D.N.; Agarwal, N.; Chaudhary, L.C. Inhibition of ruminal methanogenesis by tropical plants containing secondary compounds. *Int. Congr. Ser.*, **2006**, *1293*, 156-163.

[211] Patra, A.K.; Saxena, J. Dietary phytochemicals as rumen modifiers: a review of the effects on microbial populations. *Antonie van Leeuwenhoek*, **2009**, *96*(4), 363-375.

[212] Martin, C.; Morgavi, D.P.; Doreau, M. Methane mitigation in ruminants: from microbe to the farm scale. *Animal*, **2010**, *4*(3), 351-365.

[213] Patra, A.K.; Saxena, J. A new perspective on the use of plant secondary metabolites to inhibit methanogenesis in the rumen. *Phytochemistry*, **2010**, *71*(11-12), 1198-1222.

[214] Patra, A.K.; Saxena, J. Exploitation of dietary tannins to improve rumen metabolism and ruminant nutrition. *J. Sci. Food Agric.*, **2011**, *91*(1), 24-37.

Send Orders for Reprints to reprints@benthamscience.net
Recent Advances in Medicinal Chemistry, Vol. 1, **2014, 209-245** **209**

CHAPTER 7

Rho-GTPases and Statins: A Potential Target and a Potential Therapeutic Tool Against Tumors?

Ivana Campia, Sophie Doublier, Elisabetta Aldieri, Amalia Bosia, Dario Ghigo and Chiara Riganti*

Department of Oncology, Biology and Biochemistry, Research Center on Experimental Medicine (CeRMS), University of Torino, Via Santena 5/bis, 10126 Torino, Italy

Abstract: Rho GTPases, which control processes such as cell proliferation and cytoskeleton remodeling, are often hyperexpressed in tumors. Several members, such as RhoA/B/C, must be isoprenylated to interact with their effectors. Statins, by inhibiting the synthesis of prenyl groups, may affect RhoA/B/C activity and represent a promising tool in anticancer therapy.

Keywords: Cancer, chemotherapy, isoprenylation, rho GTPases, statins.

INTRODUCTION

Rho GTPases belong to the Ras superfamily of low molecular weight (MW 20-30 kDa) monomeric GTP-binding proteins and are found in all eukaryotic cells [1-4]. Until now, twenty mammalian genes encoding Rho GTPases have been described [4-6]. The most investigated members are Rho (Ras homologous), Rac (Ras-related C3 botulinum toxin substrate) and Cdc42 (cell division cycle 42). In this chapter we have focused our attention on the RhoA, RhoB and RhoC isoforms and on the effects of statins on Rho activity in human tumors. Similar to other regulatory GTPases, Rho proteins act as molecular switches cycling between an inactive GDP-bound state and an active GTP-bound state: in their GTP-bound form the Rho GTPases are localized at membranes and are able to interact with effector molecules initiating downstream responses. Their intrinsic GTPase activity turns the proteins back into the GDP-bound state thereby terminating signal delivery [2]. The activation of growth factor receptors and integrins can

***Address correspondence to Chara Riganti:** Department of Oncology, Via Santena 5/bis, 10126 Torino, Italy; Tel: +390116705857; Fax: +300116705845; E-mail: chiara.riganti@unito.it

Atta-ur-Rahman, Muhammad Iqbal Choudhary and George Perry (Eds)

promote the exchange of GDP for GTP on Rho proteins: among the upstream activating agonists, we can mention epidermal growth factor (EGF), hepatocyte growth factor (HGF), lysophosphatidic acid (LPA), platelet-derived growth factor (PDGF), transforming growth factor-β (TGF-β), int-1/wingless (WNT1) [7]. The cycling between the GTP- and GDP-bound states is regulated by three types of regulatory proteins: (a) guanine nucleotide exchange factors (GEFs), which catalyze the exchange of GDP for GTP to activate the switch [8]; (b) GTPase-activating proteins (GAPs), which stimulate the intrinsic GTPase activity to inactivate the switch [9]; and (c) guanine nucleotide dissociation inhibitors (GDIs), which, by binding many (but not all) Rho proteins, prevent their spontaneous activation in the cytosol [10] and favor their removal from the membranes at the end of the signaling process [11]. Besides activating Rho GTPases, GEFs participate also in the selection of downstream effectors [12]. To perform their biological functions, most Rho proteins have to dock onto cell membranes, by means of a lipid moiety, either a geranylgeranyl or farnesyl residue, attached to the cystein of the C-terminal CAAX box (C = Cys, A = aliphatic amino acid, X = any amino acid) [2, 13], a process catalyzed in the cytoplasm by either geranylgeranyltransferases or farnesyltransferases, respectively [14]. The majority of Rho family proteins (*i.e.* RhoA, RhoC, Rac1, Cdc42, Rab, Rap1A) are geranylgeranylated, while only few members, such as RhoB, RhoD, Rnd, are farnesylated. Rho B has a unique behavior amongst Rho family members, since it may be geranylgeranylated as well as farnesylated; moreover it has an additional tail of palmitic acid [5]. The attachment of the isoprenyl group to the CAAX box promotes the translocation of the GTPases to the endoplasmic reticulum, where the AAX tripeptide tail is cleaved and the new C terminus is methylated. Following full processing, GTPases are directed to their cellular location, which is often the cytoplasmic surface of cell membranes, through mechanisms that are still poorly understood [15]. The Rho-specific GDI (RhoGDI) plays an important role in this regulatory context, because it masks the isoprenyl group, thereby promoting the cytosolic sequestration of Rho [10, 16]. Finally, Rho GTPases can be regulated through direct serine phosphorylation or ubiquitination, but the meaning of these covalent modifications in normal physiology is still unclear [4].

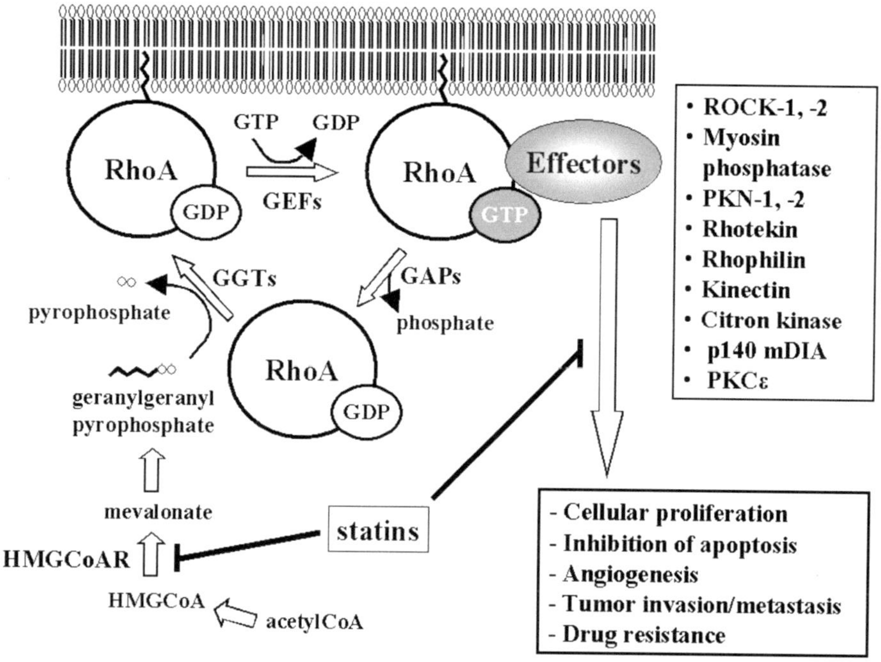

Figure 1: Schematic representation of the activation/inactivation cycle of the small GTPase RhoA, of the ultimate effects of RhoA activation and of the site of action of statins. The mechanism by which Rho GTPases lose the prenyl chain during the cycle is still poorly known. Abbreviations: GAPs: GTPase-activating proteins; GDIs: guanine nucleotide dissociation inhibitors; GEFs: guanine nucleotide exchange factors; GGT: geranylgeranyl transferase; HMGCoA: 3-hydroxy-3-methylglutaryl coenzyme A; HMGCoAR: 3-hydroxy-3-methylglutaryl coenzyme A reductase; PKC: protein kinase C; PKN: protein kinase N; ROCK: Rho-kinase.

Activated Rho GTPases interact with a large number of effector molecules that, in turn, lead to the stimulation of signaling cascades promoting general cellular responses, such as cell migration, cell adhesion, cell polarity, gene expression, cell cycle progression and transformation, cell survival, secretion, phagocytosis, endocytosis and NADPH oxidase activation [3, 4]. RhoA is ubiquitous and seems to be strongly involved in all these cellular processes (Fig. **1**). Also RhoB and RhoC proteins, which show a 85% homology with RhoA and are expressed in a great number of human tissues [5], regulate cell proliferation, polarity and migration [7, 17]. It is widely thought that Rho proteins may contribute to cancer due to their effects on cell migration (influencing invasion and metastasis) and proliferation (favoring the cell survival and growth), but, in contrast to the oncogenic Ras proteins (N-Ras, H-Ras, K-Ras), which are frequently mutated in human cancers, until now there are no reports of mutated, constitutively active forms of Rho proteins in tumors [7]. Only in haematopoietic cells of patients affected by non-Hodgkin's lymphoma it has been shown that RhoH gene is often mutated and rearranged, but it is not clear if this gene translocation may contribute to the onset and progression of the disease [17, 18]. However, recent works have shown that several Rho proteins are overexpressed in human tumors and in some cases such increased expression is associated with a poor clinical outcome [7, 18].

ROLE OF RhoA IN NORMAL AND TUMOR CELLS

RhoA is a 21-kDa protein containing 193 amino acids. Crystal structure-based comparative analysis of GDP- *versus* GTP-bound Rho revealed conformational differences in two surface regions of the N-terminal half: *Switch region 1* and *Switch region 2*. These two domains interact with GDP or GTP, as well as with Rho-specific GEF [19]: in the GDP-bound protein, the Switch 2 region is close onto Switch 1 and has a disordered conformation. The binding of Rho-GEF to Switch 2 domain causes extensive conformational changes, facilitating the loss of GDP and unmasking the binding site for GTP. Aminoacidic residues involved in GTP binding lay on both Switch 1 and Switch 2 regions [19] (Fig. **2**). The N-terminal half of RhoA contains the majority of the amino acids involved in GTP binding and hydrolysis, together with the Switch 1 and 2 regions [2]. The C-terminus of RhoA is essential for the correct localization of the protein, which is subsequent to the post-translational geranylgeranylation or farnesylation of the C-terminal cysteine [14, 15]. In addition, the C-terminal peptide of RhoA has been recently indicated as an allosteric activator of AGAP proteins, a class of GAPs that recognize Arf proteins as substrates [20]. The activation of Arf proteins by AGAP controls membrane trafficking and actin organization.

RhoA usually shuttles between cytosol and plasma membrane, RhoB may localize on plasma membrane and endosomal vesicles, RhoC may be cytosolic or associated to perinuclear structure [5]. RhoA is a target for several bacterial toxins, which modify key conserved amino acids involved in its regulation [21]. *Clostridium botulinum* exoenzyme C3 transferase specifically ADP-ribosylates RhoA at asparagine-41, inhibiting its biological activity, probably by stabilizing the Rho/GDI complex and inhibiting the GEF-mediated nucleotide exchange of RhoA [22]. The large toxins A and B from *Clostridium difficile* block the RhoA interaction with downstream effectors by glycosylating the protein at threonine-37 [21].

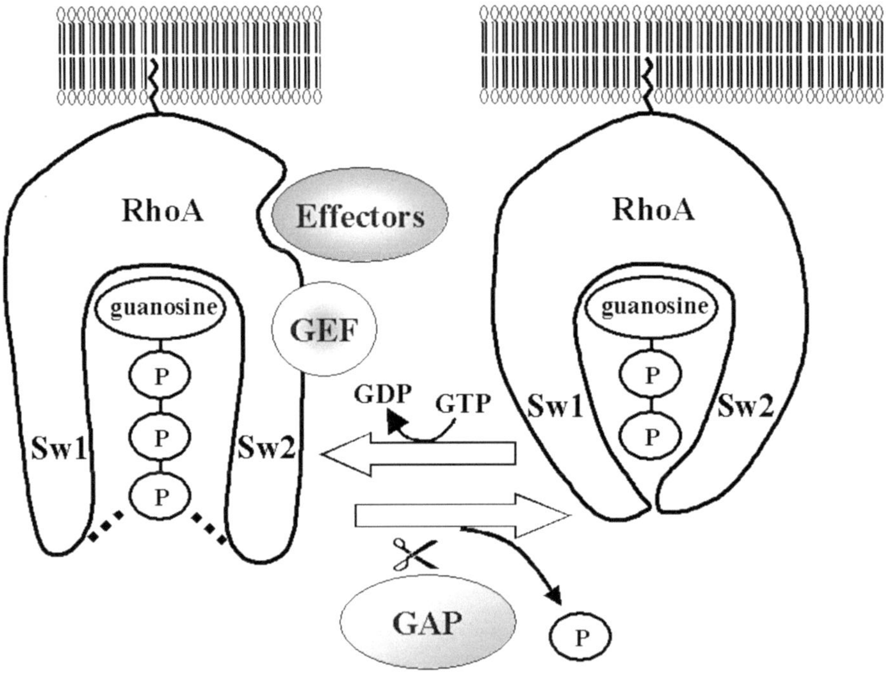

Figure 2: Role of Switch 1 and Switch 2 regions in the RhoA cycle. When bound to GDP, RhoA is in a "closed" conformation, with the Switch 2 region laying down on Switch 1 and avoiding any interaction with GTP or effectors. The binding of Rho-specific GEF to the Switch 2 domain modifies the shape of RhoA into an "open" conformation, which favors the loss of GDP and unmasks the binding site for GTP and downstream effectors. Following the action of Rho-specific GAP, GTP is hydrolysed into GDP and the protein returns in the "closed" conformation. Abbreviations: GAP: GTPase-activating protein; GEF: guanine nucleotide exchange factor; Sw1 and Sw2: Switch domains 1 and 2, respectively.

RhoA and RhoC mRNA and protein are constitutively expressed during the cell cycle; on the opposite, the amount of RhoB protein is usually low, increasing during the G1/S phase transition, and is upregulated by growth factors [5]. Activated RhoA interacts with several effector molecules including Rho-kinases (ROCK or ROK) 1 and 2, the myosin-binding subunit (MBS) of myosin phosphatase, protein kinase N (PKN) 1 and 2, rhotekin, rhophilin, kinectin, citron kinase, Lim kinase, p76RBE, protein kinase C (PKC)ε, p140 mDIA and DB1 transcription factor [2, 4, 23]. Similarly to GEFs and GAPs, effectors bind to RhoA through the Switch 1 and 2 regions, but the amino acids involved in the interaction with each target are different [2]. Although the downstream effectors of Rho proteins are often similar, slight differences exist among RhoA, RhoB and RhoC concerning their binding to specific GEF [24] or GAP proteins [25]. Furthermore, it has been reported that RhoC interacts with ROCK more efficiently than RhoA [26]. p120 β-catenin, a cytosolic effector of E-cadherin, can recruit and control the activation of ROCK1, which increases the actin polymerization, or of RhoAGAP, which turns off RhoA activity and ROCK1 effects [27]. This cycling represents the first known feed-back mechanism that controls the activity of RhoA and overcomes the schematic division into RhoA downstream and upstream effectors, since p120 β-catenin belongs to both classes. The progression of our knowledge on RhoA activation and deactivation will likely uncover other feed-back loops.

RhoA's functions in the cell are primarily related to cytoskeletal regulation. RhoA plays a central role in regulating cell shape, polarity and locomotion through its effects on actin polymerization, actomyosin contractility, cell adhesion and microtubule dynamics [2- 4]. Amongst the ascertained effects of RhoA, it is known that RhoA is required for the generation of contractile force leading to rounding of the cell body [12] and that the proper localization of RhoA in the nucleus is essential during cytokinesis [28]. In particular the activity of the RhoA effectors Citron Kinase [28] and GEF-H1 [29] is necessary for the correct control of cytokinesis in non-transformed cells. RhoA is important for cell cycle progression through G1, since it regulates the expression of cyclin D1 and cyclin-dependent kinase inhibitors [4] and it is required for processes involving cell migration [30]. RhoA regulates the activity of a variety of biochemical pathways, including the activation of MAP kinases (MAPK), in particular c-Jun-N-terminal kinases/stress-activated protein kinases (JNK/SAPK) and p38 kinase [31], as well as numerous transcription factors, such as serum response factor (SRF) [32], activator protein 1 (AP-1) [33], nuclear factor kB (NF-kB) [34], c/EBPb, FHL-2, PAX6, GATA-4, E2F, ER-α, ER-β, CREB [35, 36] and STAT proteins [37, 38].

Rho GTPases show transforming activity by their own [7, 38, 39]: indeed, the overexpression of constitutively activated Rho proteins, such as RhoA, RhoG, Rac, Cdc42 and TC10, induces tumoral transformation in non-transformed fibroblasts [7, 40, 41]. Active Rho proteins are necessary for Ras-mediated oncogenic transformation [40, 42], whereas dominant negative mutants of Rac1 and RhoA inhibit the Ras transforming activity [40]. Although at a lesser extent, also the overexpression of RhoC seems to be related to the oncogenic transformation [5, 7]. On the opposite, RhoB has been described as an oncosuppressor gene [43, 44], and the loss of RhoB expression has been shown to be involved in lung carcinogenesis [45]. Curiously, the anti-tumoral action of RhoB in murine fibroblasts is evident only when RhoB is geranylgeranylated, while it is lost if the protein is farnesylated [46].

RhoA overexpression confers to cancer cells a highly invasive phenotype. LPA, a strong activator of RhoA, promoted matrix invasion and metalloproteinase activity in ovarian cancer [47]. A highly active RhoA is necessary for the cellular motility in prostate cancer [48], where the GTPase is negatively controlled by the endocannabinoid receptors-dependent signaling [49]. RhoA favors cell motility also in tumors with aberrant activity of ephrin-B receptor [50] or E-cadherin/epidermal growth factor receptor [51]. The hyperactivity of RhoA-related proteins, such as ROCK [52] or Dia1 [53], enhanced the invasive attitude in tumors, while the overexpression of the tumor suppressor gene Deleted in Liver Cancer (DLC1) greatly reduced the cell motility in hepatocellular carcinoma because of the RhoGAP activity of DLC1 [54]. In mice injected with human pancreatic cancer cells, liver metastatic nodules were reduced when cells were transfected with the p190 RhoGAP, which slackens the RhoA signaling [55]. In normal and transformed breast epithelium, EGFR and β3-integrin control p190 RhoGAP and RhoA, which increases filopodia formation and cell migration [56]. This mechanism is important for the shape change that occurs during epithelial-mesenchymal transition and matrix invasion by breast cancer cells. RhoA GEF-H1 is another factor favoring cell invasion: in breast cancers it is under the transcriptional control of the "human pituitary tumor-transforming gene" oncogene, which up-regulates RhoA GEF-H1 and RhoA activity, increasing cell migration [57].

RhoC has a minor effect than RhoA on cell proliferation [58], but confers to cancer cells a highly invasive attitude [58, 59] and is directly related to an increased number of lung metastasis [60].

Several types of human cancers have been analyzed for Rho proteins mutations or overexpression [61]. RhoA levels are significantly increased in breast cancer, correlating with the tumor grade [62-64]. RhoA mRNA is higher in ovarian carcinoma: such an increase is particularly significant in metastatic lesions of peritoneal dissemination than in the respective primary tumors [65]. Protein expression of RhoA and its two downstream effectors ROCK1 and ROCK2 is significantly higher in testicular germ cell tumors [66]. The overexpression of RhoA GEF-H1 has been also described in aggressive cancers, where it is associated to high aneuploidy due to aberrant mitosis [29] and high invasion [57].

RhoA may control several autocrine loops in tumor cells: for instance, in transformed lung epithelium, active RhoA increases the synthesis and secretion of prostaglandin E2 [67], which is critical for epithelial tumor growth. Another attractive autocrine mechanism is the tumoral secretion of exosomes, small vesicles produced by tumor cells and carrying growth factors, cytokines, receptors, miRNA, which support or repress cancer cell proliferation. Recently, the RhoA/ROCK1/Lim Kinase pathway has been identified as a controller of secretion of exosomes with transforming activity on mitotically arrested cells [68]. RhoA also mediates the effects of endocrine messengers, as suggested by the higher responsiveness to androgens in prostate cancers overexpressing RhoA [69]. This effect is due to the RhoA/ROCK1-operated nuclear localization of transcription factors, like the so-called “serum response factor megakaryocytic acute leukemia cofactor”, which cooperates with androgen receptor in androgen-dependent prostate tumors [69].

Furthermore, RhoA has been suggested as an useful prognostic factor of the invasion and metastasis of upper urinary tract cancer: RhoA and ROCK protein levels are elevated in bladder cancer, showing higher expression in less differentiated tumors and metastatic lymph nodes [70]. The expression and activation of RhoA is greater in small cell lung carcinoma than non-small cell lung carcinoma cell lines [71]. Patients with esophageal squamous cell carcinoma overexpressing RhoA tended to have poor prognosis compared with patients with RhoA under-expression [72]. RhoA was found frequently overexpressed in gastric cancer compared with normal tissue [73]. Invasiveness of hepatocellular carcinoma is facilitated by the RhoA/ROCK pathway and is likely to be relevant to tumor progression [74]. A high proportion of colon cancers overexpresses RhoA [75] and the inhibition of RhoA activity through the introduction of dominant negative mutants completely abolishes the invasive capacity of colonic epithelial cells [76]. Plasminogen Activator Inhibitor type-1 is important for matrix invasion by colon cancer cells: its localization in the connective tissue surrounding transformed cells creates selective “hot spots” in the plasma membrane of tumor cells where RhoA and ROCK1 are activated and promote cells blebbing and epithelial-mesenchymal transition [77], one of the first steps of metastasis. The epithelial-mesenchymal transition in colon cancer is also supported by the increased activity of the mammalian Target of Rapamycin Complexes mTORC1 and mTORC2, which have RhoA and ROCK as downstream effectors [78]. These results suggest that the RhoA/ROCK axis may act as a collector of multiple signals, all promoting matrix invasion and cell migration. Furthermore, the RhoA/ROCK pathway has been implicated in the vascular endothelial growth factor (VEGF)-mediated angiogenesis [79], which is also increased in highly proliferating tumors. These evidences suggest that RhoA activation should be considered a strong marker of aggressive tumors.

As far as RhoC is concerned, its expression has been related to a more aggressive phenotype in ovarian [65], head and neck cancer [80] and in melanoma [81]. In contrast, only one contradictory study reports that RhoC enhances the tissue invasion, without affecting the directional motility of prostate cancer cells [82]. Recently, RhoC has been also proposed as a novel biomarker of tumor invasiveness, metastasis [83] and poor prognosis [84]. The selective silencing of RhoC increased the expression of oncosuppressor genes and reduced cell migration and anoikis in breast and prostate cancer cells [85]. Both RhoA and RhoC activities are necessary to explain the highly metastatic behavior of Erb2-overexpressing breast cancers, where ErB-2 oncogene recruits and phosphorylates the semaphorin receptor Plexin-1, which activates both GTPases [86].

On the other hand, RhoA and RhoC have sometimes mutually exclusive signals, due to the competition for RhoA-GDI: for instance, only when RhoC is removed by gene silencing, RhoGDIα can stabilize RhoA and promote its activation by RhoGEF, resulting in an increased activity of RhoA-operated pathways [85]. In the case of cell migration, RhoC preferentially promotes a directed and polarized migration, through the downstream effector FMNL3, which reduces the spreading of lamellipodia; the effects of RhoA are more variable, depending on which type of kinases is predominantly activated: ROCK2 promotes a polarized cell movement, whereas ROCK1 is specifically involved in the tail retraction events on the opposite side of the migratory front [87].

These and other *in vitro* and *in vivo* studies provided good evidence that RhoA and RhoC activation is highly relevant for tumor progression and invasiveness [88, 89], and have suggested that abrogation of RhoA and RhoC functions could be a promising strategy to attenuate tumor metastasis [90-93].

Synthetic compounds affecting the geranylgeranylation [94] or the post-translational modifications of RhoA [95], bacterial toxins [96] and specific anti-RhoA small interfering RNA (siRNA) [97] have shown anti-tumor activity. However, many of these strategies have dose-limiting toxicity [94] and have only been tested *in vitro* [93]. Other therapeutic tools have been addressed to inhibit the downstream RhoA effectors. Y-27632, which specifically inhibits the ROCKs [98], largely reduced metastasis in animal models [90] and the newly developed ROCK inhibitor Wf-536 reduced angiogenesis, tumor growth and metastasis *in vivo* [99, 100]. Fasudil [1-(5-isoquinolinesulfonyl)-homopiperazine, also known as HA-1077 and AT877], another ROCK inhibitor currently used in the treatment of cardiovascular [101] and neurological disorders [102], blocked the tumor progression in animal models [103] and exhibited anti-angiogenic properties [104]. A further strategy is to reduce the amount of active geranylgeranylated RhoA by statins.

STATINS INHIBIT RhoA ACTIVITY

By inhibiting the 3-hydroxy-3-methylglutaryl coenzyme A reductase (HMGCoAR), statins decrease the synthesis of cholesterol and isoprenoids molecules, such as farnesyl pyrophosphate (FPP) and geranylgeranyl pyrophosphate (GGPP) [105]. By this way, statins may impair the isoprenylation and the activity of Ras and Rho family G-proteins [92]. Nowadays, many natural and synthetic statins (Table **1**) are used in clinical practice as anti-cholesterolemic agents [105], in the prevention therapy of coronary artery disease (to view the structures of main statins, see [106]). Statins inhibit HMGCoAR by binding to the

HMGCoA pocket with a common hydrophobic bulk, whereas the other substitute groups are positioned in a non polar groove [105]. In consequence of the high number of van derWaals interactions formed with the enzyme, statins tightly bind at nanomolar concentrations, displacing the physiological substrate HMGCoA, which binds at micromolar concentrations [107]. Small differences in the chemical structure account for the different kinetic properties of each drug [108].

Factors other than the reduction of cholesterol synthesis have been invoked to justify such a variety of therapeutic properties [109]. Many statins' effects appear more related to the inhibition of RhoA activity than to the decrease of cholesterol synthesis. For instance fluvastatin prevents heart dysfunction and interstitial myocardial fibrosis in diabetic rats by inhibiting RhoA activity [110]. Using the same mechanism, statins inhibit the smooth muscle cells proliferation [111] and the cardiac remodeling [112] in hypertensive rats, and decrease the secretion of lipoprotein-associated phospholipase A2 by macrophages in atherosclerotic lesions [113]. Recently, pitavastatin has been employed as inhibitor of the accumulation of Tau protein in neurons, an effect due to the decrease in RhoA/ROCK1 pathway [114] and that opens new perspectives for the therapeutic use of statins in Alzheimer disease.

Table 1: Chemical, Pharmacodynamic and Pharmacokinetic Properties of the Most Employed Statins

Compound	Chemical Properties	Ki (nM) HMGCoAR *	IC_{50} ** (nM)	Biovailability *** (%)	Plasma $t_{1/2}$ *** (h)
Cerivastatin	- Hydrophobic drug - Entry in cells by passive diffusion	1.3	5	60	2-3
Simvastatin	- Hydrophobic drug -Administered as a lactone prodrug, which needs to be activated in liver -Entry in cells by passive diffusion - Substrate of ABC-transporters	0.1	345-1500	< 5	1.9
Atorvastatin	- Hydrophobic drug - Entry in cells by passive diffusion	0.5-1	40-100	41	12-58

Table 1: contd....

Lovastatin	- Hydrophobic drug - Administered as a lactone prodrug, which needs to be activated in liver - Substrate of ABC-transporters	0.6	24-50	< 5	1.5
Pravastatin	- Hydrophilic drug - Substrate of ABC-transporters	2.3	700-2650	10-26	1.8
Fluvastatin	- Hydrophilic drug - Substrate of ABC-transporters	0.3	30-43	25	0.5

Adapted from Moghadasian [107].
* HMGCoAR: 3-hydroxy-3-methylglutaryl coenzyme A reductase.
** Concentrations resulting in the 50% inhibition of cholesterol synthesis in HepG2 human hepatoma cells.
*** After oral administration.

STATINS AND TUMOR GROWTH/APOPTOSIS

Since the overexpression of the enzymes of mevalonate pathway cooperates with Ras to promote malignant transformation [115], drugs inhibiting this pathway have been regarded in the last years as attractive anti-cancer tools. It is conceivable that statins slacken the rate of cell proliferation by lowering the synthesis of cholesterol, a major component of cellular membranes. However, an increasing number of experimental evidences suggest that the inhibition of RhoA isoprenylation is a crucial mechanism in reducing tumor growth and eliciting apoptosis [92, 116]. Statins exert *in vitro* and *in vivo* anti-proliferative effects in solid [117, 118] and hematopoietic malignancies [119, 120]. The statin-mediated mitotic arrest was related to the reduced RhoA isoprenylation: for instance, the addition of GGPP or mevalonate, but not FPP or cholesterol, and the expression of constitutively active RhoA prevented the lovastatin-induced G1 phase cell cycle arrest and cell senescence in human prostate cancer cells [121]. The pro-apoptotic effect of statins has been related to the lowering of protein geranylgeranylation also in glioblastoma [122], melanoma [123] and acute myeloid leukemia [124]. By gene microarray approach, RhoA has been shown to be one of the genes modulated by lovastatin in cervix and head and neck squamous carcinomas cells [125]. The statin-induced apoptosis in these tumors was prevented by supplying GGPP and restoring RhoA isoprenylation [125]. The mechanism by which the reduced RhoA isoprenylation leads to growth arrest and apoptosis of tumor cells still remains to be elucidated. The lovastatin-mediated mitotic arrest in human prostate cancer cells was associated with a rapid alteration of phosphorylation state of Rb protein, a decrease in E2F-1, cyclin A and cdc2, and an accumulation of p27 protein level, leading to a significant reduction in the proportion of S phase cells [121]. Similarly, lovastatin decreased cell proliferation of anaplastic thyroid cancer cells by reducing RhoA/ROCK1 activity, which lowered cyclin A2 and cyclin D3 and increased the amounts of p27 and cyclin-dependent kinase 4, producing a G0/G1-arrest [126]. In prostate PC3 cancer cells, the cell cycle arrest induced by atorvastatin was accompanied by an increased expression of LC3-II, indicative of enhanced autophagy; this event was prevented by the addition of geranylgeraniol, suggesting that the statin inhibited a geranlygeranylated protein [127]. Such GTPase has not yet been identified.

In human breast cancer cells the simvastatin-induced apoptosis was mediated by the JNK pathway [128], while in human osteosarcoma lipophilic statins promoted apoptosis by inhibiting RhoA activity and decreasing phospho-p42/p44 levels [129]. In contrast with all these evidences, we did not find any anti-tumor activity by atorvastatin in Her2/neu-overexpressing mammary cancer: although atorvastatin decreased Ras and extracellular-regulated kinase (ERK) 1/2 activity, thus slowing down pathways that are critical for cell proliferation, it simultaneously decreased RhoA/ROCK signaling, which resulted in an increased activity of the pro-survival factor NF-kB [130]. Interestingly, Ras and RhoA displayed a differential sensitivity to atorvastatin and the latter was the most inhibited by the drug: it means that at doses compatible with those used in hypercholesterolemic patients, the proliferative signals derived from the inhibition of RhoA balanced the anti-proliferative signals derived from the inhibition of Ras [130]. To our knowledge this is the first evidence proving an antagonistic effect between Ras and RhoA in terms of tumor growth and suggests that the anti-tumor effect of statins can be highly variable and tumor-dependent.

It cannot be excluded that the anti-proliferative and pro-apoptotic effects of statins may be mediated by Rho proteins other than RhoA: for instance, the downregulation of the RhoC protein by antisense oligonucleotides [131] or siRNA [132] induced the arrest of proliferation as well as the apopotic death of cancer cells. However, no reports link the statin action to a selective inhibition of RhoC proteins.

In addition statins may also increase cellular differentiation: for instance, lovastatin was able to promote differentiation in neuroblastoma cells and in acute myeloid leukemia cells [133]. The effect of lovastatin on immature leukemia cells was similar to that evoked by retinoic acid: both drugs increased the expression of the integrins CD11b and CD18 and decreased the expression of bcl-2 protein. These changes were associated with late stage differentiation of the myeloid cells and were considered as an index of myeloid blasts maturation [133]. Lovastatin also promoted the neurite growth and immature pheochromocytoma cells, transforming them into more differentiated neuronal cells [134]. Again, such an effect was reverted by mevalonate and geranylgeraniol [134]. Not all statins exert a pro-apoptotic effect at the same extent, because of the different pharmacokinetic and pharmacodynamic properties [124]. Besides being direct pro-apoptotic agents, statins also enhanced the apoptosis induced by other chemotherapeutic drugs [118, 135]. Such effect was prevented by GGPP [118]. In several cases, statins have been also observed to exert anticancer effects independently of the mevalonate pathway [136, 137].

Recently, the overexpression of mevalonate pathway genes has been reported in breast cancers with mutated p53, where it is predictive of poor prognosis [138]: mutated p53 enhanced the transcription of mevalonate pathway genes and increased the invasive growth of cancer cells. Since simvastatin and the inhibitor of geranylgeranyltransferase fully prevented the latter event [138], it is likely that a geranylgeranylated protein of Rho family is involved.

STATINS AND ANGIOGENESIS

Both pro- and anti-angiogenic effects of statins have been widely described [139-141]. On one hand, statins increased the differentiation of endothelial progenitor cells in mice and humans [142] and stimulated the capillary formation through a hsp90- and nitric oxide (NO)-dependent mechanism [139]. On the other hand, statins blocked the proliferation and promoted the tumor necrosis factor (TNF)-α-mediated apoptosis of endothelial cells [141], inhibited the formation of vascular tubes [140], and prevented the matrix remodeling [143]. Recently it has been reported that simvastatin, fluvastatin and cerivastatin reduce the endothelial cell growth also under hypoxia [144], an environmental condition resembling that occurring in the inner core of solid tumors. The sensitivity to the anti-angiogenic effect of statins is strictly dose- and cell type-dependent [145, 146]. In human vascular smooth muscle cells and microvascular endothelial cells, which constitutively produce large amounts of VEGF, statins reduced the VEGF secretion; on the opposite, in primary macrovascular endothelial cells, which do not basally secrete VEGF, statins were pro-angiogenic at less than 1 μM and anti-angiogenic at higher concentrations [146].

In a recent screening aimed to discover new anti-angiogenic drugs for prostate cancer, four statins (mevastatin, lovastatin, simvastatin, rosuvastatin) have been identified amongst the leading anti-angiogenic compounds; rosuvastatin was the most potent *in vitro* and efficiently decreased the tumor growth in mice xenografts, thanks to a dual action, *i.e.* the reduction of microvessel density within tumor and the induction of apoptosis in tumor cells [147].

There is general agreement that most statins' anti-angiogenic effects are mediated by RhoA and RhoC inhibition. The active RhoA/ROCK pathway stimulates angiogenesis by increasing the secretion of VEGF, interleukin (IL)-6 [148] and IL-8 [149], by modulating the activity of metalloproteinase-9 [150] and by regulating the cytoskeletal remodeling and the cellular migration [143]. The ROCK inhibitor Fasudil indeed has demonstrated to possess anti-angiogenic properties in human endothelial cells [151]. The overexpression of RhoC in breast cancer cells led to increased secretion of pro-angiogenic factors, such as VEGF, basic fibroblast growth factor, IL-6 and IL-8 [152], in a MAP-kinase dependent way [153]. Both the cerivastatin-induced decrease of endothelial cell locomotion *in vitro* and the simvastatin-elicited decrease of capillary growth *in vivo* were reversed by GGPP [143, 154]. The available experimental evidences suggest that RhoA and RhoC are mainly involved in favoring angiogenesis and may be considered promising targets in the anti-angiogenic therapy. Recently RhoB expression has been shown to be crucial to regulate the endothelial survival and proliferation during the physiological vascular development [155]; however the role of RhoB in the tumor angiogenesis and the effects of statins on RhoB activity still remain to be elucidated.

STATINS AND METASTASIS

Statins inhibited the invasiveness of human colon carcinoma cells [156], human pancreatic cancer cells [157] and human anaplastic thyroid cancer cells [158]. It has been reported above that RhoA overexpression is highly relevant for tumor progression and invasiveness. In the aggressive breast cancer MDA-MB-231 cells the anti-invasive properties of statins were related to the inhibition of the RhoA/ROCK/NF-kB pathway [159]. NF-kB, whose nuclear translocation may depend on RhoA activity [48, 160, 161], in turn up-regulates the expression of genes involved in cellular invasiveness, such as urokinase-type plasminogen, tissue factor and metalloproteinase 9 [159]. Statins inhibited cell motility also by disrupting the RhoA/Focal-Adhesion-Kinase (FAK)/Akt signaling [162]: it has been reported that RhoA activity is necessary for the tyrosine phosphorylation and activation of FAK [162, 163], which is then responsible for the activation of the Akt kinase [164]. Akt may further enhance the nuclear translocation of NF-kB [162]. Interestingly, the effects of lovastatin were nearly absent in the less invasive breast cancer MCF-7 cells [159], but a differential activity of RhoA was not further investigated. Moreover, lovastatin impaired the TNF-α- and RhoA-dependent increase of E-selectin in human endothelial cells, reducing a potential mechanism of cancer cell adhesion and transendothelial migration [165]. RhoB seems responsible for the increase of E-selectin caused by TNF- α as well [165]. Statins showed good efficacy in reducing metastasis also *in vivo*: fluvastatin and lovastatin decreased the metastatic ability of renal cancer cells [166] and mammary carcinoma cells [167]. In the latter model lovastatin impaired the secretion of urokinase, a key proteolytic enzyme during tumor invasion [167]. In a murine model of melanoma, simvastatin and fluvastatin reduced the number of lung metastasis by decreasing the expression of metalloproteinases and α2-, α4-, α5-integrins, other molecules important for tumor cell invasion; these effects have been attributed to the lower amount of RhoA localized at the plasma membrane and to the lower phosphorylation of Lim kinase and myosin light chain in animals exposed to statins [168]. A similar decrease of integrins, due to the low activity of

RhoA/ROCK pathway, was reported in invasive hepatocellular carcinoma cells treated with simvastatin [169].

Due to the central role of RhoC in tumor invasion and metastasis [60], several studies pointed out a relationship between the anti-metastatic effect of statins and the specific inhibition of RhoC in human cancers: for instance atorvastatin lowered the metastatic attitude of melanoma cells by decreasing the RhoC isoprenylation [170]. By preventing the activation of both RhoA and RhoC, fluvastatin impaired the transendothelial migration of MDA-MB-231 cells [171]. Furthermore, the inhibition of both RhoA and RhoC, by specific siRNA [132, 172], prevented the matrix invasion by human breast cancer cells.

STATINS, CHEMOTHERAPY EFFICACY AND MULTIDRUG RESISTANCE

In vitro studies reported that statins synergized with γ rays [173], doxorubicin, paclitaxel and 5-fluorouracile [174] in reducing cancer growth. Besides increasing the sensitivity to doxorubicin, lovastatin also reduced the drug cardiotoxicity in mice, *via* an hypothetical lipid-lowering effect [175]. On the other hand, in a limited group of experimental works, statins and chemotherapeutic agents had no synergistic effects [176, 177]. It has been hypothesized that the p53 level may influence the efficacy of statins: indeed pravastatin and atorvastatin sensitized p53-deficient tumor cells to etoposide, doxorubicin and 5-fluorouracil, but failed in p53 wild-type cells [178]. Several evidences pointed out that the inhibition of RhoA isoprenylation is involved in modulating the response to chemotherapy. For instance, lovastatin increased the apoptotic effect of 5-fluorouracil or cisplatin in human colon cancer cells, whereas the addition of GGPP prevented the cell death [118]. Fluvastatin enhanced the pro-apoptotic effect of gemcitabine in pancreatic cancer *in vitro* and *in vivo* and such an effect was prevented by the administration of mevalonic acid [135]. Interestingly, fluvastatin increased the expression of deoxycytidine kinase, the enzyme required for the activation of gemcitabine, and simultaneously reduced the level of 5α-nucleotidase, responsible for its catabolism [135].

Multidrug resistance (MDR), an acquired or constitutive cross-resistance towards many unrelated anti-cancer drugs, is the major obstacle to a successful pharmacological therapy of tumors [179]. Many statins are substrates of ATP-binding cassette (ABC) transporters, like P-glycoprotein (Pgp) and MDR-related proteins (MRPs) [180, 181], whose overexpression mediates the enhanced efflux of chemotherapeutic agents [179]. ABC transporters are membrane pumps which bind and hydrolyze ATP, thus mediating the active efflux of endogenous metabolites and drugs [179]. Lovastatin, simvastatin, fluvastatin and pravastatin are transported out of the cells by Pgp [182], which is also responsible for the efflux of anthracyclines, Vinca alkaloids, epipodophyllotoxins, taxanes, actinomycin-D, mitoxantrone [179]. Therefore, statins might affect the accumulation of chemotherapeutics in cancer cells by competing with them for the same ABC pump-mediated transport [181]. Statins induced a selective apoptosis in drug-resistant cancer cells [183, 184]: the molecular mechanism was not fully clarified, but it has been reported that drug-resistant cells were partially protected from statins-induced apoptosis by the addition of FPP and GGPP [185]. Furthermore, a recent study implicates RhoA in MDR: hepatocellular carcinoma cells overexpressing the Rho-specific GEF Lymphoid blast crisis (Lbc) were resistant to doxorubicin, but this resistance was reverted by the C3 exotoxin from *C. Botulinum* [186]. These evidences suggest that statins could revert MDR by impairing the RhoA operation. Indeed, atorvastatin increased the doxorubicin's cytotoxic efficacy and accumulation in both sensitive and drug-resistant human colon cancer cells [187]. Interestingly, such effect of atorvastatin was mediated by its ability to induce the cellular synthesis of NO, which in turn may nitrate the ABC transporter MRP3, leading to a reduced efflux of doxorubicin [187]. The increased synthesis of NO followed by the nitration on ABC transporters was not statin-specific: indeed simvastatin produced the same sequence of events in colon cancer [188]. The molecular basis of the statins' effect was clarified in the human malignant mesothelioma, which is highly resistant to a large number of chemotherapeutic agents: both mevastatin and simvastatin corrected the doxorubicin resistance of mesothelioma cells by inhibiting the RhoA/ROCK pathway [189]. The statins' effects, reverted by mevalonic acid and mimicked by Y-27632, were NO-dependent [189]. These results led to hypothesize that the inhibition of RhoA/ROCK causes the activation of the NF-kB transcription factor and the subsequent induction of NO synthase: in mesothelioma cells the increased synthesis of NO was accompanied by the nitration of another ABC transporter, the Pgp [189]. The central role of RhoA GTPase in modulating NO synthesis and MDR was confirmed in RhoA-silenced doxorubicin-resistant colon cancer cells, where the only depletion of RhoA was sufficient to turn the drug-resistant phenotype into a drug-sensitive one [190].

A cell adhesion-mediated drug resistance (CAM-DR), dependent on Wnt3 overexpression and RhoA/Rho kinase activity [191], is often observed in myeloma cells. Also CAM-DR was totally overcome by statins and specific inhibitors of geranylgeranyltransferases and ROCKs [192].

The inhibition of RhoA does not always produce a chemosensitization: for instance, lovastatin conferred cross-resistance to doxorubicin and etoposide in human endothelial cells [177] and the expression of constitutive active RhoA induced a significant resistance to etoposide, 5-fluorouracil and taxol, but increased the sensitivity to vincristine in human prostate carcinoma cells [193]. In the light of these findings, we cannot exclude that the inhibition of RhoA by statins may modulate both chemotherapy efficacy and MDR, with different effects depending on the anti-cancer agent and on the type of tumor.

STATINS AND CHEMOPREVENTION OF TUMORS

In a small number of studies, statins exhibited a carcinogenic and genotoxic effect, but HMGCoAR inhibitors were used at concentrations higher than the common therapeutic doses [194, 195]. By inhibiting cellular proliferation and invasion, statins are likely to exert rather a cancer-preventing effect. Indeed the chemopreventive action of statins was confirmed in *in vivo* models of chemical carcinogenesis [196, 197] or pre-cancerous diseases, such as ulcerative colitis [198] and familial adenomatous polyposis [199]. The oral administration of statins, at a dose very close to that used in the treatment of cardiovascular diseases, efficiently reduced the growth of breast cancer in mice, through a MAP-kinase- and NF-kB-dependent mechanism [200].

In mice with pancreatic intraepithelial neoplasms, atorvastatin prevented the transition into invasive adenocarcinoma, reducing the activity of RhoA and of other molecules favoring survival and/or proliferation, such as PI3K, Akt, PCNA, p27, cyclin D, survivin, β-catenin [201].

Yet, when considering the cancer prevention in patients regularly taking statins, conflicting data exist: some case-control studies and randomized controlled trials found no association between the use of statins and reduced frequency of solid tumors [202, 203]. Only a long-term therapy with statins partially lowered the incidence of tumors [204]. On the opposite, other studies showed that statins efficiently reduced the incidence of pancreatic cancers [205], as well as metastasis and mortality in advanced stages of prostate cancer [206]. Randomized controlled trials for preventing cardiovascular disease indicated that statins reduced the incidence of colorectal cancer and melanoma [207].

Experimental evidences are not yet available in support of the hypothesis that the *in vivo* chemopreventive action of statins is due to the inhibition of Rho proteins. Interestingly, statins in combination with non-steroidal anti-inflammatory drugs (NSAIDs) have been shown to prevent colorectal cancer. In mice affected by adenomatous polyposis, atorvastatin and the cyclooxygenase 2 (COX2) inhibitor celecoxib synergistically prevented the development of colon adenocarcinoma [199]. Similarly, in a population-based case-control study, the association of aspirin and statins was more chemopreventive than the single drugs [208]. It has been reported that COX2 induces the activation of the RhoA/ROCK pathway, leading to the disruption of cellular adherens junctions and increased motility of colon cancer cells [209]. Since Rho and COX2 activities appear to be strictly related in colon cancer cells [210], the synergistic effect of statins and NSAIDs could be exerted by inhibiting a COX2/Rho/ROCK pathway, but this hypothesis needs to be still confirmed.

STATINS IN CANCER TREATMENT

The anti-cancer effect of statins was analyzed in different human clinical trials: the therapy with statins was well tolerated and did not enhance the adverse effects of anti-cancer drugs [211, 212] or radiotherapy [213], but conflicting results were reported about its efficacy [211]. The limited number of patients taking statins [202], the advanced stage of the disease and the too small median survival of patients [211] may affect the statistical potency of these studies. Some variability of response in hepatocellular cancer has been described: fluvastatin exerted a different anti-proliferative effect in mice, depending on the tumor stage [214], and the addition of pravastatin to the 5-fluorouracil therapy significantly prolonged the patients survival [215]. However this result was not confirmed by subsequent studies [216].

Better results have been obtained in hematological malignancies: simvastatin stabilized the disease progression in patients both sensitive and resistant to chemotherapy [217] and reversed the resistance to bortezomib and bendamustine in patients with relapsed myeloma [218]. Similarly, lovastatin improved the clinical response and the overall survival of patients with relapsed myeloma, if added to the standard therapeutic regimen (thalidomide and dexamethasone) [219].

The statins' effect in myeloma was attributed to the reduced prenylation of small G-proteins, including the Rho homologue Rap1 [217]. In a phase 1 study, pravastatin, added to idarubicin and cytarabine, obtained encouraging response rates in patients with acute myeloid leukemia [212]. In this type of tumor the exposure to cytotoxic drugs evoked an increase of cholesterol synthesis and chemoresistance, whereas statins restored the chemosensitivity by lowering the cholesterol levels [220]. Most of these experimental works provided only preliminary results and did not investigate the molecular mechanisms of the action of statins.

Other drugs targeting the mevalonate pathway, like the anti-osteoporotic drugs aminobisphosphonates, which inhibit isopentenyl diphosphate (IPP) isomerase and FPP synthase [221], showed anti-tumor activity and slackened the progression of metastasis in cancer patients [222]. Interestingly aminobisphosphonates exhibited anti-angiogenic properties by suppressing RhoA activity [223]. The association of statins and bisphosphonates was more effective than the single drugs in reducing the geranylgeranylation of proteins [224], and clinically achievable concentrations of fluvastatin and zoledronic acid synergistically induced apoptosis in cancers [225]. Another noteworthy recent study reported that simvastatin decreases the invasive attitude of p53-mutated breast cancer cells by impairing the activity of an unknown geranylgeraylated protein [138]: this result looks particularly appealing because it is the first evidence that statins treatment corrects the phenotypical consequences of a genetic mutation. RhoA, RhoB and RhoC are under intensive investigations as antitumor targets, with promising results: the C3-transferase homologue CT04, a cell-permeant inhibitor of the three Rho GTPases, but not of other geranylgeranylated or farnesylated proteins, efficiently reduced cell migration in ovarian cancer [226], one of most invasive and chemoresistant tumors. Narciclasine, a novel selective RhoA inhibitor extracted from Amaryllidaceae plants, has been tested *in vivo* against primary and metastatic brain tumors, showing good anti-tumor efficacy and few side-effects [227]. Furthermore the experiments in mice also suggest that this new inhibitor has an excellent delivery across the brain-blood barrier, which is hardly crossed by many other anti-cancer drugs.

Taken as a whole, present evidences suggest that the inhibition of RhoA might be an important anti-cancer tool *in vitro* and *in vivo*. Moreover, also the reduction of RhoC activity may decrease the tumor invasiveness and metastasis. The relative importance of the inhibition of these two isoforms in the efficacy of anti-tumor therapy with statins has to be still clarified. As to RhoB, which may have differential (enhancing or suppressive) effects on carcinogenesis, depending on the nature of its prenylation [46], the prevailing effect of statins is not known. Specific siRNA have been constructed to knock-down Rho proteins separately, but they have been only applied in mice models or in *in vitro* studies [93, 94, 97]. Presently it can be only affirmed that, by inhibiting the isoprenylation, statins lower the activity of RhoA and RhoC, and subsequently may impair the promoting effects of these GTPases in the development of many tumors. This is a stimulus to keep on investigating statins (and other inhibitors of Rho and Rho-associated regulators and effectors) as potential tools in the future anti-tumor therapy.

ACKNOWLEDGEMENTS

Declared none.

CONFLICT OF INTEREST

The authors confirm that this chapter contents have no conflict of interest.

DISCLOSURE

The chapter submitted for eBook Series entitled: "**Recent Advances in Medicinal Chemistry, Volume 1**" is an update of our article published in **Mini-Reviews in**

Medicinal Chemistry, Volume 8, Number 6, pp. 609 to 618, with additional text and references

REFERENCES

[1] Madaule, P.; Axel, R. A novel ras-related gene family. *Cell,* **1985**, *41*(1), 31-40.

[2] Bishop, A.L.; Hall, A. Rho GTPases and their effector proteins. *Biochem. J.*, **2000**, *348*(2), 241-255.

[3] Ridley, A.J. Rho family proteins: coordinating cell responses. *Trends Cell. Biol.*, **2001**, *11*(12), 471-477.

[4] Jaffe, A.B.; Hall, A. Rho GTPases: biochemistry and biology. *Annu. Rev. Cell Dev. Biol.*, **2005**, *21*, 247-269.

[5] Wennerberg, K.; Der, C.J. Rho-family GTPases: it's not only Rac and Rho (and I like it). *J. Cell. Sci.*, **2004**, *117*(8), 1301-1312.

[6] Boureux, A.; Vignal, E.; Faure; S.; Fort; P. Evolution of the Rho family of ras-like GTPases in eukaryotes. *Mol. Biol. Evol.*, **2007**, *24*(1), 203-26.

[7] Sahai, E; Marshall, C.J. RHO-GTPases and cancer. *Nat. Rev. Cancer*, **2002**, *2*(2), 133-142.

[8] Rossman, K.L.; Der, C.J; Sondek, J. GEF means go: turning on RHO GTPases with guanine nucleotide-exchange factors. *Nat. Rev. Mol. Cell Biol.*, **2005**, *6*(2), 167-180.

[9] Peck, J.; Douglas, G.[4th]; Wu, C.H.; Burbelo, P.D. Human RhoGAP domain-containing proteins: structure; function and evolutionary relationships. *FEBS Lett.*, **2002**, *528*(1-3), 27-34.

[10] Olofsson, B. Rho guanine dissociation inhibitors: pivotal molecules in cellular signalling. *Cell Signal.*, **1999**, *11*(8), 545-554.

[11] Der Mardirossian, C.; Bokoch, G.M. GDIs: central regulatory molecules in Rho GTPase activation. *Trends Cell Biol*, **2005**, *15*(7), 356-363.

[12] Malliri, A.; Collard, J.G. Role of Rho-family proteins in cell adhesion and cancer. *Curr. Opin. Cell Biol.*, **2003**, *15*(5), 583-589.

[13] Adamson, P.; Marshall, C.J.; Hall, A.; Tilbrook, P.A. Post-translational modifications of p21rho proteins. *J. Biol. Chem.*, **1992**, *267*(28), 20033-20038.

[14] Casey, P.J.; Seabra, M.C. Protein prenyltransferases. *J. Biol. Chem.*, **1996**, *271*(10), 5289-5292.

[15] Winter-Vann, A.M.; Casey, P.J. Post-prenylation-processing enzymes as new targets in oncogenesis. *Nat. Rev. Cancer*, **2005**, *5*(5),: 405-412.

[16] Gosser, Y.Q.; Nomanbhoy, T.K.; Aghazadeh, B.; Manor, D.; Combs, C.; Cerione, R.A.; Rosen, M.K. C-terminal binding domain of Rho GDP-dissociation inhibitor directs N-terminal inhibitory peptide to GTPases. *Nature*, **1997**, *387*(6635), 814-819.

[17] Ridley, A.J. Rho proteins and cancer. *Breast Cancer Res. Treat.*, **2004**, *84*(1), 13-19.

[18] Boettner, B.; Van Aelst, L. The role of Rho GTPases in disease development. *Gene*, **2002**, *286*(2), 155-174.

[19] Vetter, I.R.; Wittinghofer, A. The guanine nucleotide-binding switch in three dimensions. *Science*, **2001**, *294*(5545), 1299-1304.

[20] Luo, R.; Akpan, I.O.; Hayashi, R.; Sramko, M.; Barr, V.; Shiba, Y.; Randazzo, P.A. The GTP-binding protein-like domain of AGAP1 is a protein binding site that allosterically regulates ArfGAP catalytic activity. *J. Biol. Chem.*, **2012**, *287*(21), 17176-17185.

[21] Jank, T.; Giesemann, .T; Aktories, K. Rho-glucosylating Clostridium difficile toxins A and B: new insights into structure and function. *Glycobiology*, **2007**, *17*(4), 15R-22R.

[22] Vogelsgesang, M.; Pautsch, A.; Aktories, K. C3 exoenzymes; novel insights into structure and action of Rho-ADP-ribosylating toxins. *Naunyn Schmiedebergs Arch. Pharmacol.*, **2007**, *374*(5-6), 347-360.

[23] Kaibuchi, K.; Kuroda, S.; Amano, M. Regulation of the cytoskeleton and cell adhesion by the Rho family GTPases in mammalian cells. *Annu. Rev. Biochem.*, **1999**, *68*, 459-486.

[24] Arthur, W.T.; Ellerbroek, S.M.; Der, C.J.; Burridge, K.; Wennerberg, K. XPLN; a guanine nucleotide exchange factor for RhoA and RhoB; but not RhoC. *J. Biol. Chem.*, **2002**, *277*(45), 42964-42972.

[25] Wang, L.; Yang, L.; Luo, Y.; Zheng, Y. A novel strategy for specifically down-regulating individual Rho GTPase activity in tumor cells. *J. Biol. Chem.*, **2003**, *278*(19), 44617-44625.

[26] Sahai, E.; Marshall, C.J. RHO-GTPases and cancer. *Nat. Rev. Cancer*, **2002**, *2*(2), 133-142.

[27] Smith, A.L.; Dohn, M.R.; Brown, M.V.; Reynolds, A.B. Association of Rho-associated protein kinase 1 with E-cadherin complexes is mediated by p120-catenin. *Mol. Biol. Cell*, **2012**, *23*(1), 99-110.

[28] Bassi, Z.I.; Verbrugghe, K.J.; Capalbo, L.; Gregory, S.; Montembault, E.; Glover, D.M.; D'Avino, P.P. Sticky/Citron kinase maintains proper RhoA localization at the cleavage site during cytokinesis. *J. Cell Biol.*, **2011**, *195*(4), 595-603.

[29] Gao Y; Smith, E.; Ker, E.; Campbell, P.; Cheng, E.C.; Zou, S.; Lin, S.; Wang, L.; Halene, S.; Krause, D.S. Role of RhoA-specific guanine exchange factors in regulation of endomitosis in megakaryocytes. *Dev. Cell*, **2012**, *22*(3), 573-584.

[30] Ridley, A.J.; Schwartz, M.A.; Burridge, K.; Firtel, R.A.; Ginsberg ,M.H.; Borisy, G.; Parsons, J.T.; Horwitz, A.R. Cell migration: integrating signals from front to back. *Science*, **2003**, *302*(5651), 1704-1709.

[31] Marinissen, M.J.; Chiariello, M.; Gutkind, J.S. Regulation of gene expression by the small GTPase Rho through the ERK6 (p38 gamma) MAP kinase pathway. *Genes Dev*, **2001**, *15*(5), 535-553.

[32] Hill, C.S.; Wynne, J.; Treisman, R. The Rho family GTPases RhoA; Rac1; and CDC42Hs regulate transcriptional activation by SRF. *Cell*, **1995**, *81*(7), 1159-1170.

[33] Chang, J.H.; Pratt, J.C.; Sawasdikosol, S.; Kapeller, R.; Burakoff, S.J. The small GTP-binding protein Rho potentiates AP-1 transcription in T cells. *Mol. Cell Biol.*, **1998**, *18*(9), 4986-4993.

[34] Perona. R.; Montaner, S.; Saniger, L.; Sánchez-Pérez, I.; Bravo, R.; Lacal, J.C. Activation of the nuclear factor-kappaB by Rho; CDC42; and Rac-1 proteins. *Genes Dev.*, **1997**, *11*(4), 463-475.

[35] Marinissen, M.J.; Chiariello, M.; Tanos, T.; Bernard, O.; Narumiya, S.; Gutkind, J.S. The small GTP-binding protein RhoA regulates c-jun by a ROCK-JNK signaling axis. *Mol. Cell*, **2004**, *14*(1), 29-41.

[36] Aznar, S.; Lacal, J.C. Rho signals to cell growth and apoptosis. *Cancer Lett.*, **2001**, *165*(1), 1-10.

[37] Debidda, M.; Wang, L.; Zang, H.; Poli, V.; Zheng, Y. A role of STAT3 in Rho GTPase-regulated cell migration and proliferation. *J. Biol. Chem.*, **2005**, *280*(17),17275-17285.

[38] Benitah, S.A.; Valerón, P.F.; Rui, H.; Lacal, J.C. STAT5a activation mediates the epithelial to mesenchymal transition induced by oncogenic RhoA. *Mol. Biol. Cell*, **2003**, *14*(1), 40-53.

[39] Perona, R.; Esteve, P.; Jiménez, B.; Ballestero, R.P.; Ramón y Cajal, S.; Lacal, J.C. Tumorigenic activity of rho genes from Aplysia californica. *Oncogene*, **1993**, *8*(5), 1285-1292.

[40] Khosravi-Far, R.; Solski, P.A.; Clark, G.J.; Kinch, M.S.; Der, C.J. Activation of Rac1; RhoA; and mitogen-activated protein kinases is required for Ras transformation. *Mol. Cell Biol.*, **1995**, *15*(11), 6443-6453.

[41] Roux, P.; Gauthier-Rouvière, C.; Doucet-Brutin, S.; Fort, P. The small GTPases Cdc42Hs; Rac1 and RhoG delineate Raf-independent pathways that cooperate to transform NIH3T3 cells. *Curr. Biol.*, **1997**, *7*(9), 629-637.

[42] Prendergast, G.C.; Khosravi-Far, R.; Solski, P.A.; Kurzawa, H.; Lebowitz, P.F.; Der, C.J. Critical role of Rho in cell transformation by oncogenic Ras. *Oncogene*, **1995**, *10*(12), 2289-2296.

[43] Adnane, J.; Muro-Cacho, C.; Mathews, L.; Sebti, S.M.; Muñoz-Antonia, T. Suppression of rho B expression in invasive carcinoma from head and neck cancer patients. *Clin. Cancer Res.*, **2002**, *8*(7), 2225-2232.

[44] Jiang, K.; Delarue, F.L.; Sebti, S.M. EGFR; ErbB2 and Ras but not Src suppress RhoB expression while ectopic expression of RhoB antagonizes oncogene-mediated transformation. *Oncogene*, **2004**, *23*(5), 1136-1145.

[45] Mazieres, J.; He, B.; You, L.; Xu, Z.; Lee, A.Y.; Mikami, I.; Reguart, N.; Rosell, R.; McCormick, F.; Jablons, D.M. Wnt inhibitory factor-1 is silenced by promoter hypermethylation in human lung cancer. *Cancer Res.*, **2004**, *64*(14), 4717-4720.

[46] Mazières, J.; Tillement, V.; Allal, C.; Clanet, C.; Bobin, L.; Chen, Z.; Sebti, S.M.; Favre, G.; Pradines, A. Geranylgeranylated; but not farnesylated; RhoB suppresses Ras transformation of NIH-3T3 cells. *Exp. Cell Res.*, **2005**, *304*(2), 354-364.

[47] Fishman, D.A.; Liu, Y.; Ellerbroek, S.M.; Stack, M.S. Lysophosphatidic acid promotes matrix metalloproteinase (MMP) activation and MMP-dependent invasion in ovarian cancer cells. *Cancer Res.*, **2001**, *61*(7), 3194-3199.

[48] Hodge, J.C.; Bub, J.; Kaul, S.; Kajdacsy-Balla, A.; Lindholm, P.F. Requirement of RhoA activity for increased nuclear factor kappaB activity and PC-3 human prostate cancer cell invasion. *Cancer Res.*, **2003**, *63*(6), 1359-1364.

[49] Nithipatikom, K.; Gomez-Granados, A.D.; Tang, A.T.; Pfeiffer, A.W.; Williams, C.L.; Campbell, W.B. Cannabinoid receptor type 1 (CB1) activation inhibits small GTPase RhoA activity and regulates motility of prostate carcinoma cells. *Endocrinology*, **2012**, *153*(1), 29-41.

[50] Yang, N.Y.; Pasquale, E.B.; Owen, L.B.; Ethell, I.M. The EphB4 receptor-tyrosine kinase promotes the migration of melanoma cells through Rho-mediated actin cytoskeleton reorganization. *J. Biol .Chem.*, **2006**, *281*(43), 32574-3286.

[51] Mateus, A.R.; Seruca, R.; Machado, J.C.; Keller, G.; Oliveira, M.J.; Suriano, G.; Luber, B. EGFR regulates RhoA-GTP dependent cell motility in E-cadherin mutant cells. *Hum. Mol. Genet.*, **2007**, *16*(13), 1639-1647.

[52] Charette, S.T.; McCance, D.J. The E7 protein from human papillomavirus type 16 enhances keratinocyte migration in an Akt-dependent manner. *Oncogene*, **2007**, *26*(52), 7386-7390.

[53] Kitzing, T.M.; Sahadevan, A.S.; Brandt, D.T.; Knieling, H.; Hannemann, S.; Fackler, O.T.; Grosshans, J.; Grosse, R. Positive feedback between Dia1; LARG; and RhoA regulates cell morphology and invasion. *Genes Dev.*, **2007**, *21*(12), 1478-1483.

[54] Wong, C.M.; Yam, J.W.; Ching YP; Yau TO, Leung TH, Jin DY, Ng IO. Rho GTPase-activating protein deleted in liver cancer suppresses cell proliferation and invasion in hepatocellular carcinoma. *Cancer Res.*, **2005**, *65*(19), 8861-8868.

[55] Kusama, T.; Mukai M; Endo H; Ishikawa, O.; Tatsuta, M.; Nakamura, H.; Inoue, M. Inactivation of Rho GTPases by p190 RhoGAP reduces human pancreatic cancer cell invasion and metastasis. *Cancer Sci.*, **2006**, *97*(9), 848-853.

[56] Balanis, N.; Yoshigi, M.; Wendt, M.K.; Schiemann, W.P.; Carli, C.R. β3 integrin-EGF receptor cross-talk activates p190RhoGAP in mouse mammary gland epithelial cells. *Mol. Biol. Cell*, **2011**, *22*(22), 4288-4301.

[57] Liao, Y.C.; Ruan, J.W.; Lua, I.; Li MH, Chen WL, Wang JR, Kao RH, Chen JH. Overexpressed hPTTG1 promotes breast cancer cell invasion and metastasis by regulating GEF-H1/RhoA signalling. *Oncogene*, **2012**, *31*(25), 3086-3097.

[58] Iiizumi, M.; Bandyopadhyay, S.; Pai, S.K.; Watabe, M.; Hirota, S.; Hosobe, S.; Tsukada, T.; Miura, K.; Saito, K.; Furuta, E.; Liu, W.; Xing, F.; Okuda, H.; Kobayashi, A.; Watabe, K. RhoC promotes metastasis via activation of the Pyk2 pathway in prostate cancer. *Cancer Res.*, **2008**, *68*(18): 7613-7620.

[59] Hall, C.L.; Dubyk, C.W.; Riesenberger, T.A.; Shein, D.; Keller, E.T.; van Golen, K.L. Type I collagen receptor (alpha2beta1) signaling promotes prostate cancer invasion through RhoC GTPase. *Neoplasia*, **2008**, *10*(8), 797-803.

[60] Ikoma, T.; Takahashi, T.; Nagano, S.; Li, Y.M.; Ohno, Y.; Ando, K.; Fujiwara, T.; Fujiwara, H.; Kosai, K. A definitive role of RhoC in metastasis of orthotopic lung cancer in mice. *Clin. Cancer Res.*, **2004**, *10*(3), 1192-2000.

[61] Gómez del Pulgar, T.; Benitah, S.A.; Valerón, P.F.; Espina, C.; Lacal, J.C. Rho GTPase expression in tumourigenesis: evidence for a significant link. *Bioessays*, **2005**, *27*(6), 602-613.

[62] Burbelo, P.; Wellstein, A.; Pestell, R.G. Altered Rho GTPase signaling pathways in breast cancer cells. *Breast Cancer Res. Treat.*, **2004**, *84*(1), 43-48.

[63] Lin, M.; van Golen, K.L. Rho-regulatory proteins in breast cancer cell motility and invasion. *Breast Cancer Res. Treat.*, **2004**, *84*(1), 49-60.

[64] Fritz, G.; Brachetti, C.; Bahlmann, F.; Schmidt, M.; Kaina, B. Rho GTPases in human breast tumours: expression and mutation analyses and correlation with clinical parameters. *Br. J. Cancer*, **2002**, *87*(6), 635-644.

[65] Horiuchi, A.; Imai, T.; Wang, C.; Ohira, S.; Feng, Y.; Nikaido, T.; Konishi, I. Up-regulation of small GTPases; RhoA and RhoC; is associated with tumor progression in ovarian carcinoma. *Lab. Invest.*, **2003**, *83*(6), 861-870.

[66] Kamai, T.; Yamanishi, T.; Shirataki, H.; Takagi K, Asami H, Ito Y, Yoshida K. Overexpression of RhoA; Rac1; and Cdc42 GTPases is associated with progression in testicular cancer. *Clin. Cancer Res.*, **2004**, *10*(14), 4799-4805.

[67] Choi, H.J.; Lee, D.H.; Park, S.H.; Kim, J.; Do, K.H.; An, T.J.; Ahn, Y.S.; Park, C.B.; Moon, Y. Induction of human microsomal prostaglandin E synthase 1 by activated oncogene RhoA GTPase in A549 human epithelial cancer cells. *Biochem. Biophys. Res. Commun.*, **2011**, *413*(3), 448-453.

[68] Li, B.; Antonyak, M.A.; Zhang, J.; Cerione, R.A. RhoA triggers a specific signalling pathway that generates transforming microvesicles in cancer cells. *Oncogene*, **2012**, *31*(45):4740-4749.

[69] Schmidt, L.J.; Duncan, K.; Yadav, N.; Regan K.M.; Verone, A.R.; Lohse, C.M., Pop, E.A.; Attwood, K.; Wilding, G.; Mohler, J.L.; Sebo, T.J.; Tindall, D.J.; Heemers, H.V. RhoA as a

Mediator of Clinically Relevant Androgen Action in Prostate Cancer Cells. *Mol. Endocrinol.*, **2012**, *26*(5), 716-735.

[70] Kamai, T.; Kawakami, S.; Koga, F.; Arai, G.; Takagi, K.; Arai, K.; Tsujii ,T.; Yoshida, K.I. RhoA is associated with invasion and lymph node metastasis in upper urinary tract cancer. *BJU Int.*, **2003**, *91*(3), 234-238.

[71] Varker, K.A.; Phelps, S.H.; King, M.M.; Williams, C.L. The small GTPase RhoA has greater expression in small cell lung carcinoma than in non-small cell lung carcinoma and contributes to their unique morphologies. *Int. J. Oncol.*, **2003**, *22*(3), 671-681.

[72] Faried, A.; Nakajima, M.; Sohda, M.; Miyazaki, T.; Kato, H.; Kuwano, H. Correlation between RhoA overexpression and tumour progression in esophageal squamous cell carcinoma. *Eur. J. Surg. Oncol.*, **2005**, *31*(4), 410-414.

[73] Pan Y; Bi F; Liu N; Xue, Y.; Yao, X.; Zheng, Y.; Fan, D. Expression of seven main Rho family members in gastric carcinoma. *Biochem. Biophys. Res. Commun.*, **2004**, *315*(3), 686-691.

[74] Fukui, K.; Tamura, S.; Wada, A.; Kamada, Y.; Sawai, Y.; Imanaka, K.; Kudara, T.; Shimomura, I.; Hayashi, N. Expression and prognostic role of RhoA GTPases in hepatocellular carcinoma. *J. Cancer Res. Clin. Oncol.*, **2006**, *132*(10), 627-633.

[75] Fritz, G.; Just, I.; Kaina, B. Rho GTPases are over-expressed in human tumors. *Int. J. Cancer*, **1999**, *81*(5), 682-687.

[76] Attoub S; Noe, V.; Pirola, L.; Bruyneel, E.; Chastre, E.; Mareel, M.; Wymann, M.P.; Gespach, C. Leptin promotes invasiveness of kidney and colonic epithelial cells via phosphoinositide 3-kinase-; rho-; and rac-dependent signaling pathways. *FASEB J.*, **2000**, *14*(14), 2329-2338.

[77] Cartier-Michaud, A.; Malo, M.; Charrière-Bertrand, C.; Gadea, G.; Anguille, C.; Supiramaniam, A.; Lesne, A.; Delaplace, F.; Hutzler, G.; Roux, P.; Lawrence, D.A.; Barlovatz-Meimon, G. Matrix-bound PAI-1 supports cell blebbing via RhoA/ROCK1 signaling. *PLoS One*, **2012**, *7*(2), e32204.

[78] Gulhati P; Bowen KA; Liu J; Stevens, P.D.; Rychahou, P.G.; Chen, M.; Lee, E.Y.; Weiss, H.L.; O'Connor, K.L.; Gao, T.; Evers, B.M. mTORC1 and mTORC2 regulate EMT; motility; and metastasis of colorectal cancer via RhoA and Rac1 signaling pathways. *Cancer Res.*, **2011**, *71*(9), 3246-3256.

[79] Kumar, B.; Chile, S.A.; Ray, K.B.; Reddy, G.E.; Addepalli, M.K.; Kumar, A.S.; Ramana, V.; Rajagopal ,V. VEGF-C differentially regulates VEGF-A expression in ocular and cancer cells; promotes angiogenesis via RhoA mediated pathway. *Angiogenesis*, **2011**, *14*(3), 371-380.

[80] Kleer, C.G.; Griffith, K.A.; Sabel, M.S.; Gallagher, G.; van Golen, K.L.; Wu, Z.F.; Merajver, S.D. RhoC-GTPase is a novel tissue biomarker associated with biologically aggressive carcinomas of the breast. *Breast Cancer Res. Treat.*, 2005, *93*(2), 101-110.

[81] Ruth, M.C.; Xu, Y.; Maxwell, I.H.; Ahn, N.G.; Norris, D.A.; Shellman, Y.G. RhoC promotes human melanoma invasion in a PI3K/Akt-dependent pathway. *J. Invest. Dermatol.*, **2006**, *126*(4), 862-868.

[82] Yao, H.; Dashner, E.J.; van Golen, C.M.; van Golen, K.L. RhoC GTPase is required for PC-3 prostate cancer cell invasion but not motility. *Oncogene*, **2006**, *25*(16), 2285-2296.

[83] Kleer, C.G.; Teknos, T.N.; Islam, M.; Marcus, B.; Lee, J.S.; Pan, Q.; Merajver, S.D. RhoC GTPase expression as a potential marker of lymph node metastasis in squamous cell carcinomas of the head and neck. *Clin. Cancer Res.*, **2006**, *12*(15), 4485-4490.

[84] Zhang, H.Z.; Liu, J.G.; Wei, Y.P.; Wu, C.; Cao, Y.K.; Wang, M. Expression of G3BP and RhoC in esophageal squamous carcinoma and their effect on prognosis. *World J. Gastroenterol.*, **2007**, *13*(30), 4126-4130.

[85] Giang Ho, T.T.; Stultiens, A.; Dubail, J.; Lapière, C.M.; Nusgens, B.V.; Colige, A.C.; Deroanne, C.F. RhoGDIα-dependent balance between RhoA and RhoC is a key regulator of cancer cell tumorigenesis. *Mol. Biol. Cell.*, **2011**, *22*(17), 3263-3275.

[86] Worzfeld, T.; Swiercz, J.M.; Looso, M.; Straub, B.K.; Sivaraj, K.K.; Offermanns, S. ErbB-2 signals through Plexin-B1 to promote breast cancer metastasis. *J. Clin. Invest.*, **2012**, *122*(4), 1296-1305.

[87] Vega, F.M.; Fruhwirth, G.; Ng, T.; Ridley, A.J. RhoA and RhoC have distinct roles in migration and invasion by acting through different targets. *J. Cell Biol.*, **2011**, *193*(4), 655-665.

[88] Yoshioka, K.; Matsumura, F.; Akedo, H.; Itoh, K. Small GTP-binding protein Rho stimulates the actomyosin system; leading to invasion of tumor cells. *J. Biol. Chem.*, 1998, *273*(9), 5146-5154.

[89] Stam, J.C.; Michiels, F.; van der Kammen, R.A.; Moolenaar, W.H.; Collard, J.G. Invasion of T-lymphoma cells: cooperation between Rho family GTPases and lysophospholipid receptor signaling. *EMBO J*, **1998**, *17*(14), 4066-4074.

[90] Itoh, K.; Yoshioka, K.; Akedo, H.; Uehata, M.; Ishizaki, T.; Narumiya, S. An essential part for Rho-associated kinase in the transcellular invasion of tumor cells. *Nat. Med.*, **1999**, *5*(2), 221-225.

[91] Fritz, G; Kaina, B. Rho GTPases: promising cellular targets for novel anticancer drugs. *Curr. Cancer Drug Targets*, **2006**, *6*(1), 1-14.

[92] Fritz, G. HMG-CoA reductase inhibitors (statins) as anticancer drugs (review). *Int. J. Oncol.*, **2005**, *27*(5), 1401-1409.

[93] Walker, K.; Olson, M.F. Targeting Ras and Rho GTPases as opportunities for cancer therapeutics. *Curr. Opin. Genet. Dev.*, **2005**, *15*(1), 62-68.

[94] Lobell, R.B.; Omer, C.A.; Abrams, M.T.; Bhimnathwala, H.G.; Brucker, M.J.; Buser, C.A.; Davide, J.P.; deSolms, S.J.; Dinsmore, C.J.; Ellis-Hutchings, M.S.; Kral, A.M.; Liu, D.; Lumma, W.C.; Machotka, S.V.; Rands, E.; Williams, T.M.; Graham, S.L.; Hartman, G.D.; Oliff, A.I.; Heimbrook, D.C.; Kohl, N.E. Evaluation of farnesyl:protein transferase and geranylgeranyl:protein transferase inhibitor combinations in preclinical models. *Cancer Res.*, **2001**, *61*(24), 8758-8768.

[95] Lu, Q.; Harrington, E.O.; Newton, J.; Jankowich, M.; Rounds, S. Inhibition of ICMT induces endothelial cell apoptosis through GRP94. *Am. J. Respir .Cell Mol. Biol.*, **2007**, *37*(1), 20-30.

[96] Sheahan, K.L.; Satchell, K.J. Inactivation of small Rho GTPases by the multifunctional RTX toxin from Vibrio cholerae. *Cell Microbiol.*, **2007**, *9*(5), 1324-1335.

[97] Pillé, J.Y.; Li, H.; Blot, E.; Bertrand, J.R.; Pritchard, L.L.; Opolon, P.; Maksimenko, A.; Lu, H.; Vannier, J.P.; Soria, J.; Malvy, C.; Soria, C. Intravenous delivery of anti-RhoA small interfering RNA loaded in nanoparticles of chitosan in mice: safety and efficacy in xenografted aggressive breast cancer. *Hum. Gene Ther.*, **2006**, *17*(10), 1019-1026.

[98] Narumiya, S.; Ishizaki, T.; Uehata, M. Use and properties of ROCK-specific inhibitor Y-27632. *Methods Enzymol.*, **2000**, *325*, 273-284.

[99] Nakajima, M.; Hayashi, K.; Katayama, K.; Amano, Y.; Egi, Y.; Uehata, M.; Goto, N.; Kondo, T. Wf-536 prevents tumor metastasis by inhibiting both tumor motility and angiogenic actions. *Eur. J. Pharmacol.*, **2003**, *459*(2-3), 113-120.

[100] Somlyo, A.V.; Phelps, C.; Dipierro, C.; Eto, M.; Read, P.; Barrett, M.; Gibson, J.J.; Burnitz, M.C.; Myers, C.; Somlyo, A.P. Rho kinase and matrix metalloproteinase inhibitors cooperate to inhibit angiogenesis and growth of human prostate cancer xenotransplants. *FASEB J.*, **2003**, *17*(2), 223-234.

[101] Hirooka, Y.; Shimokawa, H. Therapeutic potential of rho-kinase inhibitors in cardiovascular diseases. *Am. J. Cardiovasc. Drugs*, **2005**, *5*(1), 31-39.

[102] Mueller, B.K.; Mack, H.; Teusch, N. Rho kinase; a promising drug target for neurological disorders. *Nat. Rev. Drug Discov.*, **2005**, *4*(5), 387-398.

[103] Ying, H.; Biroc, S.L.; Li, W.W.; Alicke, B.; Xuan, J.A.; Pagila, R.; Ohashi, Y.; Okada, T.; Kamata, Y.; Dinter, H. The Rho kinase inhibitor fasudil inhibits tumor progression in human and rat tumor models. *Mol. Cancer Ther.*, **2006**, *5*(9), 2158-2164.

[104] Yin, L.; Morishige, K.; Takahashi, T.; Hashimoto, K.; Ogata, S.; Tsutsumi, S.; Takata, K.; Ohta, T.; Kawagoe, J.; Takahashi, K.; Kurachi, H. Fasudil inhibits vascular endothelial growth factor-induced angiogenesis *in vitro* and *in vivo*. *Mol. Cancer Ther.*, 2007, *6*(5), 1517-1525.

[105] Liao, J.K.; Laufs, U. Pleiotropic effects of statins. *Annu. Rev. Pharmacol. Toxicol.*, **2005**, *45*, 89-18.

[106] Meng, C.Q. Inflammation in atherosclerosis: new opportunities for drug discovery. *Mini Rev. Med. Chem.*, **2005**, *5*(1), 33-40.

[107] Moghadasian, M.H. Clinical pharmacology of 3-hydroxy-3-methylglutaryl coenzyme A reductase inhibitors. *Life Sci.*, **1999**, *65*(13), 1329-1337.

[108] Istvan, E.S.; Deisenhofer, J. Structural mechanism for statin inhibition of HMG-CoA reductase. *Science*, **2001**, *292*(5519), 1160-164.

[109] Bellosta, S.; Ferri, N.; Bernini, F.; Paoletti, R.; Corsini, A. Non-lipid-related effects of statins. *Ann. Med.*, **2000**, *32*(3), 164-176.

[110] Dai, Q.M.; Lu, J.; Liu, N.F. Fluvastatin attenuates myocardial interstitial fibrosis and cardiac dysfunction in diabetic rats by inhibiting over-expression of connective tissue growth factor. *Chin. Med. J. (Engl).*, **2011**, *124*(1), 89-94.

[111] Ma, M.M.; Li, S.Y.; Wang, M.; Guan, Y.Y. Simvastatin Attenuated Cerebrovascular Cell Proliferation in the Development of Hypertension through Rho/Rho-kinase Pathway. *J. Cardiovasc. Pharmacol.*, **2012**, *59*(6), 576-582.

[112] Takayama, N.; Kai, H.; Kudo, H.; Yasuoka, S.; Mori, T.; Anegawa, T.; Koga, M.; Kajimoto, H.; Hirooka, Y.; Imaizumi, T. Simvastatin prevents large blood pressure variability induced aggravation of cardiac hypertrophy in hypertensive rats by inhibiting RhoA/Ras-ERK pathways. *Hypertens. Res.*, **2011**, *34*(3), 341-347.

[113] Song, J.X.; Ren, J.Y.; Chen, H. Simvastatin reduces lipoprotein-associated phospholipase A2 in lipopolysaccharide-stimulated human monocyte-derived macrophages through inhibition of the mevalonate-geranylgeranyl pyrophosphate-RhoA-p38 mitogen-activated protein kinase pathway. *J. Cardiovasc. Pharmacol.*, **2011**, *57*(2), 213-222.

[114] Amano, T.; Yen, S.H.; Gendron, T.; Ko, L.W.; Kuriyama, M. Pitavastatin decreases tau levels via the inactivation of Rho/ROCK. *Neurobiol. Aging*, **2011**, *33*(10), 2306-2320.

[115] Clendening, J.W.; Pandyra, A.; Boutros, P.C.; El Ghamrasni, S.; Khosravi, F.; Trentin, G.A.; Martirosyan, A.; Hakem, A.; Hakem, R.; Jurisica, I.; Penn, L.Z. . Dysregulation of the mevalonate pathway promotes transformation. *Proc. Natl. Acad. Sci. USA*, **2010**, *107*(34), 15051-15056.

[116] Porter, K.E.; Turner, N.A.; O'Regan, D.J.; Balmforth, A.J.; Ball, S.G. Simvastatin reduces human atrial myofibroblast proliferation independently of cholesterol lowering via inhibition of RhoA. *Cardiovasc. Res.*, **2004**, *61*(4), 745-755.

[117] Ghosh, P.M.; Ghosh-Choudhury, N.; Moyer, M.L.; Moyer, M.L.; Mott, G.E.; Thomas, C.A.; Foster, B.A.; Greenberg, N.M.; Kreisberg, J.I. Role of RhoA activation in the growth and morphology of a murine prostate tumor cell line. *Oncogene*, **1999**, *18*(28), 4120-4130.

[118] Agarwal, B.; Bhendwal, S.; Halmos, B.; Moss, S.F.; Ramey, W.G.; Holt, P.R. Lovastatin augments apoptosis induced by chemotherapeutic agents in colon cancer cells. *Clin. Cancer Res.*, **1999**, *5*(8), 2223-2229.

[119] Lewis, K.A.; Holstein, S.A.; Hohl, R.J. Lovastatin alters the isoprenoid biosynthetic pathway in acute myelogenous leukemia cells *in vivo*. *Leuk. Res.*, **2005**, *29*(5), 527-533.

[120] Clendening, J.W.; Pandyra, A.; Li, Z.; Martirosyan, A.; Lehner, R.; Jurisica, I.; Trudel, S.; Penn, L.Z. Exploiting the mevalonate pathway to distinguish statin-sensitive multiple myeloma. *Blood*, **2010**, *115*(23), 4787-4797.

[121] Lee, J.; Lee, I.; Park, C.; Kang, W.K. Lovastatin-induced RhoA modulation and its effect on senescence in prostate cancer cells. *Biochem. Biophys. Res. Commun.*, **2006**, *339*(3), 748-754.

[122] Jiang, Z.; Zheng, X.; Lytle, R.A.; Higashikubo, R.; Rich, K.M. Lovastatin-induced up-regulation of the BH3-only protein; Bim; and cell death in glioblastoma cells. *J. Neurochem.*, **2004**, *89*(1), 168-178.

[123] Shellman, Y.G.; Ribble, D.; Miller, L.; Gendall, J.; Vanbuskirk, K.; Kelly, D.; Norris, D.A.; Dellavalle, R.P. Lovastatin-induced apoptosis in human melanoma cell lines. *Melanoma Res.*, **2005**, *15*(2), 83-89.

[124] Wong, W.W.; Tan, M.M.; Xia, Z.; Dimitroulakos ,J.; Minden, M.D.; Penn, L.Z. Cerivastatin triggers tumor-specific apoptosis with higher efficacy than lovastatin. *Clin. Cancer Res.*, **2001**, *7*(7), 2067-2075.

[125] Dimitroulakos, J.; Marhin, W.H.; Tokunaga, J.; Irish, J.; Gullane, P.; Penn, L.Z.; Kamel-Reid, S. Microarray and biochemical analysis of lovastatin-induced apoptosis of squamous cell carcinomas. *Neoplasia*, **2002**, *4*(4), 337-346.

[126] Zhong, W.B.; Hsu, S.P.; Ho, P.Y.; Liang, Y.C.; Chang, T.C.; Lee, W.S. Lovastatin inhibits proliferation of anaplastic thyroid cancer cells through up-regulation of p27 by interfering with the Rho/ROCK-mediated pathway. *Biochem. Pharmacol.*, **2011**, *82*(11), 1663-72.

[127] Parikh, A.; Childress, C.; Deitrick, K.; Lin, Q.; Rukstalis, D.; Yang, W. Statin-induced autophagy by inhibition of geranylgeranyl biosynthesis in prostate cancer PC3 cells. *Prostate*, **2010**, *70*(9), 971-981.

[128] Koyuturk, M.; Ersoz, M.; Altiok, N. Simvastatin induces apoptosis in human breast cancer cells: p53 and estrogen receptor independent pathway requiring signalling through JNK. *Cancer Lett.*, **2007**, *250*(2), 220-228.

[129] Fromigué, O.; Haÿ, E.; Modrowski, D.; Bouvet, S.; Jacquel, A.; Auberger, P.; Marie, P.J. RhoA GTPase inactivation by statins induces osteosarcoma cell apoptosis by inhibiting p42/p44-MAPKs-Bcl-2 signaling independently of BMP-2 and cell differentiation. *Cell Death Differ.*, **2006**, *13*(11), 1845-1856.

[130] Riganti, C.; Pinto, H.; Bolli, E.; Belisario, D.C,; Calogero, R.A.; Bosia, A.; Cavallo, F. Atorvastatin modulates anti-proliferative and pro-proliferative signals in Her2/neu-positive mammary cancer. *Biochem. Pharmacol.*, **2011**, *82*(9), 1079-1089.

[131] Shi, Z.; Chen, M.L.; He, Q.L.; Zeng, J.H. Antisense RhoC gene suppresses proliferation and invasion capacity of human QBC939 cholangiocarcinoma cells. *Hepatobiliary Pancreat. Dis. Int.*, **2007**, *6*(5), 516-520.

[132] Sun, H.W.; Tong, S.L.; He, J.; Wang, Q.; Wang, Q.; Zou, L.; Ma, S.J.; Tan, H.Y.; Luo, J.F.; Wu, H.X. RhoA and RhoC -siRNA inhibit the proliferation and invasiveness activity of human gastric carcinoma by Rho/PI3K/Akt pathway. *World J. Gastroenterol.*, **2007**, *13*(25), 3517-3122.

[133] Dimitroulakos, J.; Thai, S.; Wasfy, G.H.; Hedley, D.W.; Minden, M.D.; Penn, L.Z. Lovastatin induces a pronounced differentiation response in acute myeloid leukemias. *Leuk, Lymphoma*, **2000**, *40*(1-2), 167-178.

[134] Fernández-Hernando, C.; Suárez, Y.; Lasunción, M.A. Lovastatin-induced PC-12 cell differentiation is associated with RhoA/RhoA kinase pathway inactivation. *Mol. Cell Neurosci.*, **2005**, *29*(4), 591-602.

[135] Bocci, G.; Fioravanti, A.; Orlandi, P.; Bernardini, N.; Collecchi, P.; Del Tacca, M.; Danesi, R. Fluvastatin synergistically enhances the antiproliferative effect of gemcitabine in human pancreatic cancer MIAPaCa-2 cells. *Br. J. Cancer*, **2005**, *93*(3), 319-330.

[136] Rao, S.; Porter, D.C.; Chen, X.; Herliczek, T.; Lowe, M.; Keyomarsi, K. Lovastatin-mediated G1 arrest is through inhibition of the proteasome; independent of hydroxymethyl glutaryl-CoA reductase. *Proc. Natl. Acad. Sci. USA*, **1999**, *96*(14), 7797-7802.

[137] Weitz-Schmidt, G.; Welzenbach, K.; Brinkmann, V.; Kamata, T.; Kallen, J.; Bruns, C.; Cottens, S.; Takada, Y.; Hommel, U. Statins selectively inhibit leukocyte function antigen-1 by binding to a novel regulatory integrin site. *Nat. Med.*, **2001**, *7*(6), 687-692.

[138] Freed-Pastor; W.A.; Mizuno; H.; Zhao; X.; Langerød, A.; Moon, S.H.; Rodriguez-Barrueco, R.; Barsotti, A.; Chicas, A.; Li, W.; Polotskaia, A.; Bissell, M.J.; Osborne, T.F.; Tian, B.; Lowe, S.W.; Silva, J.M.; Børresen-Dale, A.L.; Levine, A.J.; Bargonetti, J.; Prives, C. Mutant p53 disrupts mammary tissue architecture via the mevalonate pathway. *Cell*, **2012**; *148*(1-2), 244-258.

[139] Brouet, A.; Sonveaux, P.; Dessy, C.; Moniotte, S.; Balligand, J.L.; Feron, O. Hsp90 and caveolin are key targets for the proangiogenic nitric oxide-mediated effects of statins. *Circ. Res.*, **2001**, *89*(10), 866-873.

[140] Miura, S.; Matsuo, Y.; Saku, K. Simvastatin suppresses coronary artery endothelial tube formation by disrupting Ras/Raf/ERK signaling. *Atherosclerosis*, **2004**, *175*(2), 235-243.

[141] Tang, D.; Park, H.J.; Georgescu, S.P.; Sebti, S.M.; Hamilton, A.D.; Galper, J.B. Simvastatin potentiates tumor necrosis factor alpha-mediated apoptosis of human vascular endothelial cells via the inhibition of the geranylgeranylation of RhoA. *Life Sci.*, **2006**, *79*(15), 1484-1492.

[142] Llevadot, J.; Murasawa, S.; Kureishi, Y.; Uchida, S.; Masuda, H.; Kawamoto, A.; Walsh, K.; Isner, J.M.; Asahara, T. HMG-CoA reductase inhibitor mobilizes bone marrow-derived endothelial progenitor cells. *J. Clin. Invest.*, **2001**, *108*(3), 399-405.

[143] Vincent L; Chen W; Hong L; Mirshahi, F.; Mishal, Z.; Mirshahi-Khorassani, T.; Vannier, J.P.; Soria, J.; Soria, C. Inhibition of endothelial cell migration by cerivastatin; an HMG-CoA reductase inhibitor: contribution to its anti-angiogenic effect. *FEBS Lett.*, 2001, *495*(3), 159-166.

[144] Schaefer, C.A.; Kuhlmann, C.R.; Weiterer, S.; Fehsecke, A.; Abdallah, Y.; Schaefer, C.; Schaefer, M.B.; Mayer, K.; Tillmanns, H.; Erdogan, A. Statins inhibit hypoxia-induced endothelial proliferation by preventing calcium-induced ROS formation. *Atherosclerosis*, **2006**, *185*(2), 290-296.

[145] Weis, M.; Heeschen, C.; Glassford, A.J.; Cooke, J.P. Statins have biphasic effects on angiogenesis. *Circulation*, **2002**, *105*(6), 739-745.

[146] Frick M; Dulak J; Cisowski J; Józkowicz, A.; Zwick, R.; Alber, H.; Dichtl, W.; Schwarzacher, S.P.; Pachinger, O.; Weidinger, F. Statins differentially regulate vascular endothelial growth factor synthesis in endothelial and vascular smooth muscle cells. *Atherosclerosis*, **2003**, *170*(2), 229-236.

[147] Wang, C.; Tao, W.; Wang, Y.; Bikow, J.; Lu, B.; Keating, A.; Verma, S.; Parker, T.G.; Han, R.; Wen, X.Y. Rosuvastatin; identified from a zebrafish chemical genetic screen for antiangiogenic compounds; suppresses the growth of prostate cancer. *Eur. Urol.*, **2010**, *58*(3), 418-426.

[148] Ito, T.; Ikeda, U.; Shimpo, M.; Ohki, R.; Takahashi, M.; Yamamoto, K.; Shimada, K. HMG-CoA reductase inhibitors reduce interleukin-6 synthesis in human vascular smooth muscle cells. *Cardiovasc. Drugs Ther.*, **2002**, *16*(2), 121-126.

[149] Hippenstiel, S.; Soeth, S.; Kellas, B.; Fuhrmann, O.; Seybold, J.; Krüll, M.; Eichel-Streiber, C.; Goebeler, M.; Ludwig, S.; Suttorp, N. Rho proteins and the p38-MAPK pathway are important mediators for LPS-induced interleukin-8 expression in human endothelial cells. *Blood*, **2000**, *95*(10), 3044-3051.

[150] Watnick, R.S.; Cheng, Y.N.; Rangarajan, A.; Ince, T.A.; Weinberg, R.A. Ras modulates Myc activity to repress thrombospondin-1 expression and increase tumor angiogenesis. *Cancer Cell*, **2003**, *3*(3), 219-31.

[151] Zhang, Z.; Ren, J.H.; Li, Z.Y.; Nong, L.; Wu, G. Fasudil inhibits lung carcinoma-conditioned endothelial cell viability and migration. *Oncol. Rep.*, **2012**, *27*(5), 1561-1566.

[152] Van Golen, K.L.; Wu, Z.F.; Qiao, X.T.; Bao, L.; Merajver, S.D. RhoC GTPase overexpression modulates induction of angiogenic factors in breast cells. *Neoplasia*, **2000**, *2*(5), 418-425.

[153] Van Golen, K.L.; Bao, L.W.; Pan, Q.; Miller, F.R.; Wu, Z.F.; Merajver, S.D. Mitogen activated protein kinase pathway is involved in RhoC GTPase induced motility; invasion and angiogenesis in inflammatory breast cancer. *Clin. Exp. Metastasis*, **2002**, *19*(4), 301-311.

[154] Park, H.J.; Kong, D.; Iruela-Arispe, L.; Begley, U.; Tang, D.; Galper, J.B. 3-hydroxy-3-methylglutaryl coenzyme A reductase inhibitors interfere with angiogenesis by inhibiting the geranylgeranylation of RhoA. *Circ. Res.*, **2002**, *91*(2), 143-150.

[155] Adini, I.; Rabinovitz, I.; Sun, J.F.; Prendergast, G.C.; Benjamin, L.E. RhoB controls Akt trafficking and stage-specific survival of endothelial cells during vascular development. *Genes Dev.*, **2003**, *17*(21), 2721-2732.

[156] Kusama, T.; Mukai, M.; Tatsuta, M.; Matsumoto, Y.; Nakamura, H.; Inoue, M. Selective inhibition of cancer cell invasion by a geranylgeranyltransferase-I inhibitor. *Clin. Exp. Metastasis*, **2003**, *20*(6), 561-567.

[157] Kusama T; Mukai M; Iwasaki T; Tatsuta, M.; Matsumoto, Y.; Akedo, H.; Inoue, M.; Nakamura, H. 3-hydroxy-3-methylglutaryl-coenzyme a reductase inhibitors reduce human pancreatic cancer cell invasion and metastasis. *Gastroenterology*, **2002**, *122*(2), 308-317.

[158] Zhong, W.B.; Liang, Y.C.; Wang, C.Y.; Chang, T.C.; Lee, W.S. Lovastatin suppresses invasiveness of anaplastic thyroid cancer cells by inhibiting Rho geranylgeranylation and RhoA/ROCK signaling. *Endocr. Relat. Cancer*, **2005**, *12*(3), 615-629.

[159] Denoyelle, C.; Vasse, M.; Körner, M.; Mishal, Z,; Ganné, F.; Vannier, J.P.; Soria, J.; Soria, C. Cerivastatin; an inhibitor of HMG-CoA reductase; inhibits the signaling pathways

involved in the invasiveness and metastatic properties of highly invasive breast cancer cell lines: an *in vitro* study. *Carcinogenesis*, **2001**, *22*(8),1139-1148.

[160] Kraynack, N.C.; Corey, D.A.; Elmer, H.L.; Kelle,y T.J. Mechanisms of NOS2 regulation by Rho GTPase signaling in airway epithelial cells. *Am. J. Physiol. Lung Cell Mol. Physiol.*, **2002**, *283*(3), L604-611.

[161] Rattan, R.; Giri, S.; Singh, A.K.; Singh, I. Rho A negatively regulates cytokine-mediated inducible nitric oxide synthase expression in brain-derived transformed cell lines: negative regulation of IKKalpha. *Free Radic. Biol. Med.*, **2003**, *35*(9), 1037-1050.

[162] Denoyelle, C.; Albanese, P.; Uzan, G.; Hong, L.; Vannier, J.P.; Soria, J.; Soria, C. Molecular mechanism of the anti-cancer activity of cerivastatin; an inhibitor of HMG-CoA reductase; on aggressive human breast cancer cells. *Cell Signal.*, **2003**, *15*(3), 327-338.

[163] Clark, E.A.; King, W.G.; Brugge, J.S.; Symons, M.; Hynes, R.O. Integrin-mediated signals regulated by members of the rho family of GTPases. *J. Cell Biol.*, **1998**, *142*(2), 573-586.

[164] Reif, S.; Lang, A.; Lindquistm, J.N.; Yata, Y.; Gabele, E.; Scanga, A.; Brenner, D.A.; Rippe, R.A. The role of focal adhesion kinase-phosphatidylinositol 3-kinase-akt signaling in hepatic stellate cell proliferation and type I collagen expression. *J. Biol. Chem.*, **2003**, *278*(10), 8083-8090.

[165] Nübel, T.; Dippold, W.; Kleinert, H.; Kaina, B.; Fritz, G. Lovastatin inhibits Rho-regulated expression of E-selectin by TNFalpha and attenuates tumor cell adhesion. *FASEB J.*, **2004**, *18*(1), 140-2.

[166] Horiguchi, A.; Sumitomo, M.; Asakuma, J.; Asano, T.; Asano, T.; Hayakawa, M. 3-hydroxy-3-methylglutaryl-coenzyme a reductase inhibitor; fluvastatin; as a novel agent for prophylaxis of renal cancer metastasis. *Clin. Cancer Res.*, **2004**, *10*(24), 8648-8655.

[167] Farina, H.G.; Bublik, D.R.; Alonso, D.F.; Gomez, D.E. Lovastatin alters cytoskeleton organization and inhibits experimental metastasis of mammary carcinoma cells. *Clin. Exp. Metastasis*, **2002**, *19(6)*, 551-559.

[168] Kidera, Y.; Tsubaki, M.; Yamazoe, Y.; Shoji, K.; Nakamura, H.; Ogaki, M.; Satou, T.; Itoh, T.; Isozaki, M.; Kaneko, J.; Tanimori, Y.; Yanae, M.; Nishida, S. Reduction of lung metastasis; cell invasion; and adhesion in mouse melanoma by statin-induced blockade of the Rho/Rho-associated coiled-coil-containing protein kinase pathway. *J. Exp. Clin. Cancer Res.*, **2010**, *29*, 127.

[169] Relja, B.; Meder, F.; Wang, M.; Blaheta, R.; Henrich, D.; Marzi, I.; Lehnert, M. Simvastatin modulates the adhesion and growth of hepatocellular carcinoma cells via decrease of integrin expression and ROCK. *Int. J. Oncol.*, **2011**, *38*(3), 879-885.

[170] Collisson, E.A.; Kleer, C.; Wu, M.; De, A.; Gambhir, S.S.; Merajver, S.D.; Kolodney, M.S. Atorvastatin prevents RhoC isoprenylation; invasion; and metastasis in human melanoma cells. *Mol. Cancer Ther.*, **2003**, *2*(10), 941-948.

[171] Kusama, T.; Mukai, M.; Tatsuta, M.; Nakamura, H.; Inoue, M. Inhibition of transendothelial migration and invasion of human breast cancer cells by preventing geranylgeranylation of Rho. *Int. J. Oncol.*, **2006**, *29*(1), 217-223.

[172] Pillé, J.Y.; Denoyelle, C.; Varet, J.; Bertrand, J.R.; Soria, J.; Opolon, P.; Lu, H.; Pritchard, L.L.; Vannier, J.P.; Malvy, C.; Soria, C.; Li, H. Anti-RhoA and anti-RhoC siRNAs inhibit the proliferation and invasiveness of MDA-MB-231 breast cancer cells *in vitro* and *in vivo*. *Mol. Ther.*, **2005**, *11*(2), 267-274.

[173] Fritz, G.; Brachetti, C.; Kaina, B. Lovastatin causes sensitization of HeLa cells to ionizing radiation-induced apoptosis by the abrogation of G2 blockage. *Int. J. Radiat. Biol.*, **2003**, *79*(8), 601-610.

[174] Ahn, K.S.; Sethi, G.; Aggarwal, B.B. Reversal of chemoresistance and enhancement of apoptosis by statins through down-regulation of the NF-kappaB pathway. *Biochem. Pharmacol.*, **2008**, *75*(4), 907-913.

[175] Feleszko, W.; Mlynarczuk, I.; Balkowiec-Iskra, E.Z.; Czajka, A.; Switaj, T.; Stoklosa, T.; Giermasz, A.; Jakóbisiak, M. Lovastatin potentiates antitumor activity and attenuates cardiotoxicity of doxorubicin in three tumor models in mice. *Clin. Cancer Res.*, **2000**, *6*(5), 2044-2052.

[176] Ciocca, D.R.; Rozados, V.R.; Cuello Carrión, F.D.; Gervasoni, S.I.; Matar, P.; Scharovsky, O.G. Hsp25 and Hsp70 in rodent tumors treated with doxorubicin and lovastatin. *Cell Stress Chaperones*, **2003**, *8*(1), 26-36.

[177] Damrot, J.; Nübel, T.; Epe, B.; Roos, W.P.; Kaina, B.; Fritz, G. Lovastatin protects human endothelial cells from the genotoxic and cytotoxic effects of the anticancer drugs doxorubicin and etoposide. *Br. J. Pharmacol.*, **2006**, *149*(8), 988-997.

[178] Roudier, E.; Mistafa, O.; Stenius, U. Statins induce mammalian target of rapamycin (mTOR)-mediated inhibition of Akt signaling and sensitize p53-deficient cells to cytostatic drugs. *Mol. Cancer Ther.*, **2006**, *5*(11), 2706-2715.

[179] Gottesman, M.M.; Fojo, T.; Bates, S.E. Multidrug resistance in cancer: role of ATP-dependent transporters. *Nat. Rev. Cancer*, **2002**, *2*(1), 48-58.

[180] Kivistö, K.T.; Zukunft, J.; Hofmann, U.; Niemi, M.; Rekersbrink, S.; Schneider, S.; Luippold, G.; Schwab, M.; Eichelbaum, M.; Fromm, M.F. Characterisation of cerivastatin as a P-glycoprotein substrate: studies in P-glycoprotein-expressing cell monolayers and mdr1a/b knock-out mice. *Naunyn Schmiedebergs Arch. Pharmacol.*, **2004**, *370*(2), 124-130.

[181] Huang, L.; Wang, Y.; Grimm, S. ATP-dependent transport of rosuvastatin in membrane vesicles expressing breast cancer resistance protein. *Drug Metab. Dispos.*, **2006**, *34*(5), 738-742.

[182] Bogman, K.; Peyer, A.K.; Török, M.; Küsters, E.; Drewe, J. HMG-CoA reductase inhibitors and P-glycoprotein modulation. *Br. J. Pharmacol.*, **2001**, *132*(6), 1183-1192.

[183] Dimitroulakos, J.; Yeger, H. HMG-CoA reductase mediates the biological effects of retinoic acid on human neuroblastoma cells: lovastatin specifically targets P-glycoprotein-expressing cells. *Nat. Med.*, **1996**, *2*(3), 326-333.

[184] Maksumova, L.; Ohnishi, K.; Muratkhodjaev, F.; Zhang, W.; Pan, L.; Takeshita, A.; Ohno, R. Increased sensitivity of multidrug-resistant myeloid leukemia cell lines to lovastatin. *Leukemia*, **2000**, *14*(8), 1444-1450.

[185] Cafforio, P.; Dammacco, F.; Gernone, A.; Silvestris, F. Statins activate the mitochondrial pathway of apoptosis in human lymphoblasts and myeloma cells. *Carcinogenesis*, **2005**, *26*(5), 883-891.

[186] Sterpetti, P.; Marucci, L.; Candelaresi, C.; Toksoz, D.; Alpini, G.; Ugili, L.; Baroni, G.S.; Macarri, G.; Benedetti, A. Cell proliferation and drug resistance in hepatocellular carcinoma are modulated by Rho GTPase signals. *Am. J. Physiol. Gastrointest. Liver Physiol.*, **2006**, *290*(4), G624-632.

[187] Riganti, C.; Miraglia, E.; Viarisio, D.; Costamagna, C.; Pescarmona, G.; Ghigo, D.; Bosia, A. Nitric oxide reverts the resistance to doxorubicin in human colon cancer cells by inhibiting the drug efflux. *Cancer Res.*, **2005**, *65*(2), 516-525.

[188] Riganti, C.; Doublier, S.; Costamagna, C.; Aldieri, E.; Pescarmona, G.; Ghigo, D.; Bosia, A. Activation of nuclear factor-kappa B pathway by simvastatin and RhoA silencing

increases doxorubicin cytotoxicity in human colon cancer HT29 cells. *Mol. Pharmacol.*, **2008**, *74*(2), 476-484.

[189] Riganti, C.; Orecchia, S.; Pescarmona, G.; Betta, P.G.; Ghigo, D.; Bosia, A. Statins revert doxorubicin resistance via nitric oxide in malignant mesothelioma. *Int. J. Cancer*, **2006**, *119*(1), 17-27.

[190] Doublier, S.; Riganti, C.; Voena, C.; Costamagna, C.; Aldieri, E.; Pescarmona, G.; Ghigo, D.; Bosia, A. RhoA silencing reverts the resistance to doxorubicin in human colon cancer cells. *Mol Cancer Res.*, **2008**, *6*(10), 1607-1620.

[191] Kobune, M.; Chiba, H.; Kato, J.; Kato, K.; Nakamura, K.; Kawano, Y.; Takada, K.; Takimoto, R.; Takayama, T.; Hamada, H.; Niitsu, Y. Wnt3/RhoA/ROCK signaling pathway is involved in adhesion-mediated drug resistance of multiple myeloma in an autocrine mechanism. *Mol Cancer Ther.*, **2007**, *6*(6), 1774-1784.

[192] Schmidmaier, R.; Baumann, P.; Simsek, M.; Dayyani, F.; Emmerich, B.; Meinhardt, G. The HMG-CoA reductase inhibitor simvastatin overcomes cell adhesion-mediated drug resistance in multiple myeloma by geranylgeranylation of Rho protein and activation of Rho kinase. *Blood*, **2004**, *104*(6), 1825-1832.

[193] Kang, W.K.; Lee, I.; Ko, U.; Park, C. Differential effects of RhoA signaling on anticancer agent-induced cell death. *Oncol. Rep.*, **2005**, *13*(2), 299-304.

[194] Smith, P.F.; Grossman, S.J.; Gerson, R.J.; Gordon, L.R.; Deluca, J.G.; Majka, J.A.; Wang, R.W.; Germershausen, J.I.; MacDonald, J.S. Studies on the mechanism of simvastatin-induced thyroid hypertrophy and follicular cell adenoma in the rat. *Toxicol. Pathol.*, **1991**, *19*(3), 197-205.

[195] Lamprecht, J.; Wójcik, C.; Jakóbisiak, M.; Stoehr, M.; Schrorter, D.; Paweletz, N. Lovastatin induces mitotic abnormalities in various cell lines. *Cell Biol. Int.*, **1999**, *23*(1), 51-60.

[196] Yasuda, Y.; Shimizu, M.; Shirakami, Y.; Sakai, H.; Kubota, M.; Hata, K.; Hirose, Y.; Tsurumi, H.; Tanaka, T.; Moriwaki, H. Pitavastatin inhibits azoxymethane-induced colonic preneoplastic lesions in C57BL/KsJ-db/db obese mice. *Cancer Sci.*, **2010**, *101*(7), 1701-1707.

[197] Shimizu, M.; Yasuda, Y.; Sakai, H.; Kubota, M.; Terakura, D.; Baba, A.; Ohno, T.; Kochi, T.; Tsurumi, H.; Tanaka, T.; Moriwaki, H. Pitavastatin suppresses diethylnitrosamine-induced liver preneoplasms in male C57BL/KsJ-db/db obese mice. *BMC Cancer*, **2011**, *11*, 281.

[198] Suzuki, S.; Tajima, T.; Sassa, S.; Kudo, H.; Okayasu, I.; Sakamoto, S. Preventive effect of fluvastatin on ulcerative colitis-associated carcinogenesis in mice. *Anticancer Res.*, **2006**, *26*(6B), 4223-4228.

[199] Swamy, M.V.; Patlolla, J.M.; Steele, V.E.; Kopelovich, L.; Reddy, B.S.; Rao, C.V. Chemoprevention of familial adenomatous polyposis by low doses of atorvastatin and celecoxib given individually and in combination to APCMin mice. *Cancer Res.*, **2006**, *66*(14), 7370-7377.

[200] Campbell, M.J.; Esserman, L.J.; Zhou, Y.; Shoemaker, M.; Lobo, M.; Borman, E.; Baehner, F.; Kumar, A.S.; Adduci, K.; Marx, C.; Petricoin, E.F.; Liotta, L.A.; Winters, M.; Benz, S.; Benz, C.C. Breast cancer growth prevention by statins. *Cancer Res.*, **2006**, *66*(17), 8707-8714.

[201] Mohammed, A.; Qian, L.; Janakiram, N.B.; Lightfoot, S.; Steele, V.E.; Rao, C.V. Atorvastatin delays progression of pancreatic lesions to carcinoma by regulating PI3/AKT signaling in p48(Cre/+) LSL-Kras(G12D/+) mice. *Int. J. Cancer*, **2012**, *131*(8):1951-1962.

[202] Limburg, P.J.; Mahoney, M.R.; Ziegler, K.L.; Sontag, S.J.; Schoen, R.E; Benya, R.; Lawson, M.J.; Weinberg, D.S.; Stoffel, E.; Chiorean, M.; Heigh, R.; Levine, J.; Della'Zanna, G.; Rodriguez, L.; Richmond, E.; Gostout, C.; Mandrekar, S.J.; Smyrk, T.C.; Cancer Prevention Network. Randomized phase II trial of sulindac; atorvastatin; and prebiotic dietary fiber for colorectal cancer chemoprevention. *Cancer Prev. Res. (Phila).*, **2011**, *4*(2), 259-269.

[203] Jacobs, E.J.; Newton, C.C.; Thun, M.J.; Gapstur, S.M. Long-term use of cholesterol-lowering drugs and cancer incidence in a large United States cohort. *Cancer Res.*, **2011**, *71*(5), 1763-1771.

[204] Graaf, M.R.; Beiderbeck, A.B.; Egberts, A.C.; Richel, D.J.; Guchelaar, H.J. The risk of cancer in users of statins. *J. Clin. Oncol.*, **2004**, *22*(12), 2388-2394.

[205] Khurana, V.; Sheth, A.; Caldito, G.; Barkin, J.S. Statins reduce the risk of pancreatic cancer in humans: a case-control study of half a million veterans. *Pancreas*, **2007**, *34*(2), 260-265.

[206] Platz, E.A.; Leitzmann, M.F.; Visvanathan, K.; Rimm, E.B.; Stampfer, M.J.; Willett, W.C.; Giovannucci, E. Statin drugs and risk of advanced prostate cancer. *J. Natl. Cancer Inst.*, **2006**, *98*(24), 1819-1825.

[207] Demierre, M.F.; Higgins, P.D.; Gruber, S.B.; Hawk, E.; Lippman, S.M. Statins and cancer prevention. *Nat. Rev. Cancer*, **2005**, *5*(12), 930-942.

[208] Hoffmeister, M.; Chang-Claude, J.; Brenner, H. Individual and joint use of statins and low-dose aspirin and risk of colorectal cancer: a population-based case-control study. *Int. J. Cancer*, **2007**, *121*(6), 1325-1330.

[209] Chang, Y.W.; Marlin, J.W.; Chance, T.W.; Jakobi, R. RhoA mediates cyclooxygenase-2 signaling to disrupt the formation of adherens junctions and increase cell motility. *Cancer Res.*, **2006**, *66*(24), 11700-11708.

[210] Guruswamy, S.; Rao, C.V. Multi-Target Approaches in Colon Cancer Chemoprevention Based on Systems Biology of Tumor Cell-Signaling. *Gene Regul. Syst. Bio.*, **2008**, *2*, 163-176.

[211] Knox, J.J.; Siu, L.L.; Chen, E.; Dimitroulakos, J.; Kamel-Reid, S.; Moore, M.J.; Chin, S.; Irish, J.; LaFramboise, S.; Oza, A.M. A Phase I trial of prolonged administration of lovastatin in patients with recurrent or metastatic squamous cell carcinoma of the head and neck or of the cervix. *Eur. J. Cancer*, **2005**, *41*(4), 523-530.

[212] Kornblau, S.M.; Banker, D.E.; Stirewalt, D.; Shen, D.; Lemker, E.; Verstovsek, S.; Estrov, Z.; Faderl, S.; Cortes, J.; Beran, M.; Jackson, C.E.; Chen, W.; Estey, E.; Appelbaum, F.R. Blockade of adaptive defensive changes in cholesterol uptake and synthesis in AML by the addition of pravastatin to idarubicin + high-dose Ara-C: a phase 1 study. *Blood*, **2007**, *109*(7), 2999-3006.

[213] Larner, J.; Jane, J.; Laws, E.; Packer, R.; Myers, C.; Shaffrey, M. A phase I-II trial of lovastatin for anaplastic astrocytoma and glioblastoma multiforme. *Am. J. Clin. Oncol.*, **1998**, *21*(6), 579-583.

[214] Paragh, G.; Fóris, G.; Paragh, G. Jr.; Seres, I.; Karányi, Z.; Fülöp, P.; Balogh, Z.; Kosztáczky, B.; Teichmann, F.; Kertai, P. Different anticancer effects of fluvastatin on primary hepatocellular tumors and metastases in rats. *Cancer Lett.*, **2005**, *222*(1), 17-22.

[215] Kawata, S.; Yamasaki, E.; Nagase, T.; Inui, Y.; Ito, N.; Matsuda, Y.; Inada, M.; Tamura, S.; Noda, S.; Imai, Y.; Matsuzawa, Y. Effect of pravastatin on survival in patients with advanced hepatocellular carcinoma. A randomized controlled trial. *Br. J. Cancer*, **2001**, *84*(7), 886-891.

[216] Lersch, C.; Schmelz, R.; Erdmann, J.; Hollweck, R.; Schulte-Frohlinde, E.; Eckel, F.; Nader, M.; Schusdziarra,V. Treatment of HCC with pravastatin; octreotide; or gemcitabine--a critical evaluation. *Hepatogastroenterology*, **2004**, *51*(58), 1099-1103.

[217] Van der Spek, E.; Bloem, A.C.; van de Donk, N.W.; Bogers, L.H.; van der Griend, R.; Kramer, M.H.; de Weerdt, O.; Wittebol, S.; Lokhorst, H.M. Dose-finding study of high-dose simvastatin combined with standard chemotherapy in patients with relapsed or refractory myeloma or lymphoma. *Haematologica*, **2006**, *91*(4), 542-545.

[218] Schmidmaier, R.; Baumann, P.; Bumeder, I.; Meinhardt, G.; Straka, C.; Emmerich, B. First clinical experience with simvastatin to overcome drug resistance in refractory multiple myeloma. *Eur. J. Haematol.*, **2007**, *79*(3), 240-243.

[219] Hus, M.; Grzasko, N.; Szostek, M.; Pluta, A.; Helbig, G.; Woszczyk, D.; Adamczyk-Cioch, M.; Jawniak, D.; Legiec, W.; Morawska, M.; Kozinska, J.; Waciński, P.; Dmoszynska, A. Thalidomide; dexamethasone and lovastatin with autologous stem cell transplantation as a salvage immunomodulatory therapy in patients with relapsed and refractory multiple myeloma. *Ann. Hematol.*, **2011**, *90*(10), 1161-1166.

[220] Li, H.Y.; Appelbaum, F.R.; Willman, C.L.; Zager, R.A.; Banker, D.E. Cholesterol-modulating agents kill acute myeloid leukemia cells and sensitize them to therapeutics by blocking adaptive cholesterol responses. *Blood*, **2003**, *101*(9), 3628-3634.

[221] Van Beek, E.; Pieterman, E.; Cohen, L.; Löwik, C.; Papapoulos, S. Nitrogen-containing bisphosphonates inhibit isopentenyl pyrophosphate isomerase/farnesyl pyrophosphate synthase activity with relative potencies corresponding to their antiresorptive potencies *in vitro* and *in vivo*. *Biochem. Biophys. Res. Commun.*, **1999**, *255*(2), 491-494.

[222] Cartenì, G.; Bordonaro, R.; Giotta, F.; Lorusso, V.; Scalone, S.; Vinaccia, V.; Rondena, R.; Amadori, D. Efficacy and safety of zoledronic acid in patients with breast cancer metastatic to bone: a multicenter clinical trial. *Oncologist*, **2006**, *11*(7), 841-848.

[223] Hashimoto, K.; Morishige, K.; Sawada, K.; Tahara, M.; Shimizu, S.; Ogata, S.; Sakata, M.; Tasaka, K.; Kimura, T. Alendronate suppresses tumor angiogenesis by inhibiting Rho activation of endothelial cells. *Biochem. Biophys. Res. Commun.*, **2007**, *354*(2), 478-484.

[224] Vincenzi, B.; Santini, D.; Avvisati, G.; Baldi, A.; Cesa, A.L.; Tonini, G. Statins may potentiate bisphosphonates anticancer properties: a new pharmacological approach? *Med. Hypotheses*, **2003**, *61*(1), 98-101.

[225] Issat, T.; Nowis, D.; Legat, M.; Makowski, M.; Klejman, M.P.; Urbanski, J.; Skierski, J.; Koronkiewicz, M.; Stoklosa, T.; Brzezinska, A.; Bil, J.; Gietka, J.; Jakóbisiak, M.; Golab, J. Potentiated antitumor effects of the combination treatment with statins and pamidronate *in vitro* and *in vivo*. *Int. J. Oncol.*, **2007**, *30*(6), 1413-1425.

[226] Gest, C.; Mirshahi, P.; Li, H.; Pritchard, L.L.; Joimel, U.; Blot, E.; Chidiac, J.; Poletto, B.; Vannier, J.P.; Varin, R.; Mirshahi, M.; Cazin, L.; Pujade-Lauraine, E.; Soria, J.; Soria, C. Ovarian cancer: Stat3; RhoA and IGF-IR as therapeutic targets. *Cancer Lett.*, **2012**, *317*(2), 207-217.

[227] Van Goietsenoven, G.; Mathieu, V.; Lefranc, F.; Kornienko, A.; Evidente, A.; Kiss, R. Narciclasine as well as other Amaryllidaceae Isocarbostyrils are Promising GTP-ase Targeting Agents against Brain Cancers. *Med. Res. Rev.*, **2012**, doi: 10.1002/med.21253.

CHAPTER 8

Carotenoids that are Involved in Prostate Cancer Risk

Charles Y.F. Young[1,*], K.V. Donkena[1], H.-Q. Yuan[2], M.-L. He[3] and J.-Y. Zhang[2]

[1]Departments of Urology and Biochemistry and Molecular Biology, Mayo Clinic College of Medicine, Mayo Clinic, Rochester, Minnesota, USA; [2]Institute of Biochemistry and Molecular Biology, Medical College, Shandong University, Jinan, People's Republic of China and [3]Institute of Cancer Research, Life Science School, Tongji University, Shanghai, People's Republic of China

Abstract: Chemoprevention is presumably one of most effective means to combat prostate cancer (PCa). Patients usually require more than a decade to develop a clinically significant PCa, therefore, an ideal target for chemoprevention. This review will focus on recent findings of a group of naturally occurring chemicals, carotenoids, for potential use in reducing PCa risk.

Keywords: Antioxidant, cancer risk, carotene, carotenoids, chemoprevention, prostate cancer, retinoids, lycopene.

INTRODUCTION

Prostate cancer (PCa) is the second leading cancer killer, next to lung cancer, in male population of many highly developed countries including USA, UK and others. Inspiration from low PCa incidence in Asian countries and evidence from the more recent epidemiologic studies, nutrition and diet may indeed impact PCa as well as other types of cancer risk. It has been strongly suggested that chemoprevention would be a highly effective means to combat this disease due to its high incidence, long latency of becoming clinically significant cancer and disease-related morbidity and mortality [1-6]. It has been suggested that proper diets may eventually reduce 50-60% incidence of many types of cancer [1, 5, 6].

This review will concentrate on a class of phytochemicals, carotenoids, for their effects in prevention of PCa. Carotenoids are unique constituents of a healthy diet

***Address correspondence to Charles Y.F. Young:** Urology Research, Mayo Clinic, Rochester, MN, 55905, USA; Tel: +1 507 284 9247; Fax: +1 507 284 2384; E-mail: young.charles@mayo.edu in which they may play an important role in the network of antioxidant vitamins and phytochemicals, therefore beneficial in preventing many human diseases including cancers. Some of carotenoids may be viewed as pro-vitamin A because they can be converted to vitamin A or retinoids after consumed and absorbed by the body. However, this review will mainly focus on carotenoids. In some cases, retinoids may be also discussed side by side with carotenoids.

TYPES OF CAROTENOIDS AND THEIR DISTRIBUTION IN DIETS

Up to now, there are at least 500-600 different carotenoids being identified [7-9]. Carotenoids are fat-soluble tetraterpenoids, most of which contain a central carbon chain of alternating single and double bonds with different cyclic or acyclic end groups. The double bonds may show an array of *cis*/*trans* (E/Z) configurations in a given carotenoid [7-9]. It has been shown that the all-*trans* form is thermodynamically most stable and predominant in nature but several *cis* isomers of carotenoids can be detected in blood and tissues [10, 11]. Although carotenoids can be found in animals and plants, they are only synthesized in plants and some microorganisms including bacteria, yeasts, and molds. The major forms of carotenoids in plant foods and other food products are shown in Fig. **1** and Table **1**. Animals and humans can not synthesize carotenoids *de novo*, thus depending on dietary supply. Carotenoids may be divided into two main groups, *i.e.*, carotenes and xanthophylls [7]. β-Carotene, α-carotene, and lycopene are important members of the carotene group composed only of carbon and hydrogen atoms. The major xanthophylls including zeaxanthin, lutein, α- and β-cryptoxanthin, canthaxanthin and astaxanthin carry at least one oxygen atom. Note, because xanthophylls contain at least one hydroxyl group, they are more polar than carotenes. Moreover, as mentioned above some carotenoids, viewed as pro-vitamin A, can be converted to vitamin A or retinoic acid, including α-carotene, β-carotene and β-cryptoxanthin [7, 9]. The conversion of pro-vitamin A carotenoids like β-carotene to retinal can occur in the small bowel mucosa and in the liver by β, β-carotene 15,15' monooxygenase (previously termed beta-carotene 15,15' dioxygenase) at center of the carotene. Retinal can be further reduced by retinal resductase to retinol [8-13]. Another enzyme was found to be able to catalyze an eccentric cleavage of carotenoids into apo-carotenoids and retinal [9, 10, 12, 13]. Although retinoids possess anti-cancer activity, the non-pro-vitamin A carotenoids (*e.g.*, lycopene) also possess anti-cancer activity as demonstrated in laboratory tests and experimental models of cancer and have been associated with lower cancer risk in epidemiological studies [1, 3, 9, 13]. As mentioned above there are a relatively large number of carotenoids in natural source, however, only approximately 50 of them are present in common vegetables and fruits in US diets [8-10]. Finally, perhaps just over a dozen of these carotenoids or their metabolites can be detected in human blood and tissues which may be attributed to human health [8, 13].

Table 1: Major Carotenoids (μg/100g edible portion) in some Vegetables, Fruits and Food Products

	α-carotene	**β-Carotene**	**β-Cryptoxanthin**	**Lutein and Zeaxanthin**	**Lycopene**
Apples raw with skin	30	-	-	-	-
Apricots	-	6,640	-	-	65
Asparagus, raw	12	493	-	-	-
Avacados raw	28	53	56	-	-
Bananas	5	21	-	-	-
Bean, snap, green, raw	68	377	-	640	-
Blueberries, raw	-	35	-	-	-
Broccoli, raw	1	779	-	2,445	-
Cabbage, raw	-	65	-	310	-
Carrots, baby, raw	4,425	7,275	-	358	-
Carrots, raw	4,649	8,836	-	-	-
Celery, raw	-	150	-	232	-
Cherries, sweet, raw	-	28	-	-	-
Corn, canned	33	30	-	884	-
Grapefruit, raw, pink, red	5	603	12	12	1,462
Kale, raw	-	6,202	-	15,798	-
Lettuce, romaine, raw	-	1,272	-	2,635	-
Mangos, raw	17	445	11	-	-
Melons, cantaloupe, raw	27	1,595	-	40	-
Orange juice, raw	2	4	15	36	-
Orange, raw	16	51	122	187	-
Papayas, raw	-	276	761	75	-
Peas, green frozen	33	320	-	-	-

Table 1: contd....

Peppers, sweet, green, raw	22	198	-	-	-
Peppers, sweet, red raw	59	2,379	2,205	-	-
Pumpkin, canned, no salt	4,795	6,940	-	-	-
Spinach, raw	-	5,597	-	11,938	-
Tomato, red, ripe, raw	112	393	-	130	3,025
Tomato juice, canned	-	428	-	60	9,318
Watermelon, raw	-	295	103	17	4,868
Beef, variety meats	-	621	-	-	-
Butter	-	158	-	-	-
Cheese, cheddar	-	85	-	-	-
Egg, whole, raw	-	-	-	55	-
Margarine, regular	-	485	-	-	-
Sauce, pasta	-	440	-	160	15,990
Soup, vegetarian	410	1,500	-	160	1,930
vegetable,canned,					
condensed, commercial					
Soup, vegetable beef,	489	1,618	-	2	364
canned, condensed,					
commercial					

"-" : not available or not detectable; Adopted from (Database, 1998 Database, U.-N.C. 1998. USDA-NCC Carotenoid Database for US Foods-1998. Available from: <www.nal.usdea.gov/fnic/foodcomp/Data/car98/car98.html>).

Among carotenoids, lycopene has drawn a heavy attention to its potential role responsible for benefits to preventing chronic diseases including PCa by consumption of tomato or tomato products [1, 3, 8, 9, 13]. Interesting aspect is that the prostate is one of preferential tissues to accumulate lycopene and this preference is more towards malignant than benign prostate [14, 15]. Also lycopene detected in the prostate is predominantly cis-forms while mainly all-trans form lycopene is present in plant foods [14, 16-18]. All-trans to cis isomerization of lycopene can occur *in vitro* by heating process [19] or *in vivo* in the gastrointestinal lumen, liver, enterocytes and other tissues [20-24]. Further studies will be required to clarify the bioactivity of isoforms of lycopene in their

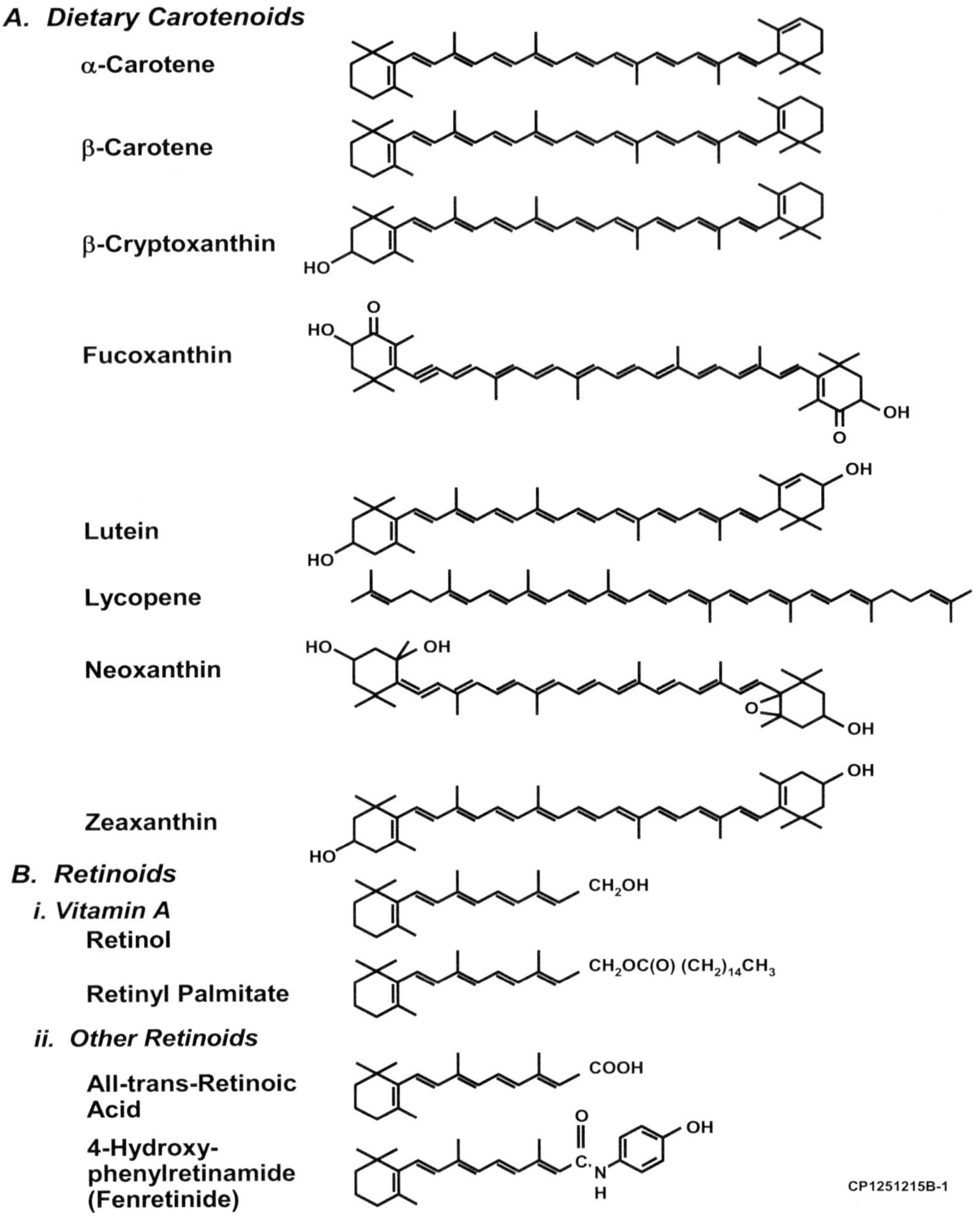

Figure 1: Chemical structures of some common carotenoids and retinoids.

anti-prostate cancer action. Recently, is has been proposed that *in vivo* lycopene metabolites or oxidation products such as apo-12'-lycopenal are also bioactive against cancer development [15, 25, 26].

Mechanisms of Anti-Prostate Cancer Activities of Carotenoids

Antioxidant Activities

Excessive reactive oxygen (ROS) and nitrogen species (RNS) can be produced during aerobic metabolism and pathological processes and then cause damages in cellular lipids, DNA or proteins [13, 27-30]. Carotenoids are usually involved in the scavenging of two types of the ROS, singlet molecular oxygen ($^1O_2$), and peroxyl radicals [29, 30]. Carotenoids can directly transfer excitation energy from $^1O_2$ and generates ground state oxygen and an excited triplet state carotenoid, which further transform the energy into heat and the ground state intact carotenoid [9, 11, 31, 32]. The potency of this kind of quench effects is closely related to the number of conjugated double bonds of a given carotenoid, thus lycopene is the top quencher among carotenoids. Carotenoids like lycopene can also trap other ROS and RNS, like $OH^{\surd}$, $NO_2^{\surd}$ or peroxynitrite, however, leading to oxidative breakdown of the carotenoid molecules [28, 33-36]. This action seems to suggest that the carotenoids may act in the front line of defense. It has been demonstrated that lycopene can reduce oxidative DNA damage in cell cultures and in animal models [37-39]. However, relatively high concenrations of lycopene may induce DNA oxidation /damage in the culture of a human PCa cell line [40]. Also, human studies showed that tomato consumption could protect human leukocytes against oxidative DNA damage [41-44]. Additional study [45] showed that lycopene-rich tomato sauce consumption could reduce oxidative DNA damage in human prostate. Nonetheless, the same group of investigators in a more recent study [46] showed increased lycopene in plasma and prostatic tissues by oral adiminstration did not reduce DNA oxidation and lipid peroxidation in a cohort of 105 African American men.

Interestingly, it has been suggested that when two structurally different antioxidant compounds are mixed together, they can generate synergistic protection activity against increased oxidative stress [8, 9]. By pairing a number of carotenoids in a multilamellar liposomes assay system for protection of oxidative damage, it was found that the pairing of other carotenoids or vitamin E with lycopene or lutein produced the most profound synergistic effect [47]. Among many individual antioxidants, vitamins E, C and β-carotene exhibit cooperative synergistic effects scavenging RNS [48, 49]. A cooperative interaction between α-tocopherol and β-carotene can be observed in a membrane model [50], showing a synergistic inhibitory effect of lipid peroxidation. The synergistic protection effect may be related to the specific positioning of different carotenoids and other lipophilic antioxidants in membranes [51]. It has been observed that lycopene at its physiological concentrations (less than 1 uM) can effectively inhibit PCa proliferation when only the presence of physiological concentrations (50 μM) of alpha-tocopherol (*i.e.*, alpha-vitamin E), but not with beta-tocopherol, ascorbic acid or probucol [52]. Similar results [53] were also obtained in xenograft of human prostate cancer cell line PC-3 in nude mice in which a daily dose of lycopene (5 or 50 mg/kg body weight), alpha-tocopheryl acetate (5 or 50 mg/kg body weight), a mixture of both, or vehicle was given. The results demonstrated that only the combined treatments of lycopene and vitamine E could significantly repress orthotopic tumor growth by 73% at day 42 as well as increase median survival time by 40% from 47 to 66 days. Furthermore, rat MatLyLu Dunning prostate tumor [54] was used to address the *in vivo* action of lycopene and vitamin E. Diet supplementation for 4 weeks with 200 ppm lycopene, 540 ppm vitamin E or both was applied to rats receiving the tumor cells orthotopically in the prostate. After 14 days tumor cell injection, tumor growth was examined by magnetic resonance imaging showing a significant increase in necrotic area in rats treated with vitamin E, lycopene or both. However, the co-treatment of both agents did not show better effects than the treatment with each agent alone Gene profiling analysis demonstrated that lycopene and vitamin may repress both prostatic anrdrogen signaling pathways and prostatic expression of interleukin 6 and insulin like growth factor-1 (IGF-1), therefore, interfering with internal autocrine or paracrine regulatory pathways for prostate tumor growth. Later, the same group of the authors [55] showed lycopene can reduce local prostatic androgen signaling, IGF-I expression, and basal inflammatory signals in normal rat prostate tissue, when young rats administered with 200 ppm lycopene in diet for up to 8 weeks. Although these studies showed the specific *in vivo* gene regulation targets by lycopene, it is still not clear if the action of lycopene is directly derived from its anti-oxidant activity. Moreover, in a subcutaneously transplanted, androgen-dependent Dunning R3327-H rat prostate tumor model fed with lycopene (250 mg/kg diet), gamma-tocopherol (200 mg/kg diet) or both, the authors [56] did not observe any tumor growth inhibitory effects from any of these treatments.

Another related antioxidant mediated anticancer activity of carotenoids is their potential of increase of antioxidant enzymes and detoxifiying enzymes [57]. However, the data varies in different experiments. A study by Gradelet *et al.,* [58] was designed to investigate if carotenoids such as canthaxanthin, astaxanthin, lycopene and lutein can alter liver drug metabolizing enzymes in male rats. The results showed after 15 days of feeding with diets containing each those carotenoids, canthaxanthin and astaxanthin indeed increased the liver content of P450, and the activities of NADH- and NADPH-cytochrome c reductase. However, lycopnene and lutein showed very little effect. Further, Breinholt *et al.*, [59] treated female rats with lycopnene at concentrations from 0.001 to 0.1 g/kg body weight/day for 2 weeks and examined alteration of drug-metabolizing (benzyloxyresorufin O-dealkylase, ethoxyresorufin O-dealkylase, hepatic quinone reductase and glutathione transferase) and antioxidant enzyme (superoxide dismutase, glutathione reductase and glutathione peroxidase) levels. The authors concluded that lycopene can enhance, although not particularly high, those enzyme activities in rat which might be relevant to the protective effects of lycopnene against human cancer. An earlier *in vitro* study [60] using Colo205 colon cancer cells showed lycopene and beta-carotene had no effect on levels of NAD(P)H:quinone reductase (QR) and GST although retinol and retinoic acid could stimulate QR activity.

Recently it has been shown [61] that carotenoids tested, especially lycopene may up-regulate phase II detoxification enzymes through activation of Nrf2 transcription factor and its binding to the antioxidant response element (ARE) of phase II detoxification enzyme genes. The study showed that lycopene induced an increase of the phase II enzymes NAD(P)H:quinone oxidoreductase and gamma-glutamylcysteine synthetase at mRNA and protein levels in human cancer cell lines tested. Further experiments [61] showed that the antioxidant actitvity of the carotenoids is not directly related to their induction ability of the phase II enzymes. In fact, it was concluded [62] that the oxidation products of lycopene and other carotenoids are responsible for the activation of the electrophilic/ARE system in prostate and breast cancer cell lines.

Biological Functions Other than Antioxidant Activities

Non-provitamin A cartenoids and retinoids may cause cell cycle retention with which IGF-I receptor signaling was diminished as well as cyclin D levels and phosphorylation of the retinoblastoma protein [63, 64] were reduced. Many biological effects pertinent to anticancer potentials of carotenoids including cell cycle/proliferation/apoptosis, cell communication and others might not totally rely on their primary antioxidant function, additional mechanisms may play a role. The studies [55, 56] in the rat system described in the above section seemed to indicate part of lycopene's anti-PCa activities may not be associated with its anti-oxidant activity. It was also found [65] that lycopene may change the mevalonate pathway by inhibiting 3-hydroxy-3-methylglutaryl-coenzyme. A reductase expression and cause Ras inactivation with subsequent cell growth inhibition of several cancer cells including human PCa cell lines. Recently, Kotake-Nara *et al.*, [66] compared 15 carotenoids for their antiproliferation effects on three human PCa cell lines and concluded that phytofluene, zeta-carotene and lycopene but not phytoene, canthaxanthin, beta-cryptoxanthin and zeaxanthin could significantly affect the growth of the PCa cells. In addition, the authors [67] found that neoxanthin and fucoxanthin, rich mainly in spinach and edible brown algae, respectively, are more effective than other carotenoids tested in anti-proliferation activity of three human PCa cell lines. These two carotenoids could decrease the expression of Bax and Bcl-2 proteins and induced apoptosis *via* caspase 3 activation.

As mentioned above, some studies [54, 55] showed lycopene may reduce androgen signaling, IGF-I expression, and basal inflammatory signals in local normal prostate or tumor tissues of rats. In mouse xenograft studies [68], lycopene seemed to reduce plasma levels of insulin-like growth factor-binding protein-3 and vascular endothelial growth factor which may be, in part, linked to tumor growth inhibitory effects of lycopene. Global gene expression in response to a 3-month intervention of lycopene supplement was analyzed on a grourp of 84 men with low risk of PCa [69]. It was conluded that there are no significant individual genes associated with the lycopene intervention, although an exploratory analysis suggested Nrf2-mediated oxidative stress response may be an *in vivo* pathway modulated by lycopene in human prostatic tissues.

Retenoids in our body system are obtained from pro-vitamin A carotenoids or preformed retinoids whose biological functions including anti-cancer effects are mainly mediated *via* two classes of nuclear receptor family, *i.e.*, retinoic acid receptors (RARs) alpha, beta, gamma and retinoid X receptors (RXRs) alpha, beta, gamma [57, 70]. All-trans-retinoic acid and the isomer 9-cisretinoic acid are ligands for RARs and RXRs, respectively. The activation of the receptors by their ligands can therefore, exert cell proliferation and differentiation. In addition, there are several hundred of structurally unrelated, synthetic retinoids exhibiting similar effects. Retinoids have been suggested to play a role in regulation of prostate growth and differentiation [70-72]. Deficiency in vitamin A can affect some early development of the prostate as also shown in the mouse knockouts of RAR gamma with metaplasia and keratinization [72]. The RAR gamma knockout [72] showed prostatic lumen defect in secretion. In laboratory studies, retinoids alone or with other agents such as vitamin D [73], can cause cell cycle arrest, inhibit PCa proliferation or induce apoptosis as well as suppress prostate carcinogenesis [70]. Retinoids may down-regulate anti-apoptotic Bcl2 and upregulate RAR beta and tissue transgluaminase in PCa cells [74]. Retinoic acid was also shown to inhibit the function of the androgen receptor [75]. Recently, epigenetic events such as promoter DNA hypermethylation have been frequently observed in many cancers including PCa and have been suggested to be involved in carcinogenesis. RARbeta2 (RARβ2) is one of frequently methylated genes whose expression is abnormally lowered in malignant cells compared to the normal counterparts [76, 77]. Whether this indicates the role of RARb2 in prostate carcinogenesis or cancer progression is not clear. Furthermore, whether reactivation of RARb2 expression by demethylating agents (*e.g.*, 5'aza-deoxycytidine) [77-79] in combination with use of retinoids can enhance the efficacy for PCa treatment may be worth an investigation.

It has been shown that both retinoids and non-provitamin A carotenoids can increase levels of a gap junctional communication (GJC) protein connexin 43 (Cx43) in several cancer cells [80-83]. Gap junctions that allow exchange of low molecular cellular components (< 1000 Da) between neighboring cells seem to be critical for maintaining normal cell phenotype [80-82]. Although the mechanisms are not well understood, GJC is implicated in controlling cell proliferation, differentiation and apoptosis. Malignant cells are usually deficient in GJC [84, 85]. Restoration of GJC proteins may resume some features of normal cell phenotype and repress cell transformation. Thus, GJC proteins or connexin family proteins may be viewed as tumor suppressors. Experiments were also performed and found that there was no correlation between the antioxidant capacity of carotenoids tested and the ability of increase in GJC proteins, suggesting that the antioxidant activities of carotenoids was not related to their effect on regulating GJC proteins [1, 86, 87]. In clinical trials, applying high doses of supplemental lycopene to PCa patients for 3 weeks prior to radical prostatectomy appeared to increase expression of Cx43 and decreased pathological severity of treated *versus* control group [88]. Similar studies were performed to show the potential benefits of lycopene to improve clinical parameters in PCa patients [89, 90]. However, an *in vitro* study showed lycopene failed to inhibit cell prolifration and to increase expression of Cx43 in a highly meatstatic PC-3 subline [91, 92]. On the other hand, Cx43 protein was not altered by lycopene and apo-12'-lycopenal even at supra-physiological concentrations, while human DU145 PCa cell proliferation was significantly reduced [25] Finally, whether this can implicate the *in vivo* heterogenecity of clonal PCa cells that can arise to lycopene resistance remains to to be determined.

Both retinoids and carotenoids up-regulate connexin mainly at the transcriptional levels [1, 82, 87]. However it seems that retenoids but not non-provitamin A carotenoids (*e.g.*, beta carotene, astaxanthin and lycopene) activate connexin *via* RARs, because RAR antagonists showed to block retinoid induced connexin expression at the transciptional and protein levels. The aforementioned non-provitamin A carotenoids have very little affinity to RARs or RXRs. In contrast, peroxisome proliferator activated receptor (PPAR) antagonist GW9662 could only block activation of connexin expression by carotenoids but not retinoids [86]. Whether the non provitamin A carotenoids can directly bind the PPAR remains to be addressed. However, it has been proposed [1, 87] that there are separate mechanisms to mediate the regulation of the connexin 43 gene expression by carotenoids or retinoids. Moreover, by transfection assays the responsive region of the connexin 43 gene promoter to both retinoids and the carotenoids was found within -158 bp and +209 bp of the transcription start site of the gene. Intriguingly this region contains no regular canonical RAR or PPAR response element but a Sp1/Sp3 GC-box. It certainly requires more studies to unravel the actual regulatory mechanisms used by the non-provitamin A carotenoids and the retinoids.

Fenretinide or N-(4-hydroxyphenyl) retinamide is a synthetic retinoid whose anti-tumor activities may be largely independent of retinoids receptors [93, 94]. Many *in vitro* and *in vivo* studies seem to suggest that fenretinide be a potential candidate for a chemopreventive and therapeutic agent [93-100]. It has been shown that fenretinide induces prostate cell death by generating reactive oxygen species [95] or mdulating mitochondrial functions [96]. Recent studies [7] showed that fenretinide can down regulate FAK, AKT phosphorylation, beta-catenin stability, VEGF and cyclin D1 expression with subsequent effects on reduced cell proliferation, cell migration and invasion and angiogenesis of PCa cells. In addition, animal studies demonstrated that fenretinide can effectively reduce prostate cancer development and progression [98-100]. However, human studies could not showed its efficacies as prostate cancer preventive or therapeutic agent [101-104].

Demonstration of Anti-Prostate Cancer Activities in Human Studies

An early study [105] investigated the relationship of diet and life style with PCa risk in a cohort of approximately 14,000 Seventh-day Adventist men who completed a detailed lifestyle questionnaire in 1976. For a 6-year follow-up, 180 histologically proven PCa were detected. One important finding was that increasing consumption of tomatoes and other fruits and vegetables was significantly associated with decreased PCa risk. Subsequently, the US Health Professionals Follow-up Study (HPFS) [106] showed there was an inverse association between high intake of tomato products and PCa risk. This study suggested that lycopene intake from tomato-derived products but not the overall intake of fruits and vegetables was associated with lowered PCa risk. The same latter investigator group conducted another analysis of the HPFS from yr1986 to 1998 consisting of 47,365 participants with 2481 men being developed PCa [107]. Although the study showed an association of lycopene and frequent consumption of tomato products with reduced PCa, the authors did acknowledge the degree of the inverse association was moderate so that a small study or one with improper assessment would miss the significance. Another case-control nested analysis of the same HPFS [108] with 450 matched PCa men showed modest inverse, but not statistically significant, associations of PCa risk with plasma levels of α-carotene, beta-carotene, and lycopene. However, there was a statistically significant inverse association between higher plasma lycopene levels and lower risk of PCa in a sub group of men with older age or whose family history showed no PCa.

A population-based case-control study (cases, n=193 *vs.* controls, n=197) [109] in Arkansas also showed that high plasma levels of lycopene, lutein/zeaxanthin, and beta-cryptoxanthin have associations with a low risk of PCa. A nested case-control study [110] was performed with 612 incident PCa cases and 612 matched controls in the Health Professionals Follow-up Study in order to determine the relationship of the MnSOD gene Ala16Val polymorphism and plasma lycopene concentrations with risk of aggressive PCa. The conclusion of this study was that men with the MnSOD Ala/Ala genotype and a low long-term lycopene status would show an increased risk of aggressive PCa.

On the other hand, regarding the protective effects of dietary or supplemental carotenoids, many observational epidemiological or clinical studies produced inconsistent results. For example, there were a few large clinical trials [111-114] as well as an observantional study with American Japanese in Hawaii [115] indicating no obvious association of beta-carotene supplementation or serum carotenoid levels with PCa risk. Yet, more recently, a relatively small case-control study with 118 non-Hispanic Caucasian with nonmetastatic PCa and 52 healthy men mainly from southeast Texas was conducted to evaluate associations between cancer risk and plasma levels of total carotenoids and several types of carotenoids [116]. The conclusion of the study suggested that higher circulating levels of alpha-cryptoxanthin, alpha-carotene, trans-beta-carotene, lutein and zeaxanthin as well as cis-lycopene isomer 1 but not total lycopenes, or other lycopene isoforms could contribute to lower PCa risk. In the Prostate, Lung, Colorectal, and Ovarian Cancer Screening Trial with 1338 cases of PCa among 29361 men for a 8 years follow-up [117], it was concluded that there was no overall association of PCa risk with intake of beta-carotene and other anti-oxidants. Similar conclusion was obtained for lycopene or tomato products [118]. In a case-control analysis (966 PCa casese *vs.* 1064 controls) of the European Prospective Investigation into Cancer and Nutrition Study [119] with a mean 6 yr follow-up was conducted to measure plasma concentrations of 7 carotenoids and retinol, and determine their associations with PCa risk. The authors found that there was no associations between plasma concentrations of carotenoids or retinol and PCa risk. However, high levels of lycopene and total carotenoids may reduce risk of advanced PCa. In a mean follow-up time of 12.6 years study [120] of 997 middle-aged Finnish men, of which 55 had PCa developed, it was found that there was no association between serum lycopene concentrations and PCa risk. Moreover, there are two studies [121, 122] resulted from the Prostate Cancer Prevention Trial investigating if serum lycopene levels or dietary supplements for lycopene can generate impact on PCa risk. Consistently, these two reports showed no significant association of lycopene in blood or lycopene supplement intake with PCa risk. However, alternative interpretations for the above results have been provided by a different viewer [123]. Moreover, a prospective nested case-control study [124] showed there was no association of serum concentrations of lycopene, β-carotene, and vitamin A with PCa development. A prospective, open phase II pilot study [125] was conducted on 18 patients with progressive hormone refractory PCa by giving 15 mg lycopene supplement daily for 6 months. The authos concluded that the participants did not show clinically related benefits. Then, a randomized, double-blind, placebo-controlled prevention trial (*i.e.*, the alpha-Tocopherol, beta-Carotene Cancer Prevention Study) [126] used beta-carotene and alpha-tocopherol supplements to invesitgate PCa survival of adult male smokers in southwestern parts of Finland. The results showed that either serum levels of beta-carotene and retinol or beta-carotene in supplements had no obvious effects on PCa survival.

Analyses of the above study [127] and other study [128] indicated that high serum beta-carotene may be associated with increased risk of aggressive PCa. In contrast, the Prostate, Lung, Colorectal, and Ovarian Cancer Screening Trial with a nested case-control study consisting of 692 PCa cases and 844 matched controls [129] concluded that increased serum levels of retinol may have effects on decreasing aggressive PCa risk. Nonetheless, the study [127] mentioned above also suggested that high supplemental beta-carotene intake *et al.*, least 2000 mug/day was associated with reduced risk of the cancer only in men with low dietary beta-carotene intakes. Another case-control study of diet and PCa in western New York with 433 men with primary PCa and 538 population-based controls [130] also revealed similar results in that only those men in the highest quartile of intake of beta-carotene, alpha-carotene, lutein, and lycopene and other phytocompounds showed reduced PCa risk when compared to men in the lowest quartile of intakes.

As already mentioned above in human studies as well as in many *in vitro* laboratory tests and animal studies, lycopene may be the one showing more promising role in preventing PCa than other carotenoids since it shows more consistent results in human studies [3] although there were some conflicting results existed [131]. Evaluating most of the studies we can still strongly argue that, overall, carotenoids especially lycopene may have protection effect against PCa development or progression. Clearly more accurate information of the carotenoid content of foods and amounts of relevant food consumption as well as absorption, metabolic, genetic and epigenetic factors of participant individuals in studies that could complicate the interpretation of results or act as confounding factors should be included, clarified or minimized in order to improve the evaluation.

Potential Harms of Use of Carotenoids in Cancer Prevention

Although there are still some debates about if higher intake of fruits and vegetables is more effective than use of supplements of dietary compounds in cancer prevention, little evidence showed higher intake of fruits and vegetables can increase cancer risk. However, as mentioned in the above section, some epidemiological studies [127, 128] seem to indicate that high-serum β-carotene levels may increase PCa risk. In addition, studies [113, 114, 132, 133] indicated that tobacco may interact with isolated forms of carotenoids and produced unexpected effects--increasing cancer risk. A double-blind, placebo-controlled trial, the Alpha-Tocopherol-Beta-Carotene (ATBC) trial [132, 134], was performed to demonstrate anti-cancer effects of two anti-oxidants, beta-carotene and vitamin E. With 29,133 male smokers receiving a daily supplement of either 20 mg of beta-carotene or 50 mg of vitamin E, both, or a placebo for a 5 to 8 year follow up, the partipicipants receiving beta-carotene had a statistically significantly higher lung cancer risk than did men who only received placebo. The intake of beta-carotene also increased the mortality from cardiovascular disease in the same study. Later, the Carotene and Retinol Efficacy Trial (CARET) [113, 114] was designed to use beta-carotene with retinol for lung cancer prevention in men and women smokers and/or asbestos workers. The study also showed similar results that lung cancer risk was statistically significantly increased in active heavy smokers. There are other studies [135-137] showed similar results in which high intake of beta-carotene can increase cancer risk in active heavy active smokers but may benefit or neutral to those never smokers. Human and animal studies showed [135, 138-142], indeed, carotene may have interaction with tobacco metabolites, the actual mechanism(s) involved is still poorly understood. However, so far, there is no indication if lycopene can enhance cancer risk in smokers yet. Although whether tobacco use can increase PCa risk is still not entirely clear and requires more studies to clarify it. Nonetheless, certain awareness needs to keep in mind that smokers should avoid the use of high dose carotenoid supplements for cancer prevention.

CONCLUSIONS

According to a variety of studies shown in the literature, carotenoids including carotenes and lycopene seem to be promising dietary compounds for reducing PCa risk. Neoxanthin and fucoxanthin may be new promising chemopreventive carotenoids for PCa that require more extensive studies in the near future. Whether that tobacco use can increase PCa risk and carotenoids can further enhance the tobacco effects on PCa is not certain at presnt time, precaution should be taken when isolated carotenoids as diet supplements are considered to be chronically used for disease or PCa prevention purposes.

ACKNOWLEDGEMENTS

This manuscript is in part supported by the NIH grant CA89000.

CONFLICT OF INTEREST

The authors confirm that this chapter contents have no conflict of interest.

DISCLOSURE

The chapter submitted for eBook Series entitled: "**Recent Advances in Medicinal Chemistry, Volume 1**" is an update of our article published in **Mini-Reviews in Medicinal Chemistry, Volume 8, Number 5, pp. 529 to 537**, with additional text and references.

REFERENCES

[1] Bertram, J.S.; Vine, A.L. Cancer prevention by retinoids and carotenoids: independent action on a common target. Biochim Biophys Acta. **2005**, *1740,* 170-178.

[2] Brand,T.C.; Canby-Hagino, E.D.; Pratap Kumar, A.; Ghosh, R.; Leach R.J.; Thompson, I.M. Chemoprevention of prostate cancer.*Hematol Oncol Clin North Am.*, **2006**, *20,* 831-843.

[3] Chan, J.M.; Gann, P.H.; Giovannucci, E.L. Role of diet in prostate cancer development and progression. J Clin Oncol. **2005**, 23, 8152-8160.

[4] Khan, N.; Afaq, F.; Saleem, M.; Ahmad, N.; Mukhtar, H. Targeting multiple signaling pathways by green tea polyphenol (-)-epigallocatechin-3-gallate.*Cancer Res.* **2006**, *66,* 2500-2505.

[5] Klein, E.A. Chemoprevention of prostate cancer. *Annu Rev Med.* **2006**, *57,* 49-63.

[6] Lippman, S.M.; Lee, J.J. Reducing the "risk" of chemoprevention: defining and targeting high risk--2005 AACR Cancer Research and Prevention Foundation Award Lecture. *Cancer Res.* 2006, *66,* 2893-2903.

[7] Olson, J.A.; Krinsky, N.I. Introduction: the colorful fascinating world of the carotenoids: important physiologic modulators. *FASEB J.* **1995**, *9,* 1547-1550.

[8] Shixian, Q.; Dai, Y.; Kakuda, Y.; Shi, J.; Mittal, G.; Yeung, D.; Jiang, Y. Synnergistic anti-oxidative effects of lycopene with other bioactive compounds. *Food Rev Internl* **2005**, *21,* 295-311.

[9] Stahl, W.; Sies, H. Bioactivity and protective effects of natural carotenoids. Biochim Biophys Acta. **2005**, *1740,* 101-7.

[10] Khachik F. Distribution and metabolism of dietary carotenoids in humans as a criterion for development of nutritional supplements. *Pure Appli. Chem.* **2006**, *78,*1551-1557.

[11] Stahl, W.; Sies, H. Antioxidant activity of carotenoids. *Mol. Aspects Med.* 2003, *24,* 345-51.

[12] Borel, P.; Drai, J.; Faure, H.; Fayol, V.; Galabert, C.; Laromiguiere, M.; Le Moel, G. Recent knowledge about intestinal absorption and cleavage of carotenoids. *Ann Biol Clin (Paris).* **2005**, *63,* 165-77.

[13] Krinsky, N.I.; Johnson, E.J. Carotenoid actions and their relation to health and disease.*Mol Aspects Med.* **2005**, *26,* 459-516.

[14] Erdman, J.W. Jr., How do nutritional and hormonal status modify the bioavailability, uptake, and distribution of different isomers of lycopene? *J. Nutr.* **2005**, *135,* 2046S-2047S.

[15] Lindshield, B.L.; Canene-Adams, K.; Erdman, J.W. Jr. Lycopenoids: Are lycopene metabolites bioactive? *Arch Biochem Biophys.* **2007**, *458,*136-140.

[16] Clinton, S.K.; Emenhiser, C.; Schwartz, S.J.; Bostwick, D.G.; Williams, A.W.; Moore, B.J.; Erdman, J.W. *cis–trans* Lycopene isomers, carotenoids, and retinol in the human prostate. *Cancer Epidemiol. Biomarkers Prev.* **1996**, *5,* 823-833.

[17] Kaplan, L.A.; Lau, J.M.; Stein, E.A. Carotenoid composition, concentrations, and relationships in various human organs. *Clin. Physiol. Biochem.* **1990**, *8,* 1-10.

[18] Yeum, K.-J.; Booth, S.; Sadowski, J.; Lin, C.; Tang, G.; Krinsky, N.I.; Russell, R.M. Human plasma carotenoid response to the ingestion of controlled diets high in fruits and vegetables. *Am. J. Clin. Nutr.* **1996**, *64,* 594-602.

[19] Boileau, T. W.; Boileau, A. C.; Erdman, J. W. Jr. Bioavailability of all-trans and cis-isomers of lycopene. *Exp Biol Med (Maywood)* **2002,** *227,* 914–919.

[20] Moraru, C.; Lee, T.C. Kinetic studies of lycopene isomerization in a tributyrin model system at gastric pH. *J Agric Food Chem.* **2005,** *53,*8997–9004.

[21] Boileau, A.C.; Merchen, N.R.; Wasson, K.; Atkinson, C.A.; Erdman, J. Cis-lycopene is more bioavailable than trans-lycopene *in vitro* and *in vivo* in lymph-cannulated ferrets. *J Nutr* **1999,** *129,* 1176–1181.

[22] Teodoro, A.J.; Perrone, D.; Martucci, R.B.; Borojevic, R. Lycopene isomerisation and storage in an *in vitro* model of murine hepatic stellate cells. *Eur J Nutr.* **2009,** *48,* 261–268.

[23] Richelle, M.; Sanchez, B.; Tavazzi, I.; Lambelet, P.; Bortlik, K.; Williamson, G. Lycopene isomerisation takes place within enterocytes during absorption in human subjects. *Br J Nutr* **2010,** *103,* 1800–1807.

[24] Ross, A.B.; Vuong le, T.; Ruckle, J.; Synal, H.A.; Schulze-König, T.; Wertz, K.; Rümbeli, R.; Liberman, R.G.; Skipper, P.L.; Tannenbaum, S.R.; Bourgeois, A.; Guy, P.A.; Enslen, M.; Nielsen, I.L.; Kochhar, S.; Richelle, M.; Fay, L.B.; Williamson, G. Lycopene bioavailability and metabolism in humans: an accelerator mass spectrometry study. *Am J Clin Nutr.* **2011,** *93,* 1263-1273.

[25] Ford, N.A.; Elsen, A.C.; Zuniga, K.; Lindshield, B.L.; Erdman, J.W. Jr. Lycopene and apo-12'-lycopenal reduce cell proliferation and alter cell cycle progression in human prostate cancer cells. *Nutr Cancer.* **2011,** *63,* 256-263.

[26] Ford, N.A.; Erdman, J.W. Jr. Are lycopene metabolites metabolically active? *Acta Biochim Pol.* **2012**, Mar 17. [Epub ahead of print]

[27] Halliwell, B. Antioxidants in human health and disease, *Annu. Rev. Nutr.* **1996**, *16,* 33-50.

[28] Mortensen, A.; Skibsted, L.H.; Truscott, T.G. Antioxidants in human health and disease. *Biochem Biophys.* **2001,** *385,* 13-19.

[29] Böhm, F.; Edge, R.; Truscott, T.G. Interactions of dietary carotenoids with activated (singlet) oxygen and free radicals: potential effects for human health. *Mol Nutr Food Res.* **2012,** *56,* 205-216.

[30] Böhm, F.; Edge, R.; Truscott, T.G. Interactions of dietary carotenoids with singlet oxygen ((1)O(2)) and free radicals: potential effects for human health. *Acta Biochim Pol.* **2012** Mar 17. [Epub ahead of print]

[31] Conn, P.F.; Schalch, W.; Truscott, T.G. The singlet oxygen carotenoid interaction. *J. Photochem. Photobiol., B Biol.* **1991**, *11,* 41-47.

[32] Schmidt, R. Deactivation of singlet oxygen by carotenoids: internal conversion of excited encounter complexes. *J. Phys. Chem.* **2004**, *108,* 5509-5513.

[33] Pannala, A.S.; Rice-Evans, C.; Sampson, J.; Singh, S. Interaction of peroxynitrite with carotenoids and tocopherols within low density lipoprotein. *FEBS Lett.* **1998**, *423,* 297-301.

[34] Sies, H. Strategies of antioxidant defense. *Eur. J. Biochem.* **1993**, *215,* 213-219.

[35] Wertz, K.; Siler, U.; Goralczyk, R. Lycopene: modes of action to promote prostate health. *Arch Biochem Biophys*. **2004**, *430,* 127-134.

[36] Woodall, A.A.; Britton, G.; Jackson, M.J. Carotenoids and protection of phospholipids in solution or in liposomes against oxidation by peroxyl radicals: Relationship between carotenoid structure and protective ability. *Biochim. Biophys. Acta* **1997**, *1336,* 575-586.

[37] Matos, H.R.; Di Mascio, P.; Medeiros, M.H. Protective Effect of Lycopene on Lipid Peroxidation and Oxidative DNA Damage in Cell Culture**.** *Arch. Biochem. Biophys.* **2000**, *383,* 56-59.

[38] Matos, H.R.; Capelozzi, V.L.; Gomes, O.F.; Mascio, P.D.; Medeiros, M.H. Lycopene Inhibits DNA Damage and Liver Necrosis in Rats Treated with Ferric Nitrilotriacetate. *Arch. Biochem. Biophys.* **2001**, *396,* 171-177.

[39] Konijeti, R.; Henning, S.; Moro, A.; Sheikh, A.; Elashoff, D.; Shapiro, A.; Ku, M.; Said, J. W.; Heber, D.; Cohen, P.; Aronson, W. J. Chemoprevention of prostate cancer with lycopene in the TRAMP model. *Prostate.* **2010,** *70,* 1547-1554.

[40] Hwang, E. S.; Bowen, P. E. Effects of lycopene and tomato paste extracts on DNA and lipid oxidation in LNCaP human prostate cancer cells. *Biofactors.* **2005,** *23,* 97-105.

[41] Pool-Zobel, B.L.; Bub, A.; Muller, H.; Wollowski I.; Rechkemmer, G. Consumption of vegetables reduces genetic damage in humans: first results of a human intervention trial with carotenoid-rich foods. *Carcinogenesis* **1997**, *18,* 1847-1850.

[42] Porrini M.; Riso. P. Lymphocyte lycopene concentration and DNA protection from oxidative damage is increased in women after a short period of tomato consumption. *J. Nutr.* **2000,** *130,* 18-192.

[43] Rehman, A.;Bourne, L.C.; Halliwell, B.; Rice-Evans, C.A. Tomato Consumption Modulates Oxidative DNA Damage in Humans. *Biochem. Biophys. Res. Commun.* **1999,** *262,* 828-831.

[44] Chen, L, Stacewicz-Sapuntzakis, M.; Duncan, C.; Sharifi, R.; Ghosh, L.; van Breemen, R.; Ashton, D.; Bowen, P. E. Oxidative DNA damage in prostate cancer patients consuming tomato sauce-based entrees as a whole-food intervention. *J Natl Cancer Inst.* **2001,** *93,* 1872-1879.

[45] Bowen, P.; Chen, L.; Stacewicz-Sapuntzakis, M.; Duncan, C.; Sharifi, R.; Ghosh, L.; Kim, H.S.; Christov-Tzelkov, K.; van Breemen, R. Tomato sauce supplementation and prostate cancer: lycopene accumulation and modulation of biomarkers of carcinogenesis. *Exp. Biol. Med. (Maywood)* **2002**, *227,* 886-93.

[46] van Breemen, R. B.; Sharifi, R.; Viana, M.; Pajkovic, N.; Zhu, D.; Yuan, L.; Yang, Y. Bowen, P. E.; Stacewicz-Sapuntzakis, M. Antioxidant effects of lycopene in African American men with prostate cancer or benign prostate hyperplasia: a randomized, controlled trial. *Cancer Prev Res (Phila).* **2011,** *4,* 711-718.

[47] Stahl, W.; Junghans, A.; de Boer, B.; Driomina, E.S.; Briviba, K.; Sies, H. Carotenoid mixtures protect multilamellar liposomes against oxidative damage: synergistic effects of lycopene and lutein. *FEBS Lett.* **1998**, *427,* 305-8.

[48] Böhm, F.; R. Edge, R.; Lange L.;Truscott, T.G. Enhanced protection of human cells against ultraviolet light by antioxidant combinations involving dietary carotenoids. *J. Photochem. Photobiol B Biol.* **1998**, *44,* 211-5.

[49] Böhm, F.; Edge, R.; McGarvey, D.J.; Truscott, T.G. Beta-carotene with vitamins E and C offers synergistic cell protection against NOx. *FEBS Lett.* **1998**, *436,* 387-389.

[50] Palozza, P.; Krinsky, N.I. Beta-carotene and alpha-tocopherol are synergistic antioxidants, *Arch. Biochem. Biophys.* **1992**, *297,* 184-187.

[51] Heber, D,; Lu, Q.Y. Overview of mechanisms of action of lycopene. *Exp Biol Med (Maywood).* 2002, *227,* 920-3.

[52] Pastori, M.; Pfander, H.; Boscoboinik, D.; Azzi, A. Lycopene in association with alpha-tocopherol inhibits at physiological concentrations proliferation of prostate carcinoma cells. *Biochem Biophys Res Commun.* **1998**, *250,* 582-5.

[53] Limpens, J.; Schroder, F.H.; de Ridder, C.M.; Bolder, C.A.; Wildhagen, M.F.; Obermuller-Jevic, U.C.; Kramer K, van Weerden WM. Combined lycopene and vitamin E treatment suppresses the growth of PC-346C human prostate cancer cells in nude mice. *J Nutr.* **2006**, *136,* 1287-93.

[54] Siler, U.; Barella. L.; Spitzer, V.; Schnorr, J.; Lein, M.; Goralczyk, R.; Wertz, K. Lycopene and vitamin E interfere with autocrine/paracrine loops in the Dunning prostate cancer model. *FASEB J.* **2004**, *18,* 1019-21.

[55] Herzog, A.; Siler, U.; Spitzer, V.; Seifert, N.; Denelavas, A.; Hunziker, P.B.; Hunziker, W.; Goralczyk, R.; Wertz, K. Lycopene reduced gene expression of steroid targets and inflammatory markers in normal rat prostate. *FASEB J.* **2005**, *19,* 272-4

[56] Lindshield, B. L.; Ford, N. A.; Canene-Adams, K.; Diamond, A. M.; Wallig, M. A.; Erdman, J. W. Jr. Selenium, but not lycopene or vitamin E, decreases growth of transplantable dunning R3327-H rat prostate tumors. *PLoS One.* **2010,** *5,* e10423.

[57] Sharoni, Y.; Danilenko, M.; Dubi, N.; Ben-Dor, A.; Levy, J. Carotenoids and transcription.*Arch Biochem Biophys.* **2004**, *430,* 89-96.

[58] Gradelet, S.; Astorg, P.; Leclerc, J.; Chevalier, J.; Vernevaut, M.F.; Siess, M.H. Effects of canthaxanthin, astaxanthin, lycopene and lutein on liver xenobiotic-metabolizing enzymes in the rat. *Xenobiotica.* **1996**, *26,* 49.

[59] Breinholt, V.; Lauridsen, S.T.; Daneshvar, B.; Jakobsen, J. Dose-response effects of lycopene on selected drug-metabolizing and antioxidant enzymes in the rat. *Cancer Lett.* **2000**, *154,* 201-10.

[60] Wang, W.; Higuchi, C.M. Induction of NAD(P)H:quinone reductase by vitamins A, E and C in Colo205 colon cancer cells.*Cancer Lett.* **1995**, *98,* 63-9.

[61] Ben-Dor, A.; Steiner, M.; Gheber, L.; Danilenko, M.; Dubi, N.; Linnewiel, K.; Zick, A.; Sharoni. Y.; Levy, J. Carotenoids activate the antioxidant response element transcription system.*Mol Cancer Ther.* **2005**, *4,* 177-86.

[62] Linnewiel, K.; Ernst, H.; Caris-Veyrat, C.; Ben-Dor, A.; Kampf, A.; Salman, H.; Danilenko, M.; Levy, J.; Sharoni, Y. Structure activity relationship of carotenoid derivatives in activation of the electrophile/antioxidant response element transcription system. *Free Radic Biol Med.* **2009,** *47,* 659-667.

[63] Karas, M.; Amir, H.; Fishman, D.; Danilenko, M.; Segal, S.; Nahum, A.; Koifmann, A.; Giat, Y.; Levy, J.; Sharoni, Y. Lycopene interferes with cell cycle progression and insulin-like growth factor I signaling in mammary cancer cells. *Nutr. Cancer.* **2000**, *36,* 101-111.

[64] Nahum, A.; Hirsch, K.; Danilenko, M.; Watts, C.K.; Prall, O.W.; Levy, J.; Sharoni. Y. Lycopene inhibition of cell cycle progression in breast and endometrial cancer cells is associated with reduction in cyclin D levels and retention of p27(Kip1) in the cyclin E-cdk2 complexes. *Oncogene.* **2001**, *20,* 3428-36.

[65] Palozza, P.; Colangelo, M.; Simone, R.; Catalano, A.; Boninsegna, A.; Lanza, P.; Monego, G.; Ranelletti, F.O. Lycopene induces cell growth inhibition by altering mevalonate pathway and Ras signaling in cancer cell lines. *Carcinogenesis.* **2010,** *31,* 1813-1821

[66] Kotake-Nara, E.; Kushiro, M.; Zhang, H.; Sugawara, T.; Miyashita, K.; Nagao, A. Carotenoids affect proliferation of human prostate cancer cells. *J Nutr.* **2001**, *131,* 3303-6.

[67] Kotake-Nara, E.; Asai, A.; Nagao, A. Neoxanthin and fucoxanthin induce apoptosis in PC-3 human prostate cancer cells. *Cancer Lett.* **2005**, *220*, 75-84.

[68] Yang, C.M.; Yen, Y.T.; Huang, C.S.; Hu, M.L. Growth inhibitory efficacy of lycopene and β-carotene against androgen-independent prostate tumor cells xenografted in nude mice. *Mol Nutr Food Res.* **2011**, *55,* 606-612.

[69] Magbanua, M. J.; Roy, R.; Sosa, E. V.; Weinberg, V.; Federman, S.; Mattie, M. D.; Hughes-Fulford, M.; Simko, J.; Shinohara, K.; Haqq, C. M.; Carroll, P. R.; Chan, J. M. Gene expression and biological pathways in tissue of men with prostate cancer in a randomized clinical trial of lycopene and fish oil supplementation. *PLoS One.* **2011,** *6,* e24004.

[70] Pasquali, D.; Rossi, V.; Bellastella, G.; Bellastella, A.; Sinisi, A.A. Natural and synthetic retinoids in prostate cancer. *Curr Pharm Des.* **2006**, *12,* 1923-1929.

[71] Aboseif, S.R.; Dahiya, R.; Narayan, P.; Cunha, G.R. Effect of retinoic acid on prostatic development. *Prostate.* **1997**, *31,* 161-167.

[72] Lohnes, D.; Kastner, P.; Dierich, A.; Mark, M.; LeMeur, M.; Chambon, P. Function of retinoic acid receptor gamma in the mouse. *Cell* **1993,** *73,* 643-658.

[73] Blutt, S.E.; Allegretto, E.A.; Pike, J.W.; Weigel, N.L. 1,25-dihydroxyvitamin D3 and 9-cis-retinoic acid act synergistically to inhibit the growth of LNCaP prostate cells and cause accumulation of cells in G1. *Endocrinology.* **1997**, *138,* 1491-1497.

[74] Pasquali, D.; Rossi, V.; Prezioso, D.; Gentile, V.; Colantuoni, V.; Lotti, T.; Bellastella, A.; Sinisi, A.A. Changes in tissue transglutaminase activity and expression during retinoic acid-induced growth arrest and apoptosis in primary cultures of human epithelial prostate cells. *J Clin Endocrinol Metab.* **1999**, *84,* 1463-1469.

[75] Young**,** C.Y.; Murtha, P.E.; Andrews, P.E.; Lindzey, J.K.; Tindall, D.J. Antagonism of androgen action in prostate tumor cells by retinoic acid. *Prostate* **1994**, *25,* 39-45.

[76] Tokumaru, Y.; Harden, S.V.; Sun, D.I.; Yamashita, K.; Epstein, J.I.; Sidransky, D. Optimal use of a panel of methylation markers with GSTP1 hypermethylation in the diagnosis of prostate adenocarcinoma.*Clin Cancer Res.* **2004**, *10,* 5518-5522.

[77] Lotan, R.; Lotan, Y. Retinoic acid receptor beta2 hypermethylation: implications for prostate cancer detection, prevention, and therapy. *Clin Cancer Res.* 2004, *10,* 3935-3936.

[78] Fang, M.Z.; Chen, D.; Sun, Y.; Jin, Z.; Christman, J.K.; Yang, C.S. Reversal of hypermethylation and reactivation of p16INK4a, RARbeta, and MGMT genes by genistein and other isoflavones from soy. *Clin Cancer Res.* **2005**. *11,* 7033-7041.

[79] Qian, D.Z.; Ren, M.; Wei, Y.; Wang, X.; van de Geijn, F.; Rasmussen, C.; Nakanishi, O.; Sacchi, N.; Pili, R. *In vivo* imaging of retinoic acid receptor beta2 transcriptional activation by the histone deacetylase inhibitor MS-275 in retinoid-resistant prostate cancer cells. *Prostate.* **2005**, *64,* 20-28.

[80] Mehta, P.P.; Bertram, J.S.; Loewenstein, W.R The actions of retinoids on cellular growth correlate with their actions on gap junctional communication, *J. Cell Biol.* **1989**, *108,* 1053-1065.

[81] Mitra, S.; Annamalai, L.; Chakraborty, S.; Johnson, K.; Song, X.H.; Batra, S.K.; Mehta, P.P. Androgen-regulated formation and degradation of gap junctions in androgen-responsive human prostate cancer cells. *Mol Biol Cell.* **2006**, *17,* 5400-5416.

[82] Oyamada, M.; Oyamada, Y.; Takamatsu, T. Regulation of connexin expression. *Biochim Biophys Acta.* **2005**, *1719,* 6-23.

[83] Zhang, L.-X.; Cooney, R.V.; Bertram, J.S. Carotenoids enhance gap junctional communication and inhibit lipid peroxidation in C3H/10T1/2 cells: relationship to their cancer chemopreventive action. *Carcinogenesis* **1991**, *12,* 2109-2114.

[84] McLachlan, E.; Shao, Q.; Wang, H.L.; Langlois, S.; Laird, D.W. Connexins act as tumor suppressors in three-dimensional mammary cell organoids by regulating differentiation and angiogenesis.*Cancer Res.* **2006**, *66,* 9886-9894.

[85] Yamasaki, H. Gap junctional intercellular communication and carcinogenesis. *Carcinogenesis* **1990**, *11,* 1051-1058.

[86] Vine, A.L.; Bertram, J.S. Upregulation of connexin 43 by retinoids but not by non-provitamin A carotenoids requires RARs. *Nutr Cancer.* **2005**, *52,* 105-113.

[87] Vine, A.L.; Bertram, J.S.; Vine, A.L.; Leung, Y.M.; Bertram, J.S. Transcriptional regulation of connexin 43 expression by retinoids and carotenoids: similarities and differences. *Mol Carcinog.* **2005**, *43,* 75-85.

[88] Kucuk, O.; Sarkar, F.H.; Sakr, W.; Djuric, Z.; Pollak, M.N.; Khachik, F.; Li, Y.W.; Banerjee, M.; Grignon, D. Bertram, J.S.; Crissman, D.; Pontes, E.J.; Wood, D.P. Jr. Phase II randomized clinical trial of lycopene supplementation before radical prostatectomy. *Cancer Epidemiol. Biomark. Prev.* **2001**, *10,* 861-868.

[89] Ansari, M.S.; Gupta, N.P. A comparison of lycopene and orchidectomy *vs.* orchidectomy alone in the management of advanced prostate cancer, *BJU Int.* **2003**, *92,* 375-378.

[90] Kim, H.S.; Bowen, P.; Chen, L.; Duncan, C.; Ghosh, L.; Sharifi, R.; Christov, K. Effects of tomato sauce consumption on apoptotic cell death in prostate benign hyperplasia and carcinoma, *Nutr. Cancer* **2003**, *47,* 40-47.

[91] Ablin, R.J. Lycopene: a word of caution. *Am J Health Syst Pharm.* **2005**, *62,* 899.

[92] Forbes, K.; Gillette, K.; Sehgal, I. Lycopene increases urokinase receptor and fails to inhibit growth or connexin expression in a metastatically passaged prostate cancer cell line: a brief communication. Exp Biol Med (Maywood). **2003**, *228,* 967-971.

[93] Hail, N. Jr.; Kim, H. J.; Lotan, R. Mechanisms of fenretinide-induced apoptosis. *Apoptosis.* **2006,** *11,* 1677-1694.

[94] Venè, R.; Arena, G.; Poggi, A.; D'Arrigo, C.; Mormino, M.; Noonan, D. M.; Albini, A.; Tosetti, F. Novel cell death pathways induced by N-(4-hydroxyphenyl)retinamide: therapeutic implications. *Mol Cancer Ther.* **2007,** *6,* 286-298.

[95] Benelli, R.; Monteghirfo, S.; Venè, R.; Tosetti, F.; Ferrari, N. The chemopreventive retinoid 4HPR impairs prostate cancer cell migration and invasion by interfering with FAK/AKT/GSK3beta pathway and beta-catenin stability. *Mol Cancer.* **2010,** *9,* 142.

[96] Hail, N. Jr.; Chen, P.; Kepa, J. J.; Bushman, L. R.; Shearn, C. Dihydroorotate dehydrogenase is required for N-(4-hydroxyphenyl)retinamide-induced reactive oxygen species production and apoptosis. *Free Radic Biol Med.* **2010,** *49,* 109-116.

[97] Hail, N. Jr.; Chen, P.; Kepa, J.J. Selective apoptosis induction by the cancer chemopreventive agent N-(4-hydroxyphenyl)retinamide is achieved by modulating mitochondrial bioenergetics in premalignant and malignant human prostate epithelial cells. *Apoptosis.* **2009,** *14,* 849-863.

[98] Takahashi, N.; Watanabe, Y.; Maitani, Y.; Yamauchi, T.; Higashiyama, K.; Ohba, T. p-Dodecylaminophenol derived from the synthetic retinoid, fenretinide: antitumor efficacy *in vitro* and *in vivo* against human prostate cancer and mechanism of action. *Int J Cancer.* **2008,** *122,* 689-698.

[99] Pienta, K.J.; Nguyen, N.M.; Lehr, J.E. Treatment of prostate cancer in the rat with the synthetic retinoid fenretinide. *Cancer Res.* **1993,** *53,* 224–226. [PubMed]

[100] Slawin, K.; Kadmon, D.; Park, S.H.; Scardino, P.T.; Anzano, M.; Sporn, M.B.; Thompson, T. C. Dietary fenretinide, a synthetic retinoid, decreases the tumor incidence and the tumor mass of ras+myc-induced carcinomas in the mouse prostate reconstitution model system. Cancer Res. 1993, *53,* 4461–4465.

[101] Cheung, E.; Pinski, J.; Dorff, T.; Groshen, S.; Quinn, D.I.; Reynolds, C.P.; Maurer, B.J.; Lara, P.N. Jr.; Tsao-Wei, D.D.; Twardowski, P.; Chatta, G.; McNamara, M.; Gandara, D.R. Oral fenretinide in biochemically recurrent prostate cancer: a California cancer consortium phase II trial. *Clin Genitourin Cancer.* **2009,** *7,* 43-50.

[102] Moore, M.M.; Stockler, M.; Lim, R.; Mok, T.S.; Millward, M.; Boyer, M.J. A phase II study of fenretinide in patients with hormone refractory prostate cancer: a trial of the Cancer Therapeutics Research Group. *Cancer Chemother Pharmacol.* **2010,** *66,* 845-850.

[103] Weiss, H.L.; Urban, D.A.; Grizzle, W.E.; Cronin, K.A.; Freedman, L.S.; Kelloff, G.J.; Lieberman, R. Bayesian monitoring of a phase 2 chemoprevention trial in high-risk cohorts for prostate cancer. *Urology.* **2001,** *57(4 Suppl 1),* 220-223.

[104] Pienta, K.J.; Esper, P.S.; Zwas, F.; Krzeminski, R.; Flaherty, L.E. Phase II chemoprevention trial of oral fenretinide in patients at risk for adenocarcinoma of the prostate. *Am J Clin Oncol.* **1997,** *20,* 36-39.

[105] Mills, P.K.; Beeson, W.L.; Phillips, R.L.; Fraser, G.E. Cohort study of diet, lifestyle, and prostate cancer in Adventist men, *Cancer* **1989**, *64,* 598-604.

[106] Giovannucci, E.; Ascherio, A.; Rimm, E.B.; Stampfer, M.J.; Colditz, G.; Willett, W.S., Intake of carotenoids and retinol in relation to risk of prostate cancer, *J. Natl. Cancer Inst.* **1995**, *87*, 1767-76.

[107] Giovannucci, E.; Rimm, E.B.; Liu, Y.; Stampfer, M.J.; Willett, W.C. A prospective study of tomato products, lycopene, and prostate cancer risk.J Natl Cancer Inst. **2002**, *94,* 391-398.

[108] Wu, K.N.; Erdman, J.W.; Schwartz, S.J.; Platz, E.A.; Leitzmann, M.; Clinton, S.K.; DeGroff, V.; Willett, W.C.; Giovannuci, E. Plasma and dietary carotenoids, and the risk of prostate cancer: A nested case-control study. *Cancer* Epidemiol Biomarkers Prev **2004**, 13, 260-269.

[109] Zhang, J.; Dhakal, I.; Stone, A.; Ning, B.; Greene, G.; Lang, N. P.; Kadlubar, F. F. Plasma carotenoids and prostate cancer: a population-based case-control study in Arkansas. *Nutr Cancer.* **2007**, *59,* 46-53.

[110] Mikhak, B.; Hunter, D.J.; Spiegelman, D.; Platz, E.A.; Wu, K.; Erdman, J.W. Jr,; Giovannucci, E. Manganese superoxide dismutase (MnSOD) gene polymorphism, interactions with carotenoid levels and prostate cancer risk. *Carcinogenesis.* **2008,** *29,* 2335-2340.

[111] Cook, N.R.; Stampfer, M.J.; Ma, J.; Manson, J.E.; Sacks, F.M.; Buring, J.E.; Hennekens, C.H. Beta-carotene supplementation for patients with low baseline levels and decreased risks of total and prostate carcinoma, *Cancer* **1999**, *86,* 1783-1792.

[112] Heinonen, O.P.; Albanes, D.; Virtamo, J.; Taylor, P.R.; Huttunen, J.K.; Hartman, A.M.; Haapakoski, J.; Malila, N.; Rautalahti, M.; Ripatti, S.; Maenpaa, H.; Teerenhovi, L.; Koss, L.; Virolainen M.; Edwards., B.K. Prostate cancer and supplementation with α-tocopherol and β-carotene: incidence and mortality in a controlled trial, *J Natl Cancer Inst* **1998**, *90,* 440-446.

[113] Omenn, G.S.; Goodman, G.E.; Thornquist, M.D.; Balmes, J.; Cullen, M.R.; Glass, A.; Keogh, J.P.; Meyskens, F.L. Jr.; Valanis, B.; Williams, J.H. Jr.; Barnhart, S.; Cherniack, M.G.; Brodkin, C.A.; Hammar, S. Risk factors for lung cancer and for intervention effects in CARET, the Beta-Carotene and Retinol Efficacy Trial. *J Natl Cancer Inst* **1996**, *88,* 1550-1559.

[114] Omenn, G.S.; Goodman, G.E.; Thornquist, M.D.; Balmes, J.; Cullen, M.R.; Glass, A.; Keogh, J.P.; Meyskens, F.L.; Valanis, B.; Williams, J.H.; Barnhart, S.; Hammar, S. Effects of a combination of beta carotene and vitamin A on lung cancer and cardiovascular disease. *N. Engl. J. Med.* **1996**, *334,* 1150-1155.

[115] Nomura, A.M.; Stemmermann, G.N.; Lee, J.; Craft, N.E. Serum micronutrients and prostate cancer in Japanese Americans in Hawaii. *Cancer Epidemiol Biomarkers Prev.* **1997**, 6, 487-491.

[116] Chang, S.; Erdman, J.W. Jr.; Clinton, S.K.; Vadiveloo, M.; Strom, S.S.; Yamamura, Y.; Duphorne, C.M.; Spitz, M.R.; Amos, C.I.; Contois, J.H.; Gu, X.; Babaian, R.J.; Scardino, P.T.; Hursting, S.D. Relationship between plasma carotenoids and prostate cancer. *Nutr Cancer.* **2005**, *53,* 127-134.

[117] Kirsh, V. A.; Hayes, R. B.; Mayne, S. T.; Chatterjee, N.; Subar, A. F.; Dixon, L. B.; Albanes, D.; Andriole, G. L.; Urban, D. A.; Peters, U. A prospective study of lycopene and tomato product intake and risk of prostate cancer. Cancer Epidemiol Biomarkers Prev. **2006**, *15,* 92-98.

[118] Kirsh, V. A.; Hayes, R. B.; Mayne, S. T.; Chatterjee, N.; Subar, A. F.; Dixon, L. B.; Albanes, D.; Andriole, G.L.; Urban, D.A.; Peters, U. Supplemental and dietary vitamin E, beta-carotene, and vitamin C intakes and prostate cancer risk. *J. Natl. Cancer Inst.* **2006**, *98,* 245-254.

[119] Key, T.J.; Appleby, P.N.; Allen, N.E.; Travis, R.C.; Roddam, A.W.; Jenab, M.; Egevad, L.; Tjønneland, A,; Johnsen, N. F.; Overvad, K.; Linseisen, J.; Rohrmann, S.; Boeing, H.; Pischon, T.; Psaltopoulou, T.; Trichopoulou, A.; Trichopoulos, D.; Palli, D.; Vineis, P.; Tumino, R.; Berrino, F.; Kiemeney, L.; Bueno-de-Mesquita, H.B,; Quirós, J.R.; González, C.A.; Martinez, C.; Larrañaga, N.; Chirlaque, M.D.; Ardanaz, E.; Stattin, P.; Hallmans, G.; Khaw, K.T.; Bingham, S.; Slimani, N.; Ferrari, P.; Rinald, S.; Riboli, E. Plasma carotenoids, retinol, and tocopherols and the risk of prostate cancer in the European Prospective Investigation into Cancer and Nutrition study. *Am J Clin Nutr.* **2007,** *86,* 672-681.

[120] Karppi, J.; Kurl, S.; Nurmi, T.; Rissanen, T. H.; Pukkala, E.; Nyyssönen, K. Serum lycopene and the risk of cancer: the Kuopio Ischaemic Heart Disease Risk Factor (KIHD) study. *Ann Epidemiol.* **2009,** *19,* 512-518.

[121] Kristal, A.R.; Till, C.; Platz, E. A.; Song, X.; King, I.B.; Neuhouser, M.L.; Ambrosone, C.B.; Thompson, I.M. Serum lycopene concentration and prostate cancer risk: results from the Prostate Cancer Prevention Trial. *Cancer Epidemiol Biomarkers Prev.* **2011,** *20,* 638-646.

[122] Kristal, A.R.; Arnold, K.B.; Neuhouser, M.L.; Goodman, P.; Platz, E.A.; Albanes, D.; Thompson, I.M. Diet, supplement use, and prostate cancer risk: results from the prostate cancer prevention trial. *Am J Epidemiol.* **2010,** *172,* 566-577.

[123] Giovannucci, E. Commentary: Serum lycopene and prostate cancer progression: a re-consideration of findings from the prostate cancer prevention trial. *Cancer Causes Control.* **2011,** *22,* 1055-1059.

[124] Beilby, J.; Ambrosini, G.L.; Rossi, E.; de Klerk, N.H.; Musk, A.W. Serum levels of folate, lycopene, β-carotene, retinol and vitamin E and prostate cancer risk. *Eur J Clin Nutr.* **2010,** *64,*1235-1238.

[125] Schwenke, C.; Ubrig, B.; Thürmann, P.; Eggersmann, C.; Roth, S. Lycopene for advanced hormone refractory prostate cancer: a prospective, open phase II pilot study. *J Urol.* **2009,** *181,* 1098-1103.

[126] Watters, J.L.; Gai, M.H.; Weinstein, S.J.; Virtamo, J.; Albanes, D. Associations between alpha-tocopherol, beta-carotene, and retinol and prostate cancer survival. *Cancer Res.* **2009,** *69,* 3833-3841.

[127] Peters, U.; Leitzmann, M. F.; Chatterjee, N.; Wang, Y.; Albanes, D.; Gelmann, E.P.; Friesen, M.D.; Riboli, E.; Hayes, R.B. Serum lycopene, other carotenoids, and prostate cancer risk: a nested case-control study in the prostate, lung, colorectal, and ovarian cancer screening trial. *Cancer Epidemiol Biomarkers Prev.* **2007**, 16, 962-968.

[128] Karppi, J.; Kurl, S.; Laukkanen, J.A.; Kauhanen, J. Serum β-Carotene in Relation to Risk of Prostate Cancer: The Kuopio Ischaemic Heart Disease Risk Factor Study. *Nutr Cancer.* **2012** Mar 16. Epub ahead of print]

[129] Schenk, J.M.; Riboli, E.; Chatterjee, N.; Leitzmann, M.F.; Ahn, J.; Albanes, D.; Reding, D. J.; Wang, Y.; Friesen, M.D.; Hayes, R.B.; Peters, U. Serum retinol and prostate cancer risk: a nested case-control study in the prostate, lung, colorectal, and ovarian cancer screening trial. *Cancer Epidemiol Biomarkers Prev.* **2009**, *18,* 1227-1231.

[130] McCann, S.E.; Ambrosone, C.B.; Moysich, K.B.; Brasure, J.; Marshall, J.R.; Freudenheim, J.L.; Wilkinson, G.S.; Graham, S. Intakes of selected nutrients, foods, and phytochemicals and prostate cancer risk in western New York. Nutr *Cancer* **2006**, 53, 33-41.

[131] Kristal, A.R. Vitamin A, retinoids and carotenoids as chemopreventive agents for prostate cancer.*J Urol.* **2004**, *171,* S54-58.

[132] Albanes, D.; Heinonen, O.P.; Taylor, P.R.; Virtamo, J.; Edwards, B.K.; Rautalahti, M.; Hartman, A.M.; Palmgren, J.; Freedman, L.S.; Haapakoski, J.; Barrett, M.J.; Pietinen, P.; Malila, N.; Tala, E.; Liippo, K.; Salomaa, E.R.; Tangrea, J.A.; Teppo, L.; Askin, F.B.; Taskinen, E.; Erozan, Y.; Greenwald, P.; Huttunen, J.K. Alpha-tocopherol and beta-carotene supplements and lung cancer incidence in the alpha-tocopherol, beta-carotene cancer prevention study: effects of base-line characteristics and study compliance, *J. Natl. Cancer Inst.* **1996**, *88,* 1560-1570.

[133] Mayne, S.T.; Lippman, S.M. Cigarettes: a smoking gun in cancer chemoprevention. *J Natl Cancer Inst.* **2005**, *97,* 1319-1321.

[134] Chatterjee, M.; Roy, K.; Janarthan, M.; Das, S.; Chatterjee, M. Biological activity of carotenoids: its implications in cancer risk and prevention. *Curr Pharm Biotechnol.* **2012,** *13,*180-190.

[135] Baron, J.A.; Cole, B.F.; Mott, L.; Haile, R.; Grau, M.; Church, T.R.; Beck, G.J.; Greenberg, E.R. Neoplastic and antineoplastic effects of beta-carotene on colorectal adenoma recurrence: results of a randomized trial, *J. Natl. Cancer Inst.* **2003**, *95,* 717-722.

[136] Greenberg, E.R.; Baron, J.A.; Stukel, T.A.; Stevens, M.M.; Mandel, J.S.; Spencer, S.K. A clinical trial of beta carotene to prevent basal-cell and squamous-cell cancers of the skin. The Skin Cancer Prevention Study Group. *N Engl J Med* **1990**, *323,*789-795.

[137] Touvier, M.; Kesse, E.; Clavel-Chapelon, F.; Boutron-Ruault, M.C. Dual association of β-carotene with risk of tobacco-related cancers in a cohort of French women.*J Natl Cancer Inst* **2005**, *97,* 1338-1344.

[138] Keijer, J.; Bunschoten, A.; Palou, A.; Franssen-van Hal, N.L. Beta-carotene and the application of transcriptomics in risk-benefit evaluation of natural dietary components. *Biochim Biophys Acta.* 2005, *1740,* 139-146.

[139] Liu, C.; Wang, X.D.; Bronson, R.T.; Smith, D.E.; Krinsky, N.I.; Russell, R.M. Effects of physiological *versus* pharmacological beta-carotene supplementation on cell proliferation and histopathological changes in the lungs of cigarette smoke-exposed ferrets. *Carcinogenesis* **2000**, *21,* 2245-2253.

[140] Paolini, M.; Cantelli-Forti, G.; Perocco, P.; Pedulli, G.F.; Abdel-Rahman, S.Z.; Legator, M.S. Co-carcinogenic effect of B-carotene. *Nature* **1999**, *398,* 760-776.

[141] Perocco, P.; Paolini, M.; Mazzullo, M.; Biagi, G.L.; Cantelli-Forti, G. B-carotene as enhancer of cell transforming activity of powerful carcinogens and cigarette-smoke condensate on BALB/c 3T3 cells *in vitro*, *Mutat. Res.* **1999**, *440,* 83-90.

[142] Wang, X.D.; Liu, C.; Bronson, R.T.; Smith, D.E.; Krinsky, N.I.; Russell, M. Retinoid signaling and activator protein-1 expression in ferrets given beta-carotene supplements and exposed to tobacco smoke. *J Natl Cancer Inst.* **1999**, *91,* 60-66.

Send Orders for Reprints to reprints@benthamscience.net
272 *Recent Advances in Medicinal Chemistry, Vol. 1*, 2014, 272-302

CHAPTER 9

Preclinical Studies of Saponins for Tumor Therapy

Christopher Bachran[1], Silke Bachran[1], Mark Sutherland[2], Diana Bachran[1] and Hendrik Fuchs[1,*]

[1]Institute of Laboratory Medicine, Clinical Chemistry and Pathobiochemistry, Charité – Universitätsmedizin Berlin, Campus Virchow-Klinikum, Berlin, Germany and [2]University of Bradford, Institute of Cancer Therapeutics, Bradford, West Yorkshire, UK

Abstract: Various saponins, plant glycosides with favorable anti-tumorigenic properties, have been used to inhibit tumor cell growth by cell cycle arrest and apoptosis with IC_{50} values of up to 0.2 µM. We describe several groups of saponins (dioscins, saikosaponins, julibrosides, soy saponins, ginseng saponins and avicins) currently investigated for their use in tumor therapy. We focus on cellular and systemic mechanisms of tumor cell growth inhibition both *in vitro* and *in vivo*, combinational approaches with saponins and conventional tumor treatment strategies, and successful syntheses of saponins. The increasing interest in saponins for tumor therapy is very promising for the future development of sophisticated anti-cancer drugs.

Keywords: Cancer, cytotoxicity, glycoside, saponin, steroid, tumor therapy, triterpenoid.

DIVERSE STRUCTURES AND FUNCTIONS OF SAPONINS

Saponins are plant glycosides and possess great diversity in their structure. They are common in a variety of higher plants and usually found in roots, tubers, leaves, blooms or seeds. They contain a steroid, steroid alkaloid or triterpene core structure, the so-called aglycone. The steroidal aglycone consists of 27 C-atoms while the triterpenoidal aglycone has 30 C-atoms. The saponins with a steroid alkaloid as aglycone are similar to those with a steroidal aglycone but contain additional nitrogen atoms within the core structure. The detailed composition of the aglycone varies between different saponins from different sources, however,

***Address correspondence to Hendrik Fuchs:** Institute of Laboratory Medicine, Clinical Chemistry and Pathobiochemistry, Charité – Universitätsmedizin Berlin, Campus Virchow-Klinikum, Augustenburger Platz 1, D-13353 Berlin, Germany; Tel.: +4930 450 569173; Fax: +4930 450 569900; E-mail: hendrik.fuchs@charite.de

Atta-ur-Rahman, Muhammad Iqbal Choudhary and George Perry (Eds)

these variations are less pronounced compared to those within the glycan structures, which are attached to the aglycone. Usually, one or more sugar chains are covalently linked to the core structure. Glucose, galactose, glucuronic acid, xylose or rhamnose are among the sugars commonly found in saponins. However, even in one source the composition of the sugars may vary, resulting in saponins with different glycosylation [1]. Thus the diversity of saponins is a result of differences in the aglycone structure and the amount and composition of the sugar side chains.

Diverse functions have been described for distinct saponins. The effects observed are often specific for certain saponins due to the great variability of their structures. They are known as foaming substances due to the combination of the non-polar aglycone and their water-soluble side chains. This property is of interest for the beverage industry. Furthermore, saponins added to the food of ruminating animals increase growth, milk or wool production by eliminating protozoa, which predate on crucial bacteria in the first stomach (reviewed in [2]). Another prominent effect of saponins is their membrane permeabilizing property [3, 4]. The pore formation is ascribed to an interaction between saponin and membrane-bound cholesterol [5]. The amount of cholesterol in the membrane has been shown to be important for this interaction [6]. Although it has been shown that hemolysis of erythrocyte membranes by certain saponins is inhibited by depletion of cholesterol, for other saponins the effect is augmented [7]. Zhao *et al.* described the formation of cholesterol-saponin complexes for the saponin platycodin D while this saponin did not interact with triglycerides [8]. It was also demonstrated by Hu *et al.* that pore-formation by saponins with two sugar side chains is independent of membrane cholesterol while those without sugars are cholesterol-dependent [9]. The number of side chains influences both hemolytic activity and membrane permeability. Woldemichael *et al.* reported that saponins possessing two side chains induce less activity than those with only one sugar side chain [10]. Taking all information into consideration, the specific effects of saponins may be due to the combination of target membrane composition, the type of the saponin side chain(s) and the nature of the aglycone [1, 11, 12]. These results underline the saponins variability in biological functions due to their different structures. Saponins are often used to permeabilize membranes in order to make intracellular compartments accessible for antibodies [13]. Interestingly Lee *et al.* demonstrated that a certain saponin from ginseng only interacted with the extracellular side of the membrane [14]. In relation to the membrane-interacting functions of saponins Morein *et al.* described saponins as adjuvant additive [15]. The saponin from the bark of the tree *Quillaja saponaria* exhibited improved adjuvant effects in formulated immunostimulating complexes in comparison to simple mixtures of saponin and immunogen [16]. The saponin forms a cage-like structure together with membrane cholesterol, demonstrating that for this saponin cholesterol is essential for its adjuvant function [17].

Saponins possess in addition many effects on tumor cells. Different cytotoxic properties have been described for a number of saponins promoting their potential as anti-cancer drugs or adjuvants. Analyses of saponins in tumor therapy from the last ten years are presented in the following section. Special emphasis is given to those studies describing a mechanistic background of saponin-mediated anti-cancer effects.

SAPONINS AS ANTI-CANCER AGENTS

One of the first studies with saponins for the treatment of cancer was described by Ebbesen *et al.* in 1976 [18]. The saponin Quil A prolonged the survival of mice, which developed spontaneous leukemia (Table **1**). While many groups used saponins to permeabilize cellular membranes for microscopic studies, an increased interest in saponins as potential drugs for the treatment of cancer took place in the 1990s. A study investigating the growth inhibitory effect of two saikosaponins and two ginsenosides, saponins from *Panax ginseng*, showed that saikosaponin-a inhibited cell proliferation of human hepatoma cells while saikosaponin-c and the ginsenosides Rb1 and Rg1 had no effect [19]. Yu *et al.* identified a triterpenoidal saponin, tubeimoside 1, from the bulb of *Bolbostemma paniculatum* [20]. This saponin had a potent anti-tumorigenic effect in a mouse skin tumor model. The number of studies on saponins as anti-tumor drugs has increased drastically in the last decade. This review focuses on the direct anti-cancer effects of certain groups of saponins while the well described adjuvant effects of saponins, *e.g.* quillaja saponin, in anti-cancer vaccines [21] are not described. This review focusses instead on the direct anti-cancer effects of saponins. Most studies were performed *in vitro* with cell culture models, however, the number of mouse studies is increasing. A general problem of several cell culture studies presented in this review is the lack of control experiments to reveal the effects of the saponins on healthy tissue. Therefore one has to be careful when considering a certain saponin whose efficacy was only demonstrated on the basis of its high toxicity on tumor cell lines.

Table 1: Anti-tumorigenic effects of saponins observed in mice

Saponin	Anti-Tumorigenic Effect	References
Quil A	prolonged survival in spontaneous leukemia model	[18]
tubeimoside 1	growth inhibitory effect on skin tumors	[20]
soyasaponin I	reduction of lung metastases	[52]
ginsenoside Rh2	reduction of human ovarian tumor cell growth suppressive effect on tumor induction enhanced anti-tumor activity of cisplatin and paclitaxel inhibition of tumor growth in combination with cyclophosphamide inhibition of tumor growth in combination with docetaxel	[61] [62] [100, 106] [110] [113]
ginsenoside Rg3	chemo-preventive and anti-mutagenic effect inhibition of tumor growth in combination with cyclophosphamide inhibition of tumor growth in combination with docetaxel inhibition of tumor growth in combination with gemcitabine	[71] [109, 111] [115] [116]
ginsenoside Rp1	prevention and growth inhibition of papillomas; increased activity of detoxifying enzymes	[76]
ginsenoside Rb1	increased radiosensitivity of tumor cells	[121]
mixture of avicins	prevention of mutations after UV radiation protective effects against papilloma-inducing chemicals	[85] [86]
formosanin-C	potentiation of growth inhibitory effect of 5-fluorouracil	[90]

DIOSCINS

Dioscin (Fig. (**1**)) is a steroidal saponin produced by many plants of different genera. In 2002 Cai *et al.* described the cellular mechanisms of dioscin, purified from the root of *Polygonatum zanlanscianense* Pamp., that lead to cell death of tumor cells with a half maximal inhibitory concentration (IC_{50}) of 4.4 μM (22]. In cell culture experiments with HeLa cervix carcinoma cells dioscin induced apoptosis *via* the mitochondrial pathway. A reduction in the expression of the anti-apoptotic protein Bcl-2 together with caspase activation was observed.

Dioscin was also isolated from rhizomes of *Smilacina atropurpurea* and induced cytotoxicity on several tumor cell lines with an IC_{50} of 1.9–6.8 μg/ml [23] indicating a high impact on tumor cells. Protodioscin and methyl protodioscin are structurally closely related to dioscin. Protodioscin from fenugreek (*Trigonella foenumgraecum*) induced cell death in the leukemic cell line HL-60 by apoptosis while the gastric cancer cell line KATO III was merely inhibited with no apoptosis observed thus demonstrating cell line-dependent results [24]. The natural derivative of protodioscin, methyl protodioscin, from the rhizome of *Dioscorea collettii var. hypoglauca* of the family *Dioscoreaceae*, inhibited many solid tumors with an IC_{50} < 10 μM while leukemia cell lines were relatively insensitive with IC_{50} values of 10–30 μM [25]. The higher cytotoxicity on solid tumors compared to leukemic cells is opposite to the effect of the related protodioscin, for which only one study with two cell lines demonstrated a higher impact on leukemic cells. Thus, further studies are necessary to determine which structural features or cell structures result in these contrasting results. The plant *Dioscorea collettii var. hypoglauca* is a traditional Chinese medicine used for the treatment of certain solid tumors, thus the observed treatment success may be attributed to the action of saponins like methyl protodioscin. Recently the mechanism of growth inhibition was studied, and cell cycle arrest and apoptosis induction observed [26, 27]. The latter was shown to be a result of upregulation of pro-apoptotic Bax and downregulation of anti-apoptotic Bcl-2 protein in HepG2 liver carcinoma cells and K562 hematopoietic malignant cells (Fig. (**2**)). Three further saponins structurally closely related to dioscin and protodioscin are gracillin, methyl protogracillin and methyl protoneogracillin, which were also isolated from rhizomes of *Dioscorea collettii var. hypoglauca*. These saponins induced cytotoxicity in the μM range on tumor cell lines as observed for methyl protodioscin [28]. However, the cytotoxicities observed on leukemia cell lines were 6 to 11-fold lower for methyl protogracillin compared to its stereoisomer (R/S configuration at C-25) methyl protoneogracillin, emphasizing distinct structural requirements for potent anti-tumorigenic activity. In 2009, Kaskiw *et al.* published a report on the synthetic modification of several diosgenyl saponins [29]. The study demonstrated that the monosaccharides analogues of this steroidal saponin are more potent than the disaccharides analogues when tested on MCF-7 breast cancer cells and HeLa cervical cancer cells.

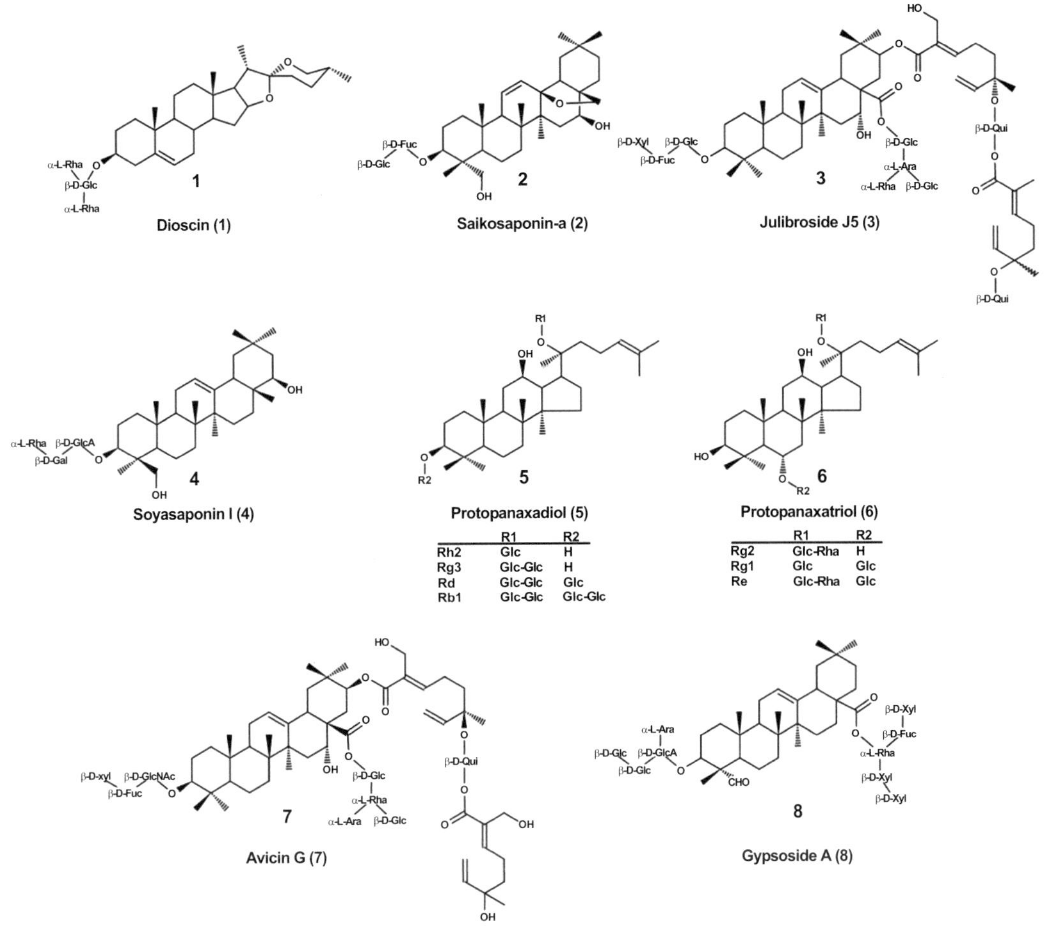

Protopanaxadiol (5)

	R1	R2
Rh2	Glc	H
Rg3	Glc-Glc	H
Rd	Glc-Glc	Glc
Rb1	Glc-Glc	Glc-Glc

Protopanaxatriol (6)

	R1	R2
Rg2	Glc-Rha	H
Rg1	Glc	Glc
Re	Glc-Rha	Glc

Figure 1: Structures of exemplary saponins. Abbreviations for sugars: α-L-arabinose (α-L-Ara), β-D-fucose (β-D-Fuc), β-D-galactose (β-D-Gal), β-D-glucose (β-D-Glc), β-D-glucuronic acid (β-D-GlcA), β-D-N-acetylglucosamine (β-D-GlcNAc), β-D-quinovose (β-D-Qui = 6-deoxy-β-D-Glc), α-L-rhamnose (α-L-Rha), β-D-xylose (β-D-Xyl). Examples for several ginsenosides varying only in the residues R1 and R2 are presented in the tables beneath the structures of protopanaxadiol and -triol.

SAIKOSAPONINS

Saikosaponins are triterpenoidal saponins produced in plants of the genus *Bupleurum* and several derivatives are described for their anti-cancer effects. The Japanese drug saiko (Chinese Chai-hu, 柴胡) contains rhizomes of *Bupleurum kaoi* and serves as source for saikosaponins. The growth-inhibiting potency of saikosaponin-a (Fig. (**1**)) in hepatoma HuH-7 cells was published as early as 1993 by Okita *et al.* while saikosaponin-c did not alter cell proliferation [19]. Surprisingly, the authors stated that growth inhibition by saikosaponin-a was independent of the cell cycle while Wu *et al.* reported in a more extensive study that saikosaponin-a inhibited cell growth of HepG2 hepatoma cells by upregulating gene expression of the cyclin dependent kinase inhibitors p-15^{INK4b} and p-16^{INK4a}, both specific inhibitors of cyclin dependent kinase 4/6 [30]. Furthermore, a link to a possible involvement of the protein kinase C pathway was also described. Wen-Sheng *et al.* reported in 2003 that the mitogen-activated protein kinase 3 (MAPK3, "ERK1") signaling pathway is involved in $p15^{INK4b}/p16^{INK4a}$ expression mediated by saikosaponin-a [31] (Fig. (**2**)). MAPK8 ("JNK1") and MAPK14 ("p38") pathways were not altered by saikosaponin-a. A pro-apoptotic effect of saikosaponin-a was described in 2003 in the human breast cancer cell lines MDA-MB-231 and MCF-7 [32]. Of special interest is that the two cell lines exhibited different cell line-specific apoptotic characteristics: Apoptosis in MCF-7 cells was dependent on the activation of p21 inhibitor by p53 whereas in MDA-MB-231 cells this process was independent of p53. The latter displayed an increase in expression of the cyclin-dependent kinase inhibitor p21 and pro-apoptotic Bax as well as activation of caspase 3. In human lung cancer cells Fas-dependent apoptosis induction was observed after treatment with a saponin-enriched fraction from *Bupleurum kaoi*, however, this fraction contained several saikosaponins [33]. Chiang *et al.* analyzed the cytotoxicity of saikosaponin-a, -c and -d on hepatoma cells. While saikosaponin-a and -d reduced cell growth with an IC_{50} of about 10 µg/ml and induced apoptosis by activating the caspases 3 and 7, saikosaponin-c was non-toxic at a concentration of 40 µg/ml [34]. However, saikosaponin-c reduced replication of hepatitis B virus in cell culture. In a further study, saikosaponin-c demonstrated a proliferating effect on endothelial cells and increased the expression of matrix metalloproteinase-2 and vascular endothelial growth factor, rendering this saponin inappropriate for cancer therapy [35]. Saikosaponin-d was intensely studied by Hsu and colleagues in 2004. This saponin induced apoptosis in two hepatoma cell lines and inhibited proliferation with IC_{50} values in the range of 2.6–4.3 µM [36]. In HepG2 cells treatment with saikosaponin-d resulted in accumulation in cell cycle phase G1, increased expression of cyclin-dependent kinase inhibitor p21 by p53 and increased expression of death receptor Fas and Fas ligand (Fig. (**2**)). Induction of apoptosis in Hep3B cells was also observed but cell cycle arrest and p21 upregulation were not detected. The ratio of pro-apoptotic Bax protein to anti-apoptotic Bcl-X_L protein increased in both cell lines, though Bax was only upregulated in HepG2 cells. A similar study on a lung tumor cell line revealed comparable effects of saikosaponin-d: Apoptosis was induced with an IC_{50} of 10.2 μM associated with p53/p21 upregulation, cell cycle arrest and Bax protein elevation [37]. The recent studies described both saikosaponin-a and -d as convenient drugs for tumor therapy by inhibiting cell growth, arresting cell cycle progression and inducing apoptosis. However, *in vivo* studies with different tumors are necessary to finally prove the anti-tumorigenic potential of saikosaponins.

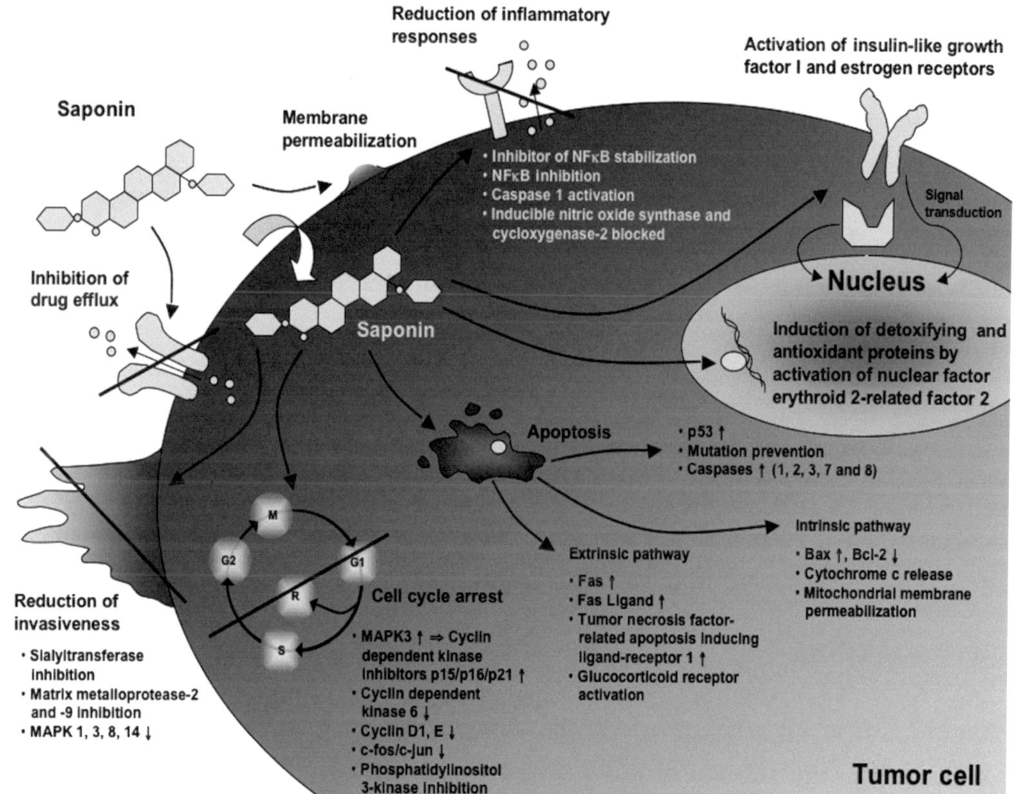

Figure 2: Cellular effects of saponins. The schematic illustration depicts the different molecular pathways contributing to the anti-tumorigenic properties of various saponins. Note that a number of pathways are observed only for certain cell lines and certain saponins.

JULIBROSIDES

The triterpenoidal julibrosides are isolated from the stem bark of *Albizia julibrissin*. These saponins contain three sugar side chains each with 2–4 sugar residues. Several derivatives have been tested on tumor cell lines in cell culture. The julibrosides J18 and J19 exhibited cytotoxicity against cervix carcinoma, hepatoma and breast cancer cell lines [38]. The julibrosides J28 and J21 exhibited cytotoxicity on a hepatoma cell line with an IC_{50} of about 10 μM [39] and 10 μg/ml, respectively [40]. The derivatives J5 (Fig. (**1**)), J8, J12 and J13 were analyzed for their cytotoxicity on a hepatocarcinoma cell line. All four saponins exhibited cytotoxicity with IC_{50} values in the range of 10 μg/ml [41]. However, J8 and J13 were most effective in killing the tumor cells since they reduced cell numbers by about 90%. Zheng *et al.* presented a more detailed study on julibroside J8, which inhibited proliferation of three cancer cell lines drastically at 100 μg/ml (46 μM) [42]. Apoptosis was observed in the cervix carcinoma cell line HeLa by induction of DNA fragmentation, upregulation of pro-apoptotic Bax, downregulation of anti-apoptotic Bcl-2 and caspase 3-activation (Fig. (**2**)). The 3 julibrosides J29, J30 and J31 exhibited significant cytotoxicity on three cancer cell lines [43]. At 10 μM all saponins inhibited growth of the cervix carcinoma cell line HeLa more than 60% while the hepatoma carcinoma cell line was inhibited by about 50%. All saponins except J31 were more cytotoxic than the chemotherapeutic drug doxorubicin (Adriamycin™). Cao *et al.* isolated 3 saponins from *Albizia gummifera*, which are structurally closely related to the julibrosides [44]. The saponins were incubated on 5 different tumor cell lines and inhibited cell growth notably with IC_{50} values of less than 1 μM. As for the saikosaponins, relevant *in vivo* data are not available for the julibrosides thus impeding estimation of their true potential.

SOY SAPONINS

Saponins from soy (*Glycine max*) have been intensely studied in recent years due to the great significance of soy on nutrition. Since Kerwin reviewed anti-cancer effects of soy saponins in 2004 [45], we will focus only on the latest publications on soy saponins. Most studies on soy saponins have been performed on human colon cancer cells, since soy saponins are expected to influence primarily the gastro-intestinal system after they are taken up as part of the diet. Kim *et al.* analyzed a crude extract of soybean containing different saponins on human colon cancer cells [46]. The saponins inhibited cell growth and reduced inflammatory responses by mediating increased inhibition of the transcription factor nuclear factor-kappa B (NFκB), which mediates expression of inflammatory proteins. These effects are the result of interference with the degradation of the inhibitor of NFκB, IκBα (Fig. (**2**)). Thus, soy saponins may present a different mechanism for anti-tumorigenic effects compared to the saponins described above. Studies with crude saponin extracts may reveal interesting results but should be considered rather preliminary. However, it is also possible that components of the crude extract interact synergistically and thus induce effects not observed for pure saponins. Therefore investigations with purified saponins are indispensable for matching a result to the molecular action of a specific saponin. In a more detailed study with purified saponins, only the aglycones soyasapogenin A and B were potent growth inhibitors of colon cancer cell lines, while the acetylated and glycosylated variants, including soyasaponin I (Fig. (**1**)) and soyasaponin III, were inactive [47]. Thus, deglycosylation of soyasaponins by intestinal bacteria in the digestive tract may help to generate potent anti-cancer drugs. While the report of Oh *et al.* supports the hypothesis that soy saponins do not induce apoptosis [48], Yanamandra *et al.* described apoptosis induction and reduced invasiveness due to the effect of the group B saponins from soy, a mixture of 4 triterpenoidal saponins possessing the same aglycone but different glycosylations [49]. However, in relation to other studies very high amounts of saponin (up to 75 μg/ml) were needed to induce the observed effects and even under these conditions activation of the caspases 3 and 9 was weak. In a further study it was shown that group B saponins induced cell cycle arrest and macroautophagy in human colon cancer cells [50]. A significant increase in autophagic vacuoles was detected by incorporation of monodansylcadaverine. Recently, the MAPK1 ("ERK2") and MAPK3 signaling pathways were identified in this process of macrophage induction [51]. Soyasaponin I from the group B saponins was described as an inhibitor of sialyltransferases, enzymes responsible for hypersialylation of cell surface proteins and are associated with highly metastatic tumors. In an *in vivo* model, soyasaponin I treatment of metastatic tumor cells drastically reduced the amount of metastases found in the lungs of mice [52] (Table **1**). The inhibition of invasiveness may block the formation of metastases and is thus very important in tumor therapy. In accordance with the previous results is the outcome of an investigation by Hsu *et al.*, who found soyasaponin I decreased cell migration thus helping to avoid metastases [53]. However, the saponin was unable to inhibit tumor cell proliferation and the cell cycle. Jun *et al.* described in an additional study a cancer protective property for soy saponins preventing DNA mutations and attachment of DNA mutagens [54]. The protective properties of soy saponins, as well as their ability to block metastases, makes them very promising drugs for tumor therapies.

GINSENG SAPONINS

The saponins produced by *Panax ginseng C. A. Meyer* (Korean ginseng, 인삼) and *Panax notoginseng* (Chinese ginseng, 三七) are called ginsenosides (Fig. (**1**)) and were analyzed in various studies to explore their anti-carcinogenic potential. The root of ginseng is a very well known drug in Korea and China for several diseases including cancer (reviewed in [55]). The ginsenosides contain a steroidal dammarane aglycone, which can either be a protopanaxadiol or a protopanaxatriol. Dammarane aglycones are found together with ginsenosides in the roots but they are also generated in the intestinal tract after saponin uptake. *Panax ginseng* produces several saponins and Wang *et al.* were able to identify 11 dammarane aglycones compounds and saponins in a preparation from the fruits [1]. All of these compounds were analyzed for their impact on apoptosis, proliferation and cell cycle progression. Of seven ginsenosides tested (Rh2, Rg1, Rg2, Rg3, Rd, Re and Rb1) only Rh2 was cytotoxic on several tumor cells with an IC_{50} in the range of 20–70 μM. However, in comparison with saponins from other groups the cytotoxicities are rather low. Two dammarane aglycones showed similar cytotoxicity while all other compounds were non-toxic. Annexin and propidium iodide staining revealed induction of apoptosis was strongest for the protopanaxadiol. The effect of the aglycones and ginsenosides on cell cycle arrest was rather small. The ginsenoside Rh2 was the focus of further studies and induction of apoptosis, cell cycle arrest and inhibition of cell growth was detected on many different tumor cell lines. The structure-activity relationship for ginsenosides, their activity *in vitro* and *in vivo* as well as their mechanisms of action are reviewed in detail by Nag *et al.* in 2012 [56]. Kim *et al.* studied Rh2 on a neuroblastoma cell line and demonstrated activation of caspases 1 and 3, increased expression of Bax (Bcl-2 remained unaltered) and activation of p53 [57] (Fig. (**2**)). In 2004, the ginsenosides Rh2 and Rg3 were identified as most potent among 11 ginsenosides to inhibit the growth of two prostate cancer cell lines with IC_{50} values of 4–14 μM [58]. In the study from Wang *et al.* described above, the authors reported lower IC_{50} values for the same ginsenosides on the same cell lines; especially for Rg3, which showed a 21 to 32-fold lower IC_{50} value [1], however, Kim *et al.* determined the inhibition of protein synthesis while Wang *et al.* detected growth inhibition. Kim *et al.* reported that both saponins induced cell detachment, which is possibly part of an apoptotic pathway. Cheng *et al.* described the induction of apoptosis by Rh2 on the human lung adenocarcinoma cell line A549 [59] that was later also analyzed by Wang *et al.* with comparable results. A more in depth study into the induction of apoptosis revealed the activation of the caspases 2, 3 and 8, cell cycle arrest, downregulation of cyclins D1, E and cyclin dependent kinase 6 and upregulation of the death receptor and tumor necrosis factor-related apoptosis-inducing ligand-receptor 1 (TRAIL-R I), which possibly plays a key role in initiating apoptosis. Jia *et al.* described the apoptosis induced by Rh2 to be mediated by glucocorticoid receptor activation [60]. As the growth inhibiting properties of Rh2 were promising, it was tested in an *in vivo* model. While Rh2 was able to reduce growth of human ovarian tumor cells in mice more effectively than cisplatin [61] and a suppressive effect on tumor induction was also shown [62] (Table **1**), the ginsenoside enhanced the metastatic potential of tumor cells in an *in vitro* assay rendering its potential use in human tumor therapy problematic [62]. Popovich *et al.* studied the ginsenosides Rh1, Rh2, Rg3, a protopanaxadiol and -triol aglycone on human leukemic cells [63]. Rh2 and the two aglycones showed high activity with IC_{50} values of about 15 μg/ml, while Rg3 and Rh1 had low cytotoxic activity with IC_{50} values greater than 200 μg/ml. Thus, Rh2 causes comparable cytotoxicities both on leukemia cells and solid tumor cells. Since Rh1 seems to be unable to inhibit tumor cell growth, other properties of the ginsenoside were investigated. Lee *et al.* identified Rh1 as a potential but weak ligand for estrogen receptors, however the binding to the receptor was specific and induced a functional signal [64] (Fig. (**2**)). Comparable results were published for ginsenoside Rg1, which stimulates growth of an estrogen receptor-positive human cell line [65]. A direct interaction between

Rg1 and the estrogen receptors does not occur and the effect of the ginsenoside seems to be mediated by stimulation of insulin-like growth factor I receptor and intracellular cross talk with estrogen receptor signaling pathways [66]. While ginsenoside Rg3 exhibited moderate toxicity in the studies presented before [1, 63] other anti-carcinogenic effects were reported. Rg3 inhibited partly papilloma formation after treatment with the phorbol ester phorbol 12-myristate 13-acetate by inhibition of NFκB [67]. The ginseng root used in this study was steamed prior to preparation, a process known to increase the activity of ginseng root extracts and leads to the production of Rg3 and Rg5 [68]. Wang *et al.* also showed the production of Rg3 and Rg5 after steaming berries of *Panax ginseng* and reported an increased cytotoxic effect of the extract [69]. Therefore the assumption is acceptable that at least one of these saponins mediates increased growth inhibition. The main ginsenoside produced was Rg3, which showed moderate growth inhibition on human tumor cells with an IC_{50} value of about 150 μM (120 μg/ml). This result is different to the result from Kim *et al.* [58], who reported a high cytotoxicity of Rg3 but measured, as stated above, protein synthesis inhibition in contrast to growth inhibition detected by Wang *et al.* who, moreover, worked only with enriched saponin extracts with lower Rg3 content. Two ginsenosides present before steaming, Rb3 and Re, were not cytotoxic at concentrations up to 1000 μM. Kim *et al.* reported that Rg3 was able to reverse drug resistance of tumor cells [70]. A concentration of 320 μM of Rg3 inhibited efflux of vinblastine and rhodamine from multidrug resistant cells, however using Rg3 at a concentration high above the IC_{50} value will make it difficult to determine whether the effects are due to the saponin or the drug. The inhibition on drug efflux was furthermore described for doxorubicin-resistant tumor cells in mice. Panwar *et al.* described a chemo-preventive and anti-mutagenic effect of Rg3 in mice although Rg3 comprised only about 3% (w/w) of the tested saponin mixture [71] (Table **1**). Rg3 inhibits furthermore the expression of CXC receptor 4 in human MDA-MB-231 breast cancer cells and reduced cell migration [72]. In another study the same ginsenoside Rg3 inhibited tumor growth in a HCT-116 colorectal cancer xenograft model by inhibiting Wnt/β-catenin signaling [73] and in a H460 lung cancer xenograft model by improving the immune response directed against the tumor cells [74]. A derivative of Rb3 mediated moderate cytotoxicity (IC_{50} 50–60 μg/ml) on human tumor cells [75], the ginsenoside Rp1 prevented development and growth of papillomas in mice after oral application and increased the activity of detoxifying enzymes [76] (Table **1**), Rd inhibited cell growth (IC_{50} 150 μg/ml) and induced apoptosis in human cervix carcinoma cells [77] but its cytotoxicity is weak compared to other saponins and apoptosis was only observed at concentrations considerably above the IC_{50}. The ginsenoside Rb2 inhibited invasiveness of human tumor cells by suppressing matrix metalloproteinase-2 activity [78] (Fig. (**2**)), however, this inhibition was not mediated by tissue inhibitors of metalloproteinases [79]. The same result was obtained for 20(S)-protopanaxadiol, which effected this inhibition at concentrations much lower than the IC_{50} value of 76 μM. Inhibition of matrix metalloproteinase-9 by another protopanaxadiol, named compound K, was demonstrated to be a result of reduced protein expression and appears to be mediated through repressing the kinases MAPK1, MAPK3, MAPK8 and MAPK14 in human glioma cells [80] (Fig. (**2**)). An influence on signal transduction by protopanaxadiols and -triols was also shown by Han *et al.*, who observed inhibition of epidermal growth factor-mediated cell proliferation with concomitant repression of c-fos and c-jun gene expression [81] (Fig. (**2**)). Notably, effects were much stronger for a total extract of ginsenosides, demonstrating that the extract may contain more potent saponins or exhibit synergistic effects. The often described effects of saponins on apoptosis-induction were also observed together with caspase activation and cytochrome c release [82, 83], as well as cell cycle arrest and inhibition of DNA synthesis with an IC_{50} of about 1 μM [84]. A multidrug resistance-reversal effect was also described for a protopanaxatriol [85], however the effect was rather small compared to the impact of Rg3 described above [70]. Two ginsenoside metabolites inhibited furthermore inflammatory processes by inactivating NFκB leading to blocked effects of inducible nitric oxide synthase and cyclooxygenase-2 [86, 87] (Fig. (**2**)). Uptake studies were performed in recent years and demonstrated that polar ginsenosides were not taken up efficiently by MCF-7 cells, while a relatively polar protopanaxadiol showed the most efficient uptake and demonstrating a potential correlation to growth inhibitory effects on tumor cells [88, 89]. The agylcone protopanaxadiol has been synthesized to gain information on structure-activity relationships for this class of ginsenosides. Some of the synthesized derivatives achieved more inhibition of proliferation than isolated protopanaxadiol [90].

Numerous effects on tumor cells have been described for ginsenosides and even more results will be revealed in the next years. While the growth inhibiting effects of ginsenosides are weaker than for other saponins, they were used to inhibit the mobility of tumor cells and may therefore be used in combination with other saponins.

AVICINS

In 2001 Haridas *et al.* identified triterpenoidal saponins from *Acacia victoriae*, an Australian tree. These saponins were named avicins and contain three sugar side chains. Avicins D and G (Fig. (**1**)) induced growth inhibition of human T lymphocytes at very low concentrations of 0.3 and 0.2 μg/ml, respectively, and promoted apoptosis as shown by activation of caspases and cytochrome c release [91] (Fig. (**2**)). The avicins are thus currently the most potent growth inhibiting saponins described. The impact on other human tumor cell lines was lower, however they still exhibited good IC_{50} values in the range of 1–2.5 μM [92]. In this context, inhibition of phosphatidylinositol 3-kinase activity especially for avicin G was reported, while the MAPK pathway seemed to be unaffected. Avicin G inhibited furthermore the activation and DNA binding of NFκB [93]. This avicin reduced the expression of inducible nitric oxide synthase and cyclooxygenase-2 as described for ginsenosides metabolites. Avicin D induced the expression of nuclear factor erythroid 2-related factor 2 (Fig. (**2**)), a transcription factor, which mediates the expression of several detoxifying and antioxidant proteins [94] (Table **1**). The protective effect was underlined by UV radiation of mice treated with avicin D, where severe damage and mutations were prevented. A mixture of avicins showed additional protective effects against papilloma-inducing chemicals 7,12-dimethylbenzanthracene and phorbol 12-myristate 13-acetate in mice and reduced the number of mutations in the oncogene H-*ras* [95] (Table **1**). The pore forming ability of the avicins D and G at concentrations of 25 μg/ml was demonstrated by Li *et al.* [96]. While pore formation was strongly cholesterol dependent for avicin G, this was not so for the close derivate avicin D. The non-selective pores are too small for proteins but allow ion flux. The authors concluded that the pores might influence the membrane potential of mitochondria. The direct influence on mitochondria was corroborated by studies on rat mitochondria, where both avicin D and G induced permeabilization leading to decreased respiratory activity [97] and ATP efflux after inhibition of the voltage dependent anion channel in the outer mitochondrial membrane [98]. This is possibly the main reason for apoptosis induction by avicins and it is likely that further saponins induce pore formation in mitochondrial membranes to induce apoptosis (Fig. (**2**)).

KNOWLEDGE, GAPS AND FUTURE

The various cellular effects of saponins described above are summarized in a schematic illustration (Fig. (**2**)). Saponins can either act extracellularly on tumor cells or influence intracellular pathways. Among the extracellular effects of saponins are plasma membrane permeabilization and inhibition of drug efflux by direct inhibition of membrane proteins. Permeabilizing effects occur at high saponin concentrations (usually > 100 μg/ml with variations for different saponins) while intracellular effects may also take place at lower concentrations. Cell cycle arrest and apoptosis induction are the best studied events. These outcomes by saponin action were shown for nearly all saponins described in this review. Other intracellular effects like the inhibition of invasiveness for metastasizing cells, the reduction of inflammatory responses, the activation of insulin-like growth factor I and estrogen receptors, and the induction of detoxifying and antioxidant proteins were only observed in a few studies and the relevance for anti-tumor therapy remains to be cleared. The most effective saponins are the avicins, which inhibit tumor cell growth at about 1 μg/ml. Due to their high impact on tumor cells, their intracellular effects were studied in great detail, especially the mechanisms leading to induction of apoptosis. However, it is notable that all described effects of saponins on tumor cells are dependent on the type of tumor and the saponin. Some activated pathways have as yet only observed for specific saponins such as the stimulation of insulin-like growth factor I receptor described for the ginsenoside Rg1, while induction of apoptosis is a common feature of nearly all saponins. More detailed studies for the diverse saponins will surely discover a higher degree of analog mechanisms for structurally related saponins. Nevertheless, the focus of future research on saponins for tumor therapy will remain the discovery of new saponins from diverse plants and the analyses of structure-activity relationships to allow identification of important structural components for the development of optimized synthetic saponins.

COMBINATIONS OF SAPONINS AND OTHER ANTI-CANCER AGENTS

The combined application of saponins with other anti-tumor drugs offers an interesting development in cancer treatment since in many reports additive or even synergistic effects between saponins and other drugs have been observed. These combinations will lead to essentially improved possibilities for the treatment of cancer. When intensified studies on saponins and their effect on cancer cells started in the 1990s, the first experiments with combinations of saponins and chemotherapeutics were initiated. Wu *et al.* reported in 1990 that formosanin-C, a saponin from *Paris formosana*, did not only increase natural killer cell activity and interferon production at a concentration of 2.5 mg/kg but potentiated additionally the growth inhibitory effect of 5-fluorouracil in mice [99] (Table **1**). When the ginsenoside Rh2 was used for treatment it did not inhibit the growth of human ovarian tumor cells in nude mice. However, a combined application with cisplatin resulted in an enhanced anti-tumor activity as a result of a synergistic action [100] (Table **1**).

Several approaches for the combined application of saponins and chemotherapeutics were investigated with the goal of achieving additional benefit of anti-tumorigenic effects on tumor cells. The well-known drug cisplatin was used in combination with different saponins. Gaidi *et al.* combined this chemotherapeutic with four triterpene jenisseensosides from *Silene* species and another triterpene saponin from *Achyranthes bidentata* [101]. While all jenisseensosides elevated the cytotoxicity of cisplatin synergistically, the *Achyranthes* saponin did not and in addition was non-toxic by itself on the colon cancer cell line used in the study. The authors concluded that the *p*-methoxycinnamoyl groups naturally present in the jenisseensosides where responsible for the enhancing potential, however, there are several further structural differences between the saponins that may be responsible for enhancing cisplatin activity. A study by Haddad *et al.* examined saponins isolated from *Albizia adianthifolia* and no potentiation of cisplatin cytotoxicity on human colon cancer cells was observed [102]. Liu *et al.* reported in 2011 a sensitizing effect of tubeimoside I for cisplatin on human ovarian cancer cells *via* effects on ERK1/2 and p38 signaling pathways [103]. Saikosaponin-a and -d have been reported to sensitize HeLa, SiHa, SKOV3, and A549 cells to cisplatin by enhancing the accumulation of reactive oxygen species [104]. A further study with four saponins from the roots of *Muraltia heisteria* revealed that only the stereoisomeric compounds 3 and 4 inhibited tumor cell growth and none were able to enhance the effect of cisplatin [105]. Thus only certain saponins seem to be able to enhance cisplatin activity due to their specific structure.

A number of further chemotherapeutics were combined with saponins to enhance the anti-tumorigenic impact. The ginsenosides described above were successfully used in many studies with chemotherapeutics. Ginsenoside Rh2 potentiated the cytotoxicity of paclitaxel (Taxol™) *in vitro* [60] and *in vivo* [106] (Table **1**). The chemotherapeutic mitoxantrone (Novantrone™) also acted synergistically with Rh2 *in vitro* but the combination failed to reduce tumor growth *in vivo* [106]. The ginsenosides Rgl, Rg3, Rhl and Rh2 and the aglycones protopanaxadiol and -triol were studied in combination with mitoxantrone and doxorubicin on drug insensitive cell lines [107]. While the enhancement of doxorubicin cytotoxicity was weak and maximal for protopanaxadiol and -triol, mitoxantrone cytotoxicity was enhanced 27 to 82-fold by the aglycones and Rh2 at a concentration of 20 μM. At high concentrations of 100 μM, the other ginsenosides enhanced mitoxantrone cytotoxicity only 2.5-fold. On tumor cells without upregulated breast cancer resistance protein all ginsenosides and aglycones were unable to increase mitoxantrone cytotoxicity, thus underlining the effect on multidrug resistance related proteins. The purified ginseng saponin panaxadiol was successfully combined with 5-fluorouracil to increase the effect on the human colon cancer cell line HCT-116 by inhibiting the cell cycle [108]. The ginsenosides Rg3 and Rh2 were recently applied together *in vivo* in mice with the alkylating agent cyclophosphamide and inhibited tumor growth synergistically [109, 110] (Table **1**). In a further study Rg3 in combination with cyclophosphamide showed only an additive effect of both substances, possibly due to the shorter treatment regimen (only 10 days compared to > 50 days) and the lower dose of Rg3 (3 mg/kg compared to 10 mg/kg) [111]. Rg3 was furthermore combined with betulinic acid to increase synergistically the apoptosis induction in human A549, HeLa, and HepG2 cells [112]. Rh2 and its aglycone aPPD were also analyzed *in vivo* for the potentiation of the tumor treatment efficacy in combination with docetaxel in four prostate xenograft models and demonstrated additive and synergistic effects [113]. The same group also studied the pharmacokinetics and biodistribution of Rh2 in a prostate PC-3 xenograft model [114]. The observed $t_{1/2}$ for the saponin at a dose of 120 mg/kg by oral gavage was 99 min and tumor tissue accumulated slightly more Rh2 than other tissues. Rg3 was studied as an sensitizer for docetaxel on several prostate tumor cell lines, however, the effects appeared to be additive rather than synergistic [115]. The same saponin was studied on lung cancer cells in combination with gemcitabine. The combination was most effective in inhibiting tumor growth and angiogenesis while reducing side effects of the gemcitabine treatment [116]. In 2010, Pokharel *et al.* described the effect of the ginsenosides Rd, Re, Rb1, and Rg1 on doxorubicin-resistant MCF-7 cells and reported decreased levels of multidrug resistance protein 1 after treatment with the ginsenosides, especially by Rd. Rd increased ubiquitination of the drug transporter, resulting in increased proteasomal degradation and reduced protein levels [117].

Chemical derivatives of diosgenyl saponins were analyzed on primary B cell chronic leukemia tumor cells in combination with the cytostatic drug cladribine (Leustatin™) and enhanced the toxic effect of cladribine only slightly [118]. The saponin mixture escin proved to be a suitable sensitizer for the tumor growth inhibition of gemcitabine in mice by inactivating NFκB [119]. Quillaja saponin is used for vaccination purposes but proved to be rather toxic when administered as a anti-cancer drug itself. Hu *et al.* combined quillaja saponin with cholesterol in order to reduce the saponin-induced toxicity [120]. Low concentrations of the complexes (around 1 µg/ml) induced apoptosis in U937 tumor cells while 30-fold higher concentrations were necessary for apoptosis induction in human monocytes. Besides the saponin-mediated enhanced chemotherapeutic cell death, a combination of saponins and radiotherapy was described by Chen *et al.* [121]. An extract of *Panax notoginseng* and ginsenoside Rb1 increased radiosensitivity of tumor cells in mice (Table **1**). The saponins had to be injected 30 minutes prior to radiation to achieve maximal tumor cell growth inhibition. Rb1 was 100-fold more effective than the saponin extract without mediating dose-dependent toxicity to the bone marrow and thus presents an encouraging approach.

Semi-synthesis and synthesis of saponins was successful in recent years. These studies open up great new opportunities for more detailed studies on structure-activity relationship and the design of more potent anti-cancer drugs. Gao *et al.* described the synthesis of derivatives of spinasaponin A [122], Zheng *et al.*

described the synthesis of highly potent OSW saponins with IC_{50} values in the low nanomolar range [123], Wang *et al.* synthesized 14 triterpenoid saponins (oleanic acid and ursonic acid saponins) and achieved potent saponins with IC_{50} values well below 10 μM after testing on HL-60, Hep-2, and BGC-823 cells [124]. Two naturally occurring saponins (28-O-beta-d-glucopyranosylbetulinic acid 3beta-O-alpha-l-arabinopyranoside) and seven bidesmosidic saponins were synthesized and the IC_{50} value of 2 μM for the most potent saponin (betulin saponin 22a) was determined on human cancer cell lines PC-3, DLD-1, A549, and MCF-7 [125].

Heisler *et al.* examined the combination of a saponin extract from *Gypsophila paniculata*, named Saponinum album, and a targeted chimeric toxin composed of human epidermal growth factor and the plant ribosome-inactivating protein saporin [126]. Saponinum album consists mainly of the saponin gypsoside A (30%) and further saponins with the identical aglycone (40%). The combination of Saponinum album at a non-permeabilizing and non-toxic concentration of 1.5 μg/ml with the targeted toxin enhanced cytotoxicity more than 3500-fold [127]. Furthermore, the highest increase in cytotoxicity of the targeted toxin was observed on epidermal growth factor receptor expressing cells, thereby increasing the therapeutic window of the targeted toxin. In relation to these observations further saponins from diverse plants and with different structures were analyzed for their ability to increase the cytotoxicity of targeted toxins [12]. Among the saponins examined only Saponinum album and quillajasaponin increased the cytotoxicity of the targeted toxin more than 1000-fold. Gypsoside A (Fig. (**1**)), the main component of Saponinum album, and quillajasaponin are both triterpene saponins with an aldehyde function at position C-4 that seems to be important for the synergistic action. Notably, only Saponinum album increased the specificity of the targeted toxin for target cells, while quillajasaponin enhanced the impact of the targeted toxin on both non-targeted and targeted cells similarly. The enhancement of cytotoxicity of several protein toxins was reported by Hebestreit *et al.* with maximal increase for the type I ribosome-inactivating proteins saporin and agrostin [128]. A biodistribution study with ^{3}H-labeled Saponium album demonstrated a very rapid distribution (within 10 min) of the saponin component without any specificity for any tissues and resulted in a fast secretion *via* the kidneys and the bladder [129]. The combination of a targeted toxin and Saponinum album in a mouse model required 50-fold lower doses of targeted toxins and resulted in a 94% reduction in tumor growth compared to untreated mice, and mild and reversible side effects [130]. A single saponin was isolated from Saponinum album and demonstrated the expected increase in the cytotoxicity of the targeted toxin [131]. Further effort has been made to develop feasible methods for the purification of biological active saponins that enhance the cytotoxicity of targeted toxins [132]. Recent studies demonstrated that among several bacterial and plant toxins saporin and its homologue dianthin are the best synergistic partners for the action of saponin [133] and that the saponin-mediated enhancement of the targeted toxin is not correlated to the target receptor density on the cell surface [134] while specificity is maintained by the targeting moiety. Mechanistic analyses revealed that saponin increases the release of the targeted toxin from endosomes [135]. The release occurs only from certain endosomes, which are involved in the endosomal uptake path followed by the targeted toxin when applied in low concentration [136]. The combination of Saponinum album and targeted toxins is to date the most powerful combination of saponins and anti-tumor drugs and presents a very strong synergism with a further studied mechanism.

CONCLUSIONS

The number of newly isolated and described saponins is increasing constantly and many further saponins will be identified due to improved methods of purification and detection. The majority of the new saponins as well as several well-known saponins possess impressive anti-cancer effects and might help to develop improved anti-cancer regimens. The avicins have currently the highest impact on cancer cells with IC_{50} values for growth inhibition in the range of 1 μg/ml. Almost all saponins induce apoptosis in tumor cells, they are preferable drugs for the treatment of cancer, because eliminating tumor cells by apoptosis is helpful to lower side effects in patients by avoiding necrosis. Other actions like the inhibition of invasiveness as mediated by ginsenoside Rb2 and the group B soy saponins is furthermore valuable in order to prevent the development of metastases. The chemical modification of saponins might be a way to further increase their activity. However, a good understanding of structure-activity relationships is a prerequisite for modifications and currently not well established. Individual studies have shown relevance of the stereochemistry of methyl protoneogracillin or higher activity for deglycosylated derivatives of soy saponin but it is impossible to determine which functional groups within a saponin would result in the highest impact on tumor cells since currently only a few studies have examined structure-activity relationships. Furthermore, no standardized experimental procedures including tumor cells and incubation time have been used to allow quantitative comparisons of different saponins. The low content of a certain saponin in an analyzed sample is a further obstacle for detecting anti-tumorigenic effects. Special attention should be given to combinations of saponins and other anti-carcinogenic drugs, since these offer very efficient treatment regimens against cancer. Most important is the saponin-mediated potentiation of tumor growth inhibition and the possibility to circumvent drug resistance. The elucidation of structure-activity relationships between different saponins in combination with conventional drugs is much more complicated than for saponins alone. Thus, it is not surprising that little mechanistic processes for these effects are known, however, detailed information on this basis is necessary for a directed improvement of saponin-based tumor therapies in the future.

ACKNOWLEDGEMENTS

We acknowledge the generous financial support of the DFG (FU 408/3-1 and FU 408/6-1), the Sonnenfeldstiftung and the Wilhelm Sander-Stiftung (2001.078.1 and 2001.078.2).

CONFLICT OF INTEREST

The authors confirm that this chapter contents have no conflict of interest.

DISCLOSURE

The chapter submitted for eBook Series entitled: "**Recent Advances in Medicinal Chemistry, Volume 1**" is an update of our article published in **Mini-Reviews in Medicinal Chemistry, Volume 8, Number 6, pp. 575 to 584**, with additional text and references.

ABBREVIATIONS

IC_{50} = Half maximal inhibitory concentration

MAPK = Mitogen-activated protein kinase

NFκB = Nuclear factor-kappa B

REFERENCES

[1] Wang W, Zhao Y, Rayburn ER, Hill DL, Wang H, Zhang R. *In vitro* anti-cancer activity and structure-activity relationships of natural products isolated from fruits of Panax ginseng. Cancer chemotherapy and pharmacology. 2006.

[2] Wina E, Muetzel S, Becker K. The impact of saponins or saponin-containing plant materials on ruminant production--a review. J Agric Food Chem. 2005;53(21):8093-105.

[3] Seeman P. Transient holes in the erythrocyte membrane during hypotonic hemolysis and stable holes in the membrane after lysis by saponin and lysolecithin. J Cell Biol. 1967;32(1):55-70.

[4] Baumann E, Stoya G, Volkner A, Richter W, Lemke C, Linss W. Hemolysis of human erythrocytes with saponin affects the membrane structure. Acta Histochem. 2000;102(1):21-35.

[5] Bangham AD, Horne RW, Glauert AM, Dingle JT, Lucy JA. Action of saponin on biological cell membranes. Nature. 1962;196:952-5.

[6] Gogelein H, Huby A. Interaction of saponin and digitonin with black lipid membranes and lipid monolayers. Biochim Biophys Acta. 1984;773(1):32-8.

[7] Segal R, Milo-Goldzweig I. The susceptibility of cholesterol-depleted erythrocytes to saponin and sapogenin hemolysis. Biochim Biophys Acta. 1978;512(1):223-6.

[8] Zhao HL, Cho KH, Ha YW, Jeong TS, Lee WS, Kim YS. Cholesterol-lowering effect of platycodin D in hypercholesterolemic ICR mice. European journal of pharmacology. 2006;537(1-3):166-73.

[9] Hu M, Konoki K, Tachibana K. Cholesterol-independent membrane disruption caused by triterpenoid saponins. Biochim Biophys Acta. 1996;1299(2):252-8.

[10] Woldemichael GM, Wink M. Identification and biological activities of triterpenoid saponins from Chenopodium quinoa. J Agric Food Chem. 2001;49(5):2327-32.

[11] Apers S, Baronikova S, Sindambiwe JB, Witvrouw M, De Clercq E, Vanden Berghe D, *et al.* Antiviral, haemolytic and molluscicidal activities of triterpenoid saponins from Maesa lanceolata: establishment of structure-activity relationships. Planta medica. 2001;67(6):528-32.

[12] Bachran C, Sutherland M, Heisler I, Hebestreit P, Melzig MF, Fuchs H. The saponin-mediated enhanced uptake of targeted saporin-based drugs is strongly dependent on the saponin structure. Exp Biol Med (Maywood). 2006;231(4):412-20.

[13] Hall JG, Birbeck MS, Robertson D, Peppard J, Orlans E. The use of detergents and immunoperoxidase reagents for the ultrastructural demonstration of internal immunoglobulin in lymph cells. J Immunol Methods. 1978;19(4):351-9.

[14] Lee BH, Jeong SM, Ha TS, Park CS, Lee JH, Kim JH, *et al.* Ginsenosides regulate ligand-gated ion channels from the outside. Mol Cells. 2004;18(1):115-21.

[15] Morein B, Sundquist B, Hoglund S, Dalsgaard K, Osterhaus A. Iscom, a novel structure for antigenic presentation of membrane proteins from enveloped viruses. Nature. 1984;308(5958):457-60.

[16] Behboudi S, Morein B, Villacres-Eriksson M. *In vivo* and *in vitro* induction of IL-6 by Quillaja saponaria molina triterpenoid formulations. Cytokine. 1997;9(9):682-7.

[17] Lovgren K, Morein B. The requirement of lipids for the formation of immunostimulating complexes (iscoms). Biotechnol Appl Biochem. 1988;10(2):161-72.

[18] Ebbesen P, Dalsgaard K, Madsen M. Prolonged survival of AKR mice treated with the saponin adjuvant Quil A. Acta Pathol Microbiol Scand [A]. 1976;84(4):358-60.

[19] Okita K, Li Q, Murakamio T, Takahashi M. Anti-growth effects with components of Sho-saiko-to (TJ-9) on cultured human hepatoma cells. Eur J Cancer Prev. 1993;2(2):169-75.

[20] Yu LJ, Ma RD, Wang YQ, Nishino H, Takayasu J, He WZ, *et al.* Potent anti-tumorigenic effect of tubeimoside 1 isolated from the bulb of Bolbostemma paniculatum (Maxim) Franquet. Int J Cancer. 1992;50(4):635-8.

[21] Ragupathi G, Damani P, Deng K, Adams MM, Hang J, George C, *et al.* Preclinical evaluation of the synthetic adjuvant SQS-21 and its constituent isomeric saponins. Vaccine. 2010;28(26):4260-7. Epub 2010/05/11.

[22] Cai J, Liu M, Wang Z, Ju Y. Apoptosis induced by dioscin in Hela cells. Biol Pharm Bull. 2002;25(2):193-6.

[23] Zhang Y, Li HZ, Zhang YJ, Jacob MR, Khan SI, Li XC, *et al.* Atropurosides A-G, new steroidal saponins from Smilacina atropurpurea. Steroids. 2006;71(8):712-9.

[24] Hibasami H, Moteki H, Ishikawa K, Katsuzaki H, Imai K, Yoshioka K, *et al.* Protodioscin isolated from fenugreek (Trigonella foenumgraecum L.) induces cell death and morphological change indicative of apoptosis in leukemic cell line H-60, but not in gastric cancer cell line KATO III. International journal of molecular medicine. 2003;11(1):23-6.

[25] Hu K, Yao X. The cytotoxicity of methyl protodioscin against human cancer cell lines *in vitro*. Cancer Invest. 2003;21(3):389-93.

[26] Wang G, Chen H, Huang M, Wang N, Zhang J, Zhang Y, *et al.* Methyl protodioscin induces G2/M cell cycle arrest and apoptosis in HepG2 liver cancer cells. Cancer Lett. 2006;241(1):102-9.

[27] Liu MJ, Yue PY, Wang Z, Wong RN. Methyl protodioscin induces G2/M arrest and apoptosis in K562 cells with the hyperpolarization of mitochondria. Cancer Lett. 2005;224(2):229-41.

[28] Hu K, Yao X. The cytotoxicity of methyl protoneogracillin (NSC-698793) and gracillin (NSC-698787), two steroidal saponins from the rhizomes of Dioscorea collettii var. hypoglauca, against human cancer cells *in vitro*. Phytother Res. 2003;17(6):620-6.

[29] Kaskiw MJ, Tassotto ML, Mok M, Tokar SL, Pycko R, Th'ng J, *et al.* Structural analogues of diosgenyl saponins: synthesis and anticancer activity. Bioorganic & medicinal chemistry. 2009;17(22):7670-9. Epub 2009/10/13.

[30] Wu WS, Hsu HY. Involvement of p-15(INK4b) and p-16(INK4a) gene expression in saikosaponin a and TPA-induced growth inhibition of HepG2 cells. Biochem Biophys Res Commun. 2001;285(2):183-7.

[31] Wen-Sheng W. ERK signaling pathway is involved in p15INK4b/p16INK4a expression and HepG2 growth inhibition triggered by TPA and Saikosaponin a. Oncogene. 2003;22(7):955-63.

[32] Chen JC, Chang NW, Chung JG, Chen KC. Saikosaponin-A induces apoptotic mechanism in human breast MDA-MB-231 and MCF-7 cancer cells. Am J Chin Med. 2003;31(3):363-77.

[33] Hsu YL, Kuo PL, Weng TC, Yen MH, Chiang LC, Lin CC. The antiproliferative activity of saponin-enriched fraction from Bupleurum Kaoi is through Fas-dependent apoptotic pathway in human non-small cell lung cancer A549 cells. Biol Pharm Bull. 2004;27(7):1112-5.

[34] Chiang LC, Ng LT, Liu LT, Shieh DE, Lin CC. Cytotoxicity and anti-hepatitis B virus activities of saikosaponins from Bupleurum species. Planta medica. 2003;69(8):705-9.
[35] Shyu KG, Tsai SC, Wang BW, Liu YC, Lee CC. Saikosaponin C induces endothelial cells growth, migration and capillary tube formation. Life Sci. 2004;76(7):813-26.
[36] Hsu YL, Kuo PL, Chiang LC, Lin CC. Involvement of p53, nuclear factor kappaB and Fas/Fas ligand in induction of apoptosis and cell cycle arrest by saikosaponin d in human hepatoma cell lines. Cancer Lett. 2004;213(2):213-21.
[37] Hsu YL, Kuo PL, Lin CC. The proliferative inhibition and apoptotic mechanism of Saikosaponin D in human non-small cell lung cancer A549 cells. Life Sci. 2004;75(10):1231-42.
[38] Zou K, Cui JR, Wang B, Zhao YY, Zhang RY. A pair of isomeric saponins with cytotoxicity from Albizzia julibrissin. J Asian Nat Prod Res. 2005;7(6):783-9.
[39] Liang H, Tong WY, Zhao YY, Cui JR, Tu GZ. An antitumor compound julibroside J28 from Albizia julibrissin. Bioorganic & medicinal chemistry letters. 2005;15(20):4493-5.
[40] Zou K, Zhao YY, Zhang RY. A cytotoxic saponin from Albizia julibrissin. Chem Pharm Bull (Tokyo). 2006;54(8):1211-2.
[41] Zou K, Tong WY, Liang H, Cui JR, Tu GZ, Zhao YY, *et al.* Diastereoisomeric saponins from Albizia julibrissin. Carbohydrate research. 2005;340(7):1329-34.
[42] Zheng L, Zheng J, Wu LJ, Zhao YY. Julibroside J8-induced HeLa cell apoptosis through caspase pathway. J Asian Nat Prod Res. 2006;8(5):457-65.
[43] Zheng L, Zheng J, Zhao Y, Wang B, Wu L, Liang H. Three anti-tumor saponins from Albizia julibrissin. Bioorganic & medicinal chemistry letters. 2006;16(10):2765-8.
[44] Cao S, Norris A, Miller JS, Ratovoson F, Razafitsalama J, Andriantsiferana R, *et al.* Cytotoxic Triterpenoid Saponins of Albizia gummifera from the Madagascar Rain Forest(,1). Journal of natural products. 2007.
[45] Kerwin SM. Soy saponins and the anticancer effects of soybeans and soy-based foods. Curr Med Chem Anti-Canc Agents. 2004;4(3):263-72.
[46] Kim HY, Yu R, Kim JS, Kim YK, Sung MK. Antiproliferative crude soy saponin extract modulates the expression of IkappaBalpha, protein kinase C, and cyclooxygenase-2 in human colon cancer cells. Cancer Lett. 2004;210(1):1-6.
[47] Gurfinkel DM, Rao AV. Soyasaponins: the relationship between chemical structure and colon anticarcinogenic activity. Nutrition and cancer. 2003;47(1):24-33.
[48] Oh YJ, Sung MK. Soybean saponins inhibit cell proliferation by suppressing PKC activation and induce differentiation of HT-29 human colon adenocarcinoma cells. Nutrition and cancer. 2001;39(1):132-8.
[49] Yanamandra N, Berhow MA, Konduri S, Dinh DH, Olivero WC, Nicolson GL, *et al.* Triterpenoids from Glycine max decrease invasiveness and induce caspase-mediated cell death in human SNB19 glioma cells. Clin Exp Metastasis. 2003;20(4):375-83.
[50] Ellington AA, Berhow M, Singletary KW. Induction of macroautophagy in human colon cancer cells by soybean B-group triterpenoid saponins. Carcinogenesis. 2005;26(1):159-67.
[51] Ellington AA, Berhow MA, Singletary KW. Inhibition of Akt signaling and enhanced ERK1/2 activity are involved in induction of macroautophagy by triterpenoid B-group soyasaponins in colon cancer cells. Carcinogenesis. 2006;27(2):298-306.
[52] Chang WW, Yu CY, Lin TW, Wang PH, Tsai YC. Soyasaponin I decreases the expression of alpha2,3-linked sialic acid on the cell surface and suppresses the metastatic potential of B16F10 melanoma cells. Biochem Biophys Res Commun. 2006;341(2):614-9.

[53] Hsu CC, Lin TW, Chang WW, Wu CY, Lo WH, Wang PH, *et al.* Soyasaponin-I-modified invasive behavior of cancer by changing cell surface sialic acids. Gynecol Oncol. 2005;96(2):415-22.
[54] Jun HS, Kim SE, Sung MK. Protective effect of soybean saponins and major antioxidants against aflatoxin B1-induced mutagenicity and DNA-adduct formation. J Med Food. 2002;5(4):235-40.
[55] Chang YS, Seo EK, Gyllenhaal C, Block KI. Panax ginseng: a role in cancer therapy? Integr Cancer Ther. 2003;2(1):13-33.
[56] Nag SA, Qin JJ, Wang W, Wang MH, Wang H, Zhang R. Ginsenosides as Anticancer Agents: *In vitro* and *in vivo* Activities, Structure-Activity Relationships, and Molecular Mechanisms of Action. Frontiers in pharmacology. 2012;3:25. Epub 2012/03/10.
[57] Kim YS, Jin SH. Ginsenoside Rh2 induces apoptosis *via* activation of caspase-1 and -3 and up-regulation of Bax in human neuroblastoma. Arch Pharm Res. 2004;27(8):834-9.
[58] Kim HS, Lee EH, Ko SR, Choi KJ, Park JH, Im DS. Effects of ginsenosides Rg3 and Rh2 on the proliferation of prostate cancer cells. Arch Pharm Res. 2004;27(4):429-35.
[59] Cheng CC, Yang SM, Huang CY, Chen JC, Chang WM, Hsu SL. Molecular mechanisms of ginsenoside Rh2-mediated G1 growth arrest and apoptosis in human lung adenocarcinoma A549 cells. Cancer chemotherapy and pharmacology. 2005;55(6):531-40.
[60] Jia WW, Bu X, Philips D, Yan H, Liu G, Chen X, *et al.* Rh2, a compound extracted from ginseng, hypersensitizes multidrug-resistant tumor cells to chemotherapy. Can J Physiol Pharmacol. 2004;82(7):431-7.
[61] Tode T, Kikuchi Y, Kita T, Hirata J, Imaizumi E, Nagata I. Inhibitory effects by oral administration of ginsenoside Rh2 on the growth of human ovarian cancer cells in nude mice. Journal of cancer research and clinical oncology. 1993;120(1-2):24-6.
[62] Tatsuka M, Maeda M, Ota T. Anticarcinogenic effect and enhancement of metastatic potential of BALB/c 3T3 cells by ginsenoside Rh(2). Jpn J Cancer Res. 2001;92(11):1184-9.
[63] Popovich DG, Kitts DD. Structure-function relationship exists for ginsenosides in reducing cell proliferation and inducing apoptosis in the human leukemia (THP-1) cell line. Arch Biochem Biophys. 2002;406(1):1-8.
[64] Lee Y, Jin Y, Lim W, Ji S, Choi S, Jang S, *et al.* A ginsenoside-Rh1, a component of ginseng saponin, activates estrogen receptor in human breast carcinoma MCF-7 cells. J Steroid Biochem Mol Biol. 2003;84(4):463-8.
[65] Chan RY, Chen WF, Dong A, Guo D, Wong MS. Estrogen-like activity of ginsenoside Rg1 derived from Panax notoginseng. J Clin Endocrinol Metab. 2002;87(8):3691-5.
[66] Chen WF, Lau WS, Cheung PY, Guo DA, Wong MS. Activation of insulin-like growth factor I receptor-mediated pathway by ginsenoside Rg1. British journal of pharmacology. 2006;147(5):542-51.
[67] Keum YS, Han SS, Chun KS, Park KK, Park JH, Lee SK, *et al.* Inhibitory effects of the ginsenoside Rg3 on phorbol ester-induced cyclooxygenase-2 expression, NF-kappaB activation and tumor promotion. Mutat Res. 2003;523-524:75-85.
[68] Kim WY, Kim JM, Han SB, Lee SK, Kim ND, Park MK, *et al.* Steaming of ginseng at high temperature enhances biological activity. Journal of natural products. 2000;63(12):1702-4.
[69] Wang CZ, Zhang B, Song WX, Wang A, Ni M, Luo X, *et al.* Steamed American ginseng berry: ginsenoside analyses and anticancer activities. J Agric Food Chem. 2006;54(26):9936-42.

[70] Kim SW, Kwon HY, Chi DW, Shim JH, Park JD, Lee YH, *et al.* Reversal of P-glycoprotein-mediated multidrug resistance by ginsenoside Rg(3). Biochem Pharmacol. 2003;65(1):75-82.

[71] Panwar M, Kumar M, Samarth R, Kumar A. Evaluation of chemopreventive action and antimutagenic effect of the standardized Panax ginseng extract, EFLA400, in Swiss albino mice. Phytother Res. 2005;19(1):65-71.

[72] Chen XP, Qian LL, Jiang H, Chen JH. Ginsenoside Rg3 inhibits CXCR4 expression and related migrations in a breast cancer cell line. International journal of clinical oncology / Japan Society of Clinical Oncology. 2011;16(5):519-23. Epub 2011/04/02.

[73] He BC, Gao JL, Luo X, Luo J, Shen J, Wang L, *et al.* Ginsenoside Rg3 inhibits colorectal tumor growth through the down-regulation of Wnt/ss-catenin signaling. International journal of oncology. 2011;38(2):437-45. Epub 2010/12/15.

[74] Park D, Bae DK, Jeon JH, Lee J, Oh N, Yang G, *et al.* Immunopotentiation and antitumor effects of a ginsenoside Rg-fortified red ginseng preparation in mice bearing H460 lung cancer cells. Environmental toxicology and pharmacology. 2011;31(3):397-405. Epub 2011/07/27.

[75] He K, Liu Y, Yang Y, Li P, Yang L. A dammarane glycoside derived from ginsenoside Rb3. Chem Pharm Bull (Tokyo). 2005;53(2):177-9.

[76] Kumar A, Kumar M, Panwar M, Samarth RM, Park TY, Park MH, *et al.* Evaluation of chemopreventive action of Ginsenoside Rp1. Biofactors. 2006;26(1):29-43.

[77] Yang ZG, Sun HX, Ye YP. Ginsenoside Rd from Panax notoginseng is cytotoxic towards HeLa cancer cells and induces apoptosis. Chem Biodivers. 2006;3(2):187-97.

[78] Fujimoto J, Sakaguchi H, Aoki I, Toyoki H, Khatun S, Tamaya T. Inhibitory effect of ginsenoside-Rb2 on invasiveness of uterine endometrial cancer cells to the basement membrane. Eur J Gynaecol Oncol. 2001;22(5):339-41.

[79] Li G, Wang Z, Sun Y, Liu K, Wang Z. Ginsenoside 20(S)-protopanaxadiol inhibits the proliferation and invasion of human fibrosarcoma HT1080 cells. Basic Clin Pharmacol Toxicol. 2006;98(6):588-92.

[80] Jung SH, Woo MS, Kim SY, Kim WK, Hyun JW, Kim EJ, *et al.* Ginseng saponin metabolite suppresses phorbol ester-induced matrix metalloproteinase-9 expression through inhibition of activator protein-1 and mitogen-activated protein kinase signaling pathways in human astroglioma cells. Int J Cancer. 2006;118(2):490-7.

[81] Han HJ, Yoon BC, Lee SH, Park SH, Park JY, Oh YJ, *et al.* Ginsenosides inhibit EGF-induced proliferation of renal proximal tubule cells *via* decrease of c-fos and c-jun gene expression *in vitro*. Planta medica. 2002;68(11):971-4.

[82] Oh SH, Lee BH. A ginseng saponin metabolite-induced apoptosis in HepG2 cells involves a mitochondria-mediated pathway and its downstream caspase-8 activation and Bid cleavage. Toxicol Appl Pharmacol. 2004;194(3):221-9.

[83] Jin YH, Yim H, Park JH, Lee SK. Cdk2 activity is associated with depolarization of mitochondrial membrane potential during apoptosis. Biochem Biophys Res Commun. 2003;305(4):974-80.

[84] Jin YH, Choi J, Shin S, Lee KY, Park JH, Lee SK. Panaxadiol selectively inhibits cyclin A-associated Cdk2 activity by elevating p21WAF1/CIP1 protein levels in mammalian cells. Carcinogenesis. 2003;24(11):1767-72.

[85] Choi CH, Kang G, Min YD. Reversal of P-glycoprotein-mediated multidrug resistance by protopanaxatriol ginsenosides from Korean red ginseng. Planta medica. 2003;69(3):235-40.

[86] Lee JY, Shin JW, Chun KS, Park KK, Chung WY, Bang YJ, *et al.* Antitumor promotional effects of a novel intestinal bacterial metabolite (IH-901) derived from the protopanaxadiol-type ginsenosides in mouse skin. Carcinogenesis. 2005;26(2):359-67.

[87] Oh GS, Pae HO, Choi BM, Seo EA, Kim DH, Shin MK, *et al.* 20(S)-Protopanaxatriol, one of ginsenoside metabolites, inhibits inducible nitric oxide synthase and cyclooxygenase-2 expressions through inactivation of nuclear factor-kappaB in RAW 264.7 macrophages stimulated with lipopolysaccharide. Cancer Lett. 2004;205(1):23-9.

[88] Ha YW, Ahn KS, Lee JC, Kim SH, Chung BC, Choi MH. Validated quantification for selective cellular uptake of ginsenosides on MCF-7 human breast cancer cells by liquid chromatography-mass spectrometry. Analytical and bioanalytical chemistry. 2010;396(8):3017-25. Epub 2010/02/20.

[89] Lee JI, Ha YW, Choi TW, Kim HJ, Kim SM, Jang HJ, *et al.* Cellular uptake of ginsenosides in Korean white ginseng and red ginseng and their apoptotic activities in human breast cancer cells. Planta medica. 2011;77(2):133-40. Epub 2010/07/30.

[90] Du GJ, Dai Q, Williams S, Wang CZ, Yuan CS. Synthesis of protopanaxadiol derivatives and evaluation of their anticancer activities. Anti-cancer drugs. 2011;22(1):35-45. Epub 2010/10/12.

[91] Haridas V, Higuchi M, Jayatilake GS, Bailey D, Mujoo K, Blake ME, *et al.* Avicins: triterpenoid saponins from Acacia victoriae (Bentham) induce apoptosis by mitochondrial perturbation. Proc Natl Acad Sci U S A. 2001;98(10):5821-6.

[92] Mujoo K, Haridas V, Hoffmann JJ, Wachter GA, Hutter LK, Lu Y, *et al.* Triterpenoid saponins from Acacia victoriae (Bentham) decrease tumor cell proliferation and induce apoptosis. Cancer Res. 2001;61(14):5486-90.

[93] Haridas V, Arntzen CJ, Gutterman JU. Avicins, a family of triterpenoid saponins from Acacia victoriae (Bentham), inhibit activation of nuclear factor-kappaB by inhibiting both its nuclear localization and ability to bind DNA. Proc Natl Acad Sci U S A. 2001;98(20):11557-62.

[94] Haridas V, Hanausek M, Nishimura G, Soehnge H, Gaikwad A, Narog M, *et al.* Triterpenoid electrophiles (avicins) activate the innate stress response by redox regulation of a gene battery. J Clin Invest. 2004;113(1):65-73.

[95] Hanausek M, Ganesh P, Walaszek Z, Arntzen CJ, Slaga TJ, Gutterman JU. Avicins, a family of triterpenoid saponins from Acacia victoriae (Bentham), suppress H-ras mutations and aneuploidy in a murine skin carcinogenesis model. Proc Natl Acad Sci U S A. 2001;98(20):11551-6.

[96] Li XX, Davis B, Haridas V, Gutterman JU, Colombini M. Proapoptotic triterpene electrophiles (avicins) form channels in membranes: cholesterol dependence. Biophys J. 2005;88(4):2577-84.

[97] Lemeshko VV, Haridas V, Quijano Perez JC, Gutterman JU. Avicins, natural anticancer saponins, permeabilize mitochondrial membranes. Arch Biochem Biophys. 2006;454(2):114-22.

[98] Haridas V, Li X, Mizumachi T, Higuchi M, Lemeshko VV, Colombini M, *et al.* Avicins, a novel plant-derived metabolite lowers energy metabolism in tumor cells by targeting the outer mitochondrial membrane. Mitochondrion. 2007;7(3):234-40.

[99] Wu RT, Chiang HC, Fu WC, Chien KY, Chung YM, Horng LY. Formosanin-C, an immunomodulator with antitumor activity. Int J Immunopharmacol. 1990;12(7):777-86.

[100] Kikuchi Y, Sasa H, Kita T, Hirata J, Tode T, Nagata I. Inhibition of human ovarian cancer cell proliferation *in vitro* by ginsenoside Rh2 and adjuvant effects to cisplatin *in vivo*. Anti-cancer drugs. 1991;2(1):63-7.

[101] Gaidi G, Correia M, Chauffert B, Beltramo JL, Wagner H, Lacaille-Dubois MA. Saponins-mediated potentiation of cisplatin accumulation and cytotoxicity in human colon cancer cells. Planta medica. 2002;68(1):70-2.

[102] Haddad M, Khan IA, Lacaille-Dubois MA. Two new prosapogenins from Albizia adianthifolia. Pharmazie. 2002;57(10):705-8.

[103] Liu HZ, Yu C, Yang Z, He JL, Chen WJ, Yin J, *et al.* Tubeimoside I sensitizes cisplatin in cisplatin-resistant human ovarian cancer cells (A2780/DDP) through down-regulation of ERK and up-regulation of p38 signaling pathways. Molecular medicine reports. 2011;4(5):985-92. Epub 2011/06/21.

[104] Wang Q, Zheng XL, Yang L, Shi F, Gao LB, Zhong YJ, *et al.* Reactive oxygen species-mediated apoptosis contributes to chemosensitization effect of saikosaponins on cisplatin-induced cytotoxicity in cancer cells. Journal of experimental & clinical cancer research : CR. 2010;29:159. Epub 2010/12/15.

[105] Elbandy M, Miyamoto T, Chauffert B, Delaude C, Lacaille-Dubois MA. Novel acylated triterpene glycosides from Muraltia heisteria. Journal of natural products. 2002;65(2):193-7.

[106] Xie X, Eberding A, Madera C, Fazli L, Jia W, Goldenberg L, *et al.* Rh2 synergistically enhances paclitaxel or mitoxantrone in prostate cancer models. J Urol. 2006;175(5):1926-31.

[107] Jin J, Shahi S, Kang HK, van Veen HW, Fan TP. Metabolites of ginsenosides as novel BCRP inhibitors. Biochem Biophys Res Commun. 2006;345(4):1308-14.

[108] Li XL, Wang CZ, Mehendale SR, Sun S, Wang Q, Yuan CS. Panaxadiol, a purified ginseng component, enhances the anti-cancer effects of 5-fluorouracil in human colorectal cancer cells. Cancer chemotherapy and pharmacology. 2009;64(6):1097-104. Epub 2009/03/12.

[109] Zhang Q, Kang X, Zhao W. Antiangiogenic effect of low-dose cyclophosphamide combined with ginsenoside Rg3 on Lewis lung carcinoma. Biochem Biophys Res Commun. 2006;342(3):824-8.

[110] Wang Z, Zheng Q, Liu K, Li G, Zheng R. Ginsenoside Rh(2) enhances antitumour activity and decreases genotoxic effect of cyclophosphamide. Basic Clin Pharmacol Toxicol. 2006;98(4):411-5.

[111] Xu TM, Xin Y, Cui MH, Jiang X, Gu LP. Inhibitory effect of ginsenoside Rg3 combined with cyclophosphamide on growth and angiogenesis of ovarian cancer. Chin Med J (Engl). 2007;120(7):584-8.

[112] Li Q, Li Y, Wang X, Fang X, He K, Guo X, *et al.* Co-treatment with ginsenoside Rh2 and betulinic acid synergistically induces apoptosis in human cancer cells in association with enhanced capsase-8 activation, bax translocation, and cytochrome c release. Molecular carcinogenesis. 2011;50(10):760-9. Epub 2011/07/14.

[113] Musende AG, Eberding A, Jia W, Ramsay E, Bally MB, Guns ET. Rh2 or its aglycone aPPD in combination with docetaxel for treatment of prostate cancer. The Prostate. 2010;70(13):1437-47. Epub 2010/08/06.

[114] Musende AG, Eberding A, Wood C, Adomat H, Fazli L, Hurtado-Coll A, *et al.* Pre-clinical evaluation of Rh2 in PC-3 human xenograft model for prostate cancer *in vivo*: formulation,

pharmacokinetics, biodistribution and efficacy. Cancer chemotherapy and pharmacology. 2009;64(6):1085-95. Epub 2009/03/21.

[115] Kim SM, Lee SY, Cho JS, Son SM, Choi SS, Yun YP, *et al.* Combination of ginsenoside Rg3 with docetaxel enhances the susceptibility of prostate cancer cells *via* inhibition of NF-kappaB. European journal of pharmacology. 2010;631(1-3):1-9. Epub 2010/01/09.

[116] Liu TG, Huang Y, Cui DD, Huang XB, Mao SH, Ji LL, *et al.* Inhibitory effect of ginsenoside Rg3 combined with gemcitabine on angiogenesis and growth of lung cancer in mice. BMC cancer. 2009;9:250. Epub 2009/07/25.

[117] Pokharel YR, Kim ND, Han HK, Oh WK, Kang KW. Increased ubiquitination of multidrug resistance 1 by ginsenoside Rd. Nutrition and cancer. 2010;62(2):252-9. Epub 2010/01/26.

[118] Myszka H, Bednarczyk D, Najder M, Kaca W. Synthesis and induction of apoptosis in B cell chronic leukemia by diosgenyl 2-amino-2-deoxy-beta-D-glucopyranoside hydrochloride and its derivatives. Carbohydrate research. 2003;338(2):133-41.

[119] Wang YW, Wang SJ, Zhou YN, Pan SH, Sun B. Escin augments the efficacy of gemcitabine through down-regulation of nuclear factor-kappaB and nuclear factor-kappaB-regulated gene products in pancreatic cancer both *in vitro* and *in vivo*. Journal of cancer research and clinical oncology. 2012. Epub 2012/01/25.

[120] Hu K, Berenjian S, Larsson R, Gullbo J, Nygren P, Lovgren T, *et al.* Nanoparticulate Quillaja saponin induces apoptosis in human leukemia cell lines with a high therapeutic index. International journal of nanomedicine. 2010;5:51-62. Epub 2010/02/18.

[121] Chen FD, Wu MC, Wang HE, Hwang JJ, Hong CY, Huang YT, *et al.* Sensitization of a tumor, but not normal tissue, to the cytotoxic effect of ionizing radiation using Panax notoginseng extract. Am J Chin Med. 2001;29(3-4):517-24.

[122] Gao J, Li X, Gu G, Liu S, Cui M, Lou HX. Facile synthesis of triterpenoid saponins bearing beta-Glu/Gal-(1-->3)-beta-GluA methyl ester and their cytotoxic activities. Bioorganic & medicinal chemistry letters. 2012;22(7):2396-400. Epub 2012/03/13.

[123] Zheng D, Zhou L, Guan Y, Chen X, Zhou W, Lei P. Synthesis of cholestane glycosides bearing OSW-1 disaccharide or its 1-->4-linked analogue and their antitumor activities. Bioorganic & medicinal chemistry letters. 2010;20(18):5439-42. Epub 2010/08/20.

[124] Wang P, Wang J, Guo T, Li Y. Synthesis and cytotoxic activity of the N-acetylglucosamine-bearing triterpenoid saponins. Carbohydrate research. 2010;345(5):607-20. Epub 2010/02/02.

[125] Gauthier C, Legault J, Lavoie S, Rondeau S, Tremblay S, Pichette A. Synthesis and cytotoxicity of bidesmosidic betulin and betulinic acid saponins. Journal of natural products. 2009;72(1):72-81. Epub 2009/01/01.

[126] Heisler I, Sutherland M, Bachran C, Hebestreit P, Schnitger A, Melzig MF, *et al.* Combined application of saponin and chimeric toxins drastically enhances the targeted cytotoxicity on tumor cells. J Control Release. 2005;106(1-2):123-37.

[127] Heisler I, Keller J, Tauber R, Sutherland M, Fuchs H. A cleavable adapter to reduce nonspecific cytotoxicity of recombinant immunotoxins. Int J Cancer. 2003;103(2):277-82.

[128] Hebestreit P, Weng A, Bachran C, Fuchs H, Melzig MF. Enhancement of cytotoxicity of lectins by Saponinum album. Toxicon. 2006;47(3):330-5.

[129] Bachran C, Weng A, Bachran D, Riese SB, Schellmann N, Melzig MF, *et al.* The distribution of saponins *in vivo* affects their synergy with chimeric toxins against tumours expressing human epidermal growth factor receptors in mice. British journal of pharmacology. 2010;159(2):345-52. Epub 2009/12/18.

[130] Bachran C, Durkop H, Sutherland M, Bachran D, Muller C, Weng A, *et al.* Inhibition of tumor growth by targeted toxins in mice is dramatically improved by saponinum album in a synergistic way. J Immunother. 2009;32(7):713-25. Epub 2009/06/30.

[131] Weng A, Jenett-Siems K, Schmieder P, Bachran D, Bachran C, Gorick C, *et al.* A convenient method for saponin isolation in tumour therapy. Journal of chromatography B, Analytical technologies in the biomedical and life sciences. 2010;878(7-8):713-8. Epub 2010/02/11.

[132] Thakur M, Weng A, Bachran D, Riese SB, Bottger S, Melzig MF, *et al.* Electrophoretic isolation of saponin fractions from Saponinum album and their evaluation in synergistically enhancing the receptor-specific cytotoxicity of targeted toxins. Electrophoresis. 2011;32(21):3085-9. Epub 2011/10/15.

[133] Weng A, Thakur M, Beceren-Braun F, Bachran D, Bachran C, Riese SB, *et al.* The toxin component of targeted anti-tumor toxins determines their efficacy increase by saponins. Molecular oncology. 2012. Epub 2012/02/09.

[134] Bachran D, Schneider S, Bachran C, Urban R, Weng A, Melzig MF, *et al.* Epidermal growth factor receptor expression affects the efficacy of the combined application of saponin and a targeted toxin on human cervical carcinoma cells. Int J Cancer. 2010;127(6):1453-61. Epub 2009/12/19.

[135] Weng A, Bachran C, Fuchs H, Krause E, Stephanowitz H, Melzig MF. Enhancement of saporin cytotoxicity by Gypsophila saponins--more than stimulation of endocytosis. Chemico-biological interactions. 2009;181(3):424-9. Epub 2009/07/21.

[136] Bachran D, Schneider S, Bachran C, Weng A, Melzig MF, Fuchs H. The endocytic uptake pathways of targeted toxins are influenced by synergistically acting Gypsophila saponins. Molecular pharmaceutics. 2011;8(6):2262-72. Epub 2011/10/11.

Send Orders for Reprints to reprints@benthamscience.net
Recent Advances in Medicinal Chemistry, Vol. 1, **2014, 303-331** **303**

CHAPTER 10

Pharmacological Neuroprotection for Acute Spinal Cord Injury

Humberto Mestre[1], Ricardo Balanza[1] and Antonio Ibarra[1,2,*]

[1]Anahuac University Avenue No. 146, Lomas Anahuac, Huixquilucan, State of Mexico, Mexico and [2]Centro de Investigación del Proyecto CAMINA A.C., Mexico City, Mexico

Abstract: Traumatic spinal cord injury (SCI) is a major problem in clinical medicine. Etiology depends on several factors such as mechanism of injury and level of injury. The result is a very heterogeneous population of SCI patients. The characteristics of this pathology make for high levels of inter-patient variability. The validation of pharmacological neuroprotective therapy in the acute phase of traumatic SCI has been a treacherous road. Today, there are no FDA-approved therapies for medical management of acute SCI. The clinician depends on recommendations from the AANS/CNS suggesting that the use of methylprednisolone or GM-1 ganglioside is permissible but no real benefits have been observed. Several poorly designed prospective randomized controlled trials have obscured the real value of these promising therapies. This review systematically revises the current treatment protocols while also analyzing the validity and feasibility of the most cutting-edge basic and clinical treatment strategies. With this in mind, the objective is to inform healthcare providers of the present state of acute SCI pharmacological neuroprotective treatment and where is it going in the future.

Keywords: Antioxidants, Apoptosis inhibitors, ATI355, Calpain inhibitors, Cethrin, Erythropoietin, Indometacin, Immunophilin ligands, Memantine, Methylprednisolone, Minocycline, Naloxone, Paraplegia, Riluzole, Steroid hormones, Tirilazad mesylate, Thyrotropin.

1. INTRODUCTION

Traumatic spinal cord injury (SCI) is a major problem in clinical practice. The spinal cord is a main component of the central nervous system (CNS) and is essential for the homeostatic function of the human body. It is a structure with a very complex anatomy and a traumatic insult to any of its components usually

***Address correspondence to Antonio Ibarra:** Av. Universidad No. 46, Col. Lomas Anahuac, Huixquilucan, Estado de Mexico, Mexico; Tel: +(52) 55 56270210 ext. 8524; Fax: +(52) 55 55735545; E-mail: iantonio65@yahoo.com

Atta-ur-Rahman, Muhammad Iqbal Choudhary and George Perry (Eds)

results in permanent injury. The etiology of SCI is highly varied and therefore results in a heterogeneous population of victims. SCI causes multisystem alterations and thus requires a multidisciplinary team of healthcare practitioners. In recent years, the growing understanding of pathophysiology and management of SCI have allowed us to reduce mortality significantly. However, growing incidence clusters around young and active age groups leaving them with a permanent disabling condition. This presents a great economic burden on society and the healthcare system, as well as a psychological and emotional tax on those who suffer from a SCI. This review has the objective of presenting the most up-to-date information on pharmacological treatment in acute SCI to healthcare practitioners involved in the everyday battle against SCI.

2. EPIDEMIOLOGY

The annual incidence of SCI worldwide is between 11.5 and 57.8 cases per million persons [1]. In the United States the yearly average is of 40 new cases per million [2]. Population projections suggest that incidence will increase to about 17,560 new cases in 2050 [2]. SCI has a bimodal age distribution affecting primarily 15-29 years of age and then 64 years of age or older, with the lowest incidence in pediatric age groups [3]. However, the mean age of injury is 37.1 years indicating the SCI affects young and productive individuals [4]. It also occurs 3-4 times more in men than it does in women [2]. Motor vehicle crashes occupy the leading cause of SCI (48.3%) in young age groups and falls are the leading cause among persons over age 60. Other causes are injuries due to acts of violence (12%) and sports-related injuries (10%) [2]. There are currently estimated to be 246,882 people living with SCI in the U.S. and that projection is said to grow to 276,281 by 2012 [5]. Due to more effective management of SCI mortality at one year after injury has been reduced by 69% from 1970 to 2007 [6]. Modern-day medicine has achieved a milestone goal by increasing life expectancy but more is needed in the field of effective SCI treatment. By innovating new therapeutic strategies that will assist in regaining partial or complete neurological integrity, patients will live longer and have a better quality of life.

3. PATHOPHYSIOLOGY

After SCI several destructive processes take place; however, these can be divided into primary and secondary injury mechanisms. The primary injury is a direct result of the traumatic force applied to the spinal cord. The injury mechanism determines the destruction sustained by the spinal cord; these can be compression, contusion, hemisection or complete transection. The secondary injury is a self-destructive phenomenon developed as a consequence of the primary injury, which further destroys neural tissue [7].

Primary injury results in the disruption of the neural circuits by structurally destroying synapses between axons and dendrites. This initial mechanism also alters regional blood flow propitiating a state of ischemia, adding to this already destructive microenvironment [8]. This primary insult causes focalized and irreversible damage to the spinal cord. Myelin breakdown is one of the most deleterious consequences of primary lesion, especially that of highly myelinated axons which are in charge of the most important functions of the spinal cord. Demyelination impedes the generation of saltatory conduction and functionally neutralizes neurons [9]. Rupture of blood vessels, primarily in the gray matter which is highly vascularized, cause hemorrhagic zones and altered blood flow leads to hypoxia and depleted nutrient supplies leading to ischemia [10, 11].

Secondary injury is caused by several self-destructive processes that are brought about by the primary injury mechanism. These increase the damaged area and begin a centrifuge craneo-caudal wave of continued neurodegeneration, radiating from the epicenter of injury [12]. The ischemic microenvironment conditioned by altered blood flow cause the failure of important metabolic mechanisms such as cellular respiration. The breakdown of these quintessential processes depletes cellular stores of energy and activates apoptotic pathways. The latter alters membrane permeability and generates lysosomal rupture. All these mechanisms result in the activation of proteases, phospholipases, ATPases and endonucleases which degrade cytoplasmic membranes as well as nuclear and cytoskeleton components [11]. Ischemic states also produce a wide range of reactive oxygen species (ROS), or free radicals. These highly volatile molecules interact with different components of the cell. One of the most harmful interactions is between ROS and membrane lipids. ROS attack the unsaturated fatty acids embedded within the cell membrane and set off a chain reaction that results in the uncontrolled release of more free radicals. This process is known as lipid peroxidation [13].

Excitotoxicity is yet another perilous phenomenon developed after the primary injury. This is defined as an excessive release of excitatory neurotransmitters, primarily glutamate. Excitotoxicity is particularly exacerbated in hypoxic environments [14]. A continuous increase in glutamate concentrations causes the overstimulation of NMDA, AMPA, and kainate receptors [15]. These ionotropic receptors allow the influx of Ca^{2+}, which activate voltage-dependent sodium channels causing a permanent depolarized state [16]. This ion imbalance produces a compensatory influx of Cl^- increasing the intracellular osmotic potential therefore inducing secondary cellular lysis. Increased intracellular calcium concentrations also result in the activation of a wide range of proteases and lipases further degrading proteins and lipids essential to cell membrane function and neurofilaments that are important in neurotransmitter transport [16]. These pathological concentrations of Ca^{2+} also inhibit the mitochondria's ability to carry out cellular respiration. The lack of energy production alters ATP-dependent Ca^{2+} uptake/sequestration further altering normal physiology. This ionic dysregulation disturbs mitochondrial membrane permeability allowing the inner membrane to be exposed and mitochondrial lysis. These destructive processes render the struggling neuron obsolete.

Mitochondrial lysis and the arrest of cellular respiration trigger the apoptotic cascade. Other triggering mechanisms are: cytokines, inflammatory response, free radical damage and excitotoxicity [17, 18]. The pathways through which apoptosis is induced causing neuronal death vary. Immune cells can activate apoptosis *via* the Fas ligand/Fas receptor pathway. ROS damage especially by the production of nitric oxide can also trigger apoptosis. Direct activation of caspase-3 proenzyme within the cell and mitochondrial damage through the pathway of cytochrome c also set off this programmed death function [10, 19]. The additional death of functional neurons contributes to the cell loss sustained after primary injury. These degenerative phenomena have important repercussions of the neurological status of the injured patient.

During the next 48 hours after injury, a series of inflammatory cell-mediated responses take place. These processes were considered to solely contribute to spinal cord damage, but have now been cast under another light. Neutrophils arrive at the site as early as one hour after injury and numbers increase up to 24 hours after [20, 21]. Peripheral circulating macrophages are observed arriving 24 hours after injury and continue to arrive 4-7 days after [21, 22]. The presence of this cellular subset has even been observed in chronic stages of SCI [23]. The resident macrophages of the CNS, microglia are activated from a resting state between three and seven days post-SCI [24]. Neutrophils and macrophages can produce ROS and increase the amount of lipid peroxidation that takes place after SCI. Studies have correlated the presence of these cells with the amount of damaged tissue after injury [21].

All these alterations, in unison, significantly increase cellular death and thereby worsen functional recovery. The secondary mechanisms of damage are currently the target of pharmacological treatment of SCI.

4. SEARCH CRITERIA

The process through which the literature was selected consisted of a three-pronged approach. The first prong included the current treatment protocols for acute SCI in adult humans. For this purpose, the most up-to-date approved clinical guidelines were employed; these are the ones co-published by the American Association of Neurological Surgeons (AANS) and the Congress of Neurological Surgery (CNS) in 2001, which were also revised in 2011.

The second prong was used to select pertinent literature for the experimental treatment of acute SCI in adult humans. In order to track down all clinical trials underway, a thorough search of the U.S. National Institutes of Health clinical trial database (www.clinicaltrials.gov) and the World Health Organization's International Clinical Trials Registry Platform Search Portal (www.apps.who.int/trialsearch/) were undertaken. Due to the nature of the review, only pharmacological interventional randomized controlled trials (RCT) were considered, this meant that all cell-based and physical (this includes electrical stimulation) therapies were discarded. Other criteria used to refine the search were: trials that had not published preliminary results or had been terminated, studies that did not evaluate functional outcome, the lack of statistical significance against a placebo or control group, interventions outside the acute setting or those that treated complications of chronic SCI.

The third prong consisted of searching for basic experimental research used in animal models of SCI. A computerized search of the National Library of Medicine and the National Institutes of Health MEDLINE database was performed using PubMed. Only published literature in English from 2001 to 2012 was taken into consideration seeing as it was deemed chronologically relevant. Since the objective of this literature revision decided to include only the most promising neuroprotective therapies a strict exclusion criteria was drafted. Parameters of exclusion were studies: not performed in *in vivo* models, that had no functional outcome analysis, that did not achieve a $P < 0.05$, that used pre-SCI treatment strategies and that used an invasive administration route that would deem it clinically unfeasible.

5. CURRENT TREATMENT OF ACUTE SPINAL CORD INJURIES

The goal of research is to find a pharmacological agent that when delivered right after sustaining a SCI would help recover neurological function and prevent permanent sequelae. Many molecules have been tested in animal models, but very few have achieved the results necessary to begin human clinical trials. Few pharmacological substances have been extensively studied and have been able to pass from the lab bench and into the clinic: methylprednisolone (MP), tirilazad mesylate, naloxone, and GM-1 ganglioside (GM-1). Other substances such as thyrotropin-releasing hormone, gacyclidine and nimodipine have also been used in human trials, but failed to show any benefit. All of these have been evaluated in randomized, controlled, blinded clinical trials in humans with SCI. Tirilazad and naloxone have been evaluated less extensively and their efficacy in the treatment of SCI is still obscure. Until now, the results from these trials have not been enough to warrant the approval by the U.S. Food and Drug Administration (FDA) of any of these molecules in human SCI. The AANS and CNS have advocated MP and GM-1 use in SCI subjects as an option but not a standard of care.

Glucocorticoid steroids have been extensively studied in SCI. The rationale for their use was based upon the theory that they would reduce post-traumatic spinal cord edema. This was later focused on the possibility that these compounds could inhibit lipid peroxidation as a result of their high lipid solubility and their ability to intercalate between membrane polyunsaturated fatty acids, stabilizing the cell membrane [25, 26]. MP is a synthetic glucocorticoid capable of inhibiting lipid peroxidation, protease-mediated neurofilament loss, phospholipase A2 activation, lactate accumulation, inflammation and post-traumatic ischemia [27]. It also increases ATP and lowers intracellular calcium concentrations [28, 29]. However, steroids also have negative properties such as increasing the incidence of pneumonia 2.6-fold [30], pulmonary embolism, sepsis, gastrointestinal hemorrhage, and wound infection [31]. In an effort to neutralize some of the adverse effects of steroids, scientist modified the steroid molecules to eliminate the glucocorticoid effect and preserve the neuroprotective effects [26]. This dawned the birth of lazaroids; the principal example of these was tirilazad mesylate. Tirilazad lacked the glucocorticoid receptor-mediated side effects that limited the clinical use of MP. Although tirilazad may induce the apparent positive effects of MP the evidence suggesting their use in acute SCI has been vague and inconclusive. At the moment, no further studies evaluating the effect of tirilazad in acute SCI have been reported. Naloxone, an opiate antagonist was also shown to improve neurological recovery and was employed as a therapeutic candidate against MP [32, 33]. This substance required further validation; however at the moment there are no reports on the topic and therefore naloxone is not considered a potential treatment in human SCI. GM-1 is a complex acidic glycolipid present in cells of the CNS. They are a major component of the cell membrane and cluster on the outer layer of the phospholipid bilayer. Although the function of these neuronal gangliosides remains unknown; they augment neurite outgrowth *in vitro*, induce regeneration and sprouting of neurons, and restore neuronal function after injury *in vivo* [34].

There have been at least four prospective randomized trials studying pharmacological therapy in acute SCI in humans. Unfortunately, these are all surrounded with controversy in the interpretation of the methods and results. The National Acute Spinal Cord Injury Study (NASCIS) I trial evaluated two sets of patients receiving MP at different doses and administration intervals [35]. The first group received 100 mg of MP followed by 25 mg every 6 hours for 10 days. The second group used 1000 mg of MP followed by 250 mg every 6 hours for 10 days, 10-fold increment in dose. After 1 year of follow-up the study failed to prove any difference between both groups. The study design was also criticized because of the lack of a placebo group. The second attempt was NASCIS II, in this study there were three groups: MP, naloxone, and a placebo. The MP group received a different dose that consisted of a 30 mg/kg bolus over an hour, followed by 5.4 mg/kg/h for the next 23 hours [36]. Patients treated with MP were analyzed separately as having received MP before 8 hours or after 8 hours post-SCI. The study concluded that the greatest benefit was observed in the group that received MP before 8 hours post-SCI. However, the study lacked information regarding concomitant surgical treatment and the statistical tools have been criticized for being confusing and difficult to replicate. One study done in Otani, Japan in 1994 was able to model the NASCIS II data and achieve mirrored results as the previous trial. The results were published as Class I-level evidence but these too have received several criticisms [37]. NASCIS III attempted to evaluate the effects of MP administration within a 24-hour or 48- hour treatment window. The comparative group consisted of 2.5 mg of tirilazad every 6 hours for 48 hours. The MP groups received a bolus of 30 mg/kg followed by 5.4 mg/kg/h either for the 24- or 48- hour window. The study observed neurological improvement at 6-weeks and at 6-months in the 48-hour group, if the first dose was given within 3-8 hours of SCI [38]. Again, deficiencies in the study design such, as randomization bias was observed further blurring the true results. Due to the controversy surrounding NASCIS I-III and he array of potential side effects, MP is used with reservations today and awaits FDA approval.

GM-1 ganglioside is the only other medication tested through prospective randomized controlled trials. An initial single-center RCT with a placebo control showed significant motor improvement after 1 year of follow-up evaluation [34]. A second, multi-center placebo-controlled RCT tried to replicate the results but failed to do so [39]. Due to the conflicting evidence on the use of GM-1 in SCI, this agent is only listed as an option by the AANS/CNS. Treatment with GM-1 at a loading dose of 300-mg and then 100 mg/day for 56 days when initiated after the administration of MP given within 8 hours of injury (NASCIS II protocol), is recommended as an option in treatment of adult patients with acute SCI [40].

Three additional pharmacological compounds have been evaluated in prospective double-blinded RCT. All three agents were considered to have neuroprotective effects and were therefore subjected to clinical trials. The first is thyrotropin-releasing hormone (TRH), the hormone responsible for the release of thyroid-stimulating hormone (TSH) causing the eventual release of bioactive thyroid hormones. This endogenous hormone was evaluated because of its antagonistic effect on damaging mediators of the secondary injury. However, only one clinical trial was taken to completion. The study demonstrated a statistically significant improvement in patients who received TRH but due to a small sample size evidence was weak [41]. The second is gacyclidine, an NMDA receptor antagonist known to compete against glutamate. Studies showed a significant benefit after 1 year of sustaining an incomplete cervical SCI. However, phase II trials for acute SCI gave disappointing results and development for this indication has been discontinued [42]. The third and last is nimodipine, studied for its ability to impede calcium-mediated injury in the secondary phase of injury. Despite animal studies showing benefit, human RCT failed to reproduce this beneficial effect [43].

6. PROMISING FUTURE PHARMACOLOGICAL THERAPIES

6.1. Human Clinical Trial Phase

The new generation of neuroprotective agents in clinical trials is still in its infancy. With most substances still in Phase I and II, most of them have yet to publish preliminary results. Most interventional studies in acute SCI that have reached clinical trials and published results are aimed at neuroregeneration; these are targeted towards axonal regeneration and remyelination. The few studies on neuroprotection are directed towards ameliorating glutamate excitotoxicity and apoptosis. The introduction of biological pharmacological agents in the form of monoclonal antibodies is an interesting and fast growing area of SCI research. Two agents are now in human clinical trials. The first, ATI355, is a monoclonal antibody directed towards neutralizing Nogo-A. Nogo-A is a key inhibitor of neuroregeneration and neurite outgrowth. Animal knockout studies show enhanced neuroregeneration and a considerable improvement in functional recovery after SCI. *In vivo* animal models of SCI demonstrated the same improvements in neuroregeneration and functional recovery after administration of ATI355 [44]. The human clinical trial enrolled 52 patients and was conducted by Novartis. It has now concluded phase I but has yet to publish its results (NCT00406016). Treatment consisted of a four-dose regimen of continuous intrathecal infusion and two regimens of repeated intrathecal bolus injections 4-14 days after SCI. The second biological is GSK249320, synthesized by Glaxo Smith Kline, this is a monoclonal antibody directed against myelin-associated glycoprotein (MAG). MAG is a protein that inhibits axonal regeneration and GSK249320 acts as a MAG antagonist [45]. This agent has just gone through phase I safety testing and results are still being analyzed (NCT00622609). The design is a placebo-controlled RCT single-blind study that enrolled 46 healthy subjects who received a single dose of intravenous infusion GSK249320. Another agent implicated in axonal regeneration is BA-210, commercialized by Alseres Pharmaceuticals as Cethrin and is a Rho protein antagonist. Several studies have validated the Rho pathway as an important mediator in the neuronal response to growth inhibitory proteins after SCI. BA-210 has shown promise in pre-clinical animal studies by increasing axonal regeneration and functional recovery. Cethrin cleared both phase I and phase IIa human clinical trials and has published its results (NCT00500812). The multi-center, open-label, dose escalation study recruited 48 patients. Intervention consisted of the transoperative administration of extradural BA-210 (dose escalating from 0.3-9 mg) while patients were receiving decompression surgery for complete ASIA A cervical or thoracic SCI. The study reported no serious adverse events but low levels of systemic exposure to the drug as well as a high inter-patient variability. ASIA motor score improvement was low across all dose groups in thoracic patients (1.8 ± 5.1) and larger in cervical patients (18.6 ± 19.3). The group with the highest motor recovery was the cervical SCI patients that received 3 mg of BA-210, these showed a (27.3 ± 13.3) point increase at 12 months postoperative. Approximately 6% of thoracic SCI patients regained an ASIA C or D compared to 31% of the cervical SCI group, highest in 3-mg dose (66%). The study was small and therefore it yielded heterogeneous results but a clear trend in neuroprotection can be seen [45]. Alseres Pharmaceuticals registered the phase IIb trial for Cethrin in 2008. It consisted of a multi-center RTC double-blind placebo-controlled trial with the same dose escalation design but reaching 18 mg of BA-210. Unfortunately, the trial was terminated prior to its completion due to the fact that Alseres would no longer develop Cethrin (NCT00610337). Other agents still undergoing phase II prospective clinical trials are: minocycline a tetracycline antibiotic and metalloproteinase inhibitor (NCT00559494), riluzole a sodium channel blocker as well as a glutamate neurotransmission blocker (NCT00876889), and SUN13837 a basic fibroblast growth factor (bFGF)-like molecule that reduces neuronal damage and improves recovery after SCI (NCT01502631). The previous are still awaiting the release of preliminary results. The gamma of pharmacological substances being evaluated in human clinical setting promises an interesting evolution in the management of acute SCI.

6.2. Animal Model Phase of Neuroprotective Agents

6.2.1. Cyclooxygenase Inhibitors

Cyclooxygenase (COX) is the enzyme responsible for eicosanoid synthesis. Such mediators include: prostaglandins, leukotrienes, prostacyclins, and thromboxane. All these are important players in the local and systemic inflammatory response. Three isoenzymes are known to date: COX-1, COX-2, COX-3 offering several pathways of intervention [46]. Inhibition of the COX pathway has been done using non-selective or isoenzyme-selective compounds.

6.2.1.1. Indomethacin

Indomethacin belongs to a family known as non-steroidal anti-inflammatory drugs (NSAID) and is a methylated indole derivative and a member of the arylalkanoic acid class of NSAIDs [47]. This agent is a non-selective inhibitor of COX, blocking both COX-1 and COX-2. This directly inhibits motility and activation of circulating polymorphic leukocytes [48]. The molecular target of this compound is on the inflammatory response developed after acute SCI. Pre-clinical animal studies demonstrated its capacity to reduce tissue damage, normalize spinal cord evoked potentials, and reduced posttraumatic edema by improving blood flow resulting in a global neurological recovery [49-51]. However, contradictory results have also been published indicating that the minimal neuroprotective dose (3 mg/kg) increases the amount of lipid peroxidation adding to the secondary phenomena [52]. The disparity in previous studies indicates the poor reliability on the neuroprotective effects of this compound.

6.2.1.2. Cyclooxygenase-2 Selective Inhibitors

COX-2 is unexpressed under normal conditions in most cells. The ability to selectively inhibit COX-2 allows us to bypass COX-1 blockade, which prevents peptic ulceration. Structural characteristics of COX-2 allow it to oxidize ester and amide derivatives of arachidonic acid produced by the enzyme phospholipase A2 from a fatty acid substrate [53]. This enzyme therefore selectively produces certain subsets of eicosanoids. There is 60% homology between COX-1 and COX-2 and they possess near-identical catalytic sites. Selective inhibition depends on a isoleucine (COX-1) substitution with valine (COX-2) at position 523 [53]. Selective inhibitors are principally known as coxibs (celecoxib, rofecoxib, parecoxib, *etc.*) but can also be named differently (*e.g.* DuP-697) [54]. Observational studies have seen that COX-2 is upregulated after SCI and this overexpression is probably resulting in additional damage to the neural tissue [55, 56]. Selective inhibition of COX-2 after SCI has resulted in modest benefits, primarily because of the oral administration routes of most coxibs [57]. SCI patients have altered gastric emptying and intestinal motility due to the neuromuscular collapse sustained resulting in deficient drug absorption [58]. For this purpose, parenteral formulas like NS-398 or parecoxib have been designed. Studies with NS-398 demonstrated significant motor recovery after SCI [59]. However, the published data does not provide a clear scenario on the usefulness of selective COX-2 inhibitors. Further research is required to develop an accurate conclusion; increasing reports of cardiovascular pathology associated with COX-2 inhibitors should also be taken into consideration [60].

6.2.2. Immunophilin Ligands

Immunophilins are a structurally diverse family of proteins abundant in CNS and immune system tissue [51]. These evolutionary-conserved proteins are receptors for immunosuppressive drugs like rapamycin, FK-506 (Tacrolimus) and cyclosporine A (CsA). The binding of these immunophilin ligands to their receptor inhibits the peptidyl-prolyl cis-trans isomerase also known as rotamase; this ligand-receptor complex also binds to calcineurin (calcium-dependent phosphoserine/phosphothreonine protein phosphatase) [51-53]. It is through these mechanisms that immunophilin ligands exert their neuroprotective effect.

6.2.2.1. Cyclosporine A

CsA is a cyclic undecapeptide that is highly lipophilic. The immunosuppressant properties of CsA reside in its ability to inhibit T helper cell proliferation. This interferes with cytokine production, neutrophil cytoskeleton motility, and expression of inducible nitric oxide synthase (iNOS) [61-64]. The broad spectrum of CsA's mechanism of action makes it an ideal treatment in SCI. Primarily because it approaches two different degenerative cascades of injury. It selectively blocks T helper cell proliferation, limiting the immune system's ability to develop an adaptive response against neural constituents; meanwhile it also prevents the expression of the main enzyme responsible for uncontrolled production of ROS (*i.e.* nitric oxide) after SCI [65-67]. Studies even suggest that CsA may promote neuroregeneration through its rotamase activity [68-71]. The effect of CsA on lipid peroxidation is comparable to that of MP without the glucocorticoid action [72, 73]. Benefits observed from CsA treatment of SCI include a reduction in demyelination, an increase in neuronal survival and functional recovery in a rat model [72, 74]. The previous studies employed a population-specific (SCI animals) dosing regimen devised from a pharmacokinetic analysis in this group of animals [75]. Contrasting studies have been published suggesting that CsA does not exert a beneficial effect after SCI; however, these studies employed a different dosing regimen and animal model of SCI [76].

6.2.2.2. FK-506

FK-506 exerts neuroprotection through calcineurin-independent mechanisms such as lowered production of leukotrienes and arachidonic acid [77, 78] and heat shock protein upregulation [79, 80]. Studies have shown that FK-506 reduces lipid peroxidation, COX-2 activity, and caspase-3 activation [81, 82]. All these improve axonal and motor neuron survival, spinal cord evoked potentials, and functional recovery [83, 84]. Although promising, further studies are needed. An initiative to isolate immunophilin ligands that do not possess the immunosuppressive effects of CsA and FK-506 should be undertaken.

6.2.3. Antioxidants

ROS-induced lipid peroxidation is by far the most detrimental processes that take place during secondary injury. It is because of this that many therapeutic interventions are aimed at scavenging these volatile free radicals before they have a chance to destroy surrounding tissue [85]. After SCI there are several ROS that are produced uncontrollably, these are nitric oxide and superoxide anion. When these come into contact they spontaneously form a radical compound called peroxynitrite. This compound has been found to be incredibly neurotoxic. Therefore, a whole area of research is dedicated to molecular scavenger that will neutralize this ROS.

6.2.3.1. Free Radical Scavengers

As mentioned previously, peroxynitrite is the result of a spontaneous reaction between two common ROS. This molecule is the most critical ROS generated after acute SCI causing increased amounts of lipid peroxidation and severely adding to neurodegenerative mechanisms [86, 87]. Tempol (4-hydroxy-2,2,6,6-tetramethylpiperidine-1-oxyl) is a membrane-permeable, superoxide dismutase mimetic that selectively inactivates superoxide anion. This ROS-specific scavenger prevents the conversion of superoxide anion into peroxynitrite. It has demonstrated to significantly improve motor recovery and neural tissue sparing when administered 40 hours after SCI [88]. Other scavengers are aimed at nitric oxide, primarily the enzyme responsible for its synthesis, iNOS. Aminoguanidine is a selective iNOS inhibitor that prevents NO formation and therefore lipid peroxidation; it also has a beneficial effect on functional recovery [89-91]. ONO-1714 has the same function as the previous, which reduces apoptosis and improves motor function after injury [91, 92]. Agmatine is also an iNOS blocker but shows NMDA receptor antagonist properties as well, this bimodal mechanism of action has made it a prime candidate in neuroprotective pharmacology. In SCI animal models it has shown to limit neuronal damage and improve the neurological recovery of injured rats [93, 94]. Safety studies should be conducted in order to understand the pleiotropic effects of these molecules.

6.2.4. Calpain Inhibitors

Calpain is a calcium-dependent cysteine protease that is responsible for the enzymatic degradation of the neuronal cytoskeleton and membrane proteins resulting in apoptosis [8]. Studies of selective calpain inhibitors have already been conducted on animal models of SCI. There are two chemical classes of calpain blockers: oxirane- and aldehyde-calpain inhibitors. The primary example of highly selective oxirane-calpain inhibitors is E-64-d, which has shown to exert neuroprotective effects *in vivo* SCI models [95-97]. Leupeptin is an aldehyde-calpain inhibitor but is less specific causing non-selective blockade of other cysteine and serine proteases, this also has shown neuroprotection after SCI [98]. Another aldehyde-calpain inhibitor, MDL28170, has no charge and is therefore highly liposoluble making it able to diffuse into the cell [8]. The chemical properties of this molecule allow it to distribute into the required site and exert its beneficial effect improving clinical outcome of SCI [99, 100].

6.2.5. Apoptosis Inhibitors

Apoptosis is also known as programmed cell death. This process is seen extensively in neurons of the spinal cord after traumatic insult. Key mediators of this chain reaction are caspase-3 and -9, they offer potential therapeutic targets to prevent apoptotic damage after SCI. zDEVD-fmk is a caspase-3 inhibitor which has shown to improve motor recovery after local administration following a SCI [101]. A similar molecule, z-LEHD-fmk is a caspase-9 inhibitor has also achieved better results after injury [102]. MAPK-selective inhibitors such as SB203580 also reduce apoptotic cells and diminish myelin damage both resulting in a better clinical outcome in treated animals [103]. Minocycline is a member of this class of agents and has now progressed to human clinical trials; this further solidifies the role of antiapoptotic therapies in the management of SCI.

6.2.6. Steroid Hormones

The pleiotropic effects of biological hormones have been studied at depth. Among these are anabolic states of protein synthesis and regeneration. This is an ideal microenvironment after a SCI, these steroidal hormones are incredibly lipophilic and are able to enter the cytoplasm easily and find their cytosolic or intranuclear receptor. The hormones that have been largely implicated as having neuroprotective properties are: progesterone and estrogen. These hormones enhance antioxidant mechanisms, increase remyelination, stimulate synaptogenesis and dendritic sprouting. They also mediate cell survival mechanisms, ameliorate excitotoxicity, and divert the destructive immune response towards an anti-inflammatory phenotype [104-111]. When administered in animal models of SCI, studies report improved neurological recovery, tissue sparing (primarily white matter), and a decrease in apoptotic cells [112-114]. This data suggests that steroid sex hormones are viable therapeutic approaches towards acute SCI; however the secondary effects of these hormones should be evaluated in a risk/benefit context.

6.2.7. Sodium Channel Blockers

Ionic dysregulation and excitotoxicity result in elevated concentrations of intracellular sodium. This unbalanced distribution of ions permanently depolarizes the neuron and triggers apoptotic mechanisms. It is this step in the neurodegenerative cascade that begs the creation of sodium channel blockers. The most recognized of these Na^+ channel blockers is tetrodotoxin, it is isolated from the puffer fish and works by binding to voltage-gated fast sodium channels and inhibiting them. Pre-clinical animal studies of SCI show significant tissue sparing and an improvement in functional recovery [115-117]. Another molecule with the same function, QX-314 showed contrasting results as it promoted neuronal sparing but failed to represent this as a functional improvement [118]. Riluzole is also a sodium channel blocker FDA-approved in the treatment of amyotrophic lateral sclerosis, which is currently undergoing human clinical trials for acute SCI. The fact that a compound from this class has reached phase II trials beckons the continued effort to research this type of therapeutic intervention.

6.2.8. Glutamate Receptor Antagonists

The most abundant excitatory neurotransmitter in the CNS is glutamate. This essential molecule has three different receptors in the mammalian nervous system: NMDA, AMPA, and kainite receptors. All of these receptors allow the influx of Ca^{2+} ions into the neuron. After SCI, normal glutamate signaling goes askew and excitotoxicity ensues. In order to decrease the pathological effect of unregulated glutamate signaling, glutamate receptor antagonists have been developed [119, 120]. Memantine is a non-competitive NMDA receptor antagonist that has already been evaluated in more than one model of SCI [121]. The published results were contradicting, in one model no benefits were achieved, probably because of the low affinity of memantine to its receptor; however in a different animal model neurological damage was substantially decreased [122]. Another non-competitive NMDA antagonist known as MK-801 has also been marred by contradictory data. Some studies have been able to see a neuroprotective effect [123-128] and others have not [129-131]. A great example of the ambiguousness of glutamate receptor antagonists is gacyclidine. In pre-clinical animal studies is showed remarkable results; however, when it was taken into clinical trials no benefit was observed at all [132, 133]. NBQX is a highly selective AMPA-kainate antagonist, when used in acute models of SCI there is a discernable protective effect on neural tissue but this does not translate into an improved functional outcome [134-140]. The data published on the use of these agents is inconclusive and confusing, if a definitive answer is desired further research is necessary. Of equal importance, are safety studies for these medications since they carry a myriad of adverse side effects at therapeutic doses.

6.2.9. Other Therapies

Erythropoietin (EPO) is a glycoprotein hormone that stimulates erythrocyte precursors in bone marrow. This hormone activates the CREB transcription pathway and increases the expression and production of brain derived growth factor (BDNF) [141]. It also decreases myeloperoxidase and caspase-3 activity, preventing apoptosis and lipid peroxidation in studies of SCI [142]. The use of a recombinant form of human EPO aided in the early recovery after SCI *via* anti-inflammatory and anti-apoptotic pathways [143]. Several studies have been published on the beneficial effect of EPO on functional outcome after SCI [144-147]. Human trials of EPO on nonacute SCI have also been conducted and resulted in a better neurological status [148]. The true efficacy of EPO in the treatment of acute SCI requires further elucidation.

The β_2-adrenoreceptor agonist, clembuterol, used as a bronchodilator has also been used in animal models of SCI. Results show greater tissue sparing which translates into less functional deficit [149]. This mechanism is probably mediated through a glutathione-dependent pathway [150]. Taurine is another agent that has received little attention. This sulfur amino acid found endogenously in humans exerts neuroprotective effects by scrubbing ROS and downregulating the expression of proinflammatory mediators such as tumor necrosis factor-alpa (TNF-α) [151, 152]. The administration of taurine after acute SCI in animal models results in less mortality in the treated group, accompanied by a regain in clinical functional recovery [153].

Citicoline has recently been evaluated as a potential neuroprotective drug. It is an essential mediator in the synthetic pathway of membrane phospholipids (*i.e.* phosphatidylcholine). Its liposoluble properties allow it to penetrate the blood-brain barrier and diffuse across cell membranes. Upon arrival to the CNS it stabilizes the neuronal membrane, increases brain metabolism, and alters the function of several neurotransmitters (*i.e.* glutamate). Citicoline reactivates ATPases especially membrane Na^+/K^+ ATPase, this regulates the concentration of intra and extracellular ions within the cell, restabilizing it. The presence of citicoline impedes the activation of several phospholipases and accelerates the resolution of edema and prevents apoptosis [154]. In SCI, citicoline attenuated lipid peroxidation and significantly improved functional recovery [155]. The true mechanisms through which citicoline and many of these experimental compounds exert their protective effects are still unknown; it is because of this that more research is needed. Continued efforts to discover these phenomena will result in more clinical trials and eventually a tangible efficacious treatment for acute SCI.

CONCLUSIONS

To date, no clinical evidence exists to definitively recommend the use of any neuroprotective pharmacologic agent, including steroids, in the treatment of acute SCI in order to improve functional recovery. MP for either 24 or 48 hours is recommended by the AANS/CNS as an option in the treatment of patients with acute SCI that should be undertaken only with the knowledge that the evidence suggesting harmful side effects is more consistent than any suggestion of clinical benefit [40, 156]. Treatment with GM-1 ganglioside is recommended as an option without demonstrated clinical benefit [40]. The problems associated with the clinical trials of acute SCI in humans have generated more questions and have limited the evidence in favor of several pharmacological agents. What is definitely clear is that more research is required. Future clinical trials should have adequate numbers of patients to achieve statistical power, appropriate control groups such as a placebo, standardized medical and surgical protocols to diminish bias, meticulous collection of relevant outcome data (this includes a standardized way of evaluating functional recovery), and the correct statistical analysis. Further initiatives should be undertaken to either solidify the evidence in favor of treatment with MP, tirilazad mesylate, naloxone, GM-1, TRH, nimodipine, and gacyclidine or push for the design of new clinical trials with the aim of evaluating the efficacy of the myriad of promising pharmacological treatments. Unfortunately, some experimental and poorly characterized SCI therapies are being offered outside a formal investigational structure, which will yield findings of limited scientific value and risk harm to patients with SCI who are understandably desperate for any intervention that might improve their function [157]. That is why a continued initiative to swiftly and accurately achieve a global consensus on the effective pharmacological treatment of traumatic acute SCI is needed.

ACKNOWLEDGEMENTS

Declared none.

CONFLICT OF INTEREST

The authors confirm that this chapter contents have no conflict of interest.

DISCLOSURE

The chapter submitted for eBook Series entitled: "**Recent Advances in Medicinal Chemistry, Volume 1**" is an update of our article published in **Mini-Reviews in Medicinal Chemistry, Volume 8, Number 3, pp. 222 to 230**, with additional text and references.

REFERENCES

[1] Ackery A, Tator C, and Krassioukov A. A global perspective on spinal cord injury epidemiology. J Neurotrauma, 21, 10 (Oct 2004), 1355-1370.

[2] Devivo MJ. Epidemiology of traumatic spinal cord injury: trends and future implications. Spinal Cord(Jan 24 2012).

[3] van den Berg ME, Castellote JM, Mahillo-Fernandez I, and de Pedro-Cuesta J. Incidence of spinal cord injury worldwide: a systematic review. Neuroepidemiology, 34, 3 2010), 184-192; discussion 192.

[4] DeVivo MJ, and Chen Y. Trends in new injuries, prevalent cases, and aging with spinal cord injury. Arch Phys Med Rehabil, 92, 3 (Mar 2011), 332-338.

[5] Lasfargues JE, Custis D, Morrone F, Carswell J, and Nguyen T. A model for estimating spinal cord injury prevalence in the United States. Paraplegia, 33, 2 (Feb 1995), 62-68.

[6] DeVivo MJ. Sir Ludwig Guttmann Lecture: trends in spinal cord injury rehabilitation outcomes from model systems in the United States: 1973-2006. Spinal Cord, 45, 11 (Nov 2007), 713-721.

[7] Dumont RJ, Okonkwo DO, Verma S, Hurlbert RJ, Boulos PT, Ellegala DB, and Dumont AS. Acute spinal cord injury, part I: pathophysiologic mechanisms. Clin Neuropharmacol, 24, 5 (Sep-Oct 2001), 254-264.

[8] Ray SK, Hogan EL, and Banik NL. Calpain in the pathophysiology of spinal cord injury: neuroprotection with calpain inhibitors. Brain Res Brain Res Rev, 42, 2 (May 2003), 169-185.

[9] Blight A. Mechanical factors in experimental spinal cord injury. J Am Paraplegia Soc, 11, 2 (Jul-Oct 1988), 26-34.

[10] Dumont, RJ, Okonkwo DO, Verma S, Hurlbert RJ, Boulos PT, Ellegala DB, and Dumont AS. Acute spinal cord injury, part I: pathophysiologic mechanisms. Clin Neuropharmacol, 24, 5 (Sep-Oct 2001), 254-264.

[11] Tator CH. Update on the pathophysiology and pathology of acute spinal cord injury. Brain Pathol, 5, 4 (Oct 1995), 407-413.

[12] Gorson KC, Ropper AH, Weinberg DH, and Weinstein R. Treatment experience in patients with anti-myelin-associated glycoprotein neuropathy. Muscle Nerve, 24, 6 (Jun 2001), 778-786.

[13] Ibarra A, Garcia E, Flores N, Martinon S, Reyes R, Campos MG, Maciel M, and Mestre H. Immunization with neural-derived antigens inhibits lipid peroxidation after spinal cord injury. Neurosci Lett, 476, 2 (May 31 2010), 62-65.

[14] Choi, D. W. Ischemia-induced neuronal apoptosis. Curr Opin Neurobiol, 6, 5 (Oct 1996), 667-672.

[15] Panter, S. S., Yum, S. W. and Faden, A. I. Alteration in extracellular amino acids after traumatic spinal cord injury. Ann Neurol, 27, 1 (Jan 1990), 96-99.

[16] Park, E., Velumian, A. A. and Fehlings, M. G. The role of excitotoxicity in secondary mechanisms of spinal cord injury: a review with an emphasis on the implications for white matter degeneration. J Neurotrauma, 21, 6 (Jun 2004), 754-774.

[17] Springer, J. E., Azbill, R. D. and Knapp, P. E. Activation of the caspase-3 apoptotic cascade in traumatic spinal cord injury. Nat Med, 5, 8 (Aug 1999), 943-946.

[18] Emery, E., Aldana, P., Bunge, M. B., Puckett, W., Srinivasan, A., Keane, R. W., Bethea, J. and Levi, A. D. Apoptosis after traumatic human spinal cord injury. J Neurosurg, 89, 6 (Dec 1998), 911-920.

[19] Citron, B. A., Arnold, P. M., Sebastian, C., Qin, F., Malladi, S., Ameenuddin, S., Landis, M. E. and Festoff, B. W. Rapid upregulation of caspase-3 in rat spinal cord after injury: mRNA, protein, and cellular localization correlates with apoptotic cell death. Exp Neurol, 166, 2 (Dec 2000), 213-226.

[20] Dusart, I. and Schwab, M. E. Secondary cell death and the inflammatory reaction after dorsal hemisection of the rat spinal cord. Eur J Neurosci, 6, 5 (May 1 1994), 712-724.

[21] Carlson, S. L., Parrish, M. E., Springer, J. E., Doty, K. and Dossett, L. Acute inflammatory response in spinal cord following impact injury. Exp Neurol, 151, 1 (May 1998), 77-88.

[22] Blight, A. R. Macrophages and inflammatory damage in spinal cord injury. J Neurotrauma, 9 Suppl 1(Mar 1992), S83-91.

[23] Guizar-Sahagun, G., Grijalva, I., Madrazo, I., Franco-Bourland, R., Salgado, H., Ibarra, A., Oliva, E. and Zepeda, A. Development of post-traumatic cysts in the spinal cord of rats-subjected to severe spinal cord contusion. Surg Neurol, 41, 3 (Mar 1994), 241-249.

[24] Popovich, P. G., Wei, P. and Stokes, B. T. Cellular inflammatory response after spinal cord injury in Sprague-Dawley and Lewis rats. J Comp Neurol, 377, 3 (Jan 20 1997), 443-464.

[25] Hall, E. D. and Braughler, J. M. Glucocorticoid mechanisms in acute spinal cord injury: a review and therapeutic rationale. Surg Neurol, 18, 5 (Nov 1982), 320-327.

[26] Hall, E. D. and Springer, J. E. Neuroprotection and acute spinal cord injury: a reappraisal. NeuroRx, 1, 1 (Jan 2004), 80-100.

[27] Braughler, J. M. and Hall, E. D. Effects of multi-dose methylprednisolone sodium succinate administration on injured cat spinal cord neurofilament degradation and energy metabolism. J Neurosurg, 61, 2 (Aug 1984), 290-295.

[28] Braughler, J. M. and Hall, E. D. Correlation of methylprednisolone levels in cat spinal cord with its effects on (Na+ + K+)-ATPase, lipid peroxidation, and alpha motor neuron function. J Neurosurg, 56, 6 (Jun 1982), 838-844.

[29] Young, W. and Flamm, E. S. Effect of high-dose corticosteroid therapy on blood flow, evoked potentials, and extracellular calcium in experimental spinal injury. J Neurosurg, 57, 5 (Nov 1982), 667-673.

[30] Gerndt, S. J., Rodriguez, J. L., Pawlik, J. W., Taheri, P. A., Wahl, W. L., Micheals, A. J. and Papadopoulos, S. M. Consequences of high-dose steroid therapy for acute spinal cord injury. J Trauma, 42, 2 (Feb 1997), 279-284.

[31] Dimar, J. R., Fisher, C., Vaccaro, A. R., Okonkwo, D. O., Dvorak, M., Fehlings, M., Rampersaud, R. and Carreon, L. Y. Predictors of complications after spinal stabilization of thoracolumbar spine injuries. J Trauma, 69, 6 (Dec 2010), 1497-1500.

[32] Faden, A. I., Jacobs, T. P. and Holaday, J. W. Opiate antagonist improves neurologic recovery after spinal injury. Science, 211, 4481 (Jan 30 1981), 493-494.

[33] Bracken, M. B. and Holford, T. R. Effects of timing of methylprednisolone or naloxone administration on recovery of segmental and long-tract neurological function in NASCIS 2. J Neurosurg, 79, 4 (Oct 1993), 500-507.

[34] Geisler, F. H., Dorsey, F. C. and Coleman, W. P. Recovery of motor function after spinal-cord injury--a randomized, placebo-controlled trial with GM-1 ganglioside. N Engl J Med, 324, 26 (Jun 27 1991), 1829-1838.

[35] Bracken, M. B., Shepard, M. J., Hellenbrand, K. G., Collins, W. F., Leo, L. S., Freeman, D. F., Wagner, F. C., Flamm, E. S., Eisenberg, H. M., Goodman, J. H. and *et al.* Methylprednisolone and neurological function 1 year after spinal cord injury. Results of the National Acute Spinal Cord Injury Study. J Neurosurg, 63, 5 (Nov 1985), 704-713.

[36] Bracken, M. B., Shepard, M. J., Collins, W. F., Holford, T. R., Young, W., Baskin, D. S., Eisenberg, H. M., Flamm, E., Leo-Summers, L., Maroon, J. and *et al.* A randomized, controlled trial of methylprednisolone or naloxone in the treatment of acute spinal-cord injury. Results of the Second National Acute Spinal Cord Injury Study. N Engl J Med, 322, 20 (May 17 1990), 1405-1411.

[37] Bracken, M. B. Steroids for acute spinal cord injury. Cochrane Database Syst Rev, 12012), CD001046.

[38] Bracken, M. B., Shepard, M. J., Holford, T. R., Leo-Summers, L., Aldrich, E. F., Fazl, M., Fehlings, M., Herr, D. L., Hitchon, P. W., Marshall, L. F., Nockels, R. P., Pascale, V., Perot, P. L., Jr., Piepmeier, J., Sonntag, V. K., Wagner, F., Wilberger, J. E., Winn, H. R. and Young, W. Administration of methylprednisolone for 24 or 48 hours or tirilazad mesylate for 48 hours in the treatment of acute spinal cord injury. Results of the Third National Acute Spinal Cord Injury Randomized Controlled Trial. National Acute Spinal Cord Injury Study. JAMA, 277, 20 (May 28 1997), 1597-1604.

[39] Geisler, F. H., Coleman, W. P., Grieco, G. and Poonian, D. The Sygen multicenter acute spinal cord injury study. Spine (Phila Pa 1976), 26, 24 Suppl (Dec 15 2001), S87-98.

[40] Pharmacological therapy after acute cervical spinal cord injury. Neurosurgery, 50, 3 Suppl (Mar 2002), S63-72.

[41] Pitts, L. H., Ross, A., Chase, G. A. and Faden, A. I. Treatment with thyrotropin-releasing hormone (TRH) in patients with traumatic spinal cord injuries. J Neurotrauma, 12, 3 (Jun 1995), 235-243.

[42] Mitha, A. P. and Maynard, K. I. Gacyclidine (Beaufour-Ipsen). Curr Opin Investig Drugs, 2, 6 (Jun 2001), 814-819.
[43] Petitjean, M. E., Pointillart, V., Dixmerias, F., Wiart, L., Sztark, F., Lassie, P., Thicoipe, M. and Dabadie, P. [Medical treatment of spinal cord injury in the acute stage]. Ann Fr Anesth Reanim, 17, 2 1998), 114-122.
[44] Schnell, L., Hunanyan, A. S., Bowers, W. J., Horner, P. J., Federoff, H. J., Gullo, M., Schwab, M. E., Mendell, L. M. and Arvanian, V. L. Combined delivery of Nogo-A antibody, neurotrophin-3 and the NMDA-NR2d subunit establishes a functional 'detour' in the hemisected spinal cord. Eur J Neurosci, 34, 8 (Oct 2011), 1256-1267.
[45] Thompson, H. J., Marklund, N., LeBold, D. G., Morales, D. M., Keck, C. A., Vinson, M., Royo, N. C., Grundy, R. and McIntosh, T. K. Tissue sparing and functional recovery following experimental traumatic brain injury is provided by treatment with an anti-myelin-associated glycoprotein antibody. Eur J Neurosci, 24, 11 (Dec 2006), 3063-3072.
[46] Chandrasekharan, N. V., Dai, H., Roos, K. L., Evanson, N. K., Tomsik, J., Elton, T. S. and Simmons, D. L. COX-3, a cyclooxygenase-1 variant inhibited by acetaminophen and other analgesic/antipyretic drugs: cloning, structure, and expression. Proc Natl Acad Sci U S A, 99, 21 (Oct 15 2002), 13926-13931.
[47] Hart, F. D. and Boardman, P. L. Indomethacin: A New Non-Steroid Anti-Inflammatory Agent. Br Med J, 2, 5363 (Oct 19 1963), 965-970.
[48] Takeuchi, K., Tanaka, A., Hayashi, Y. and Yokota, A. COX inhibition and NSAID-induced gastric damage--roles in various pathogenic events. Curr Top Med Chem, 5, 5 2005), 475-486.
[49] Simpson, R. K., Jr., Baskin, D. S., Dudley, A. W., Bogue, L. and Rothenberg, F. The influence of long-term nifedipine or indomethacin therapy on neurologic recovery from experimental spinal cord injury. J Spinal Disord, 4, 4 (Dec 1991), 420-427.
[50] Sharma, H. S., Olsson, Y. and Cervos-Navarro, J. Early perifocal cell changes and edema in traumatic injury of the spinal cord are reduced by indomethacin, an inhibitor of prostaglandin synthesis. Experimental study in the rat. Acta Neuropathol, 85, 2 1993), 145-153.
[51] Sharma, H. S., Olsson, Y., Nyberg, F. and Dey, P. K. Prostaglandins modulate alterations of microvascular permeability, blood flow, edema and serotonin levels following spinal cord injury: an experimental study in the rat. Neuroscience, 57, 2 (Nov 1993), 443-449.
[52] Guth, L., Zhang, Z., DiProspero, N. A., Joubin, K. and Fitch, M. T. Spinal cord injury in the rat: treatment with bacterial lipopolysaccharide and indomethacin enhances cellular repair and locomotor function. Exp Neurol, 126, 1 (Mar 1994), 76-87.
[53] Rouzer, C. A. and Marnett, L. J. Structural and functional differences between cyclooxygenases: fatty acid oxygenases with a critical role in cell signaling. Biochem Biophys Res Commun, 338, 1 (Dec 9 2005), 34-44.
[54] Sciulli, M. G., Capone, M. L., Tacconelli, S. and Patrignani, P. The future of traditional nonsteroidal antiinflammatory drugs and cyclooxygenase-2 inhibitors in the treatment of inflammation and pain. Pharmacol Rep, 57 Suppl2005), 66-85.
[55] Adachi, K., Yimin, Y., Satake, K., Matsuyama, Y., Ishiguro, N., Sawada, M., Hirata, Y. and Kiuchi, K. Localization of cyclooxygenase-2 induced following traumatic spinal cord injury. Neurosci Res, 51, 1 (Jan 2005), 73-80.
[56] Resnick, D. K., Graham, S. H., Dixon, C. E. and Marion, D. W. Role of cyclooxygenase 2 in acute spinal cord injury. J Neurotrauma, 15, 12 (Dec 1998), 1005-1013.

[57] Resnick, D. K., Nguyen, P. and Cechvala, C. F. Selective cyclooxygenase 2 inhibition lowers spinal cord prostaglandin concentrations after injury. Spine J, 1, 6 (Nov-Dec 2001), 437-441.

[58] Mestre, H., Alkon, T., Salazar, S. and Ibarra, A. Spinal cord injury sequelae alter drug pharmacokinetics: an overview. Spinal Cord, 49, 9 (Sep 2011), 955-960.

[59] Hains, B. C., Yucra, J. A. and Hulsebosch, C. E. Reduction of pathological and behavioral deficits following spinal cord contusion injury with the selective cyclooxygenase-2 inhibitor NS-398. J Neurotrauma, 18, 4 (Apr 2001), 409-423.

[60] Bravo, G., Guizar-Sahagun, G., Ibarra, A., Centurion, D. and Villalon, C. M. Cardiovascular alterations after spinal cord injury: an overview. Curr Med Chem Cardiovasc Hematol Agents, 2, 2 (Apr 2004), 133-148.

[61] Ibarra, A. and Diaz-Ruiz, A. Protective effect of cyclosporin-A in spinal cord injury: an overview. Curr Med Chem, 13, 22 2006), 2703-2710.

[62] Hendey, B. and Maxfield, F. R. Regulation of neutrophil motility and adhesion by intracellular calcium transients. Blood Cells, 19, 1 1993), 143-161; discussion 161-144.

[63] Diaz-Ruiz, A., Vergara, P., Perez-Severiano, F., Segovia, J., Guizar-Sahagun, G., Ibarra, A. and Rios, C. Cyclosporin-A inhibits inducible nitric oxide synthase activity and expression after spinal cord injury in rats. Neurosci Lett, 357, 1 (Feb 26 2004), 49-52.

[64] Diaz-Ruiz, A., Vergara, P., Perez-Severiano, F., Segovia, J., Guizar-Sahagun, G., Ibarra, A. and Rios, C. Cyclosporin-A inhibits constitutive nitric oxide synthase activity and neuronal and endothelial nitric oxide synthase expressions after spinal cord injury in rats. Neurochem Res, 30, 2 (Feb 2005), 245-251.

[65] Attur, M. G., Patel, R., Thakker, G., Vyas, P., Levartovsky, D., Patel, P., Naqvi, S., Raza, R., Patel, K., Abramson, D., Bruno, G., Abramson, S. B. and Amin, A. R. Differential anti-inflammatory effects of immunosuppressive drugs: cyclosporin, rapamycin and FK-506 on inducible nitric oxide synthase, nitric oxide, cyclooxygenase-2 and PGE2 production. Inflamm Res, 49, 1 (Jan 2000), 20-26.

[66] Trajkovic, V., Badovinac, V., Jankovic, V., Samardzic, T., Maksimovic, D. and Popadic, D. Cyclosporin A suppresses the induction of nitric oxide synthesis in interferon-gamma-treated L929 fibroblasts. Scand J Immunol, 49, 2 (Feb 1999), 126-130.

[67] Trajkovic, V., Badovinac, V., Jankovic, V. and Mostarica Stojkovic, M. Cyclosporin A inhibits activation of inducible nitric oxide synthase in C6 glioma cell line. Brain Res, 816, 1 (Jan 16 1999), 92-98.

[68] Sosa, I., Reyes, O. and Kuffler, D. P. Immunosuppressants: neuroprotection and promoting neurological recovery following peripheral nerve and spinal cord lesions. Exp Neurol, 195, 1 (Sep 2005), 7-15.

[69] Palladini, G., Caronti, B., Pozzessere, G., Teichner, A., Buttarelli, F. R., Morselli, E., Valle, E., Venturini, G., Fortuna, A. and Pontieri, F. E. Treatment with cyclosporine A promotes axonal regeneration in rats submitted to transverse section of the spinal cord--II--Recovery of function. J Hirnforsch, 37, 1 1996), 145-153.

[70] Sugawara, T., Itoh, Y. and Mizoi, K. Immunosuppressants promote adult dorsal root regeneration into the spinal cord. Neuroreport, 10, 18 (Dec 16 1999), 3949-3953.

[71] Ibarra, A., Hernandez, E., Lomeli, J., Pineda, D., Buenrostro, M., Martinon, S., Garcia, E., Flores, N., Guizar-Sahagun, G., Correa, D. and Madrazo, I. Cyclosporin-A enhances non-functional axonal growing after complete spinal cord transection. Brain Res, 1149(May 29 2007), 200-209.

[72] Diaz-Ruiz, A., Rios, C., Duarte, I., Correa, D., Guizar-Sahagun, G., Grijalva, I. and Ibarra, A. Cyclosporin-A inhibits lipid peroxidation after spinal cord injury in rats. Neurosci Lett, 266, 1 (Apr 30 1999), 61-64.

[73] Diaz-Ruiz, A., Rios, C., Duarte, I., Correa, D., Guizar-Sahagun, G., Grijalva, I., Madrazo, I. and Ibarra, A. Lipid peroxidation inhibition in spinal cord injury: cyclosporin-A *vs.* methylprednisolone. Neuroreport, 11, 8 (Jun 5 2000), 1765-1767.

[74] Ibarra, A., Correa, D., Willms, K., Merchant, M. T., Guizar-Sahagun, G., Grijalva, I. and Madrazo, I. Effects of cyclosporin-A on immune response, tissue protection and motor function of rats subjected to spinal cord injury. Brain Res, 979, 1-2 (Jul 25 2003), 165-178.

[75] Ibarra, A., Reyes, J., Martinez, S., Correa, D., Guizar-Sahagun, G., Grijalva, I., Castaneda-Hernandez, G., Flores-Murrieta, F. J., Franco-Bourland, R. and Madrazo, I. Use of cyclosporin-A in experimental spinal cord injury: design of a dosing strategy to maintain therapeutic levels. J Neurotrauma, 13, 10 (Oct 1996), 569-572.

[76] Rabchevsky, A. G., Fugaccia, I., Sullivan, P. G. and Scheff, S. W. Cyclosporin A treatment following spinal cord injury to the rat: behavioral effects and stereological assessment of tissue sparing. J Neurotrauma, 18, 5 (May 2001), 513-522.

[77] Gabryel, B., Chalimoniuk, M., Stolecka, A., Waniek, K., Langfort, J. and Malecki, A. Inhibition of arachidonic acid release by cytosolic phospholipase A2 is involved in the antiapoptotic effect of FK506 and cyclosporin a on astrocytes exposed to simulated ischemia *in vitro*. J Pharmacol Sci, 102, 1 (Sep 2006), 77-87.

[78] Hamasaki, Y., Kobayashi, I., Matsumoto, S., Zaitu, M., Muro, E., Ichimaru, T. and Miyazaki, S. Inhibition of leukotriene production by FK506 in rat basophilic leukemia-1 cells. Pharmacology, 50, 3 (Mar 1995), 137-145.

[79] Gold, B. G., Voda, J., Yu, X. and Gordon, H. The immunosuppressant FK506 elicits a neuronal heat shock response and protects against acrylamide neuropathy. Exp Neurol, 187, 1 (May 2004), 160-170.

[80] Oltean, M., Mera, S., Olofsson, R., Zhu, C., Blomgren, K., Hallberg, E. and Olausson, M. Transplantation of preconditioned intestinal grafts is associated with lower inflammatory activation and remote organ injury in rats. Transplant Proc, 38, 6 (Jul-Aug 2006), 1775-1778.

[81] Kaymaz, M., Emmez, H., Bukan, N., Dursun, A., Kurt, G., Pasaoglu, H. and Pasaoglu, A. Effectiveness of FK506 on lipid peroxidation in the spinal cord following experimental traumatic injury. Spinal Cord, 43, 1 (Jan 2005), 22-26.

[82] Nottingham, S., Knapp, P. and Springer, J. FK506 treatment inhibits caspase-3 activation and promotes oligodendroglial survival following traumatic spinal cord injury. Exp Neurol, 177, 1 (Sep 2002), 242-251.

[83] Bavetta, S., Hamlyn, P. J., Burnstock, G., Lieberman, A. R. and Anderson, P. N. The effects of FK506 on dorsal column axons following spinal cord injury in adult rats: neuroprotection and local regeneration. Exp Neurol, 158, 2 (Aug 1999), 382-393.

[84] Akgun, S., Tekeli, A., Kurtkaya, O., Civelek, A., Isbir, S. C., Ak, K., Arsan, S. and Sav, A. Neuroprotective effects of FK-506, L-carnitine and azathioprine on spinal cord ischemia-reperfusion injury. Eur J Cardiothorac Surg, 25, 1 (Jan 2004), 105-110.

[85] Hall, E. D., Yonkers, P. A., Andrus, P. K., Cox, J. W. and Anderson, D. K. Biochemistry and pharmacology of lipid antioxidants in acute brain and spinal cord injury. J Neurotrauma, 9 Suppl 2(May 1992), S425-442.

[86] Bao, F. and Liu, D. Peroxynitrite generated in the rat spinal cord induces neuron death and neurological deficits. Neuroscience, 115, 3 2002), 839-849.

[87] Bao, F. and Liu, D. Peroxynitrite generated in the rat spinal cord induces apoptotic cell death and activates caspase-3. Neuroscience, 116, 1 2003), 59-70.
[88] Hillard, V. H., Peng, H., Zhang, Y., Das, K., Murali, R., Etlinger, J. D. and Zeman, R. J. Tempol, a nitroxide antioxidant, improves locomotor and histological outcomes after spinal cord contusion in rats. J Neurotrauma, 21, 10 (Oct 2004), 1405-1414.
[89] Soy, O., Aslan, O., Uzun, H., Barut, S., Igdem, A. A., Belce, A. and Colak, A. Time-level relationship for nitric oxide and the protective effects of aminoguanidine in experimental spinal cord injury. Acta Neurochir (Wien), 146, 12 (Dec 2004), 1329-1335; discussion 1335-1326.
[90] Zhang, X. Y., Zhou, C. S., Jin, A. M., Tian, J., Zhang, H., Yao, W. T. and Zheng, G. Effect of aminoguanidine on the recovery of rat hindlimb motor function after spinal cord injury. Di Yi Jun Yi Da Xue Xue Bao, 23, 7 (Jul 2003), 687-689.
[91] Chatzipanteli, K., Garcia, R., Marcillo, A. E., Loor, K. E., Kraydieh, S. and Dietrich, W. D. Temporal and segmental distribution of constitutive and inducible nitric oxide synthases after traumatic spinal cord injury: effect of aminoguanidine treatment. J Neurotrauma, 19, 5 (May 2002), 639-651.
[92] Yu, Y., Matsuyama, Y., Nakashima, S., Yanase, M., Kiuchi, K. and Ishiguro, N. Effects of MPSS and a potent iNOS inhibitor on traumatic spinal cord injury. Neuroreport, 15, 13 (Sep 15 2004), 2103-2107.
[93] Yu, C. G., Marcillo, A. E., Fairbanks, C. A., Wilcox, G. L. and Yezierski, R. P. Agmatine improves locomotor function and reduces tissue damage following spinal cord injury. Neuroreport, 11, 14 (Sep 28 2000), 3203-3207.
[94] Kotil, K., Kuscuoglu, U., Kirali, M., Uzun, H., Akcetin, M. and Bilge, T. Investigation of the dose-dependent neuroprotective effects of agmatine in experimental spinal cord injury: a prospective randomized and placebo-control trial. J Neurosurg Spine, 4, 5 (May 2006), 392-399.
[95] Ray, S. K., Matzelle, D. C., Wilford, G. G., Hogan, E. L. and Banik, N. L. E-64-d prevents both calpain upregulation and apoptosis in the lesion and penumbra following spinal cord injury in rats. Brain Res, 867, 1-2 (Jun 9 2000), 80-89.
[96] Ray, S. K., Matzelle, D. D., Wilford, G. G., Hogan, E. L. and Banik, N. L. Increased calpain expression is associated with apoptosis in rat spinal cord injury: calpain inhibitor provides neuroprotection. Neurochem Res, 25, 9-10 (Oct 2000), 1191-1198.
[97] Zhang, S. X., Bondada, V. and Geddes, J. W. Evaluation of conditions for calpain inhibition in the rat spinal cord: effective postinjury inhibition with intraspinal MDL28170 microinjection. J Neurotrauma, 20, 1 (Jan 2003), 59-67.
[98] Momeni, H. R. and Kanje, M. Calpain inhibitors delay injury-induced apoptosis in adult mouse spinal cord motor neurons. Neuroreport, 17, 8 (May 29 2006), 761-765.
[99] Hung, K. S., Hwang, S. L., Liang, C. L., Chen, Y. J., Lee, T. H., Liu, J. K., Howng, S. L. and Wang, C. H. Calpain inhibitor inhibits p35-p25-Cdk5 activation, decreases tau hyperphosphorylation, and improves neurological function after spinal cord hemisection in rats. J Neuropathol Exp Neurol, 64, 1 (Jan 2005), 15-26.
[100] Arataki, S., Tomizawa, K., Moriwaki, A., Nishida, K., Matsushita, M., Ozaki, T., Kunisada, T., Yoshida, A., Inoue, H. and Matsui, H. Calpain inhibitors prevent neuronal cell death and ameliorate motor disturbances after compression-induced spinal cord injury in rats. J Neurotrauma, 22, 3 (Mar 2005), 398-406.
[101] Barut, S., Unlu, Y. A., Karaoglan, A., Tuncdemir, M., Dagistanli, F. K., Ozturk, M. and Colak, A. The neuroprotective effects of z-DEVD.fmk, a caspase-3 inhibitor, on traumatic spinal cord injury in rats. Surg Neurol, 64, 3 (Sep 2005), 213-220; discussion 220.

[102] Colak, A., Karaoglan, A., Barut, S., Kokturk, S., Akyildiz, A. I. and Tasyurekli, M. Neuroprotection and functional recovery after application of the caspase-9 inhibitor z-LEHD-fmk in a rat model of traumatic spinal cord injury. J Neurosurg Spine, 2, 3 (Mar 2005), 327-334.

[103] Horiuchi, H., Ogata, T., Morino, T., Chuai, M. and Yamamoto, H. Continuous intrathecal infusion of SB203580, a selective inhibitor of p38 mitogen-activated protein kinase, reduces the damage of hind-limb function after thoracic spinal cord injury in rat. Neurosci Res, 47, 2 (Oct 2003), 209-217.

[104] Ogata, T., Nakamura, Y., Tsuji, K., Shibata, T. and Kataoka, K. Steroid hormones protect spinal cord neurons from glutamate toxicity. Neuroscience, 55, 2 (Jul 1993), 445-449.

[105] Gonzalez, S. L., Labombarda, F., Deniselle, M. C., Mougel, A., Guennoun, R., Schumacher, M. and De Nicola, A. F. Progesterone neuroprotection in spinal cord trauma involves up-regulation of brain-derived neurotrophic factor in motoneurons. J Steroid Biochem Mol Biol, 94, 1-3 (Feb 2005), 143-149.

[106] Labombarda, F., Gonzalez, S., Roig, P., Lima, A., Guennoun, R., Schumacher, M. and De Nicola, A. F. Modulation of NADPH-diaphorase and glial fibrillary acidic protein by progesterone in astrocytes from normal and injured rat spinal cord. J Steroid Biochem Mol Biol, 73, 3-4 (Jun 2000), 159-169.

[107] Gonzalez, S. L., Labombarda, F., Gonzalez Deniselle, M. C., Guennoun, R., Schumacher, M. and De Nicola, A. F. Progesterone up-regulates neuronal brain-derived neurotrophic factor expression in the injured spinal cord. Neuroscience, 125, 3 2004), 605-614.

[108] Gonzalez Deniselle, M. C., Lopez Costa, J. J., Gonzalez, S. L., Labombarda, F., Garay, L., Guennoun, R., Schumacher, M. and De Nicola, A. F. Basis of progesterone protection in spinal cord neurodegeneration. J Steroid Biochem Mol Biol, 83, 1-5 (Dec 2002), 199-209.

[109] Labombarda, F., Gonzalez, S. L., Gonzalez, D. M., Guennoun, R., Schumacher, M. and de Nicola, A. F. Cellular basis for progesterone neuroprotection in the injured spinal cord. J Neurotrauma, 19, 3 (Mar 2002), 343-355.

[110] Sribnick, E. A., Wingrave, J. M., Matzelle, D. D., Ray, S. K. and Banik, N. L. Estrogen as a neuroprotective agent in the treatment of spinal cord injury. Ann N Y Acad Sci, 993(May 2003), 125-133; discussion 159-160.

[111] Yune, T. Y., Kim, S. J., Lee, S. M., Lee, Y. K., Oh, Y. J., Kim, Y. C., Markelonis, G. J. and Oh, T. H. Systemic administration of 17beta-estradiol reduces apoptotic cell death and improves functional recovery following traumatic spinal cord injury in rats. J Neurotrauma, 21, 3 (Mar 2004), 293-306.

[112] Thomas, A. J., Nockels, R. P., Pan, H. Q., Shaffrey, C. I. and Chopp, M. Progesterone is neuroprotective after acute experimental spinal cord trauma in rats. Spine (Phila Pa 1976), 24, 20 (Oct 15 1999), 2134-2138.

[113] Fee, D. B., Swartz, K. R., Joy, K. M., Roberts, K. N., Scheff, N. N. and Scheff, S. W. Effects of progesterone on experimental spinal cord injury. Brain Res, 1137, 1 (Mar 16 2007), 146-152.

[114] Chaovipoch, P., Jelks, K. A., Gerhold, L. M., West, E. J., Chongthammakun, S. and Floyd, C. L. 17beta-estradiol is protective in spinal cord injury in post- and pre-menopausal rats. J Neurotrauma, 23, 6 (Jun 2006), 830-852.

[115] Teng, Y. D. and Wrathall, J. R. Local blockade of sodium channels by tetrodotoxin ameliorates tissue loss and long-term functional deficits resulting from experimental spinal cord injury. J Neurosci, 17, 11 (Jun 1 1997), 4359-4366.

[116] Rosenberg, L. J., Teng, Y. D. and Wrathall, J. R. Effects of the sodium channel blocker tetrodotoxin on acute white matter pathology after experimental contusive spinal cord injury. J Neurosci, 19, 14 (Jul 15 1999), 6122-6133.

[117] Rosenberg, L. J. and Wrathall, J. R. Time course studies on the effectiveness of tetrodotoxin in reducing consequences of spinal cord contusion. J Neurosci Res, 66, 2 (Oct 15 2001), 191-202.

[118] Agrawal, S. K. and Fehlings, M. G. The effect of the sodium channel blocker QX-314 on recovery after acute spinal cord injury. J Neurotrauma, 14, 2 (Feb 1997), 81-88.

[119] Choi, D. W. Ionic dependence of glutamate neurotoxicity. J Neurosci, 7, 2 (Feb 1987), 369-379.

[120] Gentile, N. T. and McIntosh, T. K. Antagonists of excitatory amino acids and endogenous opioid peptides in the treatment of experimental central nervous system injury. Ann Emerg Med, 22, 6 (Jun 1993), 1028-1034.

[121] von Euler, M., Li-Li, M., Whittemore, S., Seiger, A. and Sundstrom, E. No protective effect of the NMDA antagonist memantine in experimental spinal cord injuries. J Neurotrauma, 14, 1 (Jan 1997), 53-61.

[122] Ehrlich, M., Knolle, E., Ciovica, R., Bock, P., Turkof, E., Grabenwoger, M., Cartes-Zumelzu, F., Kocher, A., Pockberger, H., Fang, W. C., Wolner, E. and Havel, M. Memantine for prevention of spinal cord injury in a rabbit model. J Thorac Cardiovasc Surg, 117, 2 (Feb 1999), 285-291.

[123] Kocaeli, H., Korfali, E., Ozturk, H., Kahveci, N. and Yilmazlar, S. MK-801 improves neurological and histological outcomes after spinal cord ischemia induced by transient aortic cross-clipping in rats. Surg Neurol, 64 Suppl 22005), S22-26; discussion S27.

[124] Wada, S., Yone, K., Ishidou, Y., Nagamine, T., Nakahara, S., Niiyama, T. and Sakou, T. Apoptosis following spinal cord injury in rats and preventative effect of N-methyl-D-aspartate receptor antagonist. J Neurosurg, 91, 1 Suppl (Jul 1999), 98-104.

[125] Haghighi, S. S., Johnson, G. C., de Vergel, C. F. and Vergel Rivas, B. J. Pretreatment with NMDA receptor antagonist MK801 improves neurophysiological outcome after an acute spinal cord injury. Neurol Res, 18, 6 (Dec 1996), 509-515.

[126] Hao, J. X., Watson, B. D., Xu, X. J., Wiesenfeld-Hallin, Z., Seiger, A. and Sundstrom, E. Protective effect of the NMDA antagonist MK-801 on photochemically induced spinal lesions in the rat. Exp Neurol, 118, 2 (Nov 1992), 143-152.

[127] Gomez-Pinilla, F., Tram, H., Cotman, C. W. and Nieto-Sampedro, M. Neuroprotective effect of MK-801 and U-50488H after contusive spinal cord injury. Exp Neurol, 104, 2 (May 1989), 118-124.

[128] Faden, A. I., Lemke, M., Simon, R. P. and Noble, L. J. N-methyl-D-aspartate antagonist MK801 improves outcome following traumatic spinal cord injury in rats: behavioral, anatomic, and neurochemical studies. J Neurotrauma, 5, 1 1988), 33-45.

[129] Haghighi, S. S., Agrawal, S. K., Surdell, D., Jr., Plambeck, R., Agrawal, S., Johnson, G. C. and Walker, A. Effects of methylprednisolone and MK-801 on functional recovery after experimental chronic spinal cord injury. Spinal Cord, 38, 12 (Dec 2000), 733-740.

[130] Feldblum, S., Arnaud, S., Simon, M., Rabin, O. and D'Arbigny, P. Efficacy of a new neuroprotective agent, gacyclidine, in a model of rat spinal cord injury. J Neurotrauma, 17, 11 (Nov 2000), 1079-1093.

[131] Holtz, A. and Gerdin, B. MK 801, an OBS N-methyl-D-aspartate channel blocker, does not improve the functional recovery nor spinal cord blood flow after spinal cord compression in rats. Acta Neurol Scand, 84, 4 (Oct 1991), 334-338.

[132] Gaviria, M., Privat, A., d'Arbigny, P., Kamenka, J., Haton, H. and Ohanna, F. Neuroprotective effects of a novel NMDA antagonist, Gacyclidine, after experimental contusive spinal cord injury in adult rats. Brain Res, 874, 2 (Aug 25 2000), 200-209.

[133] Gaviria, M., Privat, A., d'Arbigny, P., Kamenka, J. M., Haton, H. and Ohanna, F. Neuroprotective effects of gacyclidine after experimental photochemical spinal cord lesion in adult rats: dose-window and time-window effects. J Neurotrauma, 17, 1 (Jan 2000), 19-30.

[134] Mu, X., Azbill, R. D. and Springer, J. E. NBQX treatment improves mitochondrial function and reduces oxidative events after spinal cord injury. J Neurotrauma, 19, 8 (Aug 2002), 917-927.

[135] Gorgulu, A., Kiris, T., Unal, F., Turkoglu, U., Kucuk, M. and Cobanoglu, S. Superoxide dismutase activity and the effects of NBQX and CPP on lipid peroxidation in experimental spinal cord injury. Res Exp Med (Berl), 199, 5 (Apr 2000), 285-293.

[136] Wrathall, J. R., Teng, Y. D. and Marriott, R. Delayed antagonism of AMPA/kainate receptors reduces long-term functional deficits resulting from spinal cord trauma. Exp Neurol, 145, 2 Pt 1 (Jun 1997), 565-573.

[137] Liu, S., Ruenes, G. L. and Yezierski, R. P. NMDA and non-NMDA receptor antagonists protect against excitotoxic injury in the rat spinal cord. Brain Res, 756, 1-2 (May 9 1997), 160-167.

[138] Wrathall, J. R., Teng, Y. D. and Choiniere, D. Amelioration of functional deficits from spinal cord trauma with systemically administered NBQX, an antagonist of non-N-methyl-D-aspartate receptors. Exp Neurol, 137, 1 (Jan 1996), 119-126.

[139] Wrathall, J. R., Choiniere, D. and Teng, Y. D. Dose-dependent reduction of tissue loss and functional impairment after spinal cord trauma with the AMPA/kainate antagonist NBQX. J Neurosci, 14, 11 Pt 1 (Nov 1994), 6598-6607.

[140] Wrathall, J. R., Teng, Y. D., Choiniere, D. and Mundt, D. J. Evidence that local non-NMDA receptors contribute to functional deficits in contusive spinal cord injury. Brain Res, 586, 1 (Jul 17 1992), 140-143.

[141] Sonmez, A., Kabakci, B., Vardar, E., Gurel, D., Sonmez, U., Orhan, Y. T., Acikel, U. and Gokmen, N. Erythropoietin attenuates neuronal injury and potentiates the expression of pCREB in anterior horn after transient spinal cord ischemia in rats. Surg Neurol, 68, 3 (Sep 2007), 297-303; discussion 303.

[142] Kaptanoglu, E., Solaroglu, I., Okutan, O., Surucu, H. S., Akbiyik, F. and Beskonakli, E. Erythropoietin exerts neuroprotection after acute spinal cord injury in rats: effect on lipid peroxidation and early ultrastructural findings. Neurosurg Rev, 27, 2 (Apr 2004), 113-120.

[143] Gorio, A., Gokmen, N., Erbayraktar, S., Yilmaz, O., Madaschi, L., Cichetti, C., Di Giulio, A. M., Vardar, E., Cerami, A. and Brines, M. Recombinant human erythropoietin counteracts secondary injury and markedly enhances neurological recovery from experimental spinal cord trauma. Proc Natl Acad Sci U S A, 99, 14 (Jul 9 2002), 9450-9455.

[144] Cetin, A., Nas, K., Buyukbayram, H., Ceviz, A. and Olmez, G. The effects of systemically administered methylprednisolone and recombinant human erythropoietin after acute spinal cord compressive injury in rats. Eur Spine J, 15, 10 (Oct 2006), 1539-1544.

[145] Grasso, G., Sfacteria, A., Erbayraktar, S., Passalacqua, M., Meli, F., Gokmen, N., Yilmaz, O., La Torre, D., Buemi, M., Iacopino, D. G., Coleman, T., Cerami, A., Brines, M. and Tomasello, F. Amelioration of spinal cord compressive injury by pharmacological

preconditioning with erythropoietin and a nonerythropoietic erythropoietin derivative. J Neurosurg Spine, 4, 4 (Apr 2006), 310-318.

[146] Arishima, Y., Setoguchi, T., Yamaura, I., Yone, K. and Komiya, S. Preventive effect of erythropoietin on spinal cord cell apoptosis following acute traumatic injury in rats. Spine (Phila Pa 1976), 31, 21 (Oct 1 2006), 2432-2438.

[147] Vitellaro-Zuccarello, L., Mazzetti, S., Madaschi, L., Bosisio, P., Gorio, A. and De Biasi, S. Erythropoietin-mediated preservation of the white matter in rat spinal cord injury. Neuroscience, 144, 3 (Feb 9 2007), 865-877.

[148] Loblaw, D. A., Holden, L., Xenocostas, A., Chen, E., Chander, S., Cooper, P., Chan, P. C. and Wong, C. S. Functional and pharmacokinetic outcomes after a single intravenous infusion of recombinant human erythropoietin in patients with malignant extradural spinal cord compression. Clin Oncol (R Coll Radiol), 19, 1 (Feb 2007), 63-70.

[149] Zeman, R. J., Feng, Y., Peng, H. and Etlinger, J. D. Clenbuterol, a beta(2)-adrenoceptor agonist, improves locomotor and histological outcomes after spinal cord contusion in rats. Exp Neurol, 159, 1 (Sep 1999), 267-273.

[150] Zeman, R. J., Peng, H., Feng, Y., Song, H., Liu, X. and Etlinger, J. D. Beta2-adrenoreceptor agonist-enhanced recovery of locomotor function after spinal cord injury is glutathione dependent. J Neurotrauma, 23, 2 (Feb 2006), 170-180.

[151] Krylova, I. B., Bulion, V. V., Gavrovskaya, L. K., Selina, E. N., Kuznetzova, N. N. and Sapronov, N. S. Neuroprotective effect of a new taurinamide derivative--Taurepar. Adv Exp Med Biol, 5832006), 543-550.

[152] Gupta, R. C., Seki, Y. and Yosida, J. Role of taurine in spinal cord injury. Curr Neurovasc Res, 3, 3 (Aug 2006), 225-235.

[153] Sapronov, N. S., Bul'on, V. V., Kuznetsova, N. N. and Selina, E. N. [The neuroprotector effect of a new taurine derivative on a model of compression spinal cord trauma in rats]. Eksp Klin Farmakol, 68, 6 (Nov-Dec 2005), 45-48.

[154] Yucel, N., Cayli, S. R., Ates, O., Karadag, N., Firat, S. and Turkoz, Y. Evaluation of the neuroprotective effects of citicoline after experimental spinal cord injury: improved behavioral and neuroanatomical recovery. Neurochem Res, 31, 6 (Jun 2006), 767-775.

[155] Cakir, E., Usul, H., Peksoylu, B., Sayin, O. C., Alver, A., Topbas, M., Baykal, S. and Kuzeyli, K. Effects of citicoline on experimental spinal cord injury. J Clin Neurosci, 12, 8 (Nov 2005), 923-926.

[156] Early acute management in adults with spinal cord injury: a clinical practice guideline for health-care professionals. J Spinal Cord Med, 31, 4 2008), 403-479.

[157] Hawryluk, G. W., Rowland, J., Kwon, B. K. and Fehlings, M. G. Protection and repair of the injured spinal cord: a review of completed, ongoing, and planned clinical trials for acute spinal cord injury. Neurosurg Focus, 25, 5 2008), E14.

Send Orders for Reprints to reprints@benthamscience.net

Recent Advances in Medicinal Chemistry, Vol. 1, **2014, 332-380**

CHAPTER 11

HPLC and its Essential Role in the Analysis of Tricyclic Antidepressants in Biological Samples

V.F. Samanidou*, M.K. Nika and I.N. Papadoyannis

Laboratory of Analytical Chemistry, Department of Chemistry, Aristotle University of Thessaloniki, Thessaloniki, 541 24 Greece

Abstract: HPLC is discussed as an essential tool for the analysis of tricyclic antidepressants in biological samples, providing clinicians with efficient fast and reliable methods to define individual optimum therapeutic concentrations in treatment of depressions. Additional information on mechanism of action, structure activity relationship and metabolism is provided. Sample preparation issues are discussed.

Keywords: Biological fluids, HPLC, monitoring, medicinal chemistry, tricyclic antidepressants,.

1. INTRODUCTION

Tricyclic antidepressants (TCAs) such as amitryptiline, doxepin, imipramine, nortryptiline, trimipramine, *etc.* have been in use for many years in psychiatry for treatment of disorders. An "antidepressant" is a medication designed to treat or alleviate the symptoms of clinical depression. Many new advances in antidepressants have been made over the past half century. Generally speaking, there are three classes of antidepressant medications in use: (1) The tricyclics (or heterocyclics) (TCAs), (2) the monoamine oxidase inhibitors (MAOIs) and (3) the newer, so-called second-generation agents: Selective Serotonin Reuptake Inhibitors, (SSRIs) with first member fluoxetine, developed in the 1980's. Although there is a considerable overlap in their actions and uses, these different categories of antidepressants work by distinct mechanisms, have different side effect profiles, and may be preferred for varying indications. The first group of

***Address correspondence to V. F. Samanidou:** Laboratory of Analytical Chemistry, Department of Chemistry, Aristotle University of Thessaloniki, Thessaloniki, 541 24 Greece; Tel:+302310997698; Fax: +302310997719; E-mail: samanidu@chem.auth.gr

Atta-ur-Rahman, Muhammad Iqbal Choudhary and George Perry (Eds)

antidepressants, that were used, was MAOIs discovered in the early 1950s and their major member, iproniazid, was initially used as possible treatment for tuberculosis. Tricyclic antidepressants were the second group with imipramine as first member, discovered accidentally during a search for a new antipsychotic in the late 1950s. TCAs were as effective as MAOIs, safer than them, but dangerous in overdose. Tricyclic antidepressants being the principal pharmacological treatment for endogenous major depressions in the period between 1960 and 1980 consist in a homogeneous group of drugs, which differs mostly in their potency to inhibit presynaptic norepinephrine or serotonin uptake and in their propensity for causing variety of unwanted effects. Despite the introduction of newer and safer antidepressants the prescription of tricyclic antidepressants is still widespread as they are cheaper and are still considered to be the most effective group of antidepressants [1, 2].

Methods for the routine analysis of TCAs are required to provide clinicians with the individuals' serum levels. Analytical procedures that allow the rapid quantification of these psychotropic drugs are of paramount significance. HPLC has been proved as an efficient, fast and reliable tool in medicinal chemistry for the analysis of TCAs in biological samples.

Only a limited number of review articles can be found in literature concerning tricyclic antidepressants determination in biological matrices. These are briefly cited below.

Scoggins *et al.* in their review on the measurement of tricyclic antidepressants, published in 1980, discussed the approaches used since 1967, with emphasis on specificity, sensitivity, accuracy, precision (reproducibility), expense, convenience, and ease of sample processing [3].

In 1985, Norman, wrote a review on the analysis of TCAs in plasma and serum by chromatographic techniques, with emphasis on application to the clinical situation [4].

In 2005, Smyth presented a review on the electrospray ionisation mass spectrometric behaviour of selected nitrogen-containing drug molecules and the application of Liquid chromatography-electrospray ionisation mass spectrometry (LC-ESI-MS) to the detection and determination of TCAs in biomatrices, pharmaceutical formulations, *etc.* Analytical information on sample concentration techniques, chromatographic separation conditions, recoveries from biological media, degradation products and limits of detection (LODs) is provided, covering the period 2004-2005. Comparisons, where available, are also made with rival analytical techniques such as Gas Liquid Chromatography-Mass Spectrometry (GLC-MS), Capillary Electrophoresis-Electrospray Ionisation Mass Spectrometry (CE-ESI-MS) and Stripping Voltammetry (SV) [5].

Maurer in his review published in 2005, described multi-analyte procedures for screening and quantification of drugs in blood, plasma, or serum by Liquid Chromatography-single stage or tandem mass spectrometry (LC-MS or LC-MS/MS) relevant to clinical and forensic toxicology. TCAs are discussed among other drugs such as amphetamines, cocaine, hallucinogens, opioids, anesthetics, hypnotics, benzodiazepines, neuroleptics, antihistamines, sulfonylurea-type antidiabetics, beta-blockers, and other cardiac drugs. Basic information on the procedures is given in two tables while multi-analyte screening, identification, and quantification are illustrated in three figures. A critical discussion on the advantages and drawbacks of such LC-MS procedures is also included [6].

Kerr *et al.* in 2006 presented a review on tricyclic anti-depressant overdose, which is among the most common causes of drug poisoning seen in accident and emergency departments. This review discussed the pharmacokinetics, clinical presentation and treatment of tricyclic overdose [7].

Samanidou *et al.* in their review published in 2008 reviewed literature till 2006 focused on the recent advances on the application of HPLC in medicinal chemistry for the routine analysis of TCAs and metabolites in clinical samples. Sample preparation methods used in these methodologies were discussed. Information on medicinal chemistry of TCAs in terms of mechanism of action, etc was also provided [8].

Uddin *et al.* in 2011 presented an extended and comprehensive review focusing on sample preparation (pretreatment and extraction) and different analytical methods applied for the quantification of tricyclic antidepressants. The review covered 148 references published since 1995. It included brief information about TCA, metabolism and mechanism of action, and discussed various sample-preparation techniques and possible analytical methods with detailed description of HPLC, including the column, mobile phase and detection [9].

In the present review recent developments in the analysis of TCAs covering the time span 2006 to 2012 have been included to update the information on HPLC analysis of TCAs.

Table 1: General characteristics of antidepressant drugs [13]

TCA	Metabolites	pK_a D pK_a M	Concentration Range (mg/mL)	Half Life (h)	Time of Steady State (Days)	Doses (mg/Day)
Amitriptyline	Nortriptyline, 10- Hydroxy, *N*-oxide	9.4	500-300 (ami+Nor) 500-200 Nor only	35-50 M 20-100	7-8 M 4-10	75-150
Clomipramine	Norclomipramine 8- Hydroxy, Nor, 2- and 8-Hydroxy Clomipramine	D: 9.4 M:10.2	D: 90-300 M: 150-350	17-28 M 35	7-10	50-200
Desipramine	2- and 8-Hydroxydesipramine	9.5	20-300	2-54	14	50-200
Doxepine	Nordoxepine	8	D: 70-400 M: 75-250	12-17 28	9	50-300
Imipramine	Desipramine, Hydroxyimipramine	9.5	50-500	6-40	1.5-3	50-150
Opipramol	Dehydroethylopipramol		D: 14-64 M: 42-445	6-23		10-300
Trimipramine	Hydroxy, nor, desamino, desalkyl		50-150	7-14		50-300

Abbreviations: D drugs, M major metabolite.

2. CHEMISTRY AND ACTIVITY

2.1. Structure Chemical Characteristics

Tricyclic antidepressants bare their name due to their chemical structure, which consists of three rings of atoms. The middle ring is usually alicyclic and contains seven atoms, except for some members that possess a heterocyclic middle ring. Their side chain consists of N-alkylmethylamine or N-alkyldimethylamine. Typical TCAs are imipramine, desipramine, clomipramine, amitriptyline, nortriptyline, doxepin, trimipramine *etc.* [10-13]. General characteristics of antidepressant drugs extracted from literature are shown in Table 1. The structures of these TCAs are shown in Fig. (**1**).

Amitriptyline
3-(10,11-dihydro-5*H*-dibenzo[*a,d*] cyclohepten-5-ylidene)-*N*, *N*-dimethyl-1-propanamine

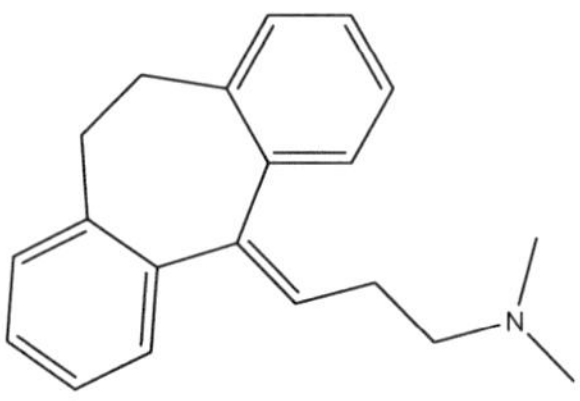

Clomipramine
[3-(3-chloro-10,11-dihydro-dibenzo[*b,f*] azepin-5-yl)-propyl]-dimethyl-amine

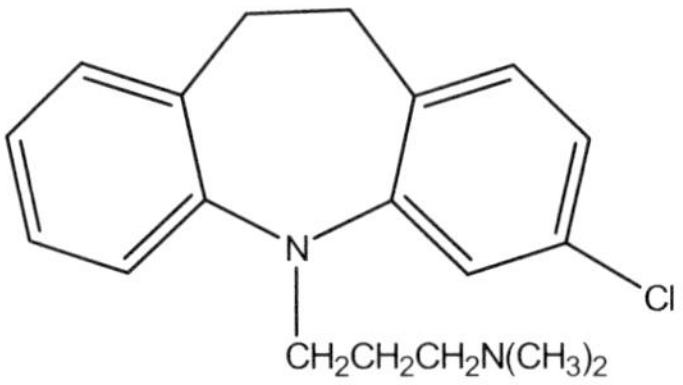

Desipramine
[3-(10,11-dihydro-dibenzo[*b,f*] azepin-5-yl)-propyl]-methyl-amine

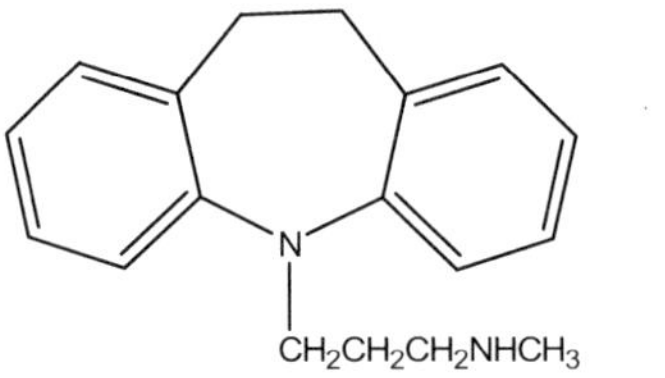

Doxepin
11-(3-(Dimethylamino) propylidene)-6H-dibenz(b,e)oxepine

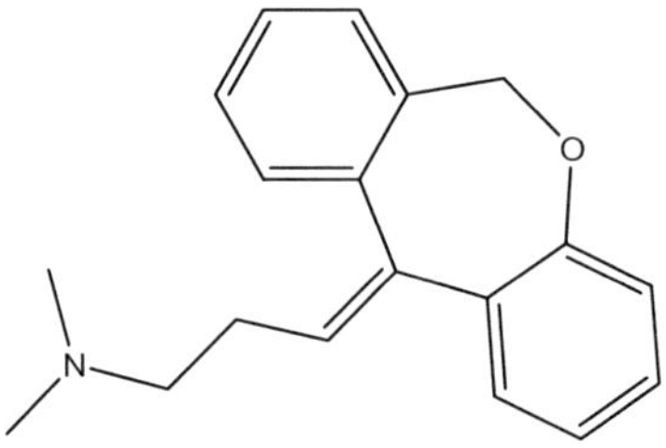

Imipramine
[3-(10,11-dihydro-dibenzo[*b,f*] azepin-5-yl)-propyl]-dimethyl-amine

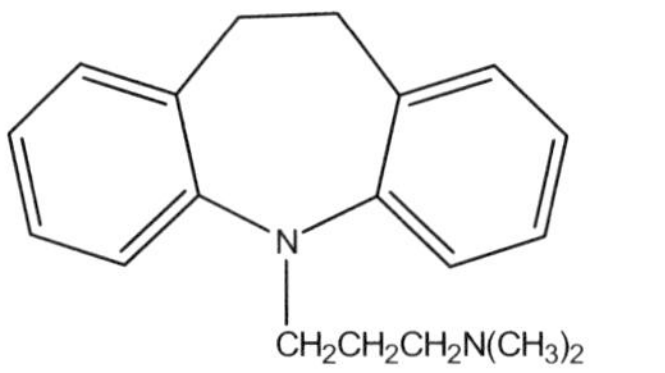

Nortriptyline
3-(10,11-dihydro-5H-dibenzo[a,d] cyclohepten-5-ylidene)- N-methyl-1-propanamine

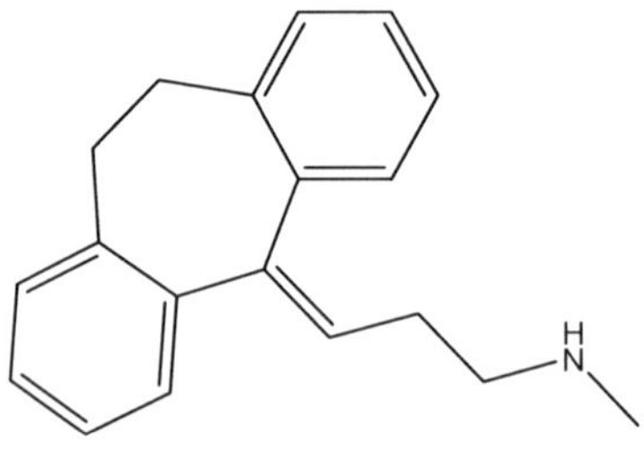

Fig. 1: contd….

Opipramol

4-[3-(5H-Dibenz[*b,f*]-azepin-5-yl)-propyl]-1-piperazineethanol

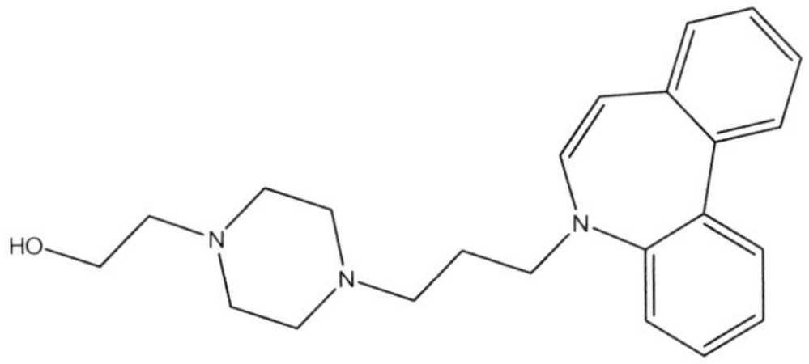

Protriptyline

[3-(5H-Dibenzo[*a,d*] cyclohepten-5-yl)-propyl]-methyl-amine

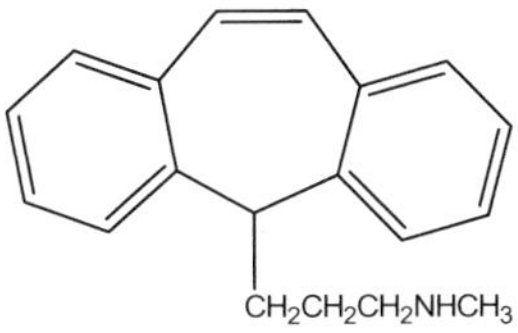

Trimipramine

[3-(10,11-dihydro-dibenzo[*b,f*] azepin-5-yl)-2-methyl-propyl]-dimethyl-amine

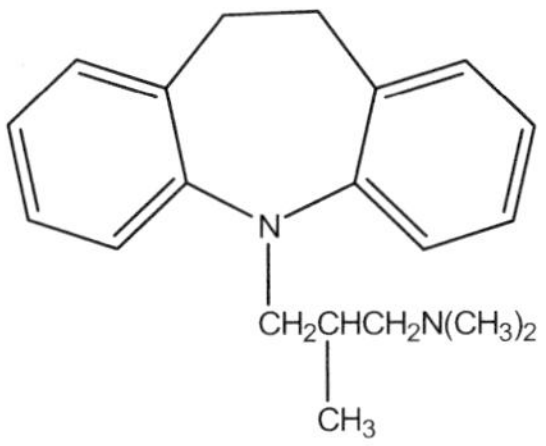

Figure 1: Chemical structures of imipramine, desipramine, clomipramine, amitriptyline, nortriptyline, doxepin, trimipramine and opipramol.

2.2. Mechanism of Action

The exact mechanism of action of TCAs in the treatment of depression is unclear and not completely understood. It is believed that their action is based on blocking the reuptake of the monoamine neurotransmitters, serotonin (5-hydroxytryptamine; 5-HT) and nor-epinephrine (NE). According to this theory these neurotransmitters are increased through inhibition of their reuptake by the presynaptic neuronal membrane.

Mood disorders are associated with reduced levels of monoamines in the brain. Neurotransmitters adjust the mood in the central nervous system. TCAs binding to 5-HT and nor-epinephrine (other name noradrenaline) reuptake transporters prevents the reuptake of these monoamines from the synaptic cleft and their subsequent degradation. As their concentration increases in the space between the nerve cells (neural synapse), by blocking the reuptake, depression relief is achieved. This reuptake blockade leads to the accumulation of 5-HT and nor-epinephrine in the synaptic cleft and the concentration returns to within the normal range. The main effect of TCAs is to block the uptake of monoamines by nerve terminals, by competing for the binding site of the carrier protein [14, 15].

Most TCAs are non-selective and inhibit nor- epinephrine and 5-HT uptake to a similar degree. Tertiary amines, such as amitriptyline, imipramine, clomipramine, doxepin and trimipramine cause reuptake inhibition of serotonin. Amitriptyline and clomipramine appear to be more potent than other tricyclics in blocking serotonin, although, through their metabolites, they become powerful inhibitors of nor-epinephrine reuptake as well. Secondary amines, such as desipramine and nortriptyline, mainly inhibit the reuptake of norepinephrine. Imipramine inhibits reuptake of nor-epinephrine and serotonin equally. Doxepin is a moderate inhibitor of nor-epinephrine and a weak inhibitor of serotonin. Mechanism of action of selective serotonin reuptake inhibitors is shown in Fig. (**2**) [14, 16].

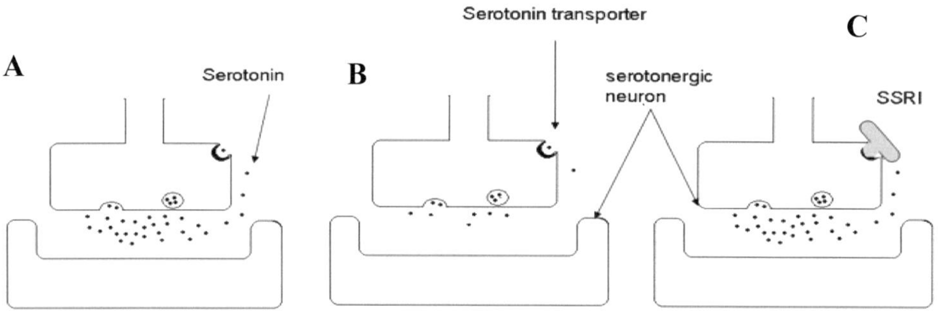

Figure 2: Mechanism of action of selective serotonin reuptake inhibitors. (**A**) Serotonergic neuron, (**B**) Short-term administration of SSRI blocks reuptake of serotonin. (**C**) Long-term administration causes downregulation of $5HT_{1A}$ autoreceptors. In the presence of blockade of reuptake, more serotonin is available to act postsynaptically.

Additionally, recent research has shown that TCAs can cause beta-adrenergic downregulation as well. Where downregulation is the phenomenon of a long-lasting increase in the availability of neurotransmitter at a synaptic receptor site resulting in a decrease in the number of receptors on the cell surface. Long-term treatment with antidepressants results in changes in postsynaptic beta-adrenergic receptor sensitivity and increased responsiveness of the adrenergic and serotonergic systems to physiologic and environmental stimuli thus contributing to the mechanism of action. Antidepressants may produce a downregulation (desensitization) of alpha 2 - or beta-adrenergic and serotonin receptors, equilibrating the noradrenergic system, and thus correcting the dysregulated monoamine output of depressed patients. Receptor changes resulting from chronic administration of TCAs appear to correlate better with antidepressant action than does the synaptic reuptake blockade of neurotransmitters, and may also account for the delay of two to four weeks in therapeutic response [17].

2.3. Pharmacokinetics and Metabolism

Pharmacokinetic studies on TCAs have shown that due to their lipophilic nature these drugs are absorbed very quickly from the small intestine and then they are metabolised in the liver (first path effect). Subsequently, the kidneys eliminate the products of the hepatic metabolism, some of which may have pharmacological action. The whole procedure is characterised as enterohepatic and enterogastric circulation. TCAs are also highly protein bound and have a long half-life, due to their large volume of distribution. Their half-life is about 24 hours, while there are some exceptions for some TCAs, which need more time to be excreted from the organism [18]. The onset of action is 3-4 weeks, because an antidepressant takes 2-4 weeks to build up its action and the maximum concentration of TCAs in plasma is reached in 2-8 hours [19].

Tertiary amines are metabolised more rapidly than secondary amines and for this reason they are preferred for therapeutic purposes [10]. Tertiary amines are metabolized in the hepatic cytochrome P450 system to secondary amines, which have pharmacological action. This metabolism is a result of a demethylation. Then, the enzyme CYP2D6 hydroxylates the secondary amines to pharmacologically inactive metabolites [20]. Although the activity of P450 system depends on genetic parameters, external factors like the use of other drugs can influence its action. Patient variabilities, such as ethnicity, age and gender also affect TCA metabolism [21]. Finally, tricyclic antidepressants (TCAs) have a narrow therapeutic index and for this reason the regulation of the dosing, the role of the metabolism and the pharmacokinetic parameters are very crucial in order to avoid toxic effects [22].

2.4. Side Effects and Toxicity

Beside their beneficial activity, TCAs cause a number of side effects as well. These are mainly due to interference with autonomic control. Some of these adverse effects relate to their anticholinergic properties. Except for atropine-like effects (muscarinic, mainly M1, cholinergic receptor block), they include postural hypotension (α1-adrenergic receptor block) and sedation (H1 histamine receptor block). Cardiotoxic side effects are also induced related to their action on cardiac Na^{+} and Ca^{2+} channels. Poor dental health, due to effects of TCAs on salivary secretion, is a common problem among middle-aged and elderly patients [15].

Possible toxic effects of TCAs are the result of their properties, namely the inhibition of norepinephrine and other neurotransmitter receptors reuptake, the adrenergic blocking, the anticholinergic action and effects on the myocardium [18, 23]. These toxic effects include dry mouth, blurred vision, constipation, urinary retention, sinus tachycardia and memory dysfunction (anticholinergic effects) [23], seizures [24], mental status changes, such as coma, respiratory failure (central nervous system effects), tachycardia, cardiac arrhytmias, orthostatic hypotension and dizziness (inhibition of nor-epinephrine reuptake, anticholinergic action, a1-adrenoreceptor antagonist properties of the tricyclics and effects on the myocardium). Also, TCAs have quinidine-like antiarrhythmic actions, which cause cardiac conduction delays and arrhythmias. These cardiovascular effects are caused by the fast sodium channel blockade and the a-1 adrenergic receptor blockade, while the anticholinergic effects are caused due to the acetylcholine inhibition.

2.5. Therapeutic Drug Monitoring

The pharmacological action of TCAs and their therapeutic results depend on their structure. Therapeutic drug monitoring is based on the measurement of serum concentration of medications to obtain optimal effective concentration. This is very crucial because many side effects of TCAs are concentration-dependent and potentially life-threatening, especially in case of drugs with narrow effective range or narrow therapeutic/toxic index such as TCAs. Obviously a possible toxic dose and a therapeutic dose are very close. Despite this, it is important to adjust the dosing according to the requirements of each patient, as indications for therapeutic drug monitoring include serious consequences for over/underdosing, narrow therapeutic/toxic index, poor relationship between drug dose and circulating concentration. Additionally good relationship between circulating concentration and therapeutic or toxic effects, alter physiologic state that may unpredictably affect circulating drug concentration, drug interactions and of course patient compliance. In order to control a possible intolerance the dosing is low at the first period of the therapy and increases gradually. It is important to understand that increasing the dosage does not normally shorten the therapy period and may increase the incidence of side effects and toxicity. Therapeutic concentrations for TCAs and their major metabolites are well established for the typical TCAs (amitriptyline, imipramine, nortriptyline, desipramine) and there is a logical correlation between drug concentration and clinical effects of these drugs. Usual dosage ranges for some antidepressant drugs (amitriptyline, clomipramine, desipramine, dothiepin, doxepin, imipramine, nortriptyline, trimipramine) from 75 to 300 mg/day [10, 14, 20, 23, 25].

3. ANALYTICAL METHODS

TCAs monitoring in biological fluids can be achieved by different analytical techniques, such as chromatographic techniques, capillary electrophoresis, voltammetry, *etc.* However HPLC is the prevalent technique in the analysis of TCAs as shown in Fig. (**3**). Both normal and reversed phase HPLC methods are applied to the analysis of TCAs. Chromatographic conditions used in the literature are extensively described below.

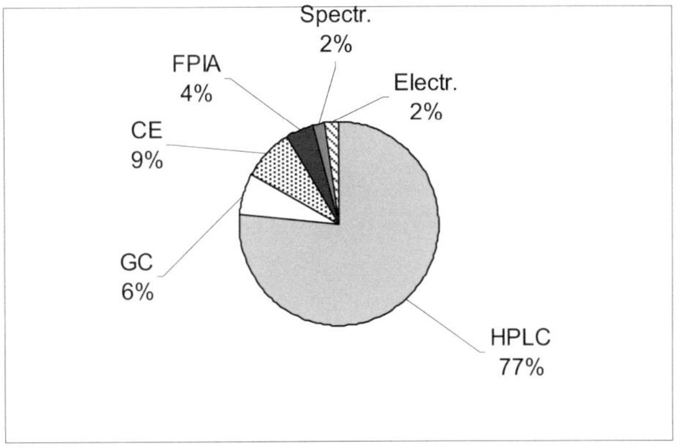

Figure 3: Analytical techniques used in TCAs analysis since 1990.

3.1. Chromatographic Conditions

In the majority of the published methods the chromatographic separation of TCAs is performed at room temperature. However in some cases elevated column temperature is also used, such as 30°C [26] and 45°C [27, 28].

Detection of separated analytes is mainly performed using ultraviolet or diode array detectors where eluted TCAs are monitored at different wavelengths within the range 200-265 nm [26, 28-52]. Other detection techniques were also used by some researchers: Electrochemical detection [45, 52-55], fluorescence detection after post-column photochemical reaction [39], Chemiluminescent detection using 0.75 mM $Ru(bipy)_3Cl_2$ solution [56] and MS/MS [57-61].

In the following paragraphs chromatographic conditions used in methods published since 1990 are cited briefly. More information upon reviewed papers is given in Table **2**.

Amitriptyline, nortriptyline, imipramine, trimipramine, clomipramine and doxepine were determined in human blood and urine using a Waters Symmetry C_8 (250 × 4.6 mm, 5 μm) analytical column with a 20 mm guard column (Waters Symmetry C_{18}). Mobile phase consisted of solvent A=phosphate buffer (pH = 3.8), solvent B=acetonitrile, used in a step gradient: 15% B for 6.5 min, then 35% until 25 min and 80% B for 3 min [26].

Amitriptyline, imipramine, trimipramine, nortriptyline, protriptyline and doxepin were determined in human serum and urine using a precolumn 10×2.1 mm and a 150 × 4.6 mm main column packed with Hypersil 5 μm octadecylsilane (C_{18}). Mobile phase was a mixture of solvent A: 50 mL/L acetonitrile and solvent B: 500 mL/L acetonitrile, both in 50 mmol/L phosphate buffer (pH 3.0), with 375 mg/L sodium octyl sulfate and 3 mL/L triethylamine. Solvent gradient conditions changed linearly from 15% B to 90% B in 20 min, stayed at 90% B for 5 min, and returned to 15% B in 3 min [27].

Table 2: Overview of HPLC methods for the determination of tricyclic antidepressants in biofluids

Analytes	Sample Type	Sample Preparation	Chromatographic Conditions	Recovery (%)
AMI, NorTRP, IMI, TRI, CLO, DOX	Human blood and urine	Add.of urine or whole blood and deion. H_2O. Mix. Centr. Removal of the org. phase, evap. Rediss. by ACN-H_2O (50:50, v/v). Vortex and centr.	Waters Symmetry C_8 (250×4.6 mm, 5 μm) with a 20 mm guard column (Waters Symmetry C_{18}). Column Temp.: 30°C. Solvent A=phosphate buff. (pH = 3.8), solvent B=ACN. Step gradient: 15% B for 6.5 min, then 35% until 25 min, 80% B for 3 min. FR: 1 mL/min for 6.5 min, then linear increase to 1.5 mL/min from 6.5 min to 25 min. Hold for 3 min.	
AMI, IMI, TRI, NorTRP, protriptyline, DOX	Human serum, urine	Serum: Add.of phosphate buff. (pH 6.0). Vortex. SPE columns wash. with MeOH and phosphate buff. (pH 6.0). Sample application. Wash with MeOH in phosphate buff. (pH 6.0) and evap. El.with MeOH and 10% ammonia (5:1 v/v). Urine: inj. of the extract directly into the chromatograph.	Precolumn 10×2.1 mm and a 150 × 4.6 mm analytical column Hypersil 5μm (C_{18}). Temp.: 45°C. MPs: Solvent A: 50 mL/L ACN and solvent B: 500 mL/L ACN, both in 50 mmol/L phosphate buff. (pH 3.0), with 375 mg/L sodium octyl sulfate and 3 mL/L TEA. Gradient changed linearly from 15% B to 90% B in 20 min, stayed at 90% B for 5 min, and returned to 15% B in 3 min. FR: 1.0 mL/min.	
AMI, NorTRP, E and Z-OH-AMI, E and Z-OH-NorTRP.	Rabbit plasma	Homog. Add. of IS. Mix. samples with NaOH, heptane and ethyl acetate. After centr. add. of H_2SO_4 to the org. layer, mix.and centr. Org. layer discarded and an aliquot of the aq. phase mix. with K_2HPO_4.	Supelco 5 μm C_{18} 250 × 4.6 mm. MP A: ACN-H_2O (10:90 v/v), 900 μL of 85% H_3PO4, 1.22 g of KH_2PO_4. MP B: ACN. A 20-50% MP B gradient in 21 min. IS: CLO (10 μg/mL). FR: 1 mL/min.	
DES	Rabbit plasma, rabbit bone marrow	Plasma sample add. of IS and saturated aq. $Na_2B_4O_7$. Vortex. Add. Hexane:isoamyl alcohol 98:2, vortex. After centr. an aliquot of the solvent phase was transferred to another centrifuge tube, dried under stream of N_2 and the cooled residue diss. with MeOH.	PE Pecosphere - 5C Silica. MP: MeOH:NH_4OH:1 M NH_4NO_3 (190:6:4 v/v/v) IS: chlorpromazine (1000 pg/mL). FR: 1.3 mL/min.	Rabbit plasma: 24.2-30.2 Rabbit bone marrow: 24.5-34.2

Table 2: contd….

DES, NorTRP, IMI, AMI, CLO	Human serum	Serum sample mix. with IS, saturated $Na_2B_4O_7$ (pH 11 with 6M NaOH) and n-hexane. Centr. Separation of the org. phase, evap. Rediss. with MP.	RP C_8 (5 μm, 150×4.6 mm) with a 2cm long Pelliguard LC-8 guard column with 40 μm packing.MP: ACN-phosphate buff. (pH 3 with H_3PO_4) in a 50:50 (v/v) mixture. IS: clobazam (20 ng/mL).FR: 1 mL/min.	93.2-110.6
AMI, NorTRP, DOX, IMI, DES, NorDOX, CLO, TRP	Human serum, liver	Vortex mix. of samples, cyanopramine, deion. H_2O. Add.of Na_2CO_3 and vortex. Add.of hexane-butan-1-ol (95:5 v/v). Agitation. Centr. of the org. layer with H_3PO_4.	Spheri-5 RP-18 (5 μm, 100×4.6 mm) and RP-18 Newguard cartridge (7 μm, 15×3.2 mm). MP: ACN 0.1M-NaH_2PO_4-diethylamine (40:57.5:2.5 v/v/v) (pH 8.0). RT.IS: cyanopramine. FR: 2.0 mL/min.	Human serum: >6[illegible] Liver: >40[illegible]
DOX, DMDOX (E- and Z- isomers)	Human serum	Add. of MeOH containing IS. Direct inj. to HPLC.	Column 1: Hypersil MOS C_8 (10×4.6 mm, 10μm). RT.MP 1: H_2O:MeOH (950:50, v/v). Column 2: Nudeosil 100 CN (250×4.6 mm, 5μm). MP: MeOH: ACN:0.01 mol/L potassium phosphate buff., pH 6.8 (188:5778:235, v/v). IS: TRP (140 μg/L, 489 nmol/L).	
AMI, CLO, demexiptiline, DES, DOX, IMI, MAPR, OPI, TRP	Human serum	Extr. with aq. NaOH sol.and a mixture of n-heptane, ethylacetate and isoamyl alcohol. The plasma or blank sample spiked of working sol.and add. of IS, NaOH, ethylacetate and n-heptane. Isoamyl alcohol was also added to the extr. mixture when the CN column was used for analysis. Vortex 20 s. Centr. The org. layer into a glass tube with methanolic HCl sol.and evapor. Rediss. in MeOH.	Columns: Kromasil C_{18} (5 μm, 150× 2.1 mm), Zorbax C_3 TMS (5 μm, 150× 4.6 mm), Nucleosil CN (5 μm, 150× 4.6 mm). RT. MP: ACN - 0.015 mol/L KH_2PO_4 buff. (adj. to the optimal pH with 0.1 mol/L H_3PO_4). MP:40/60, 30/70, 55/45 and pH of MP: 6.5, 5.8, 6.4 resp. FR: 0.35, 1.5, 1.0 mL/min resp.).	
IMI, DES	Mice serum	Add.of IS. Vortex. Add. NaOH and hexane. Vortex and centr. Org. layer dried under N_2 gas at 20°C. Rediss. with MP and inj.	Microsorb MV C_{18} column (15×0.46 cm, 5 μm). MP: 60% ACN and 40% 0.01 M TEA in distilled H_2O (pH 3.0 by 85% H3PO4). IS: CLO.FR: 1.0 mL/min.	>89.6
IMI, DES, AMI, NorTRP	Human plasma	Add.of IS, NaCl, NaOH, and hexane/isoamyl alcohol (99:1, v/v) to plasma samples. Shaking.	LiChrospher® 60 RP-select B (4×250 mm, 5 μm). MP: 50% ACN and 50% 0.25N CH_3COONa buff. (pH 5.5). IS: CLO in MeOH (5 μg/mL). FR:	89.6-103.[illegible]

Table 2: contd….

		Centr. Evapor. of the org. solvent at RT. Add. of MP to the residue. Vortex.	1.0 mL/min.	
Cis- and trans- DOX, cis- and trans-DMDOX	Human serum	Mix. with ACN, IS, 0.25N NaOH, isoamyl alcohol.Vortex. Add. of heptane. Shaking and centr. Org. layer mix. with 0.1M glycylglycine buff. (pH 3). Shake. Centr. Add.of 0.25N NaOH and n-heptane to the glycylglycine layer. Shake. Centr. Add.of n-heptane. Vortex. Evap. Rediss. with MP.	Silica column (3 µm, 6×100 mm) with a guard column 40 µm pellicular silica. MP: 0.025M Na_2HPO_4 (pH 3 with H_3PO_4)/ACN/n-nonylamine (80/20/1, v/v/v). ISs: amoxapine and loxapine (4mg/mL). FR: 1.6 mL/min.	65-75 (desmetyl OXs), 75-85 (DOXs
DES, IMI, NorTRP, MAPR, AMI, DMCLO, CLO	Human plasma	Alkal. with 2M Na_2CO_3. Add.of IS and Extr. with n-hexane. Shaking. Centr. The lower aq. layer was frozen (dry ice-acetone bath). Back Extr. of the org. layer with H_3PO_4. Shaking and centr. Inj. of acidic sol.to HPLC.	RP-C_{18} symmetry column (5 µm, 250×4.6 mm). MP: KH_2PO_4 0.067M (pH 3.0 with H_3PO_4)-ACN (65:35 v/v) IS: clovoxamine. FR: 1.2 mL/min.	>80
DOX, DMDOX	Human plasma, urine	Add of 3M ammonia (sol.A) and a mixture of n-pentane-IPA (95:5, v/v, sol.B) were added. Shaking and standing. Add. 0.1M HCl to the upper org. layer. Shaking. Aq. Layer wash. with pentane. Add sol. A and B to the wash. aq. residue, shake. Evap. org. phase and diss. with MP.	Spherisorb silica (3 µm, 150 × 4.5 mm). RT. MP:Hexane, MeOH and nonylamine (95:5:0.3 v/v/v). FR: 1.0 mL/min.	61-64 (plasma) and 63-68 (urine).
MIA, AMI, NorTRP, IMI, DES	Human plasma	Add. of Protein Releasing Reagent [PRR: an aq. sol.of 1M HCl and 25% (v:v) glycerol] and inj. into the donor ch. of the dialysis cell. Trasport of acceptor sol. (0.001M ammonium phosphate buff., pH 7.0) to the acceptor ch. Enriched analytes eluted onto the analytical column	Supelcosil LC-PCN cyanopropyl (5 µm, 150 × 4.6 mm). MP: ACN-MeOH-0.005 M ammonium phosphate buff., pH 7.0 (70:15:15, v:v). FR: 1.5 mL/min.	

Table 2: contd….

		by the HPLC MP. Both sides of dialyser wash. with the donor sol. (1 mM dodecyltrimethyl ammonium bromide in H_2O) and acceptor solution, resp. The precolumn regenerated with the acceptor sol. Next sample inj. into donor ch. of the dialysis cell.		
AMI, NorTRP, IMI, DES, CLO, DOX, TRP	Human plasma.	Mix. with IS. Alkal. by the add. of 1M NaOH. Extr. by shaking with hexane containing 1% isoamyl alcohol. Centr. Back extr. of org. phase with 0.05M HCl. Vortex. Centr. Acidic phase inj. onto the HPLC.	Ultrasphere C_8 (250×4.6 mm). MP: 35% CH_3CN-aq. phase containing 4 mM 1-octanesulphonic acid and 0.5 mM N,N,N,N-tetramethylethylene diamine (pH 2.5 with H_3PO_4). FR: 2 mL/min.IS: 200 ng AMI or 100 ng DMDOX	
AMI, NorTRP, IMI, DES, CLO, NorCLO	Human serum and plasma	Single step LLE: Mix. with IS, 0.1M $Na_2B_4O_7$ sol. (pH 11 with 30% NaOH) and hexane in a Sovirel glass tube. Shaking. Centr. Evap. Rediss. in MeOH	Nova-Pak C_{18} 4 μm, 4.6×150 mm, Waters. MP: mixing 500 mL of 5 mM aq. KH_2PO_4 buff., 500 mL ACN and 2 mL diethylamine (pH 8 with H_3PO_4).RT. IS: econazole, 90 mg mL^{-1}. FR: 0.9 mL/min.	92-105
CLO, DMCLO, 2-, 8-, and 10-OHCLO, 2-, and 8-OHDMCLO, diDMCLO	Human and rats serum and plasma	Sample clean up: Proteins and other interfering compounds wash. to waste by deion. H_2O containing 35% (v/v) ACN. Centr.	Lichrosher CN (5μm, 250×4.6 mm) and a clean up column Hypersil CN (10 μm CN, 10×4.6 mm). MP: 38% ACN and 62% $NaCLO_4$ sol. (0.02M) (pH 2.5 with $HClO_4$). FR 1.5 mL/min (5-8 min).	64-110
DOX, DES, MAPR, IMI	Human plasma	C_{18} Bond-Elut cartridges and mixtures of MeOH-aq. buff. as washing and el.solvents. SPE cartridge activ. with MeOH. Blood samples collected using sodium citrate. Wash cartridge with aq. acetate buff., ACN and washing solvent and then with el.solvent. Evap. Rediss.with MP containing IS.	Nova Pack C_{18} column (4 μm, 15 cm × 3.9 mm) and a Nova Pack C_{18} guard column (4 μm, 20×3.9 mm). MP: ACN-0.02M TEA (pH 5.5 with H_3PO_4) (35:65 v/v) IS: p-HBA n-butyl aster (butyl paraben). FR: 1.0 mL/min.	83.3

Table 2: contd….

NorTRP	Human serum	SPE on CN cartridges. LLE: add. 1M NaOH and IS containing 89 mg/mL of MAPR and 3 mL heptane/isoamylalcohol (98.5:1.5, v/v). Shaking. Centr. Org. layer evap. Diss. in MP.	IS: TRP. SPE samples: a Luna C_{18} (3 µm, 150×4.6 mm) equipped with a C_{18} guard column. MP: ACN-0.01 M TEA pH 3.0 with H_3PO_4 (34:66 v/v) in MeOH-H_2O (50:50, v/v), 6.2mg ml^{-1}.FR: 0.85 mL/min. LLE samples: MP: ACN-MeOH-conc. ammonia H_2O, 950:50:7 (v/v).FR: 1.3 mL/min.	SPE: 93-9 LLE:75 ±13
AMI, NorTRP	Human serum	Blood was collected and centrifuged. Inj. To the HPLC after dilution (1:10) in 0.15M SDS-6% pentanol at pH 7. Filtr. through 0.45µm nylon membranes.	Column: Kromasil 5 C_{18} (5 µm, 250×4.6 mm).MP: SDS (0.15M)-6% pentanol (v/v) (pH 7).FR: 1.5 mL/min.	99.8-101. (AMI) an 98.5-99.7 (NorTRP
CLO, N-DMCLO, 8-OHCLO, 2- and 8-OHDMCLO	Human plasma	Add. of phosphate buff. and IS. Vortex. Centr. SPE with an Isolute C_2. Wash columns prior sample application with MeOH/H_2O/phosphate buff. (pH 9.2) (2/1/2 v/v/v). Washing with phosphate buff./mixture of H_2O and ACN (80:20 v/v) (1/2 v/v). El.with MeOH.	Lichrospher CN, (5 µm, 250×4 mm) with a 2-cm pre-column of the same material. MP: 10 mM K_2HPO_4-ACN-MeOH (35:25:40 v/v/v). IS: MAPR. FR: 1.5 mL/min.	99.1-100.
AMI, NorTRP, IMI, DES, DOX, NorDOX, CLO, NorCLO, TRP	Human serum	SPE: 3M-Empore Extr. disk cartridges. Centr. serum. Sorbent cond. with MeOH and H_2O. The supern., melperone and 0.1M KH_2PO_4 buff. (pH 6.0) mixed. Sample extr. through disk cartridge. El. with IPA-ammmonia sol.(25%)-CH_2Cl_2 (20:2:78). Evap. Diss. in ACN-H_2O (3:7).	Nucleosil 100-5-Protect 1 (250×4.6 mm, 5 µm). MP: 25 mM KH_2PO_4 (pH 7.0)-ACN (60:40 v/v). IS: melperone (3000 ng/mL). FR: 1.0 mL/min.	75-100.2
AMI, CLO, DES, DOX, IMI, NOX, OPI	Human blood	Sample of whole blood was extr. with 0.6M NaOH and mix. Add.of hexane-isoamyl alcohol (99:1, v/v). Agitation. Centr. Evapor. of the org. layer. Reextr. of the drugs into 0.05% H_3PO_4 vortex. org. layer with diluted acid and centrif.	LiChroCART (125×4 mm) packed with octasilica LichroSpher RP Select B (5 µm). MP: H_2O with H_3PO_4 and ACN (in gradient mode). ISs: OPI and IMI.FR: 1 mL/min.	75-90. Exception AMI (67) and OPI (47)

Table 2: contd….

AMI, OPI, NOX	Urine	Sodium phosphate buff. (pH 9.0), sample sol.containing TCAs pumped into a mix.coil. Mix. solutions passed over the liquid membrane in membrane separator made of two PTFE blocks with machined spiral grooves. Impregnation of membrane by soaking in n-undecane. The membrane separated 2 ch.: the donor ch. for extr. of analytes and IS and the acceptor ch. with acidic sol. for reextr. of analytes from the membrane solvent. Sample of acceptor sol. inj. to HPLC.	Column: Shiseido RP-18, (250×4.6 mm, 5 μm). MP: 0.01M Na_3PO_4 (buff. pH 3.2)-MeOH-ACN-dimethylamine (37:55.4:7.4:0.2 v/v).IS: diethazine.FR: 0.8 mL/min.	
DOX, NorDOX, DES, IMI, NorTRP, AMI	Human breast milk	Add. of IS. Saturation with NaCl and Add.of 2M NaOH and 2% butanol in hexane. Mix. by rotation. Centr. The aq. layer was frozen (dry ice/acetone bath). Org. layer to another tube with H_2SO_4. Rotation and centr. Aspiration of org. layer and Neutr. of the samples with 1% $KHCO_3$. SPE cartridges (C_{18}) cond. with diethylamine in MeOH, followed by 1% K_2CO_3 in 10% ACN in H_2O. Sample application. Dryness. Add.of 20% ACN in H_2O. Mixing. Add.of H_2O+MeOH+ACN. El.of TCAs with diethylamine in MeOH, followed by 1% $KHCO_3$ in 10% ACN in H_2O. Evap. Rediss. with MP.	Column: MOS-2 Hypersil (C_8) (3 μm, 100×2 mm).MP: 0.02M KH_2PO_4+85 μL N,N-dimethyloctylamine/L (pH 6.5) and 34% ACN.IS: TRP (2.5 μg/mL). FR: 0.5 mL/min.	85.8-104.6 (50 mg/mL) and 90-102.3 (200 ng/mL)
AMI, DOX, CLO, TRP	Human plasma	Mixed-mode SPE(IST Isolute HCX, 80 mg/1 mL)	Phenomenex Synergi Hydro-RP HPLC (250×4.6 mm). MP: aq.	

Table 2: contd….

and their N-demethyl metabolites			CH_3COONH_4/MeOH/ACN. IS: butriptyline.FR: 1.0 mL/min.	
IMI, DES	Human serum	Direct inj. into the HPLC, after filtr. (0.45 µm nylon membranes).	Column: Kromasil 5 C_{18} (5 µm, 250×4.6 mm).RT. MP: 0.15M SDS-6% (v/v) pentanol-0.001M NaCl-0.01M NaH_2PO_4 (pH 7). FR: 1.5 mL/min.	
IMI, DES, 2- and 10-OH-IMI, 2- and 10-OH-DES	Human plasma, urine	Add. of IS and 1M sodium carbonate buff. (pH 9.6). Extr. by shaking with diethyl ether. Centr. Add. of 0.1M H_3PO_4 to the org. layer. Shaking and centr. Aliquot of acidic layer inj. to the HPLC.	(Phenomenex Bondclone 10 C_{18} 300×3.90 mm, with a pre-column (RP-18, 10µm, 40×4.6 mm). RT. MP: 30% ACN in 0.1M K_2HPO_4 buff. (with conc. H_3PO_4, pH 6.0). FR: 2 mL/min. IS: pericyazine in MeOH (200 ng).	78.6-94.3 (plasma) 10.24-28. (urine)
IMI	Human serum and plasma	One-step LLE with diethylether in presence of Na_2CO_3. Add. of IS.	Nucleosil 100 C_{18}, (5µm, 125×4 mm) (IMI) MPs: MeOH-phosphate buff. (pH 3) (30:70, v/v), 15% MeOH in 0.01M CH_3COOH, MeOH-phosphate buff. (pH 3) (60:40, v/v), MeOH-0.008M H_3PO_4 (for, IMI resp.).IS: Methyl parahydroxybenzoate, sulfathiazole, chlorpromazine hydrochloride, naproxen (IMI). FR: 1 mL/min except IMI (1.2 mL/min).	Human serum: >8
IMI, DES, CLO, AMI, NorTRP, DOX.	Human blood	Blood sample centr. Store at −20°C. Add. of $HClO_4$, vortex and centr. Neutr. of the supern. (NaOH). Filtr. Direct inj. into the HPLC.	(Inertsil ODS-3 150×4.6 mm, 5 µm, connected to the FIA system. RT. MP: mixture of ACN and phosphate buff. 0.05M (KH_2PO_4/K_2HPO_4, pH 6.9 ± 0.1) (375:625, v/v) for the detection of all TCAs, except for CLO (50:50 v/v).	DES 92.3 IMI 90.8,

Table 2: contd….

IMI, DES, AMI, NorTRP, CLO.	Human plasma	Add. of IS. Extr. by shaking with diethyl ether. Centr. Add. of H_3PO_4 to the upper org. layer. Shake. Centr. Aliquot of the acid layer inj. to HPLC.	Trimethylsilyl (TMS) (5 µm, 150 × 4.6 mm). MP: 50 mM sodium phosphate buff. (pH 7.0)-ACN (55:45, v/v).IS: CLO. FR: 1.0 mL/min.	83.0-93.8
AMI, CLO, DES, DOX, IMI, MAPR, MIA, NorCLO, NorDOX, NorTRP, OPI, TRP.	Human serum	Add of IS. Vortex. Protein precip. Centr. Supern. dil. with the MP. Inj. onto the HPLC.	Chromolith Speed ROD C_{18} (5 µm, 50×4.6 mm). RT. MP: mix.of MeOH and 5 mM acetic acid (pH 3.9). Starting at 20% MeOH and 80% buff. sol. with linear gradient to 70% MeOH in 4 min. IS: Mixture containing 50µg/L clonidine, 10µg/L dehydromethylrisperidon and 10µg/L methabenzthiazurone in ACN/MeOH (9/1 v/v). FR: 1.0 mL/min.	92-111 (average 101) exce of olanzapin (185).
DOX, DMDOX.	Human plasma	Add.of IS, 2M NaOH and 2% sol.of isoamyl alcohol in hexane. Vortex. Centr. The aq. phase was frozen and the org. phase into an ampoule with 1% formic acid solution. Vortex and the aq. phase	Phenomenex Luna C_{18} (5 µm, 150×2.1 mm). RT. MP: MeOH-H_2O-0.05% formic acid (600:400:1, v/v/v). IS: Benzoctamine-HCl. FR: 0.25 mL/min.	DOX: 90 DMDOX 5

Table 2: contd….

		inj. on the HPLC.		
CLO, NorCLO.	Blood, hair	Blood: LLE. Mix. with CH_3COONH_4 buff. (pH 5.0) Enzymatic hydrolysis by b-glucuronidase-aryl sulfatase. Neutr. with 0.1M NaOH, add. of tris buff. to pH 9.0. Mix. of the supern. with a mixture of $CHCl_3$ -IPA (3:1, v/v). Separation of the sol. on silicone treated filter paper. Org. phase, evapor. Diss. with MP. Hair: Division into three segments. Decontamination with H_2O and IPA in ultrasonic bath. Pulv. of the samples in mill-ball. Add.of diazepam-D5 and 0.1M HCl and incub. Neutr. with 0.1M NaOH. Add. $(NH_4)_2CO_3$ buff. (pH 9.3). Supern. collected, centr. Add. to the column SPE RP-18. Washing with $(NH_4)_2CO_3$ buff. and evap. El.with a mixture of 0.1% acetate acid-MeOH. Evap. Rediss. in MP.	LiChroCART column (125×3 mm, 5 μm) with Purospher RP 18 and a LiChroCART precolumn (4x4 mm, 5 μm) with LiChrospher 60 RP—select B. MP: [A] 0.1% HCOOH in H_2O and [B] 95% ACN + 5% of the phase [A]. Gradient: 95% [A] and 5% [B] for 2 min, a linear change to 30% [A] and 70% [B] in 30 min, then 30% [A] and 70% [B] for 2 min, change to 95% [A] and 5% [B] for 8 min. IS: prazepam (1.0 μg/g). FR: 0.4 mL/min.	Blood:88-90, Hair: 91-93
AMI, NorTRP, DOX, OPI.	Human plasma	Plasma samples diluted IS sol.and 0.1% formic acid. Vortex. Online Extr. on an Oasis HLB Extr. column (30μm, 1x50 mm).	Symmetry C_{18} (5 μm, 3.0×150 mm) with Sentry guard column Symmetry C_{18} (5 μm, 3.9×20 mm).RT. MP: ACN (A) and 0.1% formic acid (B) (gradient profile: 28% A for 4 min, 70% A in 1 min, isocratic at 70% A for 3 min, 28% A in 0.7 min). IS: lofepramine (10mg/L). FR: 0.6 mL/min.	>90
DOX, DES, IMI, AMI, TRI.	Human plasma.	LLE. Add. of 2 M Na_2CO_3. Extr. with hexane by rotation on a Roto-Rack mixer for 30 min. Centr. for 10 min at 3000 rpm. Org. layer evap. under N_2 at 30 °C. Rediss. in 60 μL of MP.	C_{18} column (15 × 2:1 mm). MP: 3 mM CH_3COONH_4 (pH 3.3)-ACN (66: 34). FR:1.4 ml min^{-1}. Deuterated IS, imipramine-*d*3.	75 for DES >90 for other TCA

Amitriptyline, nortriptyline, E and Z-hydroxy-amitriptyline, E and Z-hydroxy-nortriptyline were determined in rabbit plasma using an analytical column Supelco 5 μm C_{18}, 250 × 4.6 mm. The mobile phase was a mixture of A (acetonitrile-water 10/90, 900 μL of 85% phosphoric acid, 1.22 g of potassium dihydrogen phosphate) and B (acetonitrile). A 20-50% mobile phase B gradient was applied in 21 min. Clomipramine (10 μg/mL) was used as IS. Linearity was observed in the range 5 -1000 ng/mL [28].

A PE Pecosphere-5C silica column was used for the determination of desipramine in rabbit plasma and rabbit bone marrow, with a mixture of methanol:NH_4OH:1 M NH_4NO_3 (190:6:4 v/v/v) as mobile phase and chlorpromazine (1000 pg/mL) as internal standard. Linearity extended from 0.076 to 0.112 μg/mL in rabbit plasma and from 1.196 to 3.764 μg/g in rabbit bone marrow [29].

Desipramine, nortriptyline, imipramine, amitriptyline and clomipramine in human serum were separated on a RP C_8 (5μm, 150 × 4.6 mm) analytical column connected to a 2-cm long Pelliguard LC-8 guard column with 40 μm packing. Mobile phase of acetonitrile-phosphate buffer (pH 3 with phosphoric acid) in a 50:50 (v/v) mixture using clobazam (20 ng/mL) as internal standard provided a linearity from 100 to 500 ng/mL [30].

A Spheri-5 RP-18 (5 μm, 100 × 4.6 mm) analytical column and an RP-18 Newguard cartridge (7 μm, 15 × 3.2 mm) were used for the determination of amitriptyline, nortriptyline, doxepin, imipramine, desipramine, nordoxepin, clomipramine, mianserin, trimipramine in human serum and liver. A mixture of acetonitrile 0.1M-sodium dihydrogen phosphate-diethylamine (40:57.5:2.5 v/v/v) (pH 8.0) was used as mobile phase with cyanopramine as internal standard. Analysis was completed in 40 min. Linearity was observed in the range of 0.20-2.5 mg/L [31].

Doxepin, desmethyldoxepin (E-and Z-isomers) were determined in human serum using a Hypersil MOS C_8 [10 × 4.6 mm, 10 μm] column with a mixture of water:methanol (950:50, v/v) as mobile phase and a Nucleosil 100 CN analytical column (250 × 4.6 mm, 5 μm) with a mixture of methanol:acetonitrile: 0.01 mol/L potassium phosphate buffer, pH 6.8 (188:5778:235, v/v) as mobile phase.

Trimipramine was used as internal standard (140 μg/L). A linear range of 36-500 nmol/L was achieved [32].

Amitriptyline, clomipramine, desipramine, doxepine, imipramine, opipramol and trimipramine were determined in human serum using different analytical columns: Kromasil C_{18} (5 μm, 150 × 2.1 mm), Zorbax C_3 TMS (5 μm, 150 × 4.6 mm) and Nucleosil CN (5 μm, 150 × 4.6 mm). Mixtures of acetonitrile and 0.015 mol/L potassium dihydrogenphosphate buffer (adjusted to the optimal pH with 0.1 mol/L phosphoric acid) were used as mobile phases at different proportions: 40/60, 30/70, 55/45 and different pH values for each column: 6.5, 5.8, 6.4 respectively. Linearity extended from 50 to 300 ng/mL [33].

A Microsorb MV C_{18} column (15 × 0.46 cm, 5 μm) was used for the determination of imipramine and desipramine in mice serum, with a mobile phase of 60% acetonitrile and 40% 0.01 M triethylamine in distilled water (pH 3.0 by addition of 85% phosphoric acid). Clomipramine was used as internal standard. Linearity was observed within the range of 10-1000 ng/mL [34].

Imipramine, desipramine, amitriptyline and nortriptyline in human plasma were separated on a LiChrospher 60 RP-select B (4 × 250 mm, 5 μm) analytical column. Mobile phase consisted of 50% acetonitrile and 50% 0.25N sodium acetate buffer (pH 5.5). Clomipramine in methanol (5 μg/mL) was used as IS. Linear range extended from 20 to 1500 ng/mL [35].

Cis- and trans- doxepin, cis- and trans- desmethyldoxepin in human serum were separated on a Silica column (3 μm, 6 × 100 mm) with a guard column containing 40 μm pellicular silica. Mobile phase consisted of 0.025 M dibasic sodium phosphate (pH 3 with phosphoric acid)/acetonitrile/n-nonylamine (80/20/1, v/v/v). Amoxapine and loxapine (4 mg/mL) were used as internal standards.

Linear ranges were 25-250 ng/mL and 10-175 ng/mL for doxepins and desmethyldoxepins respectively [36].

An RP-C_{18} symmetry analytical column (5 μm, 250 × 4.6 mm) was used for the determination of desipramine, imipramine, nortriptyline, maprotiline, amitriptyline, desmethylclomipramine and clomipramine in human plasma.

Mobile phase consisted of KH_2PO_4 0.067 M (pH 3.0 with H_3PO_4)-acetonitrile (65:35 v/v) with clovoxamine as IS. Linearity was observed within the range 10-3000 ng/mL [37].

Doxepin and desmethyldoxepin was determined in human plasma and urine using a normal phase Spherisorb silica (3 μm, 150 × 4.5 mm) analytical column, with a mobile phase of hexane, methanol and nonylamine (95:5:0.3 v/v/v). Linearity extended from 1 to 200 ng/mL or 400 ng/mL in plasma and urine respectively [38].

Amitriptyline, nortriptyline, imipramine and desimipramine were analysed in human plasma on a cyanopropyl column Supelcosil LC-PCN (5 μm, 150 × 4.6 mm). Mobile phase was formed by acetonitrile-methanol-0.005 M ammonium phosphate buffer, pH 7.0 (70:15:15 v:v). Linearity was observed in the range from 50 to 2000 nmol/L [39].

Amitriptyline, nortriptyline, imipramine, desipramine, clomipramine, doxepin and trimipramine were determined in human plasma using an Ultrasphere C_8 (250 × 4.6 mm) analytical column. A mixture of 35 % CH_3CN mixed with an aqueous phase containing 4 mM 1-octanesulphonic acid and 0.5 mM N,N,N,N-tetramethylethylene diamine (pH 2.5 with H_3PO_4) was used as mobile phase. Amitriptyline (200 ng) was used as IS for samples from patients treated with doxepin, imipramine, or trimipramine and desmethyldoxepin (100 ng) for samples from patients treated with amitriptyline, nortriptyline, dothiepin, or clomipramine. Linearity extended from 20 to 1000 μg/L [40].

Amitriptyline, nortriptyline, imipramine, desipramine, clomipramine and norclomipramine from human serum and plasma were separated within 34 min using a reversed-phase C_{18} analytical column (Waters, Nova-Pak C_{18}, 60A, 4 μm, 4.6 × 150 mm). Mobile phase consisted of 500 mL of 5 mM aqueous KH_2PO_4 buffer, 500 mL acetonitrile and 2 mL diethylamine (pH 8 with phosphoric acid). Methanolic solution of econazole, 90 mg mL^{-1} was used as IS. Linearity extended from 20-400 ng/mL [41].

Clomipramine, demethylclomipramine, 2-, 8-, and 10-hydroxyclomipramine, 2-, and 8-hydroxydemethylclomipramine and didemethylclomipramine were determined in human and rat serum and plasma using a Lichrosher CN (5 μm, 250 × 4.6 mm) analytical column and a clean up column Hypersil CN (10 μm CN, 10 × 4.6 mm). Mobile phase consisted of 38% acetonitrile and 62% sodium perchlorate solution (0.02M) (pH 2.5 with $HClO_4$) [42].

Doxepin, desipramine and imipramine were determined in human plasma using an RP Nova Pack C_{18} column (4 μm, 15 cm × 3.9 mm) and a Nova Pack C_{18} guard column (4 μm, 20 × 3.9 mm). Mobile phase consisted of acetonitrile-0.02M triethylamine (pH 5.5 with H_3PO_4) (35:65 v/v). p-hydroxybenzoic acid n-butyl ester (butyl paraben) was used as IS. Linearity was observed within the range 0.005-2 μg/mL [43].

Nortriptyline was determined in human serum using trimipramine as IS. A Luna C_{18} (3 μm, 150 × 4.6 mm) analytical column equipped with a C_{18} guard column was used with a mobile phase consisted of acetonitrile-0.01 M triethylamine adjusted to pH 3.0 with H_3PO_4 (34:66 v/v). Linearity was observed up to1802 ng/mL [44].

Amitriptyline and nortriptyline were determined in human serum using a Kromasil 5 C_{18} (5 μm, 250 × 4.6 mm) column. Mobile phase was a mixture of SDS (0.15M)-6% pentanol (v/v) (pH 7). The method was linear up to 1000 ng/mL [45].

Clomipramine, N-desmethylclomipramine, 8-hydroxyclomipramine, 2- and 8-hydroxyl-desmethylclomipramine were determined in human plasma using a Lichrospher CN, (5 μm, 250 × 4 mm) analytical column with a 2-cm pre-column filled with the same material. Mobile phase consisted of 10 mM K_2HPO_4-acetonitrile-methanol (35:25:40 v/v/v). Maprotiline was used as IS. Linearity was observed in the range 5-500 ng/mL for clomipramine, 5-500 ng/mL for N-desmethylclomipramine, 5- 100 ng/mL for 8-hydroxyclomipramine and 5-100 ng/mL for 2- and 8- hydroxydesmethylclomipramine [46].

Amitriptyline, nortriptyline, imipramine, desipramine, doxepin, nordoxepin, clomipramine, norclomipramine, trimipramine, mianserine, maprotiline and normaprotiline were determined in human serum using a Nucleosil 100-5-Protect

1 analytical column (250 × 4.6 mm, 5 μm). Mobile phase consisted of 25 mM KH_2PO_4 (pH 7.0)-acetonitrile (60:40 v/v). Melperone (3000 ng/mL) was used as IS [47].

Amitriptyline, clomipramine, desipramine, doxepin, imipramine, noxiptyline and opipramol were analysed in human blood, using a LiChroCART (125 × 4 mm) column packed with octadecylsilica LichroSpher RP Select B (5 μm). Mobile phase consisted of water with orthophosphoric acid and acetonitrile. Opipramol and imipramine were used as IS. Linearity was observed in the range 0.125-2.0 μg/mL [48].

Amitriptyline, opipramol and noxiptyline were determined in urine using an analytical column Shiseido RP-18, (250 × 4.6 mm, 5 μm). Mobile phase consisted of 0.01M sodium phosphate (pH 3.2)-methanol-acetonitrile-dimethylamine (37: 55.4: 7.4: 0.2 v/v). Diethazine was used as IS [49].

Doxepin, nordoxepin, desipramine, imipramine, nortriptyline and amitriptyline were determined in human breast milk using a reversed-phase MOS-2 Hypersil (C_8) analytical column (3 μm, 100 × 2 mm). Mobile phase consisted of 0.02M monobasic potassium phosphate and 85 μL N,N-dimethyloctylamine/L (pH 6.5) and 34% acetonitrile. Trimipramine (2.5 μg/mL) was used as IS. The method was linear up to 800 ng/mL [50].

Amitriptyline, doxepin, clomipramine, trimipramine and their N- demethyl metabolites were determined in human plasma using a Phenomenex Synergi Hydro-RP analytical column (250 × 4.6 mm) with a mobile phase of aqueous ammonium acetate/methanol/acetonitrile. Butriptyline was used as IS. The method was linear from 10 to 550 mg/L [51].

Imipramine, desipramine in human serum were separated on a Kromasil 5 C_{18} analytical column (5 μm, 250 × 4.6 mm) using a mobile phase of 0.15M SDS-6% (v/v) pentanol-0.001M NaCl-0.01M NaH_2PO_4 (pH 7). No internal standard was used. Linearity extended from 50-1000 ng/mL [52].

Imipramine, desipramine, 2- and 10-hydroxyimipramine, as well as 2- and 10-hydroxydesipramine were determined in human plasma and urine using a reversed-phase C_{18} column (Phenomenex Bondclone 10 C_{18}, 300 × 3.90 mm) coupled to a pre-column (RP-18, 10 µm, 40 × 4.6 mm). Mobile phase consisted of 30% acetonitrile in 0.1M K_2HPO_4 buffer (with concentrated orthophosphoric acid, pH 6.0). Pericyazine in methanol (200 ng) was used as IS. Linearity was observed in the range of 15.63-500 ng/mL for imipramine and desipramine and 7.82-250 ng/mL for 2- and 10-hydroxylated metabolites [53].

Imipramine was determined in human serum and plasma using a Nucleosil 100 C_{18} analytical column (5 µm, 125 × 4 mm) using methanol-phosphate buffer (pH 3) (30:70, v/v) as mobile phase. Linearity was observed in the range 2-100 ng/mL [54].

Imipramine, desipramine, clomipramine, amitriptyline, nortriptyline and doxepin were determined in human blood using an Inertsil analytical column (ODS-3 150 × 4.6 mm, 5 µm). Mobile phase was a mixture of acetonitrile and phosphate buffer 0.05M (KH_2PO_4/K_2HPO_4, pH 6.9 ± 0.1) (375:625, v/v) for the detection of all TCAs, except for clomipramine (50:50). Linearity was observed in the range 0.05-100 µM [55].

Imipramine, desipramine, amitriptyline, nortriptyline and clomipramine were determined in human plasma using a trimethylsilyl (TMS) analytical column (5 µm, 150 × 4.6 mm) with a mobile phase of 50 mM sodium phosphate buffer (pH 7.0) and acetonitrile (55:45, v/v). Clomipramine was used as IS. Linearity was observed in the range 0.5-500 ng/mL [56].

Amitriptyline, clomipramine, desipramine, doxepin, imipramine, norclomipramine, nordoxepin, nortriptyline, opipramol and trimipramine were separated in human serum using a Chromolith Speed ROD C_{18} (5 µm, 50 × 4.6 mm) analytical column. Mobile phase was prepared by mixing of methanol and 5 mM acetic acid (pH 3.9). Starting condition was 20% methanol and 80% buffer solution with linear gradient to 70% methanol in 4 min. A mixture containing 50 µg/L clonidine, 10 µg/L dehydromethylrisperidon and 10 µg/L methabenzthiazurone in acetonitrile/methanol (9/1 v/v) was used as IS. Linearity extended from 1 to 10000 ng/mL [57].

Doxepin and desmethyldoxepin were determined in human plasma using a Phenomenex Luna C_{18} analytical column (5 µm, 150 × 2.1 mm) with a mobile phase of methanol-water-0.05% formic acid (600:400:1, v/v/v). Benzoctamine-HCl was used as IS. Linearity was observed in the range 81.1-0.320 ng/mL for doxepin and 45.1-0.178 ng/mL for desmethyldoxepin [58].

Clomipramine and norclomipramine were determined in blood and hair using a LiChroCART column (125 ×3 mm, 5 μm) filled with Purospher RP 18 and a LiChroCART pre-column (4 × 4 mm, 5 μm) filled with LiChrospher 60 RP-select B. Mobile phase was a mixture of [A] 0.1% formic acid in water and [B] 95% acetonitrile + 5% of the phase [A]. The gradient was programmed as follows: 95% [A] and 5% [B] for 2 min, a linear change to 30% [A] and 70% [B] in 30 min, then 30% [A] and 70% [B] for 2 min and then changed to 95% [A] and 5% [B] for 8 min. Prazepam (1.0 μg/g) was used as IS. Linearity was observed in the range 0.25-10μg/g in blood and 0-10 μg/g in hair [59].

The determination of amitriptyline, nortriptyline, doxepin and opipramol in human plasma was achieved on a Symmetry C_{18} (5 μm, 3.0 × 150 mm) analytical column with a Sentry guard column (Symmetry C_{18}, 5 μm, 3.9 × 20 mm). Mobile phase consisted of acetonitrile (A) and 0.1% formic acid (B) using a gradient profile: 28% A for 4 min, 70% A in 1 min, isocratic at 70% A for 3 min, 28% A in 0.7 min. Lofepramine (10 mg/L) was used as IS. Linearity extended in the range 10-800 μg/L for doxepin, amitriptyline and nortriptyline and 50-1500 μg/L for opipramol [60].

Doxepin, desipramine, imipramine, amitriptyline and trimipramine were determined in human plasma by Fast LC-API-TOF MS. The relatively short HPLC separation (18 s) was achieved using a short C_{18} column (15 × 2:1 mm i.d.) SB-C_{18} Mac Mod 15 × 2.1 mm i.d. cartridge packed with 3 mm particles (Hewlett-Packard Analytical). A high aqueous mobile phase of 3 mM ammonium acetate (pH 3.3)-acetonitrile (66:34) was used as mobile phase [61].

Amitriptyline, doxepin, clomipramine and imipramine, in pharmaceutical formulations and biological fluids were determined using a Kromasil C_8 analytical column (250 × 4 mm, 5 μm) with a mobile phase consisting of 0.05 M CH_3COONH_4 and CH_3CN (45:55 v/v) delivered at 1.5 mL/min isocratically. Quantification was performed at 238 nm, with bromazepam (1.5 ng/μL) as internal standard. The determination of TCAs in blood plasma was performed after protein precipitation. Urine analysis was performed by SPE using Lichrolut RP-18 Merck cartridges providing high absolute recoveries (from 91.0 to 114.0%). Direct analysis of urine was also performed after two-fold dilution. The developed method was validated in terms of selectivity, linearity, accuracy, precision, stability and sensitivity. The absolute detection limit of the method was calculated as 0.1 -0.6 ng in blood plasma and 0.2 -0.5 ng in extracted urine or 0.4 - 0.7 in diluted urine. The method was applied to real samples of plasma from a patient under clomipramine treatment [62].

Six benzodiazepines (BZDs) and four tricyclic antidepressants (TCAs) were determined in biological fluids by HPLC with UV detection at 240 nm. After a deproteinization step biological fluids were analyzed by direct injection. SPE on Nexus cartridges was also applied. A sequential SPE protocol has been developed to enable the effective separation of imipramine and diazepam, which were were coeluting. BZDs were eluted by a mixture of methanol/ACN (1:1 v/v), followed by the elution of TCAs with methanol. Separation was performed on a Kromasil C_8 analytical column (250 × 4 mm, 2 id, 5 μm) using a mobile phase of 0.05 M CH_3COONH_4/ACN/methanol (initial composition 55:15:30 v/v/v) at a flow rate of 1.0 mL/min delivered by a gradient program within 15 min. Colchicine was used as the internal standard (4 ng/μL). The method was linear for all analytes up to 20 ng/μL. LODs and LOQs were 0.08 -1.17 and 0.28 -3.91 ng/μL, respectively. Recovery was in the range of 92.8 -108.7% for within-day and 91.9 -109.9% for between-day assays [63].

Simultaneous determination of 1,4-benzodiazepines and tricyclic antidepressants in human saliva and pharmaceutical formulations using colchicines as internal standard was achieved by HPLC using diode array detection (DAD). The analytes and the internal standard were extracted by Nexus (Varian) SPE cartridge and separated by 15 min, utilizing a Kromasil C_8 (150 × 4.6 mm; 5 mm) column with a gradient mobile phase containing methanol, acetonitrile, and 0.05 M ammonium acetate, delivered at a flow rate of 1.0 mL/min. Calibration curves were linear up to the examined range of 20 ng/μL. The limits of detection (LOD) and quantification (LOQ) were 0.08-0.34 and 0.28-1.13 ng/μL, respectively, using 20 μL injected volumes. The mean relative recoveries of all analytes in saliva were

85-105% (n=6) and 86-84 % (n=6) for intra- and inter day, respectively. The sample preparation was very simple and the method was sensitive enough and reproducible. The method was proved to be suitable for use both in clinical analysis, as well as for quality control of pharmaceutical formulations [64, 65].

Ten frequently prescribed tricyclic and nontricyclic antidepressants: imipramine, amitriptyline, clomipramine, fluoxetine, sertraline, paroxetine, citalopram, mirtazapine, moclobemide and duloxetine were analysed by HPLC. A simple and accurate sample preparation step, consisted of liquid:liquid extraction yielded recoveries ranging between 72% and 86%, except for moclobemide (59%). Separation was obtained using a reverse phase Select B column under isocratic conditions with UV detection (230 nm). The mobile phase consisted of 35% of a mixture of acetonitrile/methanol (92:8, v/v) and 65% of 0.25 mol L^{-1} sodium acetate buffer, pH 4.5. The standard curves were linear over a working range of 2.5-1000 ng mL^{-1} for moclobemide, 5-2000 ng mL^{-1} for citalopram, duloxetine, fluoxetine, 10-2000 ng mL^{-1} for sertraline, imipramine, paroxetine, mirtazapine and clomipramine. Limits of quantification were 2.5 ng mL^{-1} for moclobemide, 5 ng mL^{-1} for citalopram, duloxetine and amitriptyline, and 10 ng mL^{-1} for mirtazapine, paroxetine, imipramine, fluoxetine, sertraline, and clomipramine. No interference of the drugs normally associated with antidepressants was observed. The method has been successfully applied to the analysis of real samples, for the drug monitoring of ten frequently prescribed tricyclic and non-tricyclic antidepressant drugs [66].

Imipramine, desipramine, amitriptyline or nortriptyline were determined in plasma by HPLC. A deproteinizing agent, with an acceptable accuracy and sensitivity, to precipitate plasma protein containing imipramine, desipramine, amitriptyline and nortriptyline was investigated along with the HPLC condition for resolving each of these compounds. The developed method was verified by performing a bio-analytical method validation, and by analyzing plasma samples from volunteers who received imipramine or amitriptyline. The mobile phase comprised of acetonitrile and 70 mM phosphate buffer (pH 6.1) (60/40 v/v). Separation was achieved on a C_{18} column, and the effluent was monitored at 251 nm or by EC detection at +1.0 V of glassy carbon against silver/silver chloride reference electrode. The chromatographic separation was excellent, without interference from endogenous plasma constituents. All compounds were resolved in a run time of 12 min. The method was suitable for quantifying drug concentrations in the ranges of 40-900 ng/mL for UV detection or 4-900 ng/mL for EC detection. The method proved to be efficient and practical for determining the plasma concentrations of each compound, following the administration of imipramine or amitriptyline to volunteers. The authors concluded that the developed method is very simple, rapid and inexpensive and is practical for routine therapeutic drug monitoring and forensic toxicological screening [67].

Amitriptyline, desipramine, imipramine, and nortriptyline were determined in serum. Following a protein precipitation from serum and dilution of supernatant with water, these analytes and their internal standards (deuterated versions of each drug) were injected, separated, and eluted from a Hypersil Gold C-18 (50 × 2.1 mm) analytical column with a gradient of water and acetonitrile each with 0.1% formic acid. Analytes were then ionized and detected over a 3.5 min analysis time by electrospray ionization mass spectrometry with multiple reaction monitoring. The limit of detection for all drugs was < 15 ng/mL and the limit quantitation for all drugs was < 22 ng/mL. Recoveries were between 97 and 131% for all examined analytes. Patient method comparison and proficiency samples were run with acceptable results. The authors concluded that the method is specific and sensitive and can be used for the rapid quantitation of tricyclic antidepressants in serum, suitable for use in the clinical laboratory [68]. Seven tricyclic antidepressants (TCAs) and seven metabolites were determined in human plasma. The analyte separation was obtained using a C_8 reversed phase column and a mobile phase composed of 68% aqueous phosphate buffer at pH 3.0 and 32% ACN. The UV detector was set at 220 nm and loxapine was used as the internal standard. A careful pre-treatment procedure for plasma samples was developed, using SPE on C_2 cartridges, which gave satisfactory extraction (>80%) and good sample purification. The LOQs were always lower than 9.1 ng/mL and the LODs always lower than 3.1 ng/mL for all analytes. The method was successfully applied to plasma samples from depressed patients undergoing therapy with one or more TCA drugs. Recovery was better than 80% and no interference from other drugs was found. Th authors concluded that the method is suitable for the therapeutic drug monitoring of patients treated with TCAs under monotherapy or polypharmacy regimens [69].

A typical chromatographic separation of four TCAs on a Kromasil C_{18} (5 μm, 250 × 4.0 mm) analytical column is presented in Fig. (**4**).

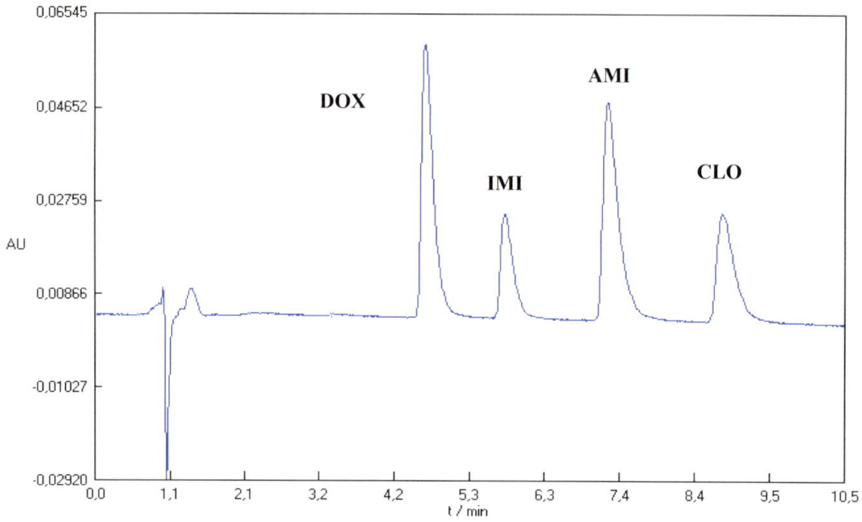

Figure 4: HPLC separation of AMI, IMI, DOX, CLO. Data from authors' laboratory. Unpublished results.

3.3. Other Techniques

Other techniques used for the analysis of TCAs in biofluids include voltammetry with lipid - coated electrodes [70], derivative spectrometry using both first- and second-derivative spectra [71], Fluorescence Polarization Immunoassay [32, 72], Gas Chromatography [73], GC - MS [74, 75], Non Aqueous Capillary Electrophoresis [76-78], Capillary Electrophoresis With Electrochemiluminescence detection based on end-column reaction of tris-(2,2_-bipyridyl)ruthenium(II) with aliphatic tertiary amino moieties [79]. These methods are briefly presented in Table **3**.

3.4. Sample Preparation

Various sample preparation techniques have been used for the isolation of TCAs from biological matrices. These include Liquid-Liquid Extraction (LLE), Solid Phase Extraction (SPE) and Solid Phase Microextraction (SPME), however direct methods after dilution, protein precipitation or dialysis have been also proposed.

Table 3: Overview of analytical methods for the determination of tricyclic antidepressants in biofluids

Analytes	Sample Type	Analytical Technique	Sample Preparation	Recovery (%)	Linear Range(LR)-LOD	Refs.
IMI, AMI, DES, NorTRP, CLO, DOX, OPI, TRI	Human serum	Fluorescence polarization immunoassay	Dilution		LR:50 - 1000 μg/L (IMI, CLOMI, DOX, AMI), 50 - 500 μg/L (DESI, NORTRP).	[32]
DOX, DES, IMI, AMI, TRI	Human plasma	Fast LC-API-TOF MS	LLE (hexane)		LR: 2 100 ng/mL (DESI), 2 – 50 ng/mL (the other four TCAs). LOD: 2 ng/mL (DESI) and 1 ng/mL (the other four TCAs).	[61]
DES, IMI, TRI	Urine	Voltammetry with lipid - coated electrodes.	Dilution		LR:1-8x10^{-7}M. LOD: 1x10^{-8}M.	[70]
IMI, AMI	Blood serum	Derivative spectrophotometry	LLE (n-hexane)		LR: 0.62 - 10.14 μg/mL (IMI) and 0.63-10.04 μg/mL (AMI).	[71]
IMI	Human serum	Fluorescence polarization immunoassay.	Dilution	97.1-102.0	LR:0.20 - 3.0 ng/mL	[72]
IMI, AMI, TRI, CLO	Urine, plasma, whole blood	GC - SID.	SPE (C_{18} cartridges).		LR: 10 - 80pg. LOD: 5 - 10pg.	[73]
IMI, DES	Human plasma	Capillary GC - MS with D4 - IMI and D4 - DES as IS. Mass selective detector at *m/z* 234 for IMI, *m/z* 238 for D4 - IMI, *m/z* 412 for DES and *m/z* 416 for D4 - DES.	LLE (*n*-heptane-isoamyl alcohol (99:1, v/v)	IMI:91-99 DES:94-103	IMI: LR:0.580-116 ng/mL DES: LR:0.545- 109 ng/mL	[74]
AMI, TRI, IMI, DOX	Hair	GC - MS.	Digestion and SPME.		AMI:LR:0.22 - 10.4 ng/mg. LOD: 0.05 ng/mg, DOX, TRIMI:0.2 ng/mg AMI: LOQ: 0.15 ng/mg, DOX,TRIMI:0.7 ng/mg	[75]
AMI, DES, norTRP, IMI	Urine and human serum	Non aqueous Capillary electrophoresis	Dialysis - SPE (20×2 mm I.D. SPE cartridge, C_2 -bonded silica)	AMI, DES: 60-70, norTRP: 80 IMI:95	LODs: in the 40 – 80 ng/mL range for urine, and the 60 – 100 ng/mL range for serum.	[76]

Table 3: contd…

IMI, AMI, norTRP	Human plasma	Non aqueous Capillary electrophoresis	LLE. (hexane/isoamyl alcohol 99:1 (v/v)	97.0-103.9.	LR:50 – 500 ng/mL. LOD: AMI and NOR: 20 ng/mL, IMI and DES: 30 ng/mL. LOQ: AMI and NOR: 30 ng/mL, IMI and DES: 50 ng/mL.	[77]
IMI, DES	Urine	Non aqueous Capillary electrophoresis	SPE (C_{18})	DESI: 95.1 ± 4.2 - 99.6 ± 4.7. IMI: 96.2 ± 4.7 - 101.2 ± 5.0.	LR:0.05 - 1.0 mg/L. LOD: DES: 15.0 µg/L, IMI: 11.0 µg/L. LOQ: DES: 50.0 µg/L, IMI: 35.0 µg/L.	[78]
AMI, DOX	Urine	Capillary Electrophoresis with electrochemiluminescence detection.	LLE (heptane:ethyl acetate = 90:10,v/v)	83 - 93.	LR:5.0 - 800 ng/mL. LODs: 0.8 ng/mL (AMI), 1.0 ng/mL (DOX)	[79]

Doxepin and desmethyldoxepin (E-and Z-isomers) were determined in human serum samples directly after addition of methanol containing the IS [32].

Clomipramine, demethylclomipramine, 2-, 8-, and 10-hydroxyclomipramine, 2-, and 8-hydroxydemethylclomipramine, didemethylclomipramine were determined in human and rat serum and plasma after removal of proteins and other interfering compounds by washing to waste using deionized water containing 35% (v/v) acetonitrile and centrifugation. Recovery ranged from 64 to 110% [42].

Amitriptyline and nortriptyline were determined in human serum directly after dilution (1:10) in 0.15 M SDS-6% pentanol at pH 7, and filtration through 0.45 µm nylon membranes. High recovery rates were obtained: 99.8-101.6% for amitriptyline and 98.5-99.7% for nortriptyline [45].

Imipramine, desipramine, clomipramine, amitriptyline, nortriptyline and doxepin were determined in human blood after protein precipitation with $HClO_4$. Average recoveries obtained after neutralization of the supernatant with NaOH, filtration and direct injection into the HPLC column were 92.3 and 90.8% for desipramine and imipramine, respectively [55].

Imipramine and desipramine were determined in human serum directly after filtration (0.45 µm nylon membranes) [52].

Amitriptyline, clomipramine, desipramine, doxepin, imipramine, norclomipramine, nordoxepin, nortriptyline, opipramol and trimipramine were determined in human serum after protein precipitation and dilution with mobile phase. Recoveries ranged from 92 to 111% [57].

Simple dilution was performed for the determination of desipramine, imipramine and trimipramine in urine [70], for determination of imipramine in human serum [72] and imipramine, amitriptyline, desipramine, nortriptyline, clomipramine, doxepin, opipramol and trimipramine in human serum [32].

The vast majority of the published methods involve LLE with different extraction media. Desipramine, nortriptyline, imipramine, amitriptyline and clomipramine were determined in human serum using n-hexane yielding recovery rates in the range from 93.2 to 110.6% [30]. Hexane was also used for imipramine and desipramine determination in mice serum (recovery >89.6%) [34], for desipramine, imipramine, nortriptyline, amitriptyline, desmethylclomipramine and clomipramine in human plasma (recovery >80%) [37], for amitriptyline, nortriptyline, imipramine, desipramine, clomipramine and norclomipramine in human serum and plasma (recovery 92-105%) [41], for imipramine and amitriptyline from blood serum [71], and for doxepin, desipramine, imipramine, amitriptyline and trimipramine in human plasma with recovery rates 75% for desipramine and >90% for the other four tricyclic amines [61].

Mixtures of hexane with isoamyl alcohol in different ratios (98:2 or 99:1) were used for the extraction of desipramine from rabbit plasma and rabbit bone marrow (with low recovery rates: 24.2-30.2% in rabbit plasma and 24.5-34.2% in rabbit bone marrow) [29], as well as for the extraction of different TCAs from human plasma: imipramine, desipramine, amitriptyline and nortriptyline (with recovery 89.6-103.2%) [35], amitriptyline, nortriptyline, imipramine, desipramine, clomipramine, doxepin and trimipramine [40], doxepin, and desmethyldoxepin using (with recovery of 90% (doxepin) and 75% (desmethyldoxepin) [58], amitriptyline, clomipramine, desipramine, doxepin, imipramine, noxiptyline and opipramol (recovery 75-90% except for amitriptyline (67%) and opipramol (47%) [48] and imipramine, amitriptyline and nortriptyline (recovery 97.0-103.9%) [77].

Amitriptyline, nortriptyline, doxepin, imipramine, desipramine, nordoxepin, clomipramine, and trimipramine were isolated from human serum and liver using hexane-butan-1-ol (95:5 v/v). More than 60% recovery was obtained for human serum and more than 40 % for liver [31].

A mixture of n-heptane, ethylacetate and isoamyl alcohol was used for the extraction of amitriptyline, clomipramine, demexiptiline, desipramine, doxepine, imipramine, opipramol and trimipramine from human serum [33], a mixture of isoamyl alcohol and heptane for cis- and trans- doxepin, cis- and trans-desmethyldoxepin isolation from human serum (65-75% recovery for desmetyldoxepins and 75-85% for doxepins) [36] and for imipramine and desipramine from human plasma after derivatization to their pentafluoropropionyl derivatives (recovery 91-99% for imipramine and 94-103% for desipramine) [74], and a mixture of heptane and ethyl acetate for doxepin, desmethyldoxepin extraction from human plasma (recovery 61-64%) and urine (recovery 63-68%) [38], for amitriptyline, nortriptyline, E and Z-hydroxy-amitriptyline, E and Z-hydroxy-nortriptyline isolation from rabbit plasma [28] and for amitriptyline and doxepin from urine (recovery 83 - 93%) [79].

Diethyl ether was used for extraction of imipramine, desipramine, 2- and 10-hydroxyimipramine, 2- and 10-hydroxydesipramine from human plasma and urine (recovery: 78.6-94.3% (plasma), 10.24-28.80% (urine) [53], for imipramine from human serum (recovery 87.6%) and plasma (recovery 86.2%) [54], and for imipramine, desipramine, amitriptyline, nortriptyline, clomipramine from human plasma with diethyl ether 83.0-93.8% [56].

Amitriptyline, nortriptyline, imipramine, trimipramine, clomipramine and doxepine were extracted from human blood and urine using Toxi-Tube A and Toxi-Tube B from Toxi-Lab (Irvine, CA, USA). These tubes are designed for use with 5-mL urine samples, containing sodium carbonate and bicarbonate to give a pH of 9.0 in a mixture of dichloromethane and dichloroethane [26].

Clomipramine and norclomipramine were determined in blood by LLE with a mixture of chloroform and isopropanol (3:1, v/v) and separation of the solution on silicone treated filter paper Whatman 1PS. Recovery obtained was 88-90% [59].

Solid Phase Extraction was applied to the isolation of TCAs from biomatrices using various sorbents. Mixed-mode SPE (IST Isolute HCX, 80 mg/1 mL) cartridges were used for the extraction of amitriptyline, doxepin, clomipramine, trimipramine and their N- demethyl metabolites from human plasma, [519] while C_{18} cartridges for extraction of imipramine, amitriptyline, trimipramine and chlorimipramine from urine, plasma, and whole blood [73].

Bond Elut Certify (Varian Sample Preparation Products, Harbor City, CA) SPE columns were used for the extraction of amitriptyline, imipramine, trimipramine, nortriptyline, protriptyline and doxepin from human serum and urine. SPE columns were washed with methanol and phosphate buffer (pH 6.0). Elution was performed with a mixture of 100% methanol and 10% ammonia (5:1 v/v). Enhancement of sensitivity was achieved by evaporation of the SPE extract after addition of a few drops of HCl [27].

C_{18} Bond-Elut cartridges were used for doxepin, desipramine, maprotiline and imipramine extraction from human plasma. Mixtures of methanol-aqueous acetate buffer were used as washing and elution solvents. The recoveries of the drugs using other sorbent materials (C_8, C_2, cyclohexyl, cyanopropyl and phenyl Bond Elut as well as copolymer HLB waters cartridges) were also examined. Mean absolute recoveries were 82-84% [43].

SPE on cyanopropyl cartridges Isolute SPE column from International Sorbent Technology LTD (Hengoed, Mid Glamorgan, UK), were used for the extraction of nortriptyline and trimipramine from human serum using ASPEC XL automatic SPE apparatus by Gilson with high recoveries (93-98 %). LLE was used in comparison using heptane/isoamylalcohol (98.5:1.5, v/v) with recovery 75 ± 13 [44].

SPE with an Isolute C_2 column was used for the extraction of clomipramine, N-desmethylclomipramine, 8-hydroxyclomipramine, 2- and 8- hydroxydesmathyl-clomipramine from human plasma. Before applying the sample, the columns were washed with methanol/water/phosphate buffer (pH 9.2) (2/1/2 v/v/v). Interference were washed out with phosphate buffer/mixture of water and acetonitrile (80:20 v/v). Elution with methanol yielded high recovery rates 99.1-100.3% [46].

Amitriptyline, nortriptyline, doxepin and opipramol were determined in human plasma by on-line extraction on an Oasis® HLB extraction column (30 μm, 1 × 50 mm) with recovery >90% [60].

RP-18 SPE cartridges were used for the extraction of clomipramine and norclomipramine from hair. The samples after division into three segments were pulverized in mill-ball and incubated with diazepam-D5 and 0.1M hydrochloric acid. After neutralization with 0.1M NaOH and treatment with ammonium carbonate buffer (pH 9.3), the supernatants were collected, centrifuged and added to the column. Washing with ammonium carbonate buffer and elution with a mixture of 0.1% acetate acid-methanol provided recovery of 91-93% [59].

Imipramine and desipramine were determined in urine on C_{18} (Waters Sep-Pak Plus, Milford, MA, USA) cartridges yielding recovery rates 95-101% [78].

A combination of LLE and SPE was applied for the extraction of doxepin, nordoxepin, desipramine, imipramine, nortriptyline and amitriptyline from human breast milk. LLE was performed using 2% butanol in hexane. SPE C_{18} cartridges were conditioned with diethylamine in methanol, followed by 1% potassium bicarbonate in 10% acetonitrile in water. After sample application benzodiazepines were washed off by a mixture of water-methanol-acetonitrile, while elution of TCAs was performed with diethylamine in methanol yielding recovery in the range 85.8-104.6% [50].

3M-Empore high-performance extraction disk cartridges (Varian, Darmstadt, Germany) were used for the extraction of amitriptyline, nortriptyline, imipramine, desipramine, doxepin, nordoxepin, clomipramine, norclomipramine and trimipramine from human serum. The sorbent was conditioned with methanol followed by water. Elution with 2-propanol-ammmonia solution (25%)-dichloromethane (20:2:78) yielded 75-100.2% recovery [47].

Membrane technologies were used for the determination of amitriptyline, nortriptyline, imipramine and desimipramine in human plasma after mixing with Protein Releasing Reagent [PRR: an aqueous solution consisting of 1M HCl and 25% (v:v) glycerol]. The mixture was injected into the donor channel of the dialysis cell while the acceptor solution (0.001M ammonium phosphate buffer, pH 7.0) was transported to the acceptor channel. The enriched analytes were eluted onto the analytical column by the HPLC mobile phase. The donor side and the acceptor side of the dialyser were simultaneously washed with the donor solution (1 mM dodecyltrimethyl ammonium bromide in water) and acceptor solution, respectively. The precolumn was regenerated with the acceptor solution and the next sample was injected into donor channel of the dialysis cell [39].

A membrane separator was also used for the extraction of amitriptyline, opipramol and noxiptyline from urine. Sample solutions containing sodium phosphate buffer (pH 9.0) were passed over the liquid membrane in membrane separator, which was made of two PTFE blocks with machined spiral grooves into a rigid construction. The membrane was impregnated in n-undecane by soaking. The membrane separated two channels: the donor channel for extraction of analytes and IS and the acceptor channel with acidic solution for re-extraction of analytes from the membrane solvent. Sample of acceptor solution was injected into the HPLC column [49].

On-line combination of Dialysis and SPE (20 × 2 mm I.D. SPE cartridge, C_2 - bonded silica was used for the determination of amitriptyline, desipramine, nortriptyline and imipramine in urine and human serum yielding recovery rates of 60-70% for amitriptylin, desipramine, 80% for nortriptyline and 95% for imipramine [76].

Solid phase microextraction was applied to the extraction of amitriptyline, trimipramine, imipramine and doxepin from hair after digestion [75].

Solid-phase microextraction (SPME)-liquid chromatography (LC) was used to analyze tricyclic antidepressant drugs desipramine, imipramine, nortriptyline, amitriptyline, and clomipramine (internal standard) in plasma samples. A Polydimethylsiloxane-divinylbenzene (60-µm film thickness) fiber was selected after the assessment of different types of coating. The chromatographic separation was performed using a C_{18} column (150 × 4.6 mm, 5-µm), ammonium acetate buffer (0.05 mol/L, pH 5.50)-acetonitrile (55:45 v/v) with 0.1% of triethylamine as mobile phase and UV-vis detection at 214 nm. Among the factorial design conditions evaluated, the best results were obtained at a pH 11.0, temperature of 30°C, and extraction time of 45 min. The proposed method, using a lab-made SPME-LC interface, allowed the determination of tricyclic antidepressants in in plasma at therapeutic concentration levels [80].

A microwave-assisted extraction (MAE) method has been developed and optimized for the extraction of six tricyclic antidepressants (TCAD; nordoxepin, nortriptyline, imipramine, amitriptyline, doxepin, dezypramine) from human serum. Optimal parameters of MAE (solvent and extraction temperature) for water solution of these drugs were defined. The microwave-assisted procedure developed was validated by extraction of serum samples at two concentration levels and then successfully applied to the analysis of reference material. Limit of quantification was in the range 0.04-0.15 μg/mL, while recovery 94-105% [81].

Microextraction in a packed syringe (MEPS) has been proposed prior to the analysis by liquid chromatography (LC) and gas chromatography-mass spectrometry (GC-MS) for the simultaneous determination of antidepressant drugs (amitriptyline, imipramine, clomipramine, mirtazapine and citalopram) in human plasma and urine samples. MEPS is a miniaturized, solid-phase extraction (SPE) technique that can be connected to LC or GC without any modifications. It is very easy to use, fully automated, inexpensive, quick and uses very small amounts (in μL) of organic solvents. Limits of detection were in the range of 0.088-0.284 ng mL^{-1} on GC-MS and between 0.133 and 0.360 ng mL^{-1} on LC-UV [82].

Microwave irradiation, for the analysis of eight tricyclic antidepressants (TCADs): nordoxepin, nortriptyline, imipramine, amitriptyline, doxepin, desipramine, clomipramine, and norclomipramine, human hair samples gas been effectively used. The developed method was based on simultaneous alkaline hair microwave-assisted hydrolysis and microwave-assisted extraction (MAH-MAE). Extracts were analyzed by high-performance liquid chromatography with diode-array detection (HPLC-DAD). A mixture of n-hexane and isoamyl alcohol (99:1, v/v) was used as extraction solvent and the process was performed at 60°C. Application of 1.0 mol L^{-1} NaOH and microwave irradiation for 40 min were found to be optimum for hair samples. Limits of detection ranged from 0.3 to 1.2 μg g^{-1} and LOQ from 0.9 to 4.0 μg g^{-1} for the different drugs. The method was suitable for analytes quantification in hair samples within average therapeutic concentration ranges [83].

4. CONCLUSIONS

Tricyclic antidepressants consist in a homogeneous group of drugs differing mostly in their potency to inhibit presynaptic norepinephrine or serotonin uptake and in their tendency for causing variety of unwanted effects. Between 1960 and 1980 tricyclic antidepressants (TCAs) represented the major pharmacological treatment for depression. Despite the introduction of newer and safer antidepressants the prescription of tricyclic antidepressants is still widespread as they are cheaper and are still considered as the most effective group of antidepressants. For this group of drugs, distinct ranges of optimal plasma concentration for therapy are required due to the wide range of inter-individual variability in metabolism and elimination. Lower concentrations are associated with sub-optimal effects, while toxicity can appear at high concentrations. They are frequently encountered in emergency toxicology screening, drug abuse testing and forensic medical examinations. Therefore, the analysis of these compounds is important not only for quality assurance in pharmaceutical preparations, but for obtaining optimum therapeutic concentrations in order to minimize the risk of toxicity as well.

Moreover, continued progress in understanding the neurobiology of antidepressant drugs will lead to further identification of the phenomenon of how the drugs act and work and development of more effective and faster acting therapeutic agents. Analytical methods presented in this review prove that HPLC is an efficient fast and reliable tool in medicinal chemistry for the multi-component analysis of TCAs in clinical samples.

ACKNOWLEDGEMENTS

Declared none.

CONFLICT OF INTEREST

The authors confirm that this chapter contents have no conflict of interest.

DISCLOSURE

The chapter submitted for eBook Series entitled: "**Recent Advances in Medicinal Chemistry, Volume 1**" is an update of our article published in **Mini-Reviews in Medicinal Chemistry, Volume 8, Number 3, pp. 256 to 275**, with additional text and references.

ABBREVIATIONS

ACN = Acetonitrile

Activ. = Activated

Add. = Addition

Adj. = Adjusted

Alkal. = Alkalisation

AMI = Amitriptyline

API = Atmospheric Pressure Ionization

buff. = Buffer Solution

CE = Capillary Electrophoresis

Centr. = Centrifugation

Ch. = Channel

CH_2Cl_2 = Methylene Chloride

CLO = Clomipramine

Conc. = Concentrated

Deion. = Deionized

Deprot.	=	Deproteinization
DES	=	Desipramine
Det.	=	Detection
Dil.	=	Dilution
Diss.	=	Dissolution
El.	=	Eluent/Elution
Electr.	=	Electrochemical
ESI	=	Electrospray Ionization
Evap.	=	Evaporation To Dryness
Extr.	=	Extraction
Filtr.	=	Filtration
FL	=	Fluorescence
FR	=	Flow Rate
Homog.	=	Homogenization
5-HT	=	5-Hydroxytryptamine
IMI	=	Imipramine
Incub.	=	Incubation
Inj.	=	Injection
IPA	=	Isopropanol
IS	=	Internal Standard

LLE = Liquid Liquid Extraction

LR = Linear Range

MAOIs = Monoamine Oxidase Inhibitors

MAPR = Maprotiline

MeOH = Methanol

Mix. = Mixing

MP = Mobile Phase

Neutr. = Neutralization

NorCLO = Norclomipramine.

NACE = Non-Aqueous Capillary Electrophoresis

NE = Nor-Epinephrine

NorTRP = Nortriptyline

Org. = Organic

OPI = Opipramol

FPIA = Fluorescence Polarization Immunoassay.

Prec. = Precipitation

ProTRP = Protriptyline

Pulv. = Pulverised

Resp. = Respectively

RT = Room Temperature

SDS = Sodium Dodecyl Sulfate

Sol. = Solution

SPE = Solid Phase Extraction

SPME = Solid Phase Microextraction

Supern. = Supernatant

SID = Surface Ionization Detection

SSRIs = Selective Serotonin Reuptake Inhibitors

TCAs = Tricyclic Antidepressants

TEA = Triethylamine

Temp. = Temperature

TRI = Trimipramine

OF = Time-Of-Flight

REFERENCES

[1] http://home.blarg.net/~charlatn/depression/tricyclic.faq.html

[2] http://en.wikipedia.org/wiki/Antidepressant, 28/10/2006.

[3] Scoggins, B.A.; Maguire K. P.; Norman, T.R.; Burrows, G.D. Measurement of tricyclic antidepressants. Part I. A review of methodology. *Clin. Chem.* **1980**, *26/1*, 5-17.

[4] Norman, T. R. ReviewAnalysis of tricyclic antidepressant drugs in plasma and serum by chromatographic techniques.*J. Chromatogr.*, **1985,** *340,* 173-197.

[5] Smyth, W.F. Recent studies on the electrospray ionisation mass spectrometric behaviour of selected nitrogen-containing drug molecules and its application to drug analysis using liquid chromatography–electrospray ionisation mass spectrometry *J. Chromatogr. B,* **2005**, *824*, 1-20.

[6] Maurer, H. H. Multi-analyte procedures for screening for and quantification of drugs in blood, plasma, or serum by liquid chromatography-single stage or tandem massspectrometry (LC-MS or LC-MS/MS) relevant to clinical and forensic toxicology. *Clin. Biochem.* **2005**, *38*, 310-318.

[7] Kerr, G.W.; McGuffie, A.C.; Wilkie, S. Tricyclic antidepressant overdose: a review. *Emerg. Med. J.* **2001**, *18*, 236-241.

[8] Samanidou, V.; Nika, M.; Papadoyannis, I. HPLC as a tool in medicinal chemistry for the monitoring of tricyclic antidepressants in biofluids. *Mini Rev. Med. Chem.*, **2008**, *8*, 256-275.

[9] Samanidou, V.F.; Uddin, MN.; Papadoyannis, I.N. Bio-sample preparation and analytical methods for the determination of tricyclic antidepressants. *Bioanalysis*, **2011**, *3*, 97-118.

[10] Ruiz-Angel, M.J.; Carda-Broch, S.; Simó – Alfonso, E.F.; Garcia – Alvarez –Coque, M.C. Optimised procedures for the reversed-phase liquid chromatographic *analysis* of formulations containing tricyclic antidepressants. *J. Pharm. and Biomed. Analysis,* **2003**, *32,* 71-84.

[11] http://www.nlm.nih.gov/medlineplus/druginfo/antidepressantstricyclicsystem202055.html, 7/05/2012.

[12] Ivandini, T.A.; Sarada, B.V; Terashima, C; Rao, T.N.; Tryk, D.A.; Ishiguro, H.; Kubota, Y.; Fujishima, A. Spectroscopic and Electrochemical Analysis of Psychotropic Drugs**.** *J. Electroanal. Chem.,***2002**, *521*,117-126.

[13] Joron, S.; Robert, H. Simultaneous determination of antidepressant drugs and metabolites by HPLC. Design and validation of a simple and reliable analytical procedure. Biomed. Chromatogr., **1994**, *8,* 158-164.

[14] http://www.psymed.com/tca.html, 27/09/2006.

[15] www.cnsforum.com/.../Drug_TCA_efficacy.png

[16] http://www.patient.co.uk/showdoc/23068678/, 7/05/2012.

[17] http://www.pharmgkb.org/do/serve?objId=1097&objCls=DrugProperties

[18] www. emj.bmjjournals.com, 27/09/2006.

[19] www.scholar.google.com, antidepressants_tricyclic 7/05/12

[20] Steimer, W.; Muller, B.; Leucht, S., Kissling, W. The CYP2D6 polymorphism in relation to the metabolism of amitriptyline and nortriptyline in the Faroese population. *Clin. Chim. Acta,* **2001**, *308*, 33-41.

[21] Rudorfer, M.V.; Potter, W.Z. Metabolism of tricyclic antidepressants. *Cell. and Mol. neurobiol.,* **1999**, *19(3),* 373-409.

[22] Poolsup, N.; Li Wan Po, A.; Knight, T.L. Pharmacogenetics of antipsychotic therapy: pivotal research issues and the prospects for clinical implementation. *J. Clin. Pharm. Ther.,* **2000**, *25,* 197-220.

[23] Burrows, G.D.; Norman, T. Antidepressants: Clinical Aspects. *Str. Med.,* **1997**, *13,* 167-172.

[24] http://pediatrics.uchicago.edu/chiefs/documents/TCA.pdf, 7/05/2012.

[25] Preskorn, S.H.; Dorey, R.C.; Jerkovich, G.S. Therapeutic drug monitoring of tricyclic antidepressants. *Clin. Chem.*, **1988***, 34(5),* 822-828.

[26] Gaillard, Y.; Pepin, G. Use of high-performance liquid chromatography with photodiode-array UV detection for the creation of a 600-compound library. Application to forensic toxicology. *J. Chromatogr. A,* **1997***, 763,* 149-163.

[27] Lai, C.; Lee, T.; Au, K.; Chan, A.Y. Uniform solid-phase extraction procedure for toxicological drug screening in serum and urine by HPLC with photodiode-array detection.*Clin. Chem.,* **1997***, 43:2,* 312-325.

[28] Abaut, A.-Y.; Chevanne, F.; Le Corre, P.; Oral bioavailability and intestinal secretion of amitriptyline: role of P-glycoprotein? *Inter. J. Pharmac.,* **2007**, *330(1-2),* 121-128.

[29] Winek, C.L.; Westwood, S.E.; Wahba, W.W. Plasma versus bone marrow desipramine: A comparative study. *For. Scien. Int.,* **1990***, 48,* 49-57.

[30] Segatti, M.; Nisi, G.; Grossi, F.; Mangiarotti, M.; Lucarelli, C. Rapid and simple high-performance liquid chromatographic determination of tricyclic antidepressants for routine and emergency serum analysis. *J. Chromatogr.*, **1991**, *536*, 319-325.

[31] Mclntyre, I.M.; King, C.V.; Skafidis, S.; Drummer, O.H. Dual ultraviolet wavelength high-performance liquid chromatographic method for the forensic or clinical analysis of seventeen antidepressants and some selected metabolites. *J. Chromatogr. Biomed. Appl.*, **1993,** *621,* 215-223.

[32] Rao, M.L.; Staberock, U.; Baumann, P.; Hiemke, C.; Deister, A.; Cuendet, C.; Amey, M.; Hartter, S.; Kraemer M. Monitoring Tricyclic Antidepressant Concentrations in Serum by Fluorescence Polarization Immunoassay Compared with Gas Chromatography and HPLC.*Clin. Chem.*, **1994**, *40(6),* 929-933.

[33] Joron, S.; Robert, H. Simultaneous determination of antidepressant drugs and metabolites by HPLC. Design and validation of a simple and reliable analytical procedure. *Biomed. Chromatogr.*,**1994**, *8,* 158-164.

[34] Yoo, S. D.; Holladay, J. W.; Fincher, T.K.; Dewey, M.J. Rapid microsample analysis of imipramine and desipramine by reversed-phase high-performance liquid chromatography with ultraviolet detection. *J. Chromatogr. B,* **1995**, *668,* 338-342.

[35] Queiroz, R.H.C.; Lanchote, V.L.; Bonato, P.S.; Carvalho, D. Simultaneous HPLC analysis of tricyclic antidepressants and metabolites in plasma samples. *Pharm. Acta Helv.,* **1995**, *70,* 181-186.

[36] Adamczyk, M.; Fishpaugh, J.R.; Harrington, C. Quantitative determination of E- and Z-doxepin and E- and Z-desmethyldoxepin by high-performance liquid chromatography.*Ther. Drug Monit.*, **1995**, *17,* 371-376.

[37] Aymard, G.; Livi, P.; Pham, Y.H.; Diquet B. Sensitive and rapid method for the simultaneous quantification of five antidepressants with their respective metabolites in plasma using high-performance liquid chromatography with diode-array detection. *J. Chromatogr. B,* **1997**, *700,* 183-189.

[38] Yan, J.; Hubbard, J.W.; MacKay, G.; Midha, K.K. Stereoselective and simultaneous measurement of cis- and trans-isomers of doxepin and N-desmethyldoxepin in plasma or urine by high-performance liquid chromatography. *J. Chromatogr. B,* **1997**, *691,* 131-138.

[39] Johansen, K.; Rasmussen, K.E. Automated on-line dialysis for sample preparation and HPLC analysis of antidepressant drugs in human plasma. Inhibition of interaction with the dialysis membrane. *J. Pharm. Biomed. Analysis,* **1998**, *16,* 1159-1169.

[40] Hackett, L.P.; Dusci, L.J.; Ilett, K.F. A comparison of high-performance liquid chromatography and fluorescence polarization immunoassay for therapeutic drug monitoring of tricyclic antidepressants.*Ther. Drug Monit.*, **1998**, *20(1),* 30-34.

[41] Theurillat, R.; Thormann W. Monitoring of tricyclic antidepressants in human serum and plasma by HPLC: characterization of a simple, laboratory developed method via external quality assessment. *J. Pharm. Biomed. Analysis,***1998**,*18,* 751-760.

[42] Weigmann, H.; Hartter, S.; Hiemke, C. Automated determination of clomipramine and its major metabolites in human and rat serum by high-performance liquid chromatography with on-line column-switching. *J. Chromatogr. B,* **1998**, *710,* 227-233.

[43] Bakkali, A.; Corta, E.; Ciria, J.I.; Berrueta, L.A.; Gallo, B.; Vicente, F. Solid-phase extraction with liquid chromatography and ultraviolet detection for the assay of antidepressant drugs in human plasma. *Talanta,* **1999**, *49,* 773-783.

[44] Vendelin Olesen, O.; Plougmann, P.; Linnet K. Determination of nortriptyline in human serum by fully automated solid-phase extraction and on-line high-performance liquid

chromatography in the presence of antipsychotic drugs. *J. Chromatogr. B*, **2000**, *746*, 233-239.

[45] Bose, D.; Durgbanshi, A.; Martinavarro-Domínguez, A.; Capella- Peiró, M.; Carda-Broch, S.; Esteve- Romeró, J.; Gil- Agustí, M. Amitriptyline and nortriptyline serum determination by micellar liquid chromatography.*J. Pharmacol. Toxicol. Methods*, **2005**, *52*, 323-329.

[46] Pirola, R.; Mundo, E.; Bellodi, L.; Bareggi, S.R. Simultaneous determination of clomipramine and its desmethyl and hydroxy metabolites in plasma of patients by high-performance liquid chromatography after solid-phase extraction. *J. Chromatogr. B*, **2002**, *772*, 205-210.

[47] Frahnert, C.; Rao, M.L.; Grasmader, K. Analysis of eighteen antidepressants, four atypical antipsychotics and active metabolites in serum by liquid chromatography: a simple tool for therapeutic drug monitoring. *J. Chromatogr. B*, **2003**, *794*, 35-47.

[48] Madej, K.; Parczewski, A.; Kała, M. HPLC/DAD screening method for selected psychotropic drugs in blood. *Tox. Mech. Meth.*, **2003**, 13, 121.

[49] Trocewicz, J.; Urine sample preparation of tricyclic antidepressants by means of a supported liquid membrane technique for high-performance liquid chromatographic analysis. *J. Chromatogr. B*, **2004**, *801*, 213-220.

[50] Hostetter, A.L.; Stowe, Z.N.; Cox, M.; Ritchie, J.C. A novel system for the determination of antidepressant concentrations in human breast milk. *Ther. Drug Monit.*, **2004**, *26*, 47-52.

[51] Morgan, P.E.; Spencer, E.P.; Flanagan, R.J. Simultaneous HPLC Of Tricyclic Antidepressants And Fluoxetine And Their N-demethyl Metabolites: 99 *Ther. Drug Monit.*, **2005**, *27(2)*, 236.

[52] Bose, D.; Martinavarro-Domínguez, A.; Gil-Agustí, M.; Carda-Broch, S.; Durgbanshi, A.; Capella-Peiró, M.; Esteve-Romeró, J. herapeutic monitoring of imipramine and desipramine by micellar liquid chromatography with direct injection and electrochemical detection. *Biomed. Chromatogr.*, **2005**, *19*, 343-349.

[53] Chen, A.G.; Wing, Y.K.; Chiu, H.; Lee, S.; Chen, C.N.; Chan, K. Simultaneous determination of imipramine, desipramine and their 2- and 10-hydroxylated metabolites in human plasma and urine by high-performance liquid chromatography. *J. Chromatogr. B*, **1997**, *693*, 153-158.

[54] Chmielewska, A.; Konieczna, L.; Plenis, A.; Lamparczyk H. Sensitive quantification of chosen drugs by reversed-phase chromatography with electrochemical detection at a glassy carbon electrode. *J. Chromatogr. B*, **2006**, *839*, 102-111.

[55] Ivandini, T.A.; Sarada, B.V.; Terashima, C.; Rao, T.N.; Tryk, D.A.; Ishiguro, H.; Kubota, Y.; Fujishima, A. Electrochemical detection of tricyclic antidepressant drugs by HPLC using highly boron-doped diamond electrodes. *J. Electroanal. Chem.*, **2002**, *521*, 117-126.

[56] Yoshida, H.; Hidaka, K.; Ishida, J.; Yoshikuni, K.; Nohta, H.; Yamaguchi, M. Highly selective and sensitive determination of tricyclic antidepressants in human plasma using high-performance liquid chromatography with post-column tris(2,2′-bipyridyl) ruthenium(III) chemiluminescence detection. *Anal. Chim. Acta*, **2000**, *413*, 137-145.

[57] Kirchherr, H.; Kuhn-Velten, W.N. Quantitative determination of forty-eight antidepressants and antipsychotics in human serum by HPLC tandem mass spectrometry: A multi-level, single-sample approach. *J. Chromatogr. B*, **2006**, *843(1)*,100-113.

[58] Badenhorst, D.; Sutherland, F.C.W.; Jager, A.D.; Scanes, T.; Hundt, H.K.L.; Swart, K.J.; Hundt, A.F. Determination of doxepin and desmethyldoxepin in human plasma using liquid chromatography-tandem mass spectrometry. *J. Chromatogr. B*, **2000**, *742*, 91-98.

[59] Kłys, M.; Scisłowski, M.; Rojek, S.; Kołodziej, A fatal clomipramine intoxication case of a chronic alcoholic patient: Application of postmortem hair analysis method of clomipramine and ethyl glucuronide using LC/APCI/MS. *J. Legal Med.,* **2005**, *7,* 319-325.

[60] Kollroser, M.; Schober, C. Simultaneous determination of seven tricyclic antidepressant drugs in human plasma by direct-injection HPLC-APCI-MS-MS with an ion trap detector. *Ther. Drug Monit.*, **2002**, 24, 537-544.

[61] Zhang, H.; Heinig, K.; Henion, J. Atmospheric pressure ionization time-of-flight mass spectrometry coupled with fast liquid chromatography for quantitation and accurate mass measurement of five pharmaceutical drugs in human plasma. *J. Mass Spectrom.,* **2000**, 35, 423-431.

[62] Samanidou, V.F.; Nika, M.K.; Papadoyannis, I.N. Development of an HPLC method for the monitoring of tricyclic antidepressants in biofluids.*J. Sep. Sci.,* **2007**, *30*, 2391-2400.

[63] Uddin, M.N.; Samanidou, V.F.; Papadoyannis, I.N. Development and validation of an HPLC method for the determination of benzodiazepines and tricyclic antidepressants in biological fluids after sequential SPE. *J. Sep. Sci.,* **2008**, *31*, 2358-2370.

[64] Uddin, M.N.; Samanidou, V.F.; Papadoyannis, I.N. HPLC method for Simultaneous Determination of 1,4-Benzodiazepines and Tricyclic Antidepressants in Pharmaceutical Formulations and Saliva After SPE. *J. Liquid Chromatogr. & Rel.Technol.*, **2009**, *32*, 1475-1504.

[65] Uddin, M.N.; Samanidou, V.F.; Papadoyannis, I.N. Simultaneous Determination of 1,4-Benzodiazepines and Tricyclic Antidepressants in Saliva after Sequential SPE Elution by the Same HPLC. *J. Chin. Chem. Soc.,* **2011**, *58*, 142-154.

[66] Malfará, W.R.; Bertucci, C.; Costa Queiroz, M.E.; Dreossi Carvalho, S.A.; de Lourdes Pires Bianchi, M.; Cesarino, E.J.; Crippa, J.A.; Costa Queiroz, R.H. Reliable HPLC method for therapeutic drug monitoring of frequently prescribed tricyclic and nontricyclic antidepressants. *J. Pharm. Biom. Anal.,* **2007**, 44, 955-962.

[67] Thongnopnua, P.; Karnjanaves, K. The rapid simultaneous tricyclic antidepressant determination by plasma deproteinization and liquid chromatography. *Asian Biomed.*, **2008**, *2*, 305-318.

[68] Breaud, A.R.; Harlan, R.; Kozak, M.; Clarke, W. A rapid and reliable method for the quantitation of tricyclic antidepressants in serum using HPLC-MS/MS. *Clin. Biochem.,* **2009**, *42*, 1300-1307.

[69] Mercolini, L.; Mandrioli, R.; Finizio, G.; Boncompagni, G.; Raggi, M.A. Simultaneous HPLC determination of 14 tricyclic antidepressants and metabolites in human plasma. *J. Sep. Sci.,* **2010**, *33,* 23-30.

[70] Wang, J.; Golden, T.; Ozsoz, M.; Lu, Z. Sensitive and selective voltammetric measurements of tricyclic antidepressants using lipid-coated electrodes*Bioelectr. Bioenerg.,* **1989**, *23,* 217-226 .

[71] Garcia Fraga, J.M.; Jimenez Abizanda, A.I., Jimenez Moreno, F.; Arias Leon, J.J. Simultaneous determination of imipramine and amitryptiline by derivative spectrophotometry. *J. Pharm. Biomed. Anal.,* **1991**, *9(2),* 109-115.

[72] Gaikwad, A.; Gomez-Hens, A.; Perez-Bendito, D. Use of stopped-flow fluorescence polarization immunoassay in drug determinations. *Anal. Chim. Acta*, **1993**, *280,* 129-135.

[73] Hattori, H.; Yamada, T.; Suzuki, O. Gas Chromatography with Surface Ionization Detection in Forensic Analysis *J. Chromatogr. A,* **1994**, *674,* 15-23.

[74] Pommier, F.; Sioufi, A.; Godbillon, J. Simultaneous determination of imipramine and its metabolite desipramine in human plasma by capillary gas chromatography with mass-selective detection. *J. Chromatogr. B,* **1997**, *703,* 147-158.

[75] Sporkert, F.; Pragst, F. Use of headspace solid-phase microextraction (HS-SPME) in hair analysis for organic compounds. *Foren. Sci. Inter.,* **2000**, *107,* 129-148.

[76] Veraart, J.A.; Brinkman, U.A.T. Dialysis–solid-phase extraction combined on-line with non-aqueous capillary electrophoresis for improved detectability of tricyclic antidepressants in biological samples. *J. Chromatogr. A,* **2001**, *922,* 339-346.

[77] Cantú, M.D.; Hillebrand, S.; Queiroz, M.E.C.; Lanças, F.M.; Carrilho, E. Validation of non-aqueous capillary electrophoresis for simultaneous determination of four tricyclic antidepressants in pharmaceutical formulations and plasma samples.*J. Chromatogr. B,* **2004**, *799,* 127-132.

[78] Flores, J.R.; Nevado, J.J.B.; Salcedo, A.M.C.; Diaz, M.P.C. Nonaqueous capillary electrophoresis method for the analysis of tamoxifen, imipramine and their main metabolites in urine. *Talanta,* **2005**, *65*, 155-162.

[79] Li, J.; Zhao, F.; Ju, H. Simultaneous determination of psychotropic drugs in human urine by capillary electrophoresis with electrochemiluminescence detection.. *Anal. Chim. Acta,* **2006**, 575, *57-61.*

[80] Alves, C.; Fernandes, C.; Dos Santos Neto, A.J.;Rodrigues, J.C.; Costa Queiroz, M.E.; Lanças, F.M. Optimization of the SPME parameters and its online coupling with HPLC for the analysis of tricyclic antidepressants in plasma samples. *J. Chrom. Sci.,* **2006**, *44,* 340-346.

[81] Woźniakiewicz, M.; Wietecha-Posłuszny, R.; Garbacik, A.; Kościelniak, P. Microwave-assisted extraction of tricyclic antidepressants from human serum followed by high performance liquid chromatography determination. *J. Chrom. A,* **2008**, 1190, 52-56.

[82] Rani, S.; Kumar, A.; Malik, A.K.; Singh, B. Quantification of Tricyclic and Nontricyclic Antidepressants in Spiked Plasma and Urine *Samples Using Microextraction in Packed Syringe and Analysis by LC and GC-MS. Chromatographia,* **2011**, 74, 235-242.

[83] Wietecha-Posłuszny, R.; Garbacik, A.; Woźniakiewicz, M.; Kościelniak, P. Microwave-assisted hydrolysis and extraction of tricyclic antidepressants from human hair. *Anal. Bioan. Chem.,* **2011,** 399, 3233-3240.

Send Orders of Reprints to reprints@benthamscience.net
Recent Advances in Medicinal Chemistry, Vol. 1, 2014, 381-403

CHAPTER 12

Gene Expression Profiles in Breast Cancer to Identify Estrogen Receptor Target Genes

Maria Aparecida Nagai[1,2,*] and Brentani M.M.[1]

[1]Center for Translational Research in Oncology, Laboratory of Molecular Genetics, Cancer Institute of the State of São Paulo (ICESP), Av Dr Arnaldo, 251, 8th Floor, CEP 01246-000, São Paulo, Brazil and [2]Laboratório de Genética Molecular do Centro de Investigação Translacional em Oncologia, Av. Dr. Arnaldo, 251, 8 andar, CEP 01246-000, São Paulo, Brazil

Abstract: The estrogens play important role in the homeostatic maintenance of several target tissues including those in the mammary gland, uterus, bone, cardiovascular system, and brain. Most of estrogen's action is thought to be mediated through its nuclear estrogen receptors, ERα and ERβ, which are members of the nuclear receptor superfamily that act as ligand-induced transcription factors. Acting *via* its receptors, estrogen also plays an essential role in the development and progression of human breast cancer. The ER and progesterone receptor (PR), which is regulated by estrogen *via* ER, have been used as prognostic markers in the clinical management of breast cancer patients. However, the prognosis of a patient with ER+/PR+ breast cancer can be highly variable and a significant proportion of hormone receptor positive breast cancers do not respond to endocrine therapy. The identification of estrogen receptor target genes may improve our understanding of the role played by estrogens in breast cancer making it possible to better tailor hormone treatments and improve a patient's response to hormonal therapy. In this review, we explore the literature for data regarding the identification of estrogen receptor-regulated genes in breast cancer cell lines and breast tumor biopsies using high throughput technologies such as serial analysis of gene expression (SAGE) and cDNA microarrays.

Keywords: Breast cancer, estrogen receptor, gene expression profiling, prognostic marker.

***Address correspondence to Maria Aparecida Nagai:** Center for Translational Research in Oncology, Laboratory of Molecular Genetics, Cancer Institute of the State of São Paulo (ICESP), Av Dr Arnaldo, 251, 8th Floor, CEP 01246-000, São Paulo, Brazil; Tel: 55-11-3893-3013; E-mail: nagai@usp.br

Atta-ur-Rahman, Muhammad Iqbal Choudhary and George Perry (Eds)

INTRODUCTION

Estrogens are small lipophilic molecules produced mainly by the ovary and carried out through the blood stream to specific target tissues. The major form and most potent natural estrogen is 17β-estradiol (E_2) followed by estrone and estriol which are less effective than estradiol (Fig. (**1**)). Besides the critical role played by estrogens in the development and maintenance of the reproductive system, there is compelling evidences that they also play important roles in regulating the physiological functions of various organs, such as bone, brain, and heart [1, 2].

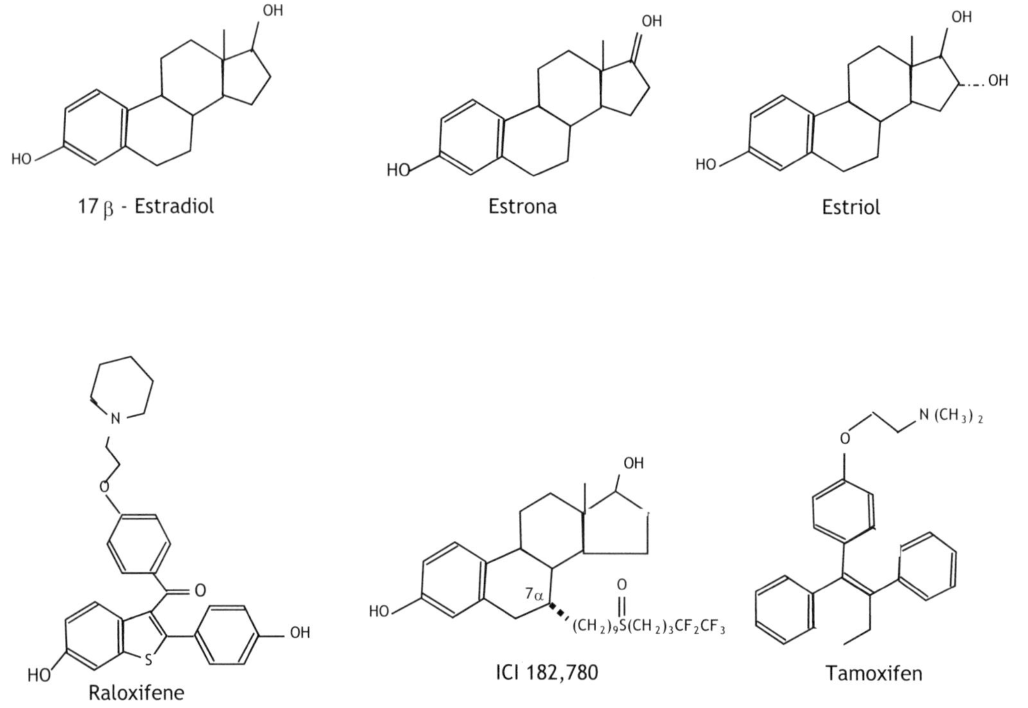

Figure 1: Chemical structure of natural estrogens (17β-estradiol, estrone and estriol); Chemical structures of selective estrogen receptor modulators (SERMs), tamoxifen, and raloxifene, and the pure antagonist ICI 182,780 ("faslodex"; fulvestrant).

In the mammary gland, the estrogens mediate key physiological processes that are essential for the normal growth and differentiation. However, there is a large body of evidence showing that estrogens, especially E_2, also play a critical role in the development and progression of breast cancer [3]. Most of the complex biological functions of the estrogens are mediated through the estrogen receptors, ERα and

ERβ, *via* the transcriptional regulation of ER target genes. Estrogen's actions can be partially blocked by selective estrogen receptor modulators (SERMs), such as tamoxifen and raloxifene, or by selective estrogen receptor down-regulators (SERDs), such as fulvestrant ("faslodex", ICI 182,780), that is a pure antagonist, which binds to ER leading to its destabilization and degradation (Fig. (**1**)) [4]. In addition, the third-generation non-steroidal aromatase inhibitors (AIs), anastrozole and letrozole, which inhibit estrogen production can be combined with fulvestrant to overcome resistance and improve patient's response to hormone therapy [5].

In fact, the ER is a useful prognostic and predictive marker for breast cancer. Approximately two-thirds of breast tumors express ER and are considered hormone-dependent, with hormone therapy widely accepted as the most important treatment for these patients. However, a proportion of the ER-positive breast cancer patients either does not respond to hormonal therapy, or become resistant to it [6]. In the last years, various molecular technologies that allow high throughput analysis of gene expression profiling have been used to identify the gene expression signature associated with the hormone-dependence of breast cancer that might improve our understanding of the ER-positive breast cancers and help select therapy for individual patients.

ESTROGEN RECEPTORS

The estrogen receptors belong to a family of structurally related and highly conserved proteins, the nuclear hormone receptor super-family, which includes receptors for progesterone, androgen, glucocorticoid, thyroid hormones, retinoic acids, and vitamin D. These steroid hormone receptors are ligand-activated transcriptional factors composed of a modular structure that includes several functional domains. These domains were designated A to F and carry out specific functions (Fig. (**2**)). The A/B domain is located in the N-terminal region and comprises the hormone-independent transcriptional activation function (AF1). The highly conserved C domain is located in the central region of the receptor molecule and corresponds to the DNA-binding domain, which contains two zinc-fingers motifs that directly interacts with the hormone response elements (HREs) in the promoter regions of the target genes. The D domain refers to a hinge region associated with receptor dimerization and interaction with co-regulatory proteins.

The E/F domain is located at the C-terminal region and contains the ligand-binding domain [7].

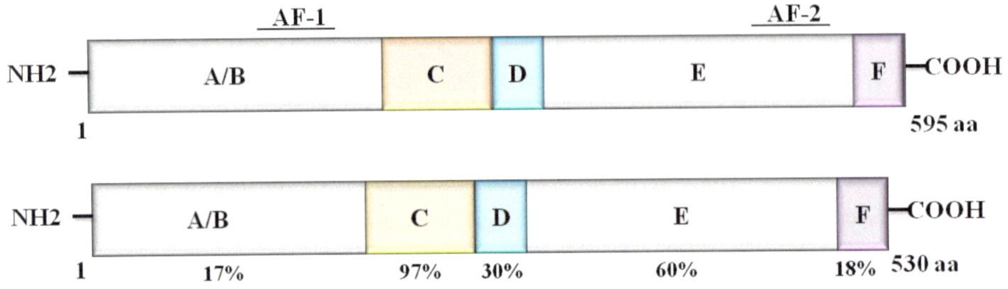

Figure 2: Diagrammatic representation of the domain structure of the human estrogen receptors, ERα and ERβ. The ERs molecules are composed of six structural and functional domains (A-F). The variable A/B domain contains the ligand-independent transcriptional-activating function (AF-1). The highly conserved DNA-binding domain (C) contains two zing finger structures that recognizes and bind to specific DNA sequences. The D domain is associated with receptor dimerization. The E/F domain contains the ligand-binding domain (LBD) and the ligand-dependent activation function (AF-2). Percentages of amino acid identity between ERα and ERβ in the corresponding functional regions are represented.

The two functional isoforms of the ERs, ERα and ERβ, are encoded by two distinct genes. The *ESR1* gene is located on chromosome 6q25.1 and encodes the 66 kDa ERα protein [8], and the *ESR2* gene is located on chromosome 14q23.2 and encodes the 54 kDa ERβ protein [9]. The expression pattern of ERα and ERβ in adults is tissue specific. ERα is ubiquitously expressed, and found predominantly in breast, uterus, cervix, and vagina, while ERβ expression is usually observed in the ovary, prostate, testis, spleen, and hypothalamus. ERβ is often co-expressed with ERα in breast carcinoma [10, 11]. These receptors share high similarity in their DNA-binding domain (97%) (Fig. (**2**)).

MECHANISMS OF ER ACTION

Due to their lipophilic properties, the estrogens can cross the cell membrane to enter into the cell cytoplasm and nucleus and bind to ERs, which dissociates from Hsp90 and acquire an activated state. Once activated, the hormone-receptors form either homo- or heterodimers, and bind to estrogen-response elements (EREs) on DNA to activate or repress gene transcription. The EREs are specific DNA sequences located in the promoter region of ER target genes to which the ERs directly interact with high affinity inducing the recruitment of co-regulatory factors and the basal core of transcriptional proteins leading to transcriptional regulation. ER transcriptional activity is enhanced by co-activators, such as the members of the SRC family (SRC1, like SRC2 and SRC3) and is repressed by the interaction of co-repressors such as NCOR1, NCOR2, and RIP140 [12]. The expression of co-regulators appears to have impact on the ER transcriptional transactivation and implications on SERMs pharmacological action [13-15]._In addition, cis-regulatory domains such as the Forkhead motif coupled with the Foxa1 transcription factor may facilitate and enhance ER-mediated transcription [16, 17]. This classical model of ER action involves direct interaction of the hormone-receptor complex with specific EREs.

The minimum consensus ERE consists of a 13 base pair perfect palindromic inverted repeat sequence separated by 3 non-conserved bases, GGTCANNNTGACC, derived from the sequence of the promoter regions of highly estrogen-induced genes [18]. These DNA binding domains are essential for the specificity of target-gene activation, and sequence deviations of the consensus ERE sequence reduces the affinity and specificity of the receptor interaction and the effectiveness of the transactivation activity [19]. Both ERα and ERβ form homo- or heterodimers and bind to specific EREs with similar specificity, but usually ERβ displays weaker transactivation activity than ERα [20, 21]. Most of the estrogen-target genes identified so far do not contain a perfect palindromic consensus sequence, which is expected to occur once every 4 million base pairs in random DNA sequences [18]. However, the ERs are able to induce transcriptional activation by binding to perfect or imperfect palindromic sequences separated by more than 3 base pairs, and also to direct repeats of half EREs, but with lower affinity [22]. A combination of experimentally defined and computationally predicted data can provide useful information about human ER target genes with putative EREs [23-25].

The classical ligand-dependent mechanism of ER action was for decades accepted as the only way through which the ER could induce transcriptional transactivation of target genes. During the last decade, it became clear that the mechanism through which the ER mediates the transcriptional regulation of gene expression is more complex. Besides the classical ligand-dependent mechanism of ER action, in which the hormone-receptor complex regulates gene transcription through its interaction with ERE consensus DNA sequences, the ER can also regulate gene transcription *via* protein-protein interaction with other transcriptional factors that bind to other promoter elements such as AP1, SP1, and CREs [21, 26, 27]. In fact, approximately one third of the ER-responsive genes have no ERE-like sequences in their promoter regions [22]. Moreover, extensive analysis of structural and functional properties of ERα and ERβ have led to additional complexity in this area, showing that these ER isoforms can transduce different hormonal signals depending on the ligand and the nature of the ERE [28, 29]. Thus, ERα and ERβ can transduce different hormonal signals depending on the ligand, and the nature of the hormone responsive element (HREs). Several studies indicated that ERß inhibits the proliferation and invasiveness of breast cancer cells causing a G2 cell cycle arrest and countering the activity of ERα [30]. Recent data demonstrated that the presence of ERβ enhances the sensitivity of breast cancer cells to the ant estrogenic effects of endoxifen (the most important tamoxifen metabolite) through the molecular actions of ERα/β heterodimers) [31].

The transactivation elicited by receptors complexes with E_2 may result in opposite signal transduction, leading to opposite biological responses in the presence of AP1 and/or CRE sites [32, 33]. In addition, several E_2-responsive genes are regulated by DNA-independent or –dependent interactions of the ERα and SP1 proteins. Otherwise, the signal transduction by growth factors and their tyrosine kinase receptors, such as EGFR, IGFR, erbB-2, and other molecules such as cAMP and dopamine, may lead to a ligand-independent ER activation, resulting from the phosphorylation of serine and tyrosine amino acid residues in the AF1 and AF2 domains in the estrogen receptor molecule [34]. ER phosphorylation promotes receptor dimerization, association with co-regulatory proteins, and transcription transactivation.

On the other hand, estrogens are also responsible for rapid biological effects, called non-genomic actions that are independent of mRNA and protein synthesis [35]. Although controversial, these non-genomic effects are thought to be displayed by E2 through a subpopulation of ER associated with the cell membrane, which induces the activation of intracellular second messengers such as calcium, nitric oxide formation, and protein kinase cascades such as ras-raf-MAPK (mitogen-activated protein kinase), ERK1 and ERK2 (extracellular-signal related kinases) or PI3K-AKT (phosphoinositide 3 kinase-protein kinase B) [36, 37]. In summary, there are at least four models of ER action: A, ligand-dependent (classical); B, ERE-independent; C, ligand-independent (cross-talk with growth factors), and D, ER cell membrane signaling (non-genomic) (Fig. (**3**)).

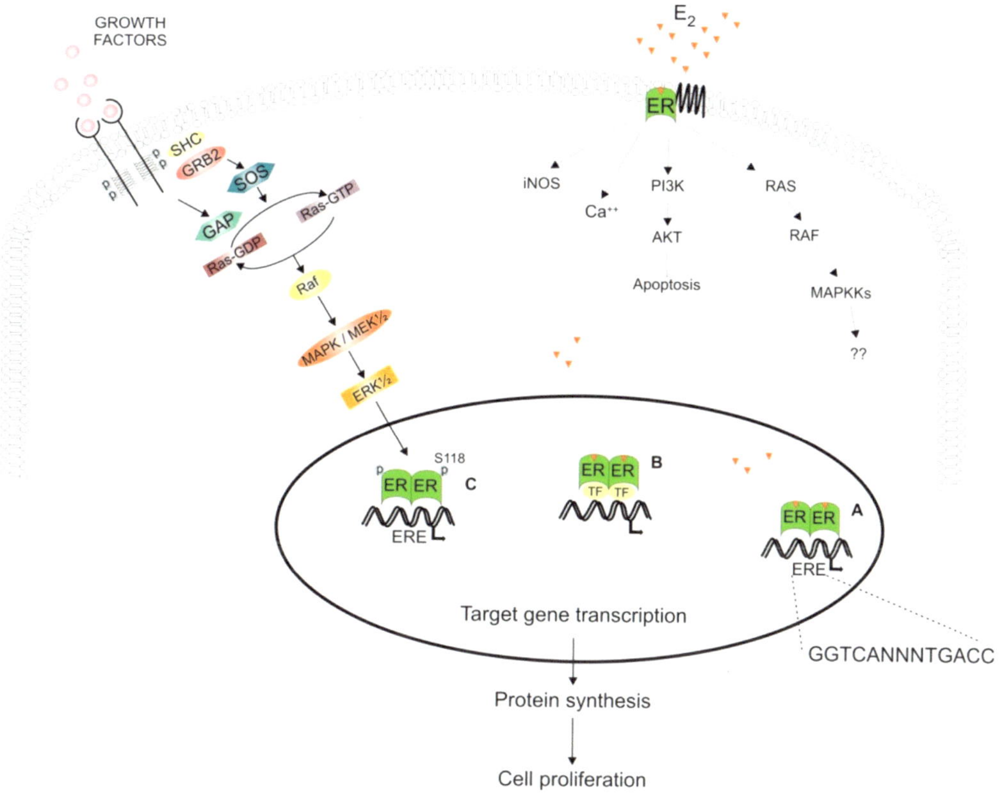

Figure 3: Schematic illustration of ER mechanisms of action. A, Classical ligand-dependent: the E_2-ER complexes as homo- or heterodimers interact with EREs in the promoter region of target genes; B, ERE-independent: the E_2-ER complexes interact with other transcription factors, such as AP1 and SP1 that bind to their cognate DNA binding sites; C, ligand-independent: growth factors, such as EGF and IGF-I, activate protein kinase cascades leading to ER phosphorylation (P) and activation; and D, non-genomic: estrogens binding to membrane–associated ERs lead to the activation of intracellular protein kinases cascades, such as MAPK and PI3K signaling pathways. E_2, estradiol (triangles); ER, estrogen receptor; GF, growth factors; TF, transcriptional factors; ERE, estrogen response element; iNOS, inducible nitric oxide synthase; PI3K, phosphoinositide-3-kinase; AKT, v-akt murine thymoma viral oncogene homologue; RAS, v-ras oncogene homologue; RAF, v-raf murine leukemia viral oncogene homologue; MAPKs, mitogen-activated protein kinases; ERK, extracellular signal-regulated kinases.

GLOBAL GENE EXPRESSION PROFILING OF ER ACTION

Although estrogens have been implicated as a major etiological factor in the tumorigenic process of breast cancer, the details of the effects of E_2 on downstream gene targets are far from fully understood. With the improvement of high-throughput experimental technologies such as SAGE and microarrays, information about estrogen signaling has been accumulating rapidly, showing that estrogen affects hundreds of genes. Gene expression profiles in response to E_2 have been carried out, particularly in single breast tumor cell lines, and most of the data have come from experiments with MCF-7, T47D, BT-474, and ZR-75 cells [38-54]. These cell lines are estrogen dependent breast cancer cells whose growth can be blocked by antiestrogens [55]. Although many investigators used similar platforms, there are differences in breast cancer cell lines expression profiles performed over a limited time course of hormone treatment (Table **1**). In particular, reports carried out using several distinct lengths of estrogen exposure revealed a diversity of temporal patterns of gene regulation by E2, with genes showing rapid changes in mRNA levels that could be sustained or not at later times, and transcripts induced or repressed beginning at later time points [41, 51, 56].

Table 1: Expression profiling of ER-regulated genes in breast cancer cell lines

Cell Line	Experimental Conditions	Platforms	Differentially Expressed Genes	References
ZR-75	5 day with 10% CSS + E_2 10^{-8}M +CHX for 6 0r 24 h;or + 4-OHT 10^{-7}M, Ral 10^{-7} M or ICI 10-7M for 24 h	Affymetrix HuGeneFL	53	Soulez and Parker, 2001 [34]
MCF-7	48 h with 5% CSS and treated with E_2 0 to 100 pM for 48 h	Affimetrix U133A	792 E_2 sensitive genes (190 up- and 602 down-regulated)	Coser et al., 2003 [38]
MCF-7	4 days with 5% CSS + E_2 10 nM for 4, 8 and 48 h (+ CHX) alone or in the presence of 4-OHT, ICI or Ral 1μM	Affimetrix Hu95A GeneChips	438 (30% up- and 70% down-regulated)	Frasor et al., 2004 [40]
T47D	4 days with CSS + E_2 10 nM for 6 and 24 h and mice tumor xenografts E2 for 7 wk	Affimetrix HG-U133A	1592 E_2 up 1277 E_2 down (cells vs. xenografts, 11% overlap	Harvell et al., 2006 [45]

Table 1: contd…

MCF-7	4 days with CSS + E_2 10 nM for 4, 8 and 24 h or + Tam 1 or 6 μM for 48 h and mice tumor xenografts E_2 for 6 wk	Affimetrix HG-U133A	1989 E_2 up 1512 E_2 down (cells vs. xenografts, >40% overlap)	Creighton et al., 2006 [46]
MCF-7, T47D and BT-474	3 days with CSS + E_2 10^{-8} M for 24 h or 4-OHT or ICI 1 μM for 24 h	Affymetrix U133A Genechip	E_2 sensitive gene in MCF-7 – 674 T47D – 140 BT0474 - 33	Rae et al., 2005 [43]
ZR-75-1	4 days with CSS + E_2 10 nM for 1 a 32h	Illumina Human WG-6 Beadchips	1488 differentially expressed genes (with 32 clusters)	Mutarelli et al., 2008 [52]
MDA-MB-231 (ERα transfected)	4 days with CSS + E_2 10 nM	Affymetrix U133A Genechip	420 ER regulated genes	Stender et al., 2010 [53]
MCF-7	4 days with CSS + E_2 10 nM for 16h	Illumina Human WG-6 Beadchips	501 E_2 up 470 E_2 down	Tolhurst et al., 2011 [54]
MCF-7	4 days with CSS + E_2 100 nM for 0, 3, 6 and 12 h	Affimetrix U133Plus 2.0 and ChIP on-chip	3 h – 275 6 h – 723 12 h – 1,023	Carroll et al., 2006 [14]
MCF-7	4 days with CSS + E_2 10 nM for 24 h	ChIP-on-chip (9000 GC-rich genomic sequence)	236	Bourdeau et al., 2004 [21]
MCF-7	3 days with 3% CSS and treated with E_2 10 nM for 24 h	ChIP-on-chip (8124 promotor sequences)	70	Jin et al., 2004 [41]
MCF-7	3 days with CSS + E2 100 nM for 45 min	ChIP-on-chip (18,668 promoter regions)	153	Langanière et al., 2005 [13]

CSS, charcoal stripped serum; CHX, cyclohexamide; Ral, reloxifene; ICI, ICI 182,780; Tam, tamoxifen.

Several previous genes identified by a variety of methodological approaches and focused on one gene or a few genes at a time such as trefoil factor-1, PR, cyclin D1, GATA-3, catepsin D, pS2/TFF-1 and c-myc were confirmed in some studies and several novel genes were described. Many of these ER up-regulated genes are important for cell proliferation and survival. However, E_2 down regulation of multiple transcriptional repressors could also contribute to increased cell proliferation. Suggested mechanisms by which E2 has been shown to repress gene expression involve the sequestering of co-activators or inhibition of NF-Kß [57]. In all those studies, the term ER was referring to ERα.

The definition of ER direct target genes was further refined and a group of genes that are responsive to E_2, sensitive to ICI 182,780, and insensitive to the protein synthesis inhibitor cycloheximide (CHX) was identified in ER+ breast cancer cell lines (T47D, ZR-75, and MCF-7). This group of genes was considerably small as compared with the total number of genes induced by E_2, suggesting the possibility of ER independent mechanisms in those cells [39, 44, 47]. Using these stringent criteria, a core set of ER direct target genes showed similar behavior in T47D and MCF-7 cells [47].

Therefore, using the initial inventory of responsive genes, one of the next questions addressed was to confirm the interaction between ER and ERE-like sites in the promoters of putative direct target genes. The Chip-on-chip technique that combines chromatin immunopreciptation with microarrays has important application for genome-wide identification of DNA binding sites for transcription factors, such as EREs in the promoter regions of ER target genes (Table **1**). Chip-on-chip experiments were performed to monitor recruitment of ERα to the EREs in several genes known to be regulated directly by E2 in MCF-7 cells and binding was observed for practically all tested genes [24]. Similar results in T47D and MCF-7 cells were observed by Lin *et al.* 2004 [47], suggesting that the ERE is the major response element mediating the specific regulation of ER direct target genes. In the report of Carrol *et al.* 2006 [17], it was also verified that E_2 up-regulated genes which have adjacent estrogen receptor binding sites are more likely to contain EREs, and that genes down regulated generally contain AP1 sites. In summary, in comparison to the overall number of E_2 regulated genes in ERα positive breast cancer cell lines, the E_2 direct response pathway accounts for only a portion of the molecular signature and a significant enrichment of EREs in the regulatory regions of these direct target genes was observed [47, 58, 59].

Many studies have shown distinct patterns of gene expression related to ER status in breast cancer biopsies and identified genes related to ER signaling [59-70]. The results provided evidence that ER+ and ER- or ER+PR+, ER-PR- tumors display remarkably different gene-expression phenotypes, but the association between ER discriminator genes and genes regulated by E2 are unclear.

A key question to be answered is whether the *in vitro* observations in cell lines reflect biological significance *in vivo*. Lin *et al.* [47] compared the E_2-induced expression profiles of MCF-7 cells and the behavior of these genes in ER-positive tumor samples and observed that the number of direct estrogen responsive genes was small in comparison to the overall number of genes that define the ER+ breast tumors. According to the work of Harvell *et al.* 2006 [50], a comparison of *in vivo* estrogen-regulated genes in a model of human breast tumor xenografts compared with the identical cell grown *in vitro* revealed only an 11% of overlap. In spite of differences in individual genes, similar functions were maintained in general. On the other hand, another report found a good agreement between the estrogen-regulated pattern in MCF-7 cells *in vitro* and that obtained in the same cells grown as xenografts (over 40%). Interestingly, a significant number of genes induced by E_2 *in vitro* were correlated with tumor profiles in ERα breast cancers from patients within a narrow age range of 41-44 years [51].

Abba and co-workers [59] used SAGE to analyze the gene profile of breast carcinomas based on ERα status coupled with the identification of putative high affinity EREs in the promoter regions of the SAGE-identified up modulated genes. Approximately 31% of ERα associated transcripts were involved in biological process related to cell growth. The authors suggested that many of these genes were transcriptionally regulated by non-ERE mediated mechanisms such as those involving ER-binding to the AP1 or SP1 transcription factors. Comparison of the *in vitro* transcripts (MCF-7) and *in vivo* profiles revealed that only few transcripts behaved similarly in both studies, confirming the observation of Meltzer and co-workers that the majority of genes regulated in cell culture do not predict ER status in breast cancer [44, 60]. The combination of genome-wide analysis for gene expression and global ERE binding sites identification are important to discriminate ER target genes that are direct or indirectly regulated by estrogens, however the expression profile of ER-target genes *in vitro* or *in vivo* will depend on the cellular context and the expression of co-activators and co-repressors [15, 71].

As most of the data were based on expression of mRNAs isolated from tumor masses which includes fibroblasts and lymphocytes and the proportion of tumor cells in clinical samples varies significantly, the multiple cell population may compromise the gene expression data associated with ER that is expressed on epithelial cells. In the report on Yang *et al.* [69], epithelial tumor cells obtained by laser capture micro dissection that allows one to isolated nearly pure cell populations from a heterogeneous environment retained only 43% of the genes, unique to this category. Several genes classified in this category have been demonstrated in ERα tumors such as trefoil factors 1 and 3, GATA 3, GREB1, XBP1 and keratin 18. Other cause of discrepancy might be the presence of ERß, which seems to have a significant impact on the pattern of gene expression in breast cancer cells that contain ERα [11]. However, it was demonstrated that a complex pattern of genes (not including ER) could also identify most ER-negative from ER-positive breast tumors and could be used to predict clinical ER status, suggesting that the differences between ER- and ER+ cancers may not simply be attributable to the absence or presence of ER function but rather reflect different molecular phenotypes [64].

In accordance, a novel molecular taxonomy has been proposed which stratified breast cancer into several clinical relevant subtypes. Among them, ER positive cancers were classified as luminal tumors, whereas ER negative comprised HER-2 rich, triple negative (TNP) and normal breast like cancers. Approximately 80% of TNP tumors were classified as presented a basal like phenotype [72]. In the past years additional ER negative molecular subtypes have been described such as the claudin-low subgroup which comprises tumors that have transcriptome features suggestive of a cancer stem cell like subtype [73]. Luminal cancers were next subclassified into 2 groups (luminal A and B) according to the expression levels of proliferating related genes. Luminal A expressed the highest levels of ER and ER-related genes and the lowest level of proliferation-linked genes whereas luminal B cancers presented an opposite profile [74]. Only approximately 30% of luminal B are HER-2 positive indicating that this clinical marker alone is not sensitive enough to identify most luminal B tumors. However, the major biological distinction between luminal A and B is the proliferation signature including genes such as CCNB1, MK167 and MYBL2 which have higher expression in luminal B tumors. However a recent meta analyze indicated that levels of proliferation related genes in ER positive cancers are a continuum ranging from those lesions with low proliferation to those with high proliferation and therefore the sub classifications of luminal cancers seems to be arbitrary [74-76].

The luminal A type tumors are also characterized by low histological grades and for being responsive to adjuvant hormonal treatment and associated with improved survival whereas luminal B carcinomas presented more often high histological grades [70, 74]. Proliferation related gene sets were also associated with increased chemotherapy sensitivity in ER positive breast cancer [77].

Many more pathways were associated with prognosis in ER positive than in ER negative cancers. Indeed many prognostic markers have been developed for this group of patients including several that are being used in clinical practice including Oncotype SX, Mammaprint and others reviewed by Weigelt *et al.* 2010 [75]. In ER positive cancers, gene sets that were involved in mitosis, cell cycle, DNA replication and chromosome duplication pathways were associated with poor prognosis, and gene sets involved with immune and inflammatory responses were associated with good prognosis.

Desmedt *et al.* 2008 [78] demonstrated that most of the signatures associated with prognosis are composed of proliferation markers and have limited discriminatory power in ER negative cancers which seem to constitute a collection of distinct entities whose prognosis is not dependent of proliferation but of other biological characteristics [75]. Part of the genes that co-clustered with ER was previously identified in E2-responsive carcinoma cell lines [79]. However its is possible that these expression cassettes segregating with ER status are reflecting other tumor aspects such as slower growth, enhanced differentiation that are only partly related to the presence of ER, reflecting different molecular phenotypes perhaps arisen from different precursors in the breast. All the studies outlined above emphasize the molecular complexicity of the mechanisms by which estrogen receptor dictates tumor status and showed that other molecular events could influence sensitivity to hormonal therapy and clinical outcome.

The high cost of gene expression profiling has limited its incorporation into most clinical trials. The PAM50 method has provided an approach for breast cancer molecular classification using RT-PCR [80]. Several groups have attempted to develop surrogate immunohistocemical markers for the molecular subtypes. It was suggested that ER, PR and HER-2 determined by immunohistochemistry can be used to approximate the subtypes defined by gene expression profiling. Tumors that are ER and or PR positive and HER-2 negative are most likely luminal A, those that are ER positive and or PR positive and HER-2 positive are luminal B, and those that are ER negative and HER-2 positive are considered as HER-2 rich and those that are ER, PR and HER-2 negative are triple negative. Within the triple negative group expression of basal cytokeratins (CK 5/6, CK14) and or EGFR have been proposed to define basal-like breast cancers [72]. The study of Cheang *et al.* 2009 [81] defined the luminal B subtype by selecting a Ki-67 index of 14% or more as the best cut point for identifying luminal B cancers. Two recent meta analysis have reported an statistically significant association between high Ki67 expression and increased risk of breast cancer relapse and death [82, 83]. A recent publication suggested that quantitative measurements of HER-2, ER and AURKA (aurora kinase) can provide relevant clinical molecular classification [84].

Since its introduction more than 30 years ago tamoxifen has been the most widely used drug in endocrine therapy for the treatment of women with advanced breast cancer. However, almost all patients with metastatic disease and as many as 40% receiving adjuvant tamoxifen eventually relapse due to intrinsic (*de novo*) or acquired resistance and need further treatment options [85]. Some multigene prognostic predictors of tamoxifen response are already being proposed [86-89] and interestingly although experimental and computational studies revealed that a large number of genes is potentially regulated by ER signaling pathways, the discriminatory signatures consist of a relatively small number of genes.

Tumors consist of cancer cells and of diverse stromal cells that influence the behavior of cancer [90]. Much of the tumor stroma consists of cancer associated fibroblasts that express αSMA, vimentin or S100A4 and biologically are different from fibroblasts associated to normal breast. Recent studies also reported that stromal gene signatures are associated with breast prognosis. Bianchini *et al.* 2010 [91] assessed the ability of stromal genes to further risk stratify patients within the ER positive groups in low and high proliferative cancers

The mechanistic bases of the different antiestrogens described are not yet fully understood. Microarray analysis used to identify transcriptional programs regulated by tamoxifen, raloxifen and ICI in ER positive cells, showed a very low degree of overlapping indicating that each individual compound exhibited a very specific gene expression profile [45, 89, 92]. The identification of antiestrogenic effects on gene regulation *in vivo* may be expected to provide a better understanding of the mechanisms that lead to a poor antiproliferative sequential response. Two recent reports used a neodjuvante protocol with aromatase inhibitors and analyzed gene expression in individual biopsies taken before and after treatments highlighting that decreased expression of proliferation-related genes were particularly prominent [93, 94].

As illustrated in this review, several studies have been published reporting interesting results of estrogen responsive gene profiles in breast cancer cell lines and tumor biopsies and they clearly pointed that only the analysis of multiple genetic elements could classify tumors in terms of response to therapy more accurately than conventional biomarkers. The main challenge in the area is to establish a group of limited number of genes that are suitable for determining sensitivity to anti-hormone therapy.

FUTURE CONSIDERATIONS

Breast cancer is not one disease but a collection of several biologically different diseases. Different prognostic and treatment markers may be necessary for the different molecular types of breast cancer [95]. Although the gene expression profile technology is promising, several problems elicited by the poor association between bioinformatics and biology has to be solved. The advent of massively DNA sequences may help to established more homogenous molecular subgroups [96]. Further optimization and standardization of methodology including mathematical and statistical analyses, similar criteria of patients and protocols selection and properly designed clinical trials [97, 98] will be required to improve our understanding of breast cancer molecular-phenotype and its relationship with anti-cancer agents response.

ACKNOWLEDGEMENTS

Declared none.

CONFLICT OF INTEREST

The authors confirm that this chapter contents have no conflict of interest.

DISCLOSURE

The chapter submitted for eBook Series entitled: "**Recent Advances in Medicinal Chemistry, Volume 1**" is an update of our article published in **Mini-Reviews in Medicinal Chemistry, Volume 8, Number 5, pp. 448 to 454**, with additional text and references.

REFERENCES

[1] Baldini, V.; Mastropasqua M.; Francucci, C.M.; D'Erasmo, E. Cardiovascular disease and osteoporosis. *J. Endocrinol. Invest.*, **2005**, *28*, 69-72.

[2] Thakur, M.K.; Sharma, P.K. Aging of brain: role of estrogen. *Neurochem. Res.*, **2006**, *31*, 1389-1398.

[3] Medina, D. Mammary developmental fate and breast cancer risk. *Endocr. Relat. Cancer.*, **2005**, *12*, 483-495.

[4] Lewis, J.S.; Jordan, V.C. Selective estrogen receptor modulators (SERMs): mechanisms of anticarcinogenesis and drug resistance. *Mutat. Res.*, **2005**, *591*, 247-263.

[5] Adamo, V.; Iorfida, M.; Montalto, E.; Festa, V.; Garipoli, C.; Scimone, A.; Zanghì, M.; Caristi, N. Overview and new strategies in metastatic breast cancer (MBC) for treatment of tamoxifen-resistant patients. Ann Oncol., **2007**, *18*: 53-57.

[6] Ring, A.; Dowsett, M. Mechanisms of tamoxifen resistance. *Endocr. Relat. Cancer.*, **2004**, *11*, 643-658.

[7] Aranda, A.; Pascual, A. Nuclear hormone receptors and gene expression. *Physiol. Rev.*, **2001**, *81*, 1269-1304.

[8] Green, S.; Walter, P.; Kumar, V.; Krust, A.; Bornert, J.M.; Argos, P.; Chambon, P. Human oestrogen receptor cDNA: sequence, expression and homology to v-erb-A. *Nature,* **1986**, *320*, 134-139.

[9] Kuiper, G.G.; Enmark, E.; Pelto-Huikko, M.; Nilsson, S.; Gustafsson, J.A. Cloning of a novel receptor expressed in rat prostate and ovary. *Proc. Natl. Acad. Sci. U.S.A.,* **1996**, *93*, 5925-5930.

[10] Couse, J.F.; Lindzey, J.; Grandien, K.; Gustafsson, J.A.; Korach, K.S. Tissue distribution and quantitative analysis of estrogen receptor-alpha (ERalpha) and estrogen receptor-beta (ERbeta) messenger ribonucleic acid in the wild-type and ERalpha-knockout mouse. *Endocrinology*, **1997**, *138*, 4613-4621.

[11] Chang, E.C.; Frasor, J.; Komm, B.; Katzenellenbogen, B.S. Impact of estrogen receptor beta on gene networks regulated by estrogen receptor alpha in breast cancer cells. *Endocrinolog*, **2006**, *147*, 4831-4842.

[12] Smith, C.L.; O'Malley, B.W. Coregulator function: a key to understanding tissue specificity of selective receptor modulators. *Endocr. Rev.*, **2004**, *25*, 45-71.

[13] Brisken, C.; O'Malley, B. Hormone action in the mammary gland. *Cold Spring Harb Perspect Biol.*, **2010,** *2*, a003178.

[14] McDonnell, D.P.; Wardell, S.E. The molecular mechanisms underlying the pharmacological actions of ER modulators: implications for new drug discovery in breast cancer. *Curr Opin Pharmacol., 2010, 10,* 620-628.

[15] Zwart, W.; Theodorou, V.; Kok, M.; Canisius, S.; Linn, S.; Carroll, J.S. Oestrogen receptor-co-factor-chromatin specificity in the transcriptional regulation of breast cancer. *EMBO J.*, *2011*, *30,* 4764-4776.

[16] Laganiere, J.; Deblois, G.; Lefebvre, C.; Bataille, A.R.; Robert, F.; Giguere, V. From the Cover: Location analysis of estrogen receptor alpha target promoters reveals that FOXA1 defines a domain of the estrogen response. *Proc .Natl. Acad. Sci. U.S.A.*, **2005**, *102*, 11651-11656.

[17] Carroll, J.S.; Meyer, C.A.; Song, J.; Li, W.; Geistlinger. T.R.; Eeckhoute, J.; Brodsky, A.S.; Keeton, E.K.; Fertuck, K.C.; Hall, G.F.; Wang, Q.; Bekiranov, S.; Sementchenko, V.; Fox, E.A.; Silver, P.A.; Gingeras, T.R.; Liu, X.S.; Brown, M. Genome-wide analysis of estrogen receptor binding sites. *Nat. Genet.*, **2006**, *38*, 1289-1297.

[18] Gruber, C.J.; Gruber, D.M.; Gruber, I.M.; Wieser, F.; Huber, J.C. Anatomy of the estrogen response element. *Trends Endocrinol. Metab.*, **2004**, *15*, 73-78.

[19] Sanchez, R.; Nguyen, D.; Rocha, W.; White, J.H.; Mader, S. Diversity in the mechanisms of gene regulation by estrogen receptors. *Bioessays.*, **2002**, *24*, 244-254.

[20] Hyder, S.M.; Chiappetta, C.; Stancel, G.M. Interaction of human estrogen receptors alpha and beta with the same naturally occurring estrogen response elements. *Biochem. Pharmacol.*, **1999**, *57*, 597-601.

[21] Kushner, P.J.; Agard, D.A.; Greene, G.L.; Scanlan, T.S.; Shiau, A.K.; Uht, R.M.; Webb, P. Estrogen receptor pathways to AP-1. *J. Steroid Biochem. Mol. Biol.*, **2000**, *74*, 311- 307.

[22] O'Lone, R.; Frith, M.C.; Karlsson, E.K.; Hansen, U. Genomic targets of nuclear estrogen receptors. *Mol. Endocrinol.*, **2004**, *18*, 1859-1875.

[23] Bajic, V.B.; Tan, S.L.; Chong, A.; Tang, S.; Strom, A.; Gustafsson, J.A.; Lin, C.Y.; Liu, E.T. Dragon ERE Finder version 2: A tool for accurate detection and analysis of estrogen response elements in vertebrate genomes. *Nucleic Acids Res.*, **2003**, *31*, 3605-3607.

[24] Bourdeau, V.; Deschenes, J.; Metivier, R.; Nagai, Y.; Nguyen, D.; Bretschneider, N.; Gannon, F.; White, J.H.; Mader, S. Genome-wide identification of high-affinity estrogen response elements in human and mouse. *Mol. Endocrinol.*, **2004**, *18*, 1411-1427.

[25] Jin, V.X.; Sun, H.; Pohar, T.T.; Liyanarachchi, S.; Palaniswamy, S.K.; Huang, T.H.; Davuluri, R.V. ERTargetDB: an integral information resource of transcription regulation of estrogen receptor target genes. *J. Mol. Endocrinol.*, **2005**, *35*, 225-230.

[26] Sabbah, M.; Courilleau, D.; Mester, J.; Redeuilh, G. Estrogen induction of the cyclin D1 promoter: involvement of a cAMP response-like element. *Proc. Natl. Acad. Sci. USA.*, **1999**, *96*, 11217-11222.

[27] Saville, B.; Wormke, M.; Wang, f.; Nguyen, T.; Enmark, E.; Kuiper, G.; Gustafsson, J.A.; Safe, S. Ligand-, cell-, and estrogen receptor subtype (alpha/beta)-dependent activation at GC-rich (Sp1) promoter elements. *J. Biol. Chem.*, **2000**, *275*, 5379-5387.

[28] Nilsson, S.; Makela, S.; Treuter, E.; Tujague, M.; Thomsen, J.; Andersson, G.; Enmark, E.; Pettersson, K.; Warner, M.; Gustafsson, J.A. Mechanisms of estrogen action. *Physiol. Rev.*, **2001**, *81*, 1535-1565.

[29] Pike, A.C. Lessons learnt from structural studies of the oestrogen receptor. *Best. Pract. Res. Clin. Endocrinol .Metab.*, **2006**, *20*, 1-14.

[30] Hartman, J.; Lindberg, K.; Morani, A.; Inzunza, J.; Ström, A.; Gustafsson, J.A. Estrogen receptor beta inhibits angiogenesis and growth of T47D breast cancer xenografts. *Cancer Res.*, **2006**, *66*, 11207-11213.

[31] Wu, X.; Subramaniam, M.; Grygo, S.B.; Sun, Z.; Negron, V.; Lingle, W.L.; Goetz, M.P.; Ingle, J.N.; Spelsberg, T.C.; Hawse, J.R. Estrogen receptor-beta sensitizes breast cancer cells to the anti-estrogenic actions of endoxifen. *Breast Cancer Res.*, **2011**, *13*, R27.

[32] Paech, K.; Webb, P.; Kuiper, G.G.; Nilsson, S.; Gustafsson, J.; Kushner, P.J.; Scanlan, T.S. Differential ligand activation of estrogen receptors ERalpha and ERbeta at AP1 sites. *Science*, **1997**, *277*, 1508-1510.

[33] Castro-Rivera, E.; Samudio, I.; Safe, S. Estrogen regulation of cyclin D1 gene expression in ZR-75 breast cancer cells involves multiple enhancer elements. *J. Biol. Chem.*, **2001**, *276*, 30853-30861.

[34] Kato, S.; Kitamoto, T.; Masuhiro, Y.; Yanagisawa, J. Molecular mechanism of a cross-talk between estrogen and growth-factor signaling pathways. *Oncology*, **1998**, *55*, 5-10.

[35] Acconcia, F.; Kumar, R. Signaling regulation of genomic and nongenomic functions of estrogen receptors. *Cancer Lett.*, **2006**, *238*, 1-14.

[36] Kelly, M.J.; Levin, E.R. Rapid actions of plasma membrane estrogen receptors. *Trends Endocrinol. Metab.*, **2001**, *12*, 152-156.

[37] Bjornstrom, L.; Sjoberg, M. Mechanisms of estrogen receptor signaling: convergence of genomic and nongenomic actions on target genes. *Mol. Endocrinol.*, **2005**, *19*, 833-842.

[38] Charpentier, A.H.; Bednarek, A.K.; Daniel, R.L.; Hawkins, K.A.; Laflin, K.J.; Gaddis S.; MacLeod, M.C.; Aldaz, C.M. Effects of estrogen on global gene expression: identification of novel targets of estrogen action. *Cancer Res.*, **2000**, *60*, 5977-5983.

[39] Soulez, M.; Parker, M.G. Identification of novel oestrogen receptor target genes in human ZR75-1 breast cancer cells by expression profiling. *J. Mol. Endocrinol.*, **2001**, *27*, 259-274.

[40] Seth, P.; Krop, I.; Porter, D.; Polyak, K. Novel estrogen and tamoxifen induced genes identified by SAGE (Serial Analysis of Gene Expression). *Oncogene.*, **2002**, *21*, 836-842.

[41] Inoue, A.; Yoshida, N.; Omoto, Y.; Oguchi, S.; Yamori, T.; Kiyama, R.; Hayashi, S. Development of cDNA microarray for expression profiling of estrogen-responsive genes. *J. Mol. Endocrinol.*, **2002**, *29*, 175-192.

[42] Levenson, A.S.; Kliakhandler, I.L.; Svoboda, K.M.; Pease, K.M.; Kaiser, S.A.; Ward, J.E. Molecular classification of selective oestrogen receptor modulators on the basis of gene expression profiles of breast cancer cells expressing oestrogen receptor alpha. *Br. J. Cancer*, **2002**, *87*, 449-456.

[43] Coser, K.R.; Chesnes, J.; Hur, J.; Ray, S.; Isselbacher, K.J.; Shioda, T. Global analysis of ligand sensitivity of estrogen inducible and suppressible genes in MCF7/BUS breast cancer cells by DNA microarray. *Proc. Natl. Acad. Sci. U.S.A.*, **2003**, *100*, 13994-13999.

[44] Cunliffe, H.E.; Ringner, M.; Bilke, S.; Walker, R.L.; Cheung, J.M.; Chen, Y.; Meltzer P.S. The gene expression response of breast cancer to growth regulators: patterns and correlation with tumor expression profiles. *Cancer Res.*, **2003**, *63*, 7158-7166.

[45] Frasor, J.; Stossi, F.; Danes, J.M.; Komm, B.; Lyttle, C.R.; Katzenellenbogen, B.S. Selective estrogen receptor modulators: discrimination of agonistic versus antagonistic activities by gene expression profiling in breast cancer cells. *Cancer Res.*, **2004**, *64*, 1522-1533.

[46] Jin, V.X.; Leu, Y.W.; Liyanarachchi, S.; Sun, H.; Fan, M.; Nephew, K.P.; Huang, T.H.; Davuluri, R.V. Identifying estrogen receptor alpha target genes using integrated computational genomics and chromatin immunoprecipitation microarray. *Nucleic Acids Res.*, **2004**, *32*, 6627-6635.

[47] Lin, C.Y.; Strom, A.; Vega, V.B.; Kong, S.L.; Yeo, A.L.; Thomsen, J.S.; Chan, W.C.; Doray, B.; Bangarusamy, D.K.; Ramasamy, A.; Vergara, L.A.; Tang, S.; Chong, A.; Bajic, V.B.; Miller, L,D,; Gustafsson, J.A.; Liu, E.T. Discovery of estrogen receptor alpha target genes and response elements in breast tumor cells. *Genome Biol.*, **2004**, *5*, R66.

[48] Rae, J.M.; Johnson, M.D.; Scheys, J.O.; Cordero, K.E., Larios, J.M.; Lippman, M.E. GREB 1 is a critical regulator of hormone dependent breast cancer growth. *Breast Cancer Res. Treat.*, **2005**, *92*, 141-149.

[49] Scafoglio, C.; Ambrosino, C.; Cicatiello, L.; Altucci, L.; Ardovino, M.; Bontempo, P.; Medici, N.; Molinari, A.M.; Nebbioso, A.; Facchiano, A.; Calogero, R.A.; Elkon, R.; Menini, N.; Ponzone, R.; Biglia, N.; Sismondi, P.; Bortoli, M.D.; Weisz, A. Comparative gene expression profiling reveals partially overlapping but distinct genomic actions of different antiestrogens in human breast cancer cells. *J. Cell Biochem.*, **2006**, *98*, 1163-1184.

[50] Harvell, D.M.; Richer, J,K.; Allred, D.C.; Sartorius, C.A.; Horwitz, K.B. Estradiol regulates different genes in human breast tumor xenografts compared with the identical cells in culture. *Endocrinology*, **2006**, *147*, 700-713.

[51] Creighton, C.J.; Cordero, K.E.; Larios, J.M.; Miller, R.S.; Johnson, M.D.; Chinnaiyan A.M.; Lippman, M.E.; Rae, J.M. Genes regulated by estrogen in breast tumor cells *in vitro* are similarly regulated *in vivo* in tumor xenografts and human breast tumors. *Genome Biol.*, **2006**, *7*, R28.

[52] Mutarelli, M.; Cicatiello, L.; Ferraro, L.; Grober, O.M.; Ravo, M.; Facchiano, A.M.; Angelini, C.; Weisz, A. Time-course analysis of genome-wide gene expression data from hormone-responsive human breast cancer cells. *BMC Bioinformatics.*, **2008**, 26 (9 Suppl 2), S12.

[53] Stender, J.D.; Kim, K.; Charn, T.H.; Komm, B.; Chang, K.C.; Kraus, W.L.; Benner, C.; Glass, C.K.; Katzenellenbogen BS. Genome-wide analysis of estrogen receptor alpha DNA binding and tethering mechanisms identifies Runx1 as a novel tethering factor in receptor-mediated transcriptional activation. *Mol Cell Biol.*, **2010,** *30*, 3943-3955.

[54] Tolhurst, R.S.; Thomas, R.S.; Kyle, F.J.; Patel, H.; Periyasamy, M.; Photiou, A.; Thiruchelvam, P.T.; Lai, C.F.; Al-Sabbagh, M.; Fisher, R.A.; Barry, S.; Crnogorac-Jurcevic, T.; Martin, L.A.; Dowsett, M.; Charles Coombes, R.; Kamalati, T.; Ali, S.; Buluwela, L. Transient over-expression of estrogen receptor-α in breast cancer cells promotes cell survival and estrogen-independent growth. *Breast Cancer Res Treat.* **2011**, *128,* 357-368.

[55] Lacroix, M.; Leclercq, G. Relevance of breast cancer cell lines as models for breast tumours: an update. *Breast Cancer Res. Treat.*, **2004**, *83*, 249-289.

[56] Frasor J, Danes, J.M.; Komm, B.; Chang, K.C.; Lyttle, C.R.; Katzenellenbogen, B.S. Profiling of estrogen up- and down-regulated gene expression in human breast cancer cells: insights into gene networks and pathways underlying estrogenic control of proliferation and cell phenotype. *Endocrinology*, **2003**, *144*, 4562-4574.

[57] McKay, L.I.; Cidlowski, J.A. Molecular control of immune/inflammatory responses: interactions between nuclear factor-kappa B and steroid receptor-signaling pathways. *Endocr. Rev.*, **1999**, *20*, 435-459.

[58] Vega, V.B.; Lin, C.Y.; Lai, K.S.; Li Kong, S.; Xie, M.; Su, X.; The, H.F.; Thomsen, J.S.; Li Yeo, A.; Sung, W.K.; Bourque, G.; Liu, E.T. Multiplatform genome-wide identification and modeling of functional human estrogen receptor binding sites. *Genome Biol.*, **2006**, *7*, R82.

[59] Abba, M.C.; Hu, Y.; Sun, H.; Drake, J.A.; Gaddis, S.; Baggerly, K.; Sahin, A.; Aldaz, C.M. Gene expression signature of estrogen receptor alpha status in breast cancer. *BMC Genomics*, **2005**, *6*, 37.

[60] Gruvberger, S.; Ringner, M.; Chen, Y.; Panavally, S.; Saal, L.H.; Borg, A.; Ferno, M.; Peterson, C.; Meltzer, P.S. Estrogen receptor status in breast cancer is associated with remarkably distinct gene expression patterns. *Cancer Res.*, **2001**, *61*, 5979-5984.

[61] West, M.; Blanchette, C.; Dressman, H.; Huang, E.; Ishida, S.; Spang, R.; Zuzan, H.; Olson, J.A. Jr; Marks, J.R.; Nevins, J.R. Predicting the clinical status of human breast cancer by using gene expression profiles. *Proc. Natl. Acad. Sci. U.S.A.*, **2001**, *98*, 11462-11467.

[62] Porter, D.A.; Krop, I.E.; Nasser, S.; Sgroi, D.; Kaelin, C.M.; Marks, J.R.; Riggins, G.; Polyak, K. A SAGE (serial analysis of gene expression) view of breast tumor progression. *Cancer Res.*, **2001**, *61*, 5697-5702.

[63] Sorlie, T.; Tibshirani, R.; Parker, J.; Hastie, T.; Marron, J.S.; Nobel, A.; Deng, S.; Johnsen, H.; Pesich, R.; Geisler, S.; Demeter, J.; Perou, C.M.; Lonning, P.E.; Brown, P.O.; Borresen-Dale, A.L.; Botstein, D. Repeated observation of breast tumor subtypes in independent gene expression data sets. *Proc. Natl. Acad. Sci. U.S.A.*, **2003**, *100*, 8418-8423.

[64] Pusztai, L.; Ayers, M.; Stec, J.; Clark, E.; Hess, K.; Stivers, D.; Damokosh, A.; Sneige, N.; Buchholz, T.A.; Esteva, F.J.; Arun, B.; Cristofanilli, M.; Booser, D.; Rosales, M.; Valero, V.; Adams, C.; Hortobagyi, G.N.; Symmans, W.F. Gene expression profiles obtained from fine-needle aspirations of breast cancer reliably identify routine prognostic markers and reveal large-scale molecular differences between estrogen-negative and estrogen-positive tumors. *Clin. Cancer Res.*, **2003**, *9*, 2406-2415.

[65] Nagai, M.A.; Da Ros, N.; Neto, M.M.; de Faria Junior, S.R.; Brentani, M.M.; Hirata, R. Jr; Neves, E.J. Gene expression profiles in breast tumors regarding the presence or absence of estrogen and progesterone receptors. *Int. J. Câncer*, **2004**, *111*, 892-899.

[66] Sotiriou, C.; Neo, S.Y.; McShane, L.M.; Korn, E.L.; Long, P.M.; Jazaeri, A.; Martiat, P.; Fox, S.B.; Harris, A.L.; Liu, E.T. Breast cancer classification and prognosis based on gene expression profiles from a population-based study. *Proc. Natl. Acad. Sci. U.S.A.*, **2003**, *100*, 10393-10398.

[67] Gruvberger-Saal, S.K.; Eden, P.; Ringner, M.; Baldetorp, B.; Chebil, G.; Borg, A.; Ferno, M.; Peterson, C.; Meltzer, P.S. Predicting continuous values of prognostic markers in breast cancer from microarray gene expression profiles. *Mol Cancer Ther.*, **2004**, *3*,161-168.

[68] Oh, D.S.; Troester, M.A.; Usary, J.; Hu, Z.; He, X.; Fan, C.; Wu, J.; Carey, L.A.; Perou, C.M. Estrogen-regulated genes predict survival in hormone receptor-positive breast cancers. *J. Clin. Oncol.*, **2006**, *24*, 1656-1664.

[69] Yang, F.; Foekens, J.A.; Yu, J.; Sieuwerts, A.M.; Timmermans, M.; Klijn, J.G.; Atkins, D.; Wang, Y.; Jiang, Y. Laser microdissection and microarray analysis of breast tumors reveal ER-alpha related genes and pathways. *Oncogene*, **2006**, *25*, 1413-1419.

[70] Sorlie, T.; Perou, C.M.; Fan, C.; Geisler, S.; Aas, T.; Nobel, A.; Anker, G. Akslen, L.A.; Botstein, D.; Borresen-Dale, A.L.; Lonning, P.E. Gene expression profiles do not consistently predict the clinical treatment response in locally advanced breast cancer. *Mol. Cancer Ther.*, **2006**, *5*, 2914-2918.

[71] Gao, H.; Dahlman-Wright, K. The gene regulatory networks controlled by estrogens. *Mol Cell Endocrinol.*, **2011**, *334*, 83-90.

[72] Badve, S.; Dabbs, D.J.; Schnitt, S.J.; Baehner, F.L.; Decker, T.; Eusebi, V.; Fox, S.B.; Ichihara, S.; Jacquemier, J.; Lakhani, S.R.; Palacios, J.; Rakha, E.A.; Richardson, A.L.; Schmitt, F.C,; Tan, P.H.; Tse, G.M.; Weigelt, B.; Ellis, I.O.; Reis-Filho, J.S. Basal-like and

triple-negative breast cancers: a critical review with an emphasis on the implications for pathologists and oncologists. *Mod Pathol.* **2011,** *24*,157-167.

[73] Hennessy, B.T.; Gonzalez-Angulo, A.M.; Stemke-Hale, K.; Gilcrease, M.Z.; Krishnamurthy, S.; Lee, J.S.; Fridlyand, J.; Sahin, A.; Agarwal, R.; Joy, C.; Liu, W.; Stivers, D.; Baggerly, K.; Carey, M.; Lluch, A.; Monteagudo, C.; He, X.; Weigman, V.; Fan, C.; Palazzo, J.; Hortobagyi, G.N.; Nolden, L.K.; Wang, N.J,; Valero, V.; Gray, J.W.; Perou, C.M.; Mills, G.B. Characterization of a naturally occurring breast cancer subset enriched in epithelial-to-mesenchymal transition and stem cell characteristics. *Cancer Res.*, **2009**, *69*, 4116-4124.

[74] Wirapati, P.; Sotiriou, C.; Kunkel, S.; Farmer, P.; Pradervand, S.; Haibe-Kains, B.; Desmedt, C.; Ignatiadis, M.; Sengstag, T.; Schütz, F.; Goldstein, D.R.; Piccart, M.; Delorenzi, M. Meta-analysis of gene expression profiles in breast cancer: toward a unified understanding of breast cancer subtyping and prognosis signatures. *Breast Cancer Res.,* **2008,** *10*, R65.

[75] Weigelt, B.; Baehner, F.L.; Reis-Filho, J.S. The contribution of gene expression profiling to breast cancer classification, prognostication and prediction: a retrospective of the last decade. *J Pathol.*, **2010**, *220,* 263-280.

[76] Haibe-Kains, B.; Desmedt, C.; Loi, S.; Culhane, A.C.; Bontempi, G.; Quackenbush, J.; Sotiriou, C. A three-gene model to robustly identify breast cancer molecular subtypes. *J Natl Cancer Inst.*, **2012**, *104*, 311-325.

[77] Iwamoto, T.; Bianchini, G.; Booser, D.; Qi, Y.; Coutant, C.; Shiang, C.Y.; Santarpia, L.; Matsuoka, J.; Hortobagyi, G.N.; Symmans, W.F.; Holmes, F.A.; O'Shaughnessy, J.; Hellerstedt, B.; Pippen, J.; Andre, F.; Simon, R.; Pusztai, L. Gene pathways associated with prognosis and chemotherapy sensitivity in molecular subtypes of breast cancer. *J Natl Cancer Inst.*, **2011,** *103*, 264-272.

[78] Desmedt, C.; Haibe-Kains, B.; Wirapati, P.; Buyse, M.; Larsimont, D.; Bontempi, G.; Delorenzi, M.; Piccart, M.; Sotiriou, C. Biological processes associated with breast cancer clinical outcome depend on the molecular subtypes. *Clin Cancer Res.*, **2008**, *14*, 5158-5165.

[79] Charafe-Jauffret, E.; Ginestier, C.; Monville, F.; Finetti, P.; Adelaide, J.; Cervera, N.; Fekairi, S.; Xerri, L.; Jacquemier, J.; Birnbaum, D.; Bertucci, F. Gene expression profiling of breast cell lines identifies potential new basal markers. *Oncogene,* **2006**, *25*, 2273-2284.

[80] Nielsen, T.O.; Parker, J.S.; Leung, S.; Voduc, D.; Ebbert, M.; Vickery, T.; Davies, S.R.; Snider, J.; Stijleman, I.J.; Reed, J.; Cheang, M.C.; Mardis, E.R.; Perou, C.M.; Bernard, P.S.; Ellis, MJ. A comparison of PAM50 intrinsic subtyping with immunohistochemistry and clinical prognostic factors in tamoxifen-treated estrogen receptor-positive breast cancer. *Clin Cancer Res.*, **2010**, *16,* 5222-5232.

[81] Cheang, M.C.; Chia, S.K.; Voduc, D.; Gao, D.; Leung, S.; Snider, J.; Watson, M.; Davies, S.; Bernard, P.S.; Parker, J.S.; Perou, C.M.; Ellis, M.J.; Nielsen, T.O. Ki67 index, HER2 status, and prognosis of patients with luminal B breast cancer. *J Natl Cancer Inst.*, **2009**; *101*, 736-750.

[82] De Azambuja, E.; Cardoso, F.; de Castro, G. Jr.; Colozza, M.; Mano, M.S.; Durbecq, V.; Sotiriou, C.; Larsimont, D.; Piccart-Gebhart, M.J.; Paesmans, M. Ki-67 as prognostic marker in early breast cancer: a meta-analysis of published studies involving 12,155 patients. *Br J Cancer.*, **2007,** *96*, 1504-1513.

[83] Stuart-Harris, R.; Caldas, C.; Pinder, S.E.; Pharoah, P. Proliferation markers and survival in early breast cancer: a systematic review and meta-analysis of 85 studies in 32,825 patients. *Breast.*, **2008,** *17,* 323-334.

[84] Haibe-Kains, B.; Olsen, C.; Djebbari, A.; Bontempi, G.; Correll, M.; Bouton, C.; Quackenbush, J. Predictive networks: a flexible, open source, web application for integration and analysis of human gene networks. *Nucleic Acids Res.*, **2012,** *40,* D866-875.

[85] Early Breast Cancer: Trialists' Collaborate Group: Effects of chemotherapy and hormonal therapy for early breast cancer on recurrence and 15-year survival. Lancet, **2005**, *365*, 1687-1717.

[86] Paik, S.; Shak, S.; Tang, G.; Kim, C.; Baker, J.; Cronin, M.; Baehner, F.L.; Walker, M.G.; Watson, D.; Park, T.; Hiller, W.; Fisher, E.R.; Wickerham, D.L.; Bryant, J.; Wolmark, N. A multigene assay to predict recurrence of tamoxifen-treated, node-negative breast cancer. *N. Engl. J. Med.*, **2004**, *351*, 2817-2826.

[87] Ma, X.J.; Wang, Z.; Ryan, P.D.; Isakoff, S.J.; Barmettler, A.; Fuller, A.; Muir, B.; Mohapatra, G.; Salunga, R.; Tuggle, J.T.; Tran, Y.; Tran, D.; Tassin, A.; Amon, P.; Wang, W.; Wang, W.; Enright, E.; Stecker, K.; Estepa-Sabal, E.; Smith, B.; Younger, J.; Balis, U.; Michaelson, J.; Bhan, A.; Habin, K.; Baer, T.M.; Brugge, J.; Haber, D.A.; Erlander, M.G.; Sgroi, D.C. A two-gene expression ratio predicts clinical outcome in breast cancer patients treated with tamoxifen. *Cancer Cell*, **2004**, *5*, 607-616.

[88] Jansen, M.P.; Foekens, J.A.; van Staveren, I.L.; Dirkzwager-Kiel, M.M.; Ritstier, K.; Look, M.P.; Meijer-van Gelder, M.E.; Sieuwerts, A.M.; Portengen, H.; Dorssers, L.C.; Klijn, J.G.; Berns, E.M. Molecular classification of tamoxifen-resistant breast carcinomas by gene expression profiling. *J. Clin. Oncol.*, **2005**, *23*, 732-740.

[89] Frasor, J.; Chang, E.C.; Komm, B.; Lin, C.Y.; Vega, V.B.; Liu, E.T.; Miller, L.D.; Smeds, J.; Bergh, J.; Katzenellenbogen, B.S. Gene expression preferentially regulated by tamoxifen in breast cancer cells and correlations with clinical outcome *Cancer Res.*, **2006**, *66*, 7334-7340.

[90] Kalluri, R.; Zeisberg, M. Fibroblasts in cancer. *Nat Rev Cancer.*, **2006**, *6*, 392-401.

[91] Bianchini, G.; Qi, Y.; Alvarez, R.H.; Iwamoto, T.; Coutant, C.; Ibrahim, N.K.; Valero, V.; Cristofanilli, M.; Green, M.C.; Radvanyi, L.; Hatzis, C.; Hortobagyi, G.N.; Andre, F.; Gianni, L.; Symmans, W.F.; Pusztai, L. Molecular anatomy of breast cancer stroma and its prognostic value in estrogen receptor-positive and -negative cancers. *J Clin Oncol.* **2010,** *28*, 4316-4323.

[92] Sismondi, P.; Biglia, N.; Ponzone, R.; Fuso, L.; Scafoglio, C.; Cicatiello, L.; Ravo, M.; Weisz, A.; Cimino, D.; Altobelli, G.; Friard, O.; De Bortoli, M. Influence of estrogens and antiestrogens on the expression of selected hormone-responsive genes. *Maturitas*, **2007**, *57*, 50-55.

[93] Mackay, A.; Urruticoechea, A.; Dixon, J.M.; Dexter, T.; Fenwick, K.; Ashworth, A.; Drury, S.; Larionov, A.; Young, O.; White, S.; Miller, W.R.; Evans, D.B.; Dowsett, M. Molecular response to aromatase inhibitor treatment in primary breast cancer. *Breast Cancer Res.*, **2007**, *9*, R37.

[94] Miller, W.R.; Larionov, A.; Renshaw, L.; Anderson, T.J.; White, S.; Hampton, G.; Walker, J.R.; Ho, S.; Krause, A.; Evans, D.B.; Dixon, J.M. Aromatase inhibitors--gene discovery. *J Steroid Biochem Mol Biol.*, **2007**, *106*,130-142.

[95] Pusztai, L,; Iwamoto, T. Breast cancer prognostic markers in the post-genomic era. *Breast Cancer Res Treat.*, **2011,** *125,* 647-650.

[96] Sun, Z.; Asmann, Y.W,; Kalari, K.R.; Bot, B.; Eckel-Passow, J.E.; Baker, T.R.; Carr, J.M.; Khrebtukova, I.; Luo, S.; Zhang, L.; Schroth, G.P.; Perez, E.A.; Thompson, E.A. Integrated analysis of gene expression, CpG island methylation, and gene copy number in breast cancer cells by deep sequencing. *PLoS One*., **2011**, *6,* e17490.

[97] Simon, R.; Radmacher, M.D.; Dobbin, K.; McShane, L.M. Pitfalls in the use of DNA microarray data for diagnostic and prognostic classification. *J Natl Cancer Inst*., **2003**, *95*, 14-18.

[98] Pusztai, L. The use of microarray technology in the management of breast cancer. *Clin Adv Hematol Oncol*., **2007**, *5*,193-197.

Send Orders for Reprints to reprints@benthamscience.net
 Recent Advances in Medicinal Chemistry, Vol. 1, 2014, 404-432

CHAPTER 13

Trans-Plasma Membrane Electron Transport in Human Blood Platelets: An Update

Luciana Avigliano*, **I. Savini, M. V. Catani and D. Del Principe**

Department of Experimental Medicine & Surgery, University of Rome, Tor Vergata, Rome, Italy

Abstract: The plasma membrane redox (PMR) system is important for cell metabolism and survival; it is also crucial for blood coagulation and thrombosis. This review will give an update on the PMR system, with a particular regard to platelets, and on the role of antioxidant vitamins belonging to this system.

Keywords: Antioxidants, free radicals, nicotinamide, plasma membrane redox system, platelets, ubiquinone, vitamin E, vitamin C.

PLASMA MEMBRANE REDOX SYSTEM

Enzymes involved in electron transport are usually associated to the respiratory chain, localized in the mitochondrial inner membrane and responsible for aerobic ATP production. However, the existence of a plasma membrane electron transport (PMET) or plasma membrane redox (PMR) system has been known since 1960, although its physiological relevance has only recently been investigated.

The primary biological function of the PMET system, which is ubiquitous in every living cell (including bacteria, yeast, plants and animals), is the maintenance of cytoplasmic NAD^+/NADH ratio, thus regulating energy levels and redox homeostasis. Cellular ATP usually comes from oxidative phosphorylation, but when mitochondrial activity is depressed (*e.g.* in the presence of mitochondrial dysfunction or strenuous physical activity) cells can survive by using alternative pathways, including cytoplasmic glycolysis. In such conditions, the NAD^+/NADH balance, needed for sustaining ATP levels, is maintained by

***Address correspondence to Luciana Avigliano:** Department of Experimental Medicine & Surgery, University of Rome, Tor Vergata, Via Montpellier 1, 00133 Rome, Italy; Tel: + 39 6 72596472; Fax: + 39 6 72596379; E-mail: avigliano@uniroma2.it

Atta-ur-Rahman, Muhammad Iqbal Choudhary and George Perry (Eds)

compensatory mechanisms, such as the pyruvate/lactate couple and enhanced PMET activity. Indeed, higher activity of the PMR system has been observed in human ρ^0 cells (devoid of mitochondrial DNA), where mitochondrial respiration is impaired [1], and in lymphocytes derived from diabetes mellitus patients, lacking functional mitochondria [2]. So, the PMR system guarantees glycolytic metabolism, thus ensuring survival of cells deficient in mitochondrial electron transport (*e.g.* aged or tumour cells).

Components of the PMR System

The basic components of the PMR system include electron donors and acceptors, antioxidants and reductases. These systems may have operational flexibility, since electrons can be transferred from different donors to several acceptors.

a) Electron Donors and Acceptors

The intracellular sources of reducing equivalents are mainly represented by the pyridine coenzymes NADH and NADPH, derived from the hydrophilic vitamin niacin. NAD^+ is converted to NADH during the catabolism of carbohydrates, fats and proteins; on the contrary, the NADPH/$NADP^+$ system functions in anabolic reactions, including synthesis of fatty acids and cholesterol, as well as in hydroxylation and detoxification reactions. Oxidation/reduction cycle, fluctuating in response to metabolic changes, determines the pyridine derivative ratio, which plays a crucial role in regulating the intracellular redox state [3, 4].

Especially in red blood cells, the PMR system can also use intracellular vitamin C (named ascorbate in the reduced form) to reduce extracellular oxidants, with a mechanism that is dependent on the NADH content of the cell [5]. Ascorbate-mediated electron transfer occurs through two main mechanisms: (i) enzyme-mediated electron transport, where vitamin C is the electron donor for trans-membrane oxidoreductases, and (ii) shuttle electron transfer, where ascorbate, released from cells, directly acts as reducing agent, thus being oxidized to dehydroascorbate (DHA) *via* the intermediate ascorbyl free radical (AFR) [5-7]. DHA can be taken up by cells and reduced back to ascorbate thanks to reducing equivalents coming from cell metabolism [6, 8, 9].

Other recently discovered, intracellular substrates are some flavonoids (quercetin, myricetin, fisetin); their ability to function as electron donors is linked to the presence of cathecol structure of the B ring (responsible for reducing activity) and the double bond and 4-oxo function of the C ring (responsible for cellular uptake) [10].

Although the membrane-impermeant ferricyanide has been routinely used to probe the activity of different enzymes, nonetheless several oxidants are physiologically reduced by the PMR system. Ubiquinone (or coenzyme Q; CoQ) is the direct acceptor of intracellular electrons in the plasma membrane. It can be transformed to the ubisemiquinone radical ($CoQ^{·-}$) or to hydroquinone ($CoQH_2$), depending on the membrane enzyme involved, such as the NADH:cytochrome b5 reductase or the NAD(P)H: ubiquinone oxidoreductase [11, 12]. Then, CoQ is a key membrane redox player, which can directly or indirectly (through the participation of intermediate carriers, including b cytochromes, flavin and vitamin E) transfer electron(s) to final acceptors. Membrane proteins also contribute to this "plasma membrane electron transport chain", since they may be responsible for the disulfide-thiol interchange activity ascribed to some enzymes belonging to the PMR system [13].

Molecular oxygen is the final electron acceptor for membrane NAD(P)H oxidases [14], thus generating water; if partially reduced, it generates superoxide anions ($O_2^{·-}$) and hydrogen peroxide (H_2O_2). These reactive oxygen species (ROS) modulate specific bio-molecules involved in cellular functions and signal transduction. Other extracellular acceptors are the ascorbyl free radical that is reduced to ascorbate, and the ferric iron, that is reduced to ferrous ion needed for intestinal iron absorption [15].

b) Antioxidants

Thanks to their ability to exist as stable radicals, some of the electron carriers described above (CoQ, vitamin E and vitamin C) also exert antioxidant effects.

The major lipophilic antioxidants are ubiquinone and alpha tocopherol, which protect cells against lipid peroxidation. Ubiquinone directly scavenges ROS which can damage unsaturated lipid chains; demonstrations of this phenomenon come from the findings that ubiquinone-deficient yeast mutants show high levels of lipid radicals [16]. In the quinol form, it is also necessary for regenerating tocopherol from tocopheryl radical and ascorbate outside the cell from ascorbyl radical [17]. Alpha-tocopherol (vitamin E) is also a lipophilic chain-breaking antioxidant; its radical can be reduced by ubiquinol, ascorbate and glutathione [12]. The antioxidant activity of alpha-tocopherol plays a key role in erythrocyte membranes, which are highly susceptible to peroxidation because of the presence of many polyunsaturated fatty acids, the continuous exposure to high concentrations of oxygen and the presence of a transition metal catalyst [18].

Although being water-soluble, vitamin C exerts some important antioxidant functions on the plasma membrane, since it is capable of neutralizing an array of ROS and regenerating alpha-tocopherol from its radical [19].

c) Enzymes

The PMR enzymes include the: (i) NADH:ferricyanide reductase, (ii) NAD(P)H:ubiquinone oxidoreductase 1, (iii) cytochrome b5 reductase, (iv) superoxide-generating NADPH oxidases, (v) disulfide-thiol exchangers and (vi) duodenal cytochrome b. The major components of the PMR system are summarized in Fig. (**1**).

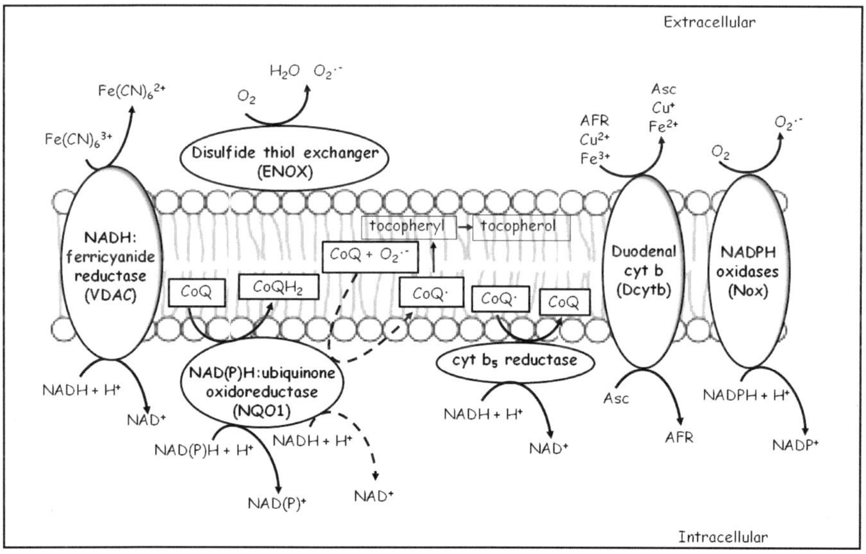

Figure 1: Key enzymes of the PMR system. Membrane localisation and catalysed reactions are shown in the scheme. Alternative reactions are depicted with hatched lines. CoQ regenerates transplasma membrane tocopherol, thus protecting platelets from lipid peroxidation and external oxidative insults. Asc: ascorbate. AFR: ascorbyl free radical.

The NADH:ferricyanide reductase [known as Voltage-Dependent Anion-selective Channel (VDAC) or porin] was originally discovered in the outer mitochondrial membrane, but it is also present in the plasma membrane [20]. In mammals, three highly conserved genes have been shown to encode distinct isoforms, named VDAC1, VDAC2 and VDAC3, which mainly differ in their channel-forming activity. The tertiary structure of human VDAC1, the best-characterized member of this family, has recently been solved [21]: it consists of 19 β-barrel, membrane-spanning strands, with the N-terminal α-helix folded horizontally midway within the pore. Similar folding properties are likely shared by other VDAC proteins, as they have >80% sequence identity [21].

In mice, VDAC1 exists as two different proteins, generated by the use of alternative first exons; one isoform, expressing a leader peptide in its N-terminus, is directed to the plasma membrane, while the second one, lacking the pre-sequence, is targeted to mitochondria. Thus, the same polypeptide may be localized in two membranes, where it exerts different functions; these could be explained assuming that (i) the protein changes its conformation when it is targeted to the plasma or mitochondrial membrane, or (ii) some not yet identified effector proteins control VDAC activity.

In the mitochondrion, VDAC1 forms pores freely permeable to low molecular-weight molecules (including ADP/ATP, succinate and citrate), thus controlling metabolite trafficking between the cytosol and mitochondria. It has also been suggested a role for this pore-forming protein in the release of apoptogenic proteins from mammalian mitochondria [22]. Nonetheless, data on VDAC1 involvement in formation of permeability transition pore, during apoptosis, are still controversial [23], since VDAC-deficient mice are indistinguishable from their wild-type counterparts in the ability to undergo cell death in response to different inducers of apoptosis [24, 25].

Conversely, the protein exerts a NADH:ferricyanide reductase activity in the plasma membrane: indeed, the enzyme is capable of reducing the cell-impermeant ferricyanide by using NADH as the intracellular electron donor [26]. The reductase activity is strictly dependent on NADH, as it could not be replaced by NADPH; in addition, it can not be inhibited by anion channel inhibitors, while it is sensitive to thiol chelators, suggesting a critical role for two cysteine residues, present in the molecule, in electron transfer. The natural substrate(s) of VDAC1 remains to be established, since ferricyanide is not a physiological molecule; however, recent data suggest that CoQ may act as an acceptor for the enzyme [26]. However, it should be recalled that other options may exist: the effects attributed to VDAC may arise from VDAC itself acting as (i) a NADH-dependent reductase or (ii) a channel directly releasing intracellular reducing equivalents into the medium, or from (iii) VDAC assembly with other proteins that drive specific functions [27]. Anyway, VDAC1 appears to be necessary for sensing the cellular redox state and regulating the NAD^+/NADH ratio, thus enabling cell growth and apoptosis [22].

The NAD(P)H:ubiquinone oxidoreductase (NQO1, also named DT-diaphorase) is a largely cytosolic, homodimeric FAD-containing protein of 33 kDa, which is translocated into the plasma membrane under oxidative stress. It catalyzes obligatory NADH or NADPH-dependent two-electron reductions of quinones (in particular, it converts oxidised CoQ in the hydroquinone reduced form), but it also reduces quinone-imines, nitro and azo compounds [28]. These reactions prevent the cytotoxic and carcinogenic effects of xenobiotics, that can produce free-radicals leading to DNA and cell damage. An isoenzyme, lacking 43 residues at the C-terminus (called NQO2) and encoded by a different gene, has been found; unlike NQO1, it preferentially uses dihydronicotinamide riboside [instead of NAD(P)H], as a source of reducing equivalents [29], and it is not involved in PMET. Both genes are up-regulated in response to oxidative stress and are over-expressed in some tumours. The crystal structure of NQO1 reveals that the catalytic site is at the dimer interface and that the enzyme works through a ping-pong mechanism. Indeed, the FAD prosthetic group is always bound to each subunit, while NADH or NADPH are bound and released during catalytic cycling, in order to allow substrate binding: thus, the reaction consists of hydride transfer from NAD(P)H to FAD and, then, from $FADH_2$ to the quinone [30]. Conversely, NQO1 can also catalyze a NADH-driven, superoxide-dependent, one-electron reduction of CoQ, thus generating the semiquinone radical, which in turn leads to regeneration of alpha-tocopherol [31]. Dicoumarol and dicoumarol analogues are competitive inhibitors of NQO1 activity [32] and are currently under investigation for their ability to suppress the malignant phenotype of pancreatic cancer cells, where NQO1 expression is upregulated [33, 34].

The cytochrome b5 reductase is a FAD-containing monomer of 32 kDa, existing in two isoforms produced by alternative promoter usage. The first one is a membrane-bound enzyme catalyzing the one-electron transfer from NADH to CoQ, thus generating the radical CoQ^{-} in the plasma membrane [12]. The second isoform produces the soluble form found in erythrocytes, which transfers electrons derived from NAD(P)H to the heme protein. In red blood cells, hemoglobin can be converted in methemoglobin, if the iron of the heme group is oxidized from the ferrous (Fe^{2+}) to the ferric (Fe^{3+}) state, thus impairing oxygen delivery. Cytochrome b5 reductase is the main endogenous enzymatic mechanism responsible for hemoglobin reduction; indeed, deficiency of this enzyme is the most common cause of congenital type I and II methemoglobinemias, characterized by increased levels of methemoglobin above its steady state [35].

The membrane-bound isoform also reduces AFR back to ascorbate [36], especially in synaptic plasma membrane vesicles [37]; the reductase forms redox centers, localized in the lipid rafts, at the interneuronal contact sites, thus suggesting that alterations of these redox centers may affect neuron–neuron interactions and cell survival [37].

The superoxide-generating NADPH oxidases represent a protein family, including at least seven proteins (Nox1 to Nox5, Duox1 and Duox2) [14, 38, 39], whose best characterized member is Nox2. This is an inducible, membrane-bound cytochrome b_{558}, composed of two subunits (named $p22^{phox}$ and $gp91^{phox}$) and mainly expressed in phagocytes of the immune system, as well as in platelets. The enzyme catalyzes the transfer of one electron from NADPH to molecular oxygen, thus generating $O_2^{\cdot -}$ anions. The Rac1 and Rac2 GTPases and three cytosolic factors ($p47^{phox}$, $p67^{phox}$, $p40^{phox}$) are needed for the catalytic activity; in particular, membrane translocation of these cytosolic proteins promote the production of superoxide. ROS generation by Nox2 represents the first mechanism of defence against pathogens: indeed, ROS act as microbiocidal agents in the phenomenon known as "respiratory burst" [40].

Other Nox family members are all homologs of the $gp91^{phox}$ subunit of Nox2. Like Nox2, they generate tightly controlled superoxide anions, but, showing a tissue-specific distribution, they mediate distinct and specific functions [41]. Nox1 is primarily expressed in colon epithelial cells, but it can be detected in vascular smooth muscle cells, where it has a role in angiotensin II- and growth factor-induced cell hypertrophy or proliferation [42]. Several findings also suggest that Nox1 could participate in the host defence of different cell types [43]. Nox3 is expressed in several foetal tissues, including kidney, liver, lung, and spleen [44]; it seems involved in tumour cell proliferation, but more studies are needed to provide better insight. Because of the peculiar expression in the inner ear, Nox3 has been proposed to play an essential role for normal vestibular function; indeed, impairment of Nox3 leads to balance defects due to absence of otoconia [45]. Nox4 is expressed at high levels in the kidney, where it appears to play a role in oxygen sensing and regulation of erythropoietin synthesis [46]. Nox5 was originally found in testis, but it is also present in T- and B-lymphocytes [47]. This oxidase, which is activated by intracellular calcium-dependent conformational changes, drives sperm capacitation, a process strictly dependent on ROS-mediated signal transduction; Nox5-induced ROS production is also important for acrosome reaction and sperm-oocyte fusion during fertilization. Finally, dual oxidases (Duox1 and Duox2) are peculiar, because they possess peroxidase and superoxide-generating NADPH oxidase activities. Furthermore, they are unusual, in that they produce H_2O_2 rather than $O_2^{\cdot-}$. They are expressed in the membrane of thyroid glands and serve in iodide oxidation during thyroxine synthesis [48,49].

The disulfide-thiol exchangers are a family of cell surface proteins, located on the outer side of the plasma membrane and whose members exhibit a time-keeping $CoQH_2$ (NADH) oxidase and protein disulfide-thiol interchange activities [50]. At least three members of this family have been described so far: the constitutive NADH oxidase (ENOX1 or cNOX), expressed in normal cells [51], the tumour-associated NADH oxidase (also known as ENOX2 or tNOX) [52], which is present on the surface of invasive cancer cells and in the sera of cancer patients, and the age-related NADH oxidase (also named arNOX), which seems to be related to the aging process, as it is present only in individuals after the age of 30, in late passage cultured cells and in senescent plants [53].

ENOX proteins contain an NADH-binding site, a disulfide-thiol interchange site and potential copper-binding sites, as well as two cysteine residues critical for catalytic activity [52, 54]. In addition, ENOX2 is specifically inhibited by natural and synthetic quinone analogues with anticancer activity (including capsaicin and adriamycin), whereas ENOX1 is not drug-responsive, but hormone- and growth factor-responsive [55], and arNOX is inhibited by CoQ10 [56]. The exact biological role of these oxidases is not well defined, although they have been implicated in aging, the enlargement phase of cell growth, and regulation of circadian oscillatory system [57-59].

The duodenal cytochrome b (Dcytb, also known as Cybrd1) was originally identified in duodenal enterocytes as a ferrireductase involved in dietary iron uptake [60], but, more recently, it has been found in the plasma membrane of other cell types (such as astrocytes, erythrocytes, hepatocytes and epithelial cells) [61-63]. Although the tertiary structure has not yet been solved, a partial characterization of a recombinant form of human Dcytb has been performed: it is a monomeric protein binding two heme groups, as previously suggested by the phylogenetic relationship with the cytochrome b_{561} family [64]. Dcytb may act as a trans-plasma membrane ascorbate:AFR oxidoreductase [65], thus maintaining physiological ascorbate concentrations within biological fluids (plasma, interstitial, and cerebrospinal fluid). Alternative electron acceptors are extracellular ferric or cupric ions [66, 67]; this finding, together with Dcytb expression on the brush border of duodenal enterocytes alongside the divalent metal ion transporter 1 (DMT1), indicate that this enzyme may play a key role in the uptake of dietary copper and nonheme iron [60].

Biological Functions of the PMR System

All the PMET pathways described so far have a general role in the control of redox homeostasis; thus, the intracellular redox state and, accordingly, the PMR system efficaciously modulate cell growth and survival.

Besides its above-mentioned role in the control of NAD^+/NADH ratio and bioenergetics, this system is also correlated with modulation of internal pH and redox homeostasis. Indeed, the activity of the PMR system is usually associated with a Na^+/H^+ antiport system, which pumps protons outside cells, thus buffering changes in intracellular pH and leading to acidification of the medium [17, 68]. Proton influx and efflux across the plasma membrane, not only contribute to maintain a constant intracellular pH, but also provide a mechanism modulating the polarization of biological membranes and regulating growth control of normal and transformed cells [69].

In addition, the PMR system exerts a modulating action on oxidative stress-induced apoptosis; indeed, CoQ sustains growth of cells in serum-limiting conditions characterized by mild oxidative stress and lipid peroxidation, whereas inhibitors of the PMR system induce cell growth arrest and apoptosis. The PMR system is constitutively activated in tumour cells and, thanks to its unique plasma membrane localisation, can be a novel target for the growth-inhibitory effects of several anti-cancer drugs (vanilloids, antracyclines, cis-platinum, bleomycin and phenoxodiol) [12, 70]; blocking its activity will compromise the cellular $NADH/NAD^+$ ratio, thus compromising the viability of transformed cells [71]. PMR, for example, is a primary site of action of phenoxodiol, since the drug is able to bind and inactivate the plasma membrane oxidoreductase tNOX, thus inducing G1 arrest and apoptosis [72, 73]. Nonetheless, phenoxodiol does not act specifically on tNOX or cancer cells, as it can inhibit proliferation and trigger cell death also in primary cells, including endothelial cells and activated primary T lymphocytes or peripheral blood mononuclear cells [74,75]. These findings suggest a potential harmful role for some drugs, which alter the PMR system: pharmacological concentrations of phenoxodiol can induce immunosuppressive effects, as already reported in Phase I clinical trials where 19 out of 21 patients with solid cancer experienced lymphocytopenia at doses up to 30 mg/kg [76].

Parallel to the ubiquitous functions described above, some specific biological roles of the PMR system can be put in evidence, depending on the cell type considered. For shortness, only three significative examples are reported below.

a) Iron Uptake

Dietary iron uptake from the intestinal lumen is mediated by DMT1 that transports ferrous iron. As described above, Dcytb is the ferrireductase driving the ferric to ferrous iron reduction. The mechanism of action is still unclear, since Gunshin *et al.*, [77] found that loss of the ferrireductase in mice has no impact on body iron stores. In addition, a nonenzymatic ferrireduction by ascorbate has been proposed by Lane and Lawen [78], which demonstrated that DHA uptake and ascorbate release are enough for extracellular ferrireduction.

b) Cellular Defence

Because of its location, the PMR system represents a mechanism of cellular defence. It exerts a dual role, as it can be both pro-oxidant and anti-oxidant. Indeed, some PMR enzymes maintain adequate antioxidant levels into the plasma membrane to counteract oxidative stress. On the other hand, plasma membrane NADPH oxidases can produce localized ROS (such as superoxide, hydrogen peroxide and singlet oxygen), involved in death of invading microorganisms. Recently, Huang *et al.,* reported that Nox2-mediated ROS generation mediates recruitment of the autophagy machinery, a defence mechanism responsible for clearance of cellular aberrant proteins and invading pathogens [79-81]. Finally, plasma membrane ROS generation is also involved in the oxidative unmasking of auto-antibodies in healthy individuals [82]: it has been suggested, for example, that oxidative modification of antibodies can occur through the action of both NADH and NADPH plasma membrane oxidases [83]. This finding should be useful in treatment of autoimmune diseases, including rheumatoid arthritis and systemic lupus erythematosus, characterized by chronic ROS production that leads to generation of neo-antigenic determinants able to activate T and/or B cells [84, 85].

c) Fertilization

Another biological function concerns the involvement of the PMR system in fertilization. We have previously mentioned the presence of sperm-specific redox enzymes that control male fertility. Also oocytes possess a NADPH-specific oxidase needed for generating protein coats, which provide a structural block to polyspermy, after fertilization; this enzyme, homolog to the dual oxidase Duox1, accumulates at the cell surface of each zygote, generating hydrogen peroxide that is necessary for the physical block to polyspermy [86].

Finally, several components of the PMR system are modulated during the aging process, which is characterized by decreased levels of alpha-tocopherol and increased levels of lipid peroxidation. Lowered mitochondrial respiratory activity due to the aging process can be compensated by up-regulating the plasma membrane NADH oxidizing systems, to maintain NAD^+ levels and glycolytic activity [87]. Thereby, mechanisms enhancing the PMR system could delay aging processes; this is the case of caloric restriction, which up-regulates the plasma membrane redox activity, that otherwise declines with age [88]. Similarly, as arNOX activity may be a crucial player in the onset of aged-related pathologies, its inhibition by CoQ10 supplementation provides a rational basis for developing new anti-aging drugs, especially in highly susceptible tissues [89-92].

From these findings, it appears clear that the PMR system plays a crucial role in body homeostasis. In particular, PMR-derived ROS are necessary for normal cellular functions, metabolism and survival, but the upside-down of the medal is that compromising this system and overproducing ROS can contribute to pathological diseases, including neurotoxicity, endothelial dysfunction and cardiovascular diseases [93]. What determines the cellular biological response initiated by ROS? Different levels of ROS induce distinct responses within a cell and, moreover, there is considerable variation between cells in the concentration required to initiate a particular biological effect. For example, superoxide anion and hydrogen peroxide employed for the killing of bacteria are produced by macrophage plasma membrane NADPH oxidase in the mM range, while non-toxic nM concentrations are produced by the not-phagocytic NADPH oxidases. Furthermore, the ability of these Nox systems to be modulated by extracellular effectors (growth factors and hormones) may allow a tailored ROS production in order to respond to the needs of the various tissues [94]. All aerobic forms of life use intracellular H_2O_2 concentrations ranging from 0.001 to 0.5-0.7 μM for signaling purposes. As long as the intracellular H_2O_2 level is maintained below 0.7 μM cells will undergo proliferation, above 0.7 μM oxidative stress and apoptotic death will occur [95]. Recently, it has been also suggested that the local antioxidant capacity contributes to the susceptibility of a cellular target to oxidative damage or signaling [95].

Plasma Membrane Redox System in Platelets

In the seventies, it has been shown that platelets possess an enzymatic armamentarium linked to the plasma membrane. Early papers [96, 97] suggested the presence of superoxide in the medium of platelet suspensions, based on the ability of resting and stimulated cells to reduce the nitroblue tetrazolium salt (NBT); indeed, the product of NBT reduction (formazan), being insoluble and unable to cross membranes, can give indications about $O_2^{\bullet-}$ localization. Later on,

Marcus *et al.,* [98] demonstrated the presence of two reducing activities in platelets. The first one led to production of extracellular superoxide, with a mechanism independent from the aggregation process; the second one was a superoxide-independent reducing activity, which increased with aggregation and seemed to be membrane-bound. Then, the burst in oxygen consumption, observed in stimulated platelets, was shown to be dependent on NADH or NADPH [99]. A plasma membrane activity was hypothesized, since pyridine nucleotides are membrane-impermeant and, moreover, the O_2 consumption was cyanide insensitive. In addition, the presence of multiple plasma membrane-associated activities was postulated, based on different results obtained with NADH or NADPH [99]. The platelet membrane redox activity, as well as its involvement in the aggregation response, is now widely recognized, thanks to the use of last generation tetrazolium salts, which are membrane-impermeant [100].

The first evidence of a NADH oxidase activity producing hydrogen peroxide (H_2O_2), in platelet plasma membrane, derived from an ultrastructural-cytochemical demonstration [101]; after that, several platelet agonists (thrombin, collagen, immunological stimuli) has been shown to stimulate H_2O_2 production [102, 103]. More recently, hydrogen peroxide produced by platelets has been implicated in specific signal transduction pathways, such as tyrosine phosphorylation [104]. The presence of an enzyme similar to ENOX1 or ENOX2 has been suggested from the study of Peter *et al.,* [105], which demonstrated a periodic and light responsive oxidation of NADH by human buffy coats (a mixture of white cells and platelets). Recently, we demonstrated that human platelets express ENOX1: furthermore, its expression was translationally modulated by capsaicin, through activation of the transient receptor potential cation channel, subfamily V, member 1 (TRPV1) and ROS generation [106]. The finding that ENOX1 is a redox-sensitive protein opens a new research field in terms of electron movement during platelet activation and cell-to-cell interactions [107].

Platelets also possess a NADPH oxidase complex, similar to that identified in phagocytic cells. The cytosolic ($p47^{phox}$, $p67^{phox}$) and membrane-bound ($p22^{phox}$ and $gp91^{phox}$) components of this enzymatic complex have been identified in platelets [108-110]. On the other hand, platelets are considered "covercytes" and not true phagocytic cells and, most importantly, they do not kill bacteria [111];

thus the function of NADPH oxidase in platelets is different. NADPH oxidase plays a key role in intracellular signaling leading to α_{IIb}/β_3-integrin activation [112]; furthermore, O_2^- (or the derivative H_2O_2) enhances the ADP release, resulting in increased platelet recruitment [110]. In addition, in host defence response, platelet NADPH oxidase causes the release of thromboxane A2, which, in turn, enhances ROS production by neutrophils and their cytotoxic action [113].

It is noteworthy that hydrogen peroxide has been disclosed as the primary second messenger in several cell types, affecting the activity of protein kinases and phosphatases [114]. Thus, the NADPH oxidase-derived O_2^- is likely converted to H_2O_2 which can diffuse inside cells. The temporal and spatial organization of ROS production appears to be important for activation of specific redox signaling pathways, as recently reported [115]. Since ROS generation by platelets may occur either inside the cells [112, 116, 117] or on the external surface, the platelet PMR activity may induce autocrine and paracrine effects, as shown in Fig. (**2**).

Although the NAD(P)H oxidases are the main enzymes of the PMR system in platelets, nonetheless other components have been identified, including NADH-diaphorase, cytochrome b5 reductase [118] and, more importantly, at least three thiol-related enzymes. In particular, two thiol isomerases and a glutathione reductase activity appear crucial for the rearrangement of disulfide bonds [119, 120]. Platelet membrane proteins contain redox-sensitive sulfhydryl groups needed for platelet aggregation, secretion and post-aggregation events through the activation of integrin receptors. The thiol/disulfide balance is fine-tunely regulated by low molecular weight thiols (glutathione or homocysteine) and by nitric oxide (NO), derived from S-nitrosothiols, which can change in several diseases [121]. Among the thiol isomerase family, protein disulphide isomerase (PDI) and endoplasmic reticulum protein 5 (ERP5) have been identified on the platelet surface; PDI interacts with $\alpha_2\beta_1$ and $\alpha_{IIb}\beta_3$ integrins, whereas ERP5 is associated only with the β3 integrin subunit. In addition, PDI concentrates at fibrin and thrombus formation sites of vascular injury, thus regulating coagulation factor ligation to thrombin-stimulated platelets and subsequent feedback activation of thrombin generation [122]. Importantly, PDI itself is activated by changes in the sulfhydryl state of its active site. So far, it is not clear which mechanism mediates PDI activation; Essex [119] suggested the involvement of external reducing compound(s) (such as glutathione), maintained in its active form by either NADPH oxidase or glutahione reductase. On the other hand, the presence, on platelet surface, of NADH oxidases with protein disulfide-thiol interchange activity can raise the possibility of an alternative mechanism of action, which directly modulates PDI (or integrins). Finally, a relationship between PDI and NADPH oxidases has been recently discovered: in this case, PDI regulates the activity of NADPH oxidase [123], so that a bi-directional mode of action may lead to reciprocal regulation of these two enzymes.

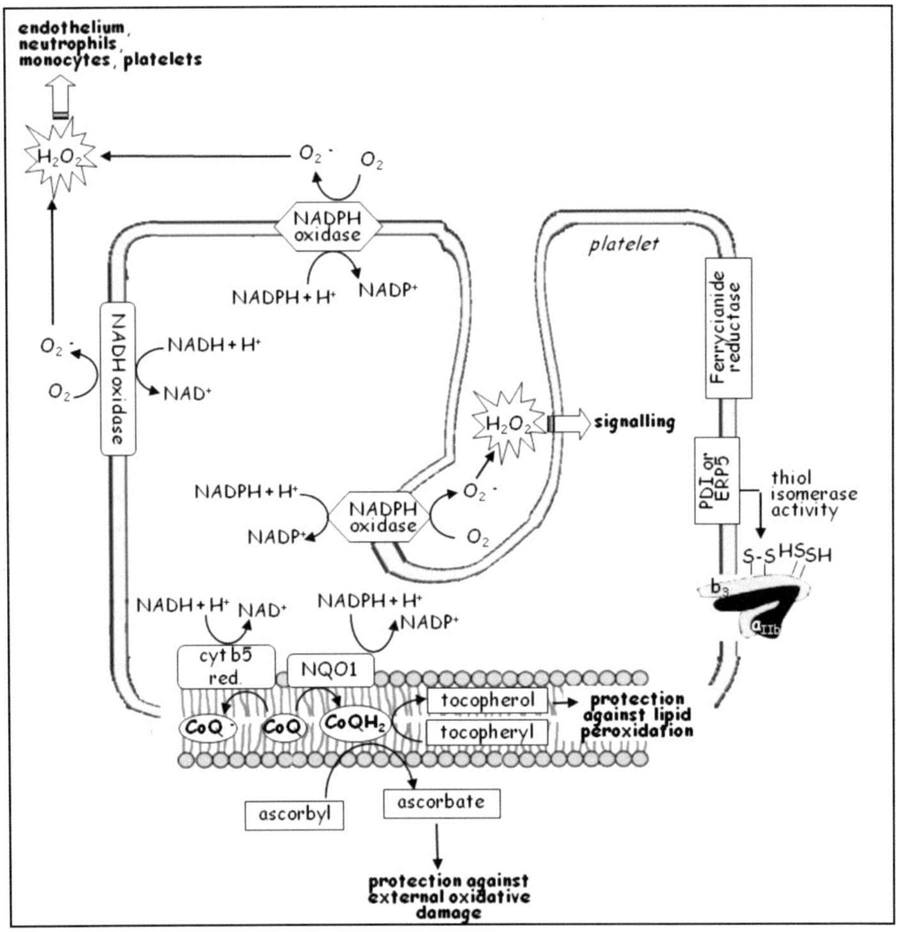

Figure 2: The PMR system in platelets. Plasma membrane enzymes are present on membranes, looking outside or inside (channels of the open canalicolar system) the cell. The main components are NADH- and NADPH-related enzymes (NADH and NADPH oxidases), which mantain the levels of $NAD(P)H/NAD(P)^+$ and reduced coenzyme Q (CoQ), while generating reactive oxygen species ($\mathbf{O_2^-}$ and H_2O_2).

MEDICINAL EFFECTS ON PLATELET PMR SYSTEM

From the literature data reported above, it is conceivable that increased levels of ROS play a key role in the pathogenesis of several diseases, including diabetes, essential hypertension, sickle cell disease, preeclampsia, thalassemia and cardiovascular disorders [124-126]. In blood, alteration on redox-state leads to dramatic changes in platelet functions, such as the activation/aggregation process; thus, platelets are involved in thrombotic vascular occlusion, and chronic platelet activation may play a role in thromboembolic complications. Just an example, a cross-sectional study including 51 cirrhotic patients demonstrated that, in these patients, platelets are iper-activated due to dysfunction of the PMR system (and, in particular, to increased Nox2 activity) [127].

Vitamin C has physiological relevance in platelets, because it can modulate the redox state of sulfhydryl groups. Indeed, we have demonstrated that intracellular ascorbate concentrations dose-dependently increased the amounts of surface thiols, while vitamin depletion decreased them. This effect appears to be specific for surface sulphydryl groups, since intracellular thiols and glutathione content (the main non-protein thiol in platelets) are unaffected by ascorbate supplementation or depletion [128]. Beside its role during aggregation, vitamin C seems to have also important effects on platelet-fibrin clot strength and stability that are crucial in pathophysiological conditions. Thrombus strength generated during post-clotting events is significantly reduced in ascorbate-depleted platelets with respect to control platelets and this parameter is in part restored if depleted platelets are supplemented with ascorbate [128]. Due to its relevance in platelet physiology, vitamin C uptake is strictly controlled, platelets compensate for fluctuations in ascorbate levels by modulating (at translational level) the expression of the Na^+-dependent transporter SVCT2. The control of ascorbate uptake, through regulation of its carrier, also occurs during platelet activation, characterized by vitamin C deprivation and alteration in redox state [128]. The platelet PMR system is capable to prevent extracellular ascorbate autoxidation (our unpublished data), thus contributing to stabilization of reduced vitamin C and, consequently, to modulation of the redox state in the micro-environment; indeed, changes on redox potential of blood may regulate activation of α_{IIb}/β_3 integrin in platelets. The ascorbate/dehydroascorbate redox couple (instead of ascorbate *per se*) seems to be important in those events accompanying platelet aggregation, as dehydroascorbate can be a substrate for PDI present on platelet surface [129].

Accumulating evidences show that vitamin C is also indirectly involved in platelet functions, by protecting membrane components susceptible to free radical damage and by regulating membrane systems sensitive to ROS-mediated signalling. Vitamin C reduces platelet CD40L expression, a transmembrane pro-inflammatory and pro-thrombotic protein implicated in initiation and progression of atherosclerotic disease through scavenging the O_2^- generated by NADPH oxidase activation [130].

By these data, it appears clear that a rational use of vitamin C supplementation may have beneficial effects on platelet-related diseases. It can decrease levels of platelet-derived microparticles, exerting a protective effect during myocardial infarction in patients with high thrombotic risk [131], and ameliorate the oxidative stress-mediated hypercoagulable state of patients with sickle cell disease and thalassaemia [124,132]. Vitamin C, whose levels (together with those of vitamins E) are lowered in preeclamptic women, can counteract platelet activation and oxidative stress, involved in early pregnancy and pathogenesis of preeclampsia [133]. Ascorbate supplementation suppresses platelet nitric oxide and $O_2^{\cdot-}$ production, as well as NADH oxidase activity, often associated with nitrate tolerance observed after continuous administration of organic nitrates [134]. Finally, in chronic smokers, oral vitamin C administration reverted the NADPH oxidase-mediated oxidative stress of platelets; this is achieved by restoring nitric oxide release, intraplatelet cGMP levels and platelet aggregability [135]. It is worthmentioning that other antioxidants (such as epicatechin present in dark chocolate) enhance artery dilatation in smokers, by lowering Nox2 activation: in a crossover, single-blind study enrolling 20 smokers and 20 healthy subjects, it has been found that platelets from smokers showed lower p47(phox) translocation to platelet membrane when incubated with oncreasing concentrations of epicatechin [136].

Vitamin E supplementation has been shown to play a protective effect against several cardiovascular disorders, through its ability to influence platelet activation and aggregation. Platelet vitamin E, whose levels strictly depend on dietary intake, has a dual role, exerting antioxidant as well as signalling effects unrelated with its antioxidant function [137]. Indeed, vitamin E supplementation reverts biochemical abnormalities (such as lipid peroxidation) observed in platelets from haemodialysis patients [138]; it also inhibits prostaglandin production and platelet aggregation, thus reducing vascular complications in diabetic patients [139], and, like vitamin C, reduces the platelet reactivity of thalassemic patients or individuals with sickle cell disease [133, 140]. Among the antioxidant-unrelated mechanisms of action, there is the vitamin E-dependent modulation of protein kinase C (PKC) activity. Indeed, the vitamin, by interfering with the PKC signalling pathway(s), inhibits platelet adhesion, activation and aggregation [141, 142]. In addition, mixed tocopherols have been shown to affect platelet aggregation by increasing NO release and decreasing superoxide production, through regulation of NOS and SOD activities [143]. Finally, the finding that vitamin E can impair NADPH-oxidase activation, in human subjects and animal models, suggests an additional mechanism by which tocopherols prevent coronary diseases [144].

The importance of nicotinamide supplementation in the maintenance of NAD^+ intracellular levels (thus influencing cellular life span) has been recently demonstrated [145, 146]. Beside its action as $NAD^+/NADP^+$ precursor, nicotinamide is also a substrate for three classes of enzymes: mono-ADP-ribosyltransferases, poly-ADP-ribose polymerase and ADP-ribosyl cyclase. In platelets, the multifunctional CD38 membrane enzyme has been shown to possess ADP-ribosyl cyclase and NAD^+ hydrolase activities [147]. The resulting cyclic ADP-ribose leads to calcium mobilization from intracellular stores, thus playing a crucial role during platelet aggregation [147]; this is further supported by the findings that thrombin stimulation leads to α_{IIb}/β_3 integrin-dependent association of the enzymatic activity with cytoskeleton [148]. Also mono ADP ribosylation of proteins has been demonstrated in platelets and a functional role in platelet activation has been suggested [149]. So, ADP ribosylation and NAD^+ glycohydrolase activity [147, 150], leading to NAD^+ depletion, influence the activity of platelet PMR system. It has recently been shown that nicotinic acid-adenine dinucleotide phosphate (NAADP) is a calcium-releasing second messenger contributing to thrombin-mediated platelet activation; the finding that a novel cell-permeant NAADP receptor antagonist has an inhibitory effect on platelet aggregation, secretion and spreading provides a potential avenue for platelet-targeted therapy and the regulation of thrombosis [151]. Finally, since exogenous nicotinamide has been shown to be effective when an unbalance in the NAD^+/NADH ratio occurs (at least in some cell types) [145, 146], further studies on the effects of supplementation on platelet functions should be hoped.

In conclusion, the PMR system is crucial in platelets, keeping in mind that the activation of coagulation pathways strictly depends on membrane activities, which are also required for the cross-talk among cells (platelets, monocytes, neutrophils, and endothelium). Further knowledge in regards to the role of the PMR system in platelet functions should promote the development of therapeutic approaches against several diseases, including diabetes, inflammation and thromboembolic pathologies.

ACKNOWLEDGEMENTS

This work was supported by grants from the Italian MIUR.

CONFLICT OF INTEREST

The authors confirm that this chapter contents have no conflict of interest.

ABBREVIATIONS

PMR	=	Plasma membrane redox system
PMET	=	Plasma membrane electron transport
CoQ	=	Ubiquinone coenzyme Q
ROS	=	Reactive oxygen species
NQO	=	NAD(P)H:ubiquinone oxidoreductase
VDAC	=	Voltage-Dependent Anion-selective Channel
NOX (1-5)	=	NADPH oxidases

tNOX = Tumour-associated NADH oxidase

cNOX = Constitutive NADH oxidase

PDI = Protein disulphide isomerase

ERP5 = Endoplasmic reticulum protein.

REFERENCES

[1] Scarlett, D.J.; Herst, P.; Tan, A.; Prata, C.; Berridge, M. Mitochondrial gene-knockout (rho0) cells: a versatile model for exploring the secrets of trans-plasma membrane electron transport. *Biofactors*, **2004**, *20*, 199-206.

[2] Lenaz, G.; Paolucci, U.; Fato, R.; D'Aurelio, M.; Parenti Castelli G.; Sgarbi, G.; Bigini, G.; Ragni, L.; Salardi, S.; Cacciari, E. Enhanced activity of the plasma membrane oxidoreductase in circulating lymphocytes from insulin-dependent diabetes mellitus patients. *Biochem. Biophys. Res. Commun.*, **2002**, *290*, 1589-1592.

[3] Lin, S.J.; Guarente, L. Nicotinamide adenine dinucleotide, a metabolic regulator of transcription, longevity and disease. *Curr. Opinion. Cell. Biol.*, **2003**, *15*, 241-246.

[4] Houtkooper, R.H.; Cantò, C.; Wanders, R.J.; Auwerx, The secret life of NAD+: an old metabolite controlling new metabolic signaling pathways. J. *Endocr. Rev.*, **2010**, *31*, 194-223.

[5] VanDuijn, M.M.; Van der Zee, J.; Van den Broek, P.J. The ascorbate-driven reduction of extracellular ascorbate free radical by the erythrocyte is an electrogenic process. *FEBS Lett.*, **2001**, *491*, 67-70.

[6] Lane, D.J.; Lawen, A. Transplasma membrane electron transport comes in two flavors. *Biofactors*, **2009**, *34*, 191-200.

[7] Lane, D.J.R.; Lawen, A. Non-transferrin iron reduction and uptake are regulated by transmembrane ascorbate cycling in K562 cells. *J. Biol. Chem.*, **2008**, *283*, 12701-12708.

[8] Himmelreich, U.; Drew, K.N.; Serianni, A.S.; Kuchel, P.W. 13C NMR studies of vitamin C transport and its redox cycling in human erythrocytes. *Biochemistry*, **1998**, *37*, 7578-7588.

[9] Lane, D.J.; Lawen, A. Ascorbate and plasma membrane electron transport--enzymes vs efflux. *Free Radic. Biol. Med.*, **2009**, *47*, 485-495.

[10] Fiorani, M.; Accorsi, A. Dietary flavonoids as intracellular substrates for an erythrocyte trans-plasma membrane oxidoreductase activity. *Br J Nutr.*, **2005**, *94*, 338-345.

[11] Villalba, J.M.; Navarro, F.; Gomez-Diaz, C.; Arroyo, A.; Bello, R.I.; Navas, P.; Role of cytochrome b5 reductase on the antioxidant function of coenzyme Q in the plasma membrane. *Mol. Aspects Med.*, **1997**, *18* (*Suppl.*), S7-13.

[12] Villalba, J.M.; Navas, P. Plasma membrane redox system in the control of stress-induced apoptosis. *Antioxid. Redox Signal.*, **2000**, *2*, 213-230.

[13] Morré, D.J.; Jacobs, E.; Sweeting, M.; de Cabo, R.; Morré, D.M. A protein disulfide-thiol interchange activity of HeLa plasma membranes inhibited by the antitumor sulfonylurea N-(4-methylphenylsulfonyl)-N'-(4-chlorophenyl) urea (LY181984). *Biochim. Biophys. Acta,* **1997**, *1325*, 117-125.

[14] Geiszt, M.; Leto, T.L. The Nox family of NAD(P)H oxidases: host defense and beyond. *J. Biol. Chem.* **2004**, *279*, 51715-51718.

[15] McKie, A.T.; Barrow, D.; Latunde-Dada, G.O.; Rolfs, A.; Sager, G.; Mudaly, E.; Mudaly, M.; Richardson, C.; Barlow, D.; Bomford, A.; Peters, T.J.; Raja, K.B.; Shirali, S.; Hediger, M.A.; Farzaneh, F.; Simpson, R.J. An iron-regulated ferric reductase associated with the absorption of dietary iron. *Science,* **2001**, *291*, 1755-1759.

[16] Poon, W.W.; Do, T.Q.; Marbois, B.N.; Clarke, C.F. Sensitivity to treatment with polyunsaturated fatty acids is a general characteristic of the ubiquinone-deficient yeast coq mutants. *Mol. Aspect. Med.* **1997**,18(Suppl), S121-127.

[17] Crane, F.L. Biochemical functions of coenzyme Q10. *J. Am. Coll. Nutr.,* **2001**, *20*, 591-598.

[18] Clemens, M.R.; Waller, H.D. Lipid peroxidation in erythrocytes. *Chem. Phys. Lipids,* **1987**, *45*, 251-268.

[19] Huang, J.; May, J.M. Ascorbic acid spares alpha-tocopherol and prevents lipid peroxidation in cultured H4IIE liver cells. *Mol. Cell. Biochem.,* **2003**, *247*,171-176.

[20] Lawen, A.; Ly, J.D.; Lane, D.J.; Zarschler, K.; Messina, A.; De Pinto, V. Voltage-dependent anion-selective channel 1 (VDAC1)--a mitochondrial protein, rediscovered as a novel enzyme in the plasma membrane. *Int. J. Biochem. Cell. Biol.,* **2005**, *37*, 277-282.

[21] Bayrhuber, M.; Meins, T.; Habeck, M.; Becker, S.; Giller, K.; Villinger, S.; Vonrhein, C.; Griesinger, C.; Zweckstetter, M.; Zeth, K. Structure of the human voltage-dependent anion channel. *Proc. Natl. Acad. Sci. U.S.A.*, **2008**, *105*, 15370-15375.

[22] Lawen, A. Apoptosis-an introduction. *BioEssays*, **2003**, *25*, 888-896.

[23] Berridge, M.V.; Herst, P.M.; Lawen, A. Targeting mitochondrial permeability in cancer drug development. *Mol. Nutr. Food Res.*, **2009**, *53*, 76-86.

[24] Baines, C.P.; Kaiser, R.A.; Sheiko, T.; Craigen, W.J.; Molkentin, J.D. Voltage-dependent anion channels are dispensable for mitochondrial-dependent cell death. *Nat. Cell Biol.*, **2007**, *9*, 550-555.

[25] Krauskopf, A.; Eriksson, O.; Craigen, W.J.; Forte, M.A.; Bernardi, P. Properties of the permeability transition in VDAC1(-/-) mitochondria. *Biochim. Biophys. Acta*, **2006**, *1757*, 590-595.

[26] Baker, M.A.; Lane, D.J.; Ly, J.D.; De Pinto, V.; Lawen, A. VDAC1 is a transplasma membrane NADH-ferricyanide reductase. *J. Biol. Chem.*, **2004**, *279*, 4811-4819.

[27] De Pinto, V.; Messina, A.; Lane, D.J.; Lawen, A. Voltage-dependent anion-selective channel (VDAC) in the plasma membrane. *FEBS Lett.*, **2010**, *584*, 1793-1799.

[28] Li, R.; Bianchet, M.A.; Talalay, P.; Amzel, L.M. The three-dimensional structure of NAD(P)H:quinone reductase, a flavoprotein involved in cancer chemoprotection and chemotherapy: mechanism of the two-electron reduction. *Proc. Natl. Acad. Sci.*, **1995**, *92*, 8846-8850.

[29] Wu, K.; Knox, R.; Sun, X.; Joseph, P.; Jaiswal, A.; Zhang, D.; Deng, P.; Chen, S. Catalytic properties of NAD(P)H:quinone oxidoreductase-2 (NQO2), a dihydronicotinamide riboside dependent oxidoreductase. *Arch. Biochem. Biophys.*, **1997**, *347*, 221-228.

[30] Bianchet, M.A.; Faig, M.; Amzel, L.M. Structure and mechanism of NAD[P]H:quinone acceptor oxidoreductases (NQO). *Methods in Enzymol.*, **2004**, *382*, 144-174.

[31] Kagan, V.E.; Arroyo, A.; Tyurin, V.A.; Tyurinaa, Y.Y.; Villalba, J.M.; Navas, P. Plasma membrane NADH-coenzyme Q0 reductase generates semiquinone radicals and recycles vitamin E homologue in a superoxide-dependent reaction. *FEBS Lett.* **1998**, *428*, 43-46.

[32] Hosoda, S.; Nakamura, W.; Hayashi, K. Properties and reaction mechanism of DT diaphorase from rat liver. *J. Biol. Chem.*, **1974**, *249*, 6416-6423.

[33] Cullen, J.J.; Hinkhouse, M.M.; Grady, M.; Gaut, A.W.; Liu, J.; Zhang, Y.P.; Darby Weydert, C.J.; Domann, F.E.; Oberley, L.W. Dicumarol inhibition of NADPH:quinone oxidoreductase induces growth inhibition of pancreatic cancer *via* a superoxide-mediated mechanism. *Cancer Res.*, **2003**, *63*, 5513-5520.

[34] Nolan, K.A.; Doncaster, J.R.; Dunstan, M.S.; Scott, K.A.; Frenkel, A.D.; Siegel, D.; Ross, D.; Barnes, J.; Levy, C.; Leys, D.; Whitehead, R.C.; Stratford, I.J.; Bryce, R.A. Synthesis and biological evaluation of coumarin-based inhibitors of NAD(P)H: quinone oxidoreductase-1 (NQO1). *J. Med. Chem.*, **2009**, *52*, 7142-7156.

[35] Percy, M.J.; Lappin, T.R. Recessive congenital methaemoglobinaemia: cytochrome b(5) reductase deficiency. *Br. J. Haematol.*, **2008**, *141*, 298-308.

[36] Shirabe, K.; Landi, M.T.; Takeshita, M.; Uziel, G.; Fedrizzi, E.; Borgese, N. A novel point mutation in a 3' splice site of the NADH-cytochrome b5 reductase gene results in immunologically undetectable enzyme and impaired NADH-dependent ascorbate regeneration in cultured fibroblasts of a patient with type II hereditary methemoglobinemia. *Am. J. Hum. Genet.*, **1995**, *57*, 302-310.

[37] Samhan-Arias, A.K.; Garcia-Bereguiain, M.A.; Martin-Romero, F.J.; Gutierrez-Merino, C. Clustering of plasma membrane-bound cytochrome b5 reductase within 'lipid raft' microdomains of the neuronal plasma membrane. *Mol. Cell. Neurosci.*, **2009**, *40*, 14-26.

[38] Nauseef, W.M. Biological roles for the NOX family NADPH oxidases. *J. Biol. Chem.*, **2008**, *283*, 16961-16965.

[39] Sumimoto, H. Structure, regulation and evolution of Nox-family NADPH oxidases that produce reactive oxygen species. *FEBS J.*, **2008**, *275*, 3249-3277.

[40] Dang, P.M.; Cross, A.R.; Babior, B.M. Assembly of the neutrophil respiratory burst oxidase: a direct interaction between p67PHOX and cytochrome b558. *Proc. Natl. Acad. Sci.*, **2001**, *98*, 3001-3005.

[41] Brown, D.I.; Griendling, K.K. Nox proteins in signal transduction. *Free Radic. Biol. Med.*, **2009**, *47*, 1239-1253.

[42] Lassegue, B.; Sorescu, D.; Szocs, K.; Yin, Q.; Akers, M.; Zhang, Y.; Grant, S.L.; Lambeth, J.D.; Griendling, K.K. Novel gp91(phox) homologues in vascular smooth muscle cells : nox1 mediates angiotensin II-induced superoxide formation and redox-sensitive signaling pathways. *Circ. Res.*, **2001**, *88*, 888-894.

[43] Kawahara, T.; Kuwano, Y.; Teshima-Kondo, S.; Takeya, R.; Sumimoto, H.; Kishi, K.; Tsunawaki, S.; Hirayama, T.; Rokutan, K. Role of nicotinamide adenine dinucleotide phosphate oxidase 1 in oxidative burst response to Toll-like receptor 5 signaling in large intestinal epithelial cells. *J. Immunol.*, **2004**, *172*, 3051-3058.

[44] Cheng, G.; Cao, Z.; Xu, X.; van Meir, E.G.; Lambeth, J.D. Homologs of gp91phox: cloning and tissue expression of Nox3, Nox4, and Nox5. *Gene*, **2001**, *269*, 131-140.

[45] Paffenholz, R.; Bergstrom, R.A.; Pasutto, F.; Wabnitz, P.; Munroe, R.J.; Jagla, W.; Heinzmann, U.; Marquardt, A.; Bareiss, A.; Laufs, J.; Russ, A.; Stumm, G.; Schimenti, J.C.; Bergstrom, D.E. Vestibular defects in head-tilt mice result from mutations in Nox3, encoding an NADPH oxidase. *Genes Dev.*, **2004**, *18*, 486-491.

[46] Shiose, A.; Kuroda, J.; Tsuruya, K.; Hirai, M.; Hirakata, H.; Naito, S.; Hattori, M.; Sakaki, Y.; Sumimoto, H. A novel superoxide-producing NAD(P)H oxidase in kidney. *J. Biol. Chem.*, **2001**, *276*, 1417-1423.

[47] Banfi, B.; Molnar, G.; Maturana, A.; Steger, K.; Hegedus, B.; Demaurex, N.; Krause, K.H. A Ca(2+)-activated NADPH oxidase in testis, spleen, and lymph nodes. *J. Biol. Chem.*, **2001**, *276*, 37594-37601.
[48] De Deken, X.; Wang, D.; Many, M.C.; Costagliola, S.; Libert, F.; Vassart, G.; Dumont, J.E.; Miot, F. Cloning of two human thyroid cDNAs encoding new members of the NADPH oxidase family. *J. Biol. Chem.*, **2000**, *275*, 23227-23233.
[49] Song, Y.; Ruf, J.; Lothaire, P.; Dequanter, D.; Andry, G.; Willemse, E.; Dumont, J.E.; Van Sande, J.; De Deken, X. Association of duoxes with thyroid peroxidase and its regulation in thyrocytes. *J. Clin. Endocrinol. Metab.*, **2010**, *95*, 375-382.
[50] Morré, D.J.; Morré, D.M. Cell surface NADH oxidases (ECTO-NOX proteins) with roles in cancer, cellular time-keeping, growth, aging and neurodegenerative diseases. *Free Radic. Res.*, **2003**, *37*, 795-808.
[51] Wang, S.; Pogue, R.; Morré, D.M.; Morré, D.J. NADH oxidase activity (NOX) and enlargement of HeLa cells oscillate with two different temperature-compensated period lengths of 22 and 24 minutes corresponding to different NOX forms. *Biochim. Biophys. Acta*, **2001**,*1539*, 192-204.
[52] Chueh, P.J.; Kim, C.; Cho, N.; Morré, D.M.; Morré, D.J. Molecular cloning and characterization of a tumor-associated, growth-related, and time-keeping hydroquinone (NADH) oxidase (tNOX) of the HeLa cell surface. *Biochemistry*, **2002**, *41*, 3732-3741.
[53] Morrè, D.M.; Gu,o F.; and Morrè, D.J. An aging-related cell surface NADH oxidase (arNOX) generates superoxide and is inhibited by coenzyme Q. *Mol. Cell. Biochem.*, **2003**, *254*, 101-109.
[54] Jiang, Z.; Gorenstein, N.M.; Morrè, D.M.; Morrè, D.J. Molecular cloning and characterization of a candidate human growth-related and time-keeping constitutive cell surface hydroquinone (NADH) oxidase. *Biochemistry*, **2008**, *47*, 14028-14038.
[55] Brightman, A.O.; Wang, J.; Miu, R.K.; Sun, I.L.; Barr, R.; Crane, F.L.; Morrè, D.J. A growth factor- and hormone-stimulated NADH oxidase from rat liver plasma membrane. *Biochim. Biophys. Acta*, **1992**, *1105*, 109-117.
[56] Morrè, D.M.; Morrè, D.J.; Rehmus, W.; Kern, D. Supplementation with CoQ10 lowers age-related (ar) NOX levels in healthy subjects. *Biofactors*, **2008**, *32*, 221-230.
[57] Wang, S.; Pogue, R.; Morrè, D.M.; Morrè, D.J. NADH oxidase activity (NOX) and enlargement of HeLa cells oscillate with two different temperature-compensated period lengths of 22 and 24 minutes corresponding to different NOX forms. *Biochim. Biophys. Acta*, **2001**, *1539*, 192-204.
[58] Morrè, D.J.; Chueh, P-J.; Pletcher, J.; Tang, X.; Wu, L-Y.; Morrè, D.M. Biochemical basis for the biological clock. *Biochemistry*, **2002**, *41*, 11941-11945.
[59] Morrè, D.M.; Guo, F.; Morrè, D.J. An aging-related cell surface NADH oxidase (arNOX) generates superoxide and is inhibited by coenzyme Q. *Mol. Cell. Biochem.*, **2003**, *254*, 101-109.
[60] McKie, A.T.; Barrow, D.; Latunde-Dada, G.O.; Rolfs, A.; Sager, G.; Mudaly, E.; Mudaly, M.; Richardson, C.; Barlow, D.;, Bomford, A.; Peters, T.J.; Raja, K.B.; Shirali, S.; Hediger, M.A.; Farzaneh, F.; Simpson, R.J. An iron-regulated ferric reductase associated with the absorption of dietary iron. *Science*, **2001**, *291*, 1755-1759.
[61] Balusikova, K.; Neubauerova, J.; Dostalikova-Cimburova, M.; Horak, J.; Kovar, Differing expression of genes involved in non-transferrin iron transport across plasma membrane in

various cell types under iron deficiency and excess. J. *Mol. Cell. Biochem.*, **2009**, *321*, 123-133.

[62] Su, D.; May, J.M.; Koury, M.J.; Asard, H. Human erythrocyte membranes contain a cytochrome b561 that may be involved in extracellular ascorbate recycling. *J. Biol. Chem.*, **2006**, *281*, 39852-39859.

[63] Turi, J.L.; Wang, X.; McKie, A.T.; Nozik-Grayck, E.; Mamo, L.B.; Crissman, K.; Piantadosi, C.A.; Ghio, A.J. Duodenal cytochrome b: a novel ferrireductase in airway epithelial cells. *Am. J. Physiol. Lung Cell. Mol. Physiol.*, **2006**, *291*, L272-280.

[64] Ludwiczek, S.; Rosell, F.I.; Ludwiczek, M.L.; Mauk, A.G. Recombinant expression and initial characterization of the putative human enteric ferric reductase Dcytb. *Biochemistry*, **2008**, *47*, 753-761.

[65] Oakhill, J.S.; Marritt, S.J.; Gareta, E.G.; Cammack, R.; McKie, A.T. Functional characterization of human duodenal cytochrome b (Cybrd1): Redox properties in relation to iron and ascorbate metabolism. *Biochim. Biophys. Acta*, **2008**, *1777*, 260-268.

[66] McKie, A.T. The role of Dcytb in iron metabolism: an update. *Biochem. Soc. Trans.*, **2008**, *36*, 1239-1241.

[67] Wyman, S.; Simpson, R.J.; McKie, A.T.; Sharp, P.A. Dcytb (Cybrd1) functions as both a ferric and a cupric reductase *in vitro*. *FEBS Lett.*, **2008**, *582*, 1901-1906.

[68] Berridge, M.V.; Tan, A.S. High-capacity redox control at the plasma membrane of mammalian cells: trans-membrane, cell surface, and serum NADH-oxidases. *Antioxid. Redox Signal.*, **2000**, *2*, 231-242.

[69] Chueh, P.J. Cell membrane redox systems and transformation. *Antioxid. Redox Signal.*, **2000**, *2*, 177-187.

[70] Navas, P.; Fernandez-Ayala, D.M.; Martin, S.F.; Lopez-Lluch, G.; De Caboa, R.; Rodriguez-Aguilera J.C.; Villalba, J.M. Ceramide-dependent caspase 3 activation is prevented by coenzyme Q from plasma membrane in serum-deprived cells. *Free Radic. Res.* **2002**, *36*, 369-374.

[71] Herst, P.M.; Berridge, M.V. Plasma membrane electron transport: a new target for cancer drug development. *Curr. Mol. Med.,* **2006**, *6*, 895-904.

[72] Morré, D.J.; Chueh, P.J.; Yagiz, K.; Balicki, A.; Kim, C.; Morré, D.M. ECTO-NOX target for the anticancer isoflavene phenoxodiol. *Oncol. Res.,* **2007**, *16*, 299-312.

[73] De Luca, T.; Morré, D.M.; Zhao, H.; Morré, D.J. NAD+/NADH and/or CoQ/CoQH2 ratios from plasma membrane electron transport may determine ceramide and sphingosine-1-phosphate levels accompanying G1 arrest and apoptosis. *Biofactors,* **2005**, *25*, 43-60.

[74] Gamble, J.R.; Xia, P.; Hahn, C.N.; Drew, J.J.; Drogemuller, C.J.; Brown, D.; Vadas, M.A. Phenoxodiol, an experimental anticancer drug, shows potent antiangiogenic properties in addition to its antitumour effects. *Int. J. Cancer.* **2006**, *118*, 2412-2420.

[75] Herst, P.M.; Petersen, T.; Jerram, P.; Baty, J.; Berridge, M.V. The antiproliferative effects of phenoxodiol are associated with inhibition of plasma membrane electron transport in tumour cell lines and primary immune cells. *Biochem. Pharmacol.* **2007**, *74*, 1587-1595.

[76] Choueiri, T.K.; Mekhail, T.; Hutson, T.E.; Ganapathi, R.; Kelly, G.E.; Bukowski, R.M. Phase I trial of phenoxodiol delivered by continuous intravenous infusion in patients with solid cancer. *Ann. Oncol.*, **2006**, *17*, 860-865.

[77] Gunshin, H.; Starr, C.N.; Direnzo, C.; Fleming, M.D.; Jin, J.; Greer, E.L.; Sellers, V.M.; Galica, S.M.; Andrews, N.C. Cybrd1 (duodenal cytochrome b) is not necessary for dietary iron absorption in mice. *Blood*, **2005**, *106*, 2879-2883.

[78] Lane, D.J.R.; Lawen, A. Non-transferrin iron reduction and uptake are regulated by transmembrane ascorbate cycling in K562 cells. *J. Biol. Chem.*, **2008**, *283*, 12701-12708.

[79] Huang, J.; Brumell, J.H. NADPH oxidases contribute to autophagy regulation. *Autophagy*, **2009**, *5*, 887-889.

[80] Huang, J.; Canadien, V.; Lam, G.Y.; Steinberg, B.E.; Dinauer, M.C.; Magalhaes, M.A.; Glogauer, M.; Grinstein, S.; Brumell, J.H. Activation of antibacterial autophagy by NADPH oxidases. *Proc. Natl. Acad. Sci. U.S.A.*, **2009**, *106*, 6226-6231.

[81] Mizushima, N.; Levine, B.; Cuervo, A.M.; Klionsky, D.J. Autophagy fights disease through cellular self-digestion. *Nature*, **2008**, *451*, 1069-1075.

[82] McIntyre, J.A.; Wagenknecht, D.R.; Faulk, W.P. Autoantibodies unmasked by redox reactions. *J. Autoimmun.*, **2005**, *24*, 311-317.

[83] Crane, F.L.; Löw, H. Reactive oxygen species generation at the plasma membrane for antibody control. *Autoimmun. Rev.*, **2008**, *7*, 518-522.

[84] Griffiths, H.R. Is the generation of neo-antigenic determinants by free radicals central to the development of autoimmune rheumatoid disease? *Autoimmun. Rev.*, **2008**, *7*, 544-549.

[85] Margutti, P.; Matarrese, P.; Conti, F.; Colasanti, T.; Delunardo, F.; Capozzi, A.; Garofalo, T.; Profumo, E.; Rigano, R.; Siracusano, A.; Alessandri, C.; Salvati, B.; Valesini, G.; Malorni, W.; Sorice, M.; Ortona, E. Autoantibodies to the C-terminal subunit of RLIP76 induce oxidative stress and endothelial cell apoptosis in immune-mediated vascular diseases and atherosclerosis. *Blood*, **2008**, *111*, 4559-4570.

[86] Wong, J.L.; Creton, R.; Wessel, G.M. The oxidative burst at fertilization is dependent upon activation of the dual oxidase Udx1. *Dev. Cell.*, **2004**, *7*, 801-814.

[87] Kopsidas, G.; Kovalenko, S.A.; Heffernan, D.R.; Yarovaya, N.; Kramarova, L.; Stojanovski, D.; Borg, J.; Islam, M.M.; Caragounis, A.; Linnane, A.W. Tissue mitochondrial DNA changes. A stochastic system. *Ann. N. Y. Acad. Sci.*, **2000**, *908*, 226-243.

[88] Gedik, C.M.; Grant, G.; Morrice, P.C.; Wood, S.G.; Collins, A.R. Effects of age and dietary restriction on oxidative DNA damage, antioxidant protection and DNA repair in rats. *Eur. J. Nutr.*, **2005**, *44*, 263-272.

[89] Kern, D.G.; Draelos, Z.D.; Meadows, C.; Morrè, D.J.; Morrè, D.M. Controlling reactive oxygen species in skin at their source to reduce skin aging. *Rejuvenation Res.*, **2010**, *13*, 165-167.

[90] Morrè, D.J.; Morrè, D.M. Aging-related cell surface ECTO-NOX protein, arNOX, a preventive target to reduce atherogenic risk in the elderly. *Rejuvenation Res.*, **2006**, *9*, 231-236.

[91] Morrè, D.M.; Guo, F.; Morrè, D.J. An aging-related cell surface NADH oxidase (arNOX) generates superoxide and is inhibited by coenzyme Q. *Mol. Cell. Biochem.*, **2003**, *254*, 101-109.

[92] Morrè, D.M.; Morrè, D.J.; Rehmus, W.; Kern, D. Supplementation with CoQ10 lowers age-related (ar) NOX levels in healthy subjects. *Biofactors*, **2008**, *32*, 221-230.

[93] Cave, A.C.; Brewer, A.C.; Narayanapanicker, A.; Ray, R.; Grieve, D.J.; Walker, S.; Shah, A.M. NADPH oxidases in cardiovascular health and disease. *Antioxid. Redox Signal.*, **2006**, *8*, 691-728.

[94] Linnane, A.W.; Kios, M.; Vitetta, L. Healthy aging: regulation of the metabolome by cellular redox modulation and prooxidant signaling systems: the essential roles of superoxide anion and hydrogen peroxide. *Biogerontology*, **2007**, *8*, 445-467.

[95] Stone, J.R.; Yang, S. Hydrogen peroxide: a signaling messenger. *Antioxid. Red. Signaling.*, **2006**, *8*, 243-270.

[96] Dejesus, M. Jr; Fikrig, S.; Detwiler, T.J. Phagocytosis-stimulated nitroblue tetrazolium reduction by platelets. *Lab. Clin. Med.*, **1972**, *80*, 117-124.

[97] Del Principe, D.; Calducci, L.; Sabetta, G. Letter: NBT test in newborn platelets. *Thromb. Diath. Haemorrh.*, **1974**, *31*, 368-369.

[98] Marcus, A.J.; Silk, S.T.; Safier, L.B.; Ullman, H.L. Superoxide production and reducing activity in human platelets. *J. Clin. Invest.*, **1977**, *59*,149-158.

[99] Del Principe D.; Mancuso, G.; Menichelli, A.; Maretto, G.; Sabetta, G. Oxygen consumption in platelets of newborn infants before and after stimulation by thrombin. *Thromb. Haemost.*, **1976**, *35*, 712-716.

[100] Maekawa, Y.; Yagi, K.; Nonomura, A.; Kuraoku, R.; Nishiura, E.; Uchibori, E.; Takeuchi, K. A tetrazolium-based colorimetric assay for metabolic activity of stored blood platelets. *Thromb. Res.*, **2003**, *109*, 307-314.

[101] Del Principe, D.; Mancuso, G.; Menichelli, A.; Gabbiotti, M.; Cosmi, E.V.; Ghepardi, G. Production of hydrogen peroxide in phagocyting human platelets: an electron microscopic cytochemical demonstration. *Biol. Cell.*, **1980**, *38*, 135-140.

[102] Finazzi-Agrò, A.; Menichelli, A.; Persiani, M.; Biancini, G.; Del Principe, D. Hydrogen peroxide release from human blood platelets. *Biochim. Biophys. Acta*, **1982**, *718*, 21-25.

[103] Leoncini, G.; Maresca, M.; Colao, C.; Piana, A.; Armani, U. Increased hydrogen peroxide formation in platelets of patients affected with essential thrombocythaemia (ET). *Blood Coagul. Fibrinolysis*,**1992**, *3*, 271-277.

[104] Krotz, F.; Sohn, H.Y.; Pohl, U. Reactive oxygen species: players in the platelet game. *Arterioscler. Thromb. Vasc. Biol.*, **2004**, *24*, 1988-1996.

[105] Peter, A.D.; Morré, D.J, Morré, D.M. A light-responsive and periodic NADH oxidase activity of the cell surface of Tetrahymena and of human buffy coat cells. *Antioxid. Redox. Signal.*, **2000**, *2*, 289-300.

[106] Savini, I.; Arnone, R.; Rossi, A.; Catani, M.V.; Del Principe, D.; Avigliano, L. Redox modulation of Ecto-NOX1 in human platelets. *Mol. Membr. Biol.*, **2010**, *27*, 160-169.

[107] Del Principe, D.; Frega, G.; Savini, I.; Catani, M.V.; Rossi, A.; Avigliano, L. The plasma membrane redox system in human platelet functions and platelet-leukocyte interactions. *Thromb. Haemost.*, **2009**, *101*, 284-289.

[108] Seno, T.; Inoue, N.; Gao, D.; Okuda, M.; Sumi, Y.; Matsui, K.; Yamada, S.; Hirata, K.I.; Kawashima, S.; Tawa, R.; Imajoh-Ohmi, S.; Sakurai, H.; Yokoyama, M. Involvement of NADH/NADPH oxidase in human platelet ROS production. *Thromb. Res.*, **2001**, *103*, 399-409.

[109] Pignatelli, P.; Sanguigni, V.; Lenti, L.; Ferro, D.; Finocchi, A.; Rossi, P.; Violi, F. gp91phox-dependent expression of platelet CD40 ligand. *Circulation*, **2004**, *110*, 1326-1329.

[110] Krotz, F.; Sohn, H.Y.; Gloe, T.; Zahler, S.; Riexinger, T.; Schiele, T.M.; Becker, B.F.; Theisen, K.; Klauss, V.; Pohl, U. NAD(P)H oxidase-dependent platelet superoxide anion release increases platelet recruitment. *Blood*, **2002**, *100*, 917-924.

[111] White, J.G. Platelets are covercytes, not phagocytes: uptake of bacteria involves channels of the open canalicular system. *Platelets*, **2005**, *16*, 121-131.

[112] Begonja, A.J.; Gambaryan, S.; Geiger, J.; Aktas, B.; Pozgajova, M.; Nieswandt, B.; Walter, U. Platelet NAD(P)H-oxidase-generated ROS production regulates alphaIIbbeta3-integrin activation independent of the NO/cGMP pathway. *Blood*, **2005**, *106*, 2757-2760.

[113] Chlopicki, S.; Olszanecki, R.; Janiszewski, M.; Laurindo, F.R.; Panz, T.; Miedzobrodzki, J. Functional role of NADPH oxidase in activation of platelets. *Antioxid. Redox. Signal.*, **2004**, *6*, 691-698.

[114] Rhee, S.G.; Bae, Y.S.; Lee, S.R.; Kwon, J. Hydrogen peroxide: a key messenger that modulates protein phosphorylation through cysteine oxidation. *Sci STKE*, **2000**, *2000*, PE1.

[115] Ushio-Fukai, M. Localizing NADPH oxidase-derived ROS. *Sci STKE*, **2006**, *2006*, re8.

[116] Del Principe, D.; Menichelli, A.; De Matteis, W.; Di Giulio, S.; Giordani, M.; Savini, I.; Finazzi Agrò, A. Hydrogen peroxide is an intermediate in the platelet activation cascade triggered by collagen, but not by thrombin. *Thromb. Res.*, **1991**, *62*, 365-375.

[117] Rosado, J.A.; Nunez, A.M.; Lopez, J.J.; Pariente, J.A.; Salido, G.M. Intracellular Ca2+ homeostasis and aggregation in platelets are impaired by ethanol through the generation of H2O2 and oxidation of sulphydryl groups. *Arch. Biochem. Biophys.*, **2006**, *452*, 9-16.

[118] Takeshita, M.; Matsuki, T.; Tanishima, K.; Yubisui, T.; Yoneyama, Y.; Kurata, K.; Hara, N.; Igarashi, T. Alteration of NADH-diaphorase and cytochrome b5 reductase activities of erythrocytes, platelets, and leucocytes in hereditary methaemoglobinaemia with and without mental retardation. *J. Med. Genet.*,**1982**, *19*, 204-209.

[119] Essex, D.W.; Li, M.; Feinman, R.D.; Miller, A. Platelet surface glutathione reductase-like activity. *Blood*, **2004**, *104*, 1383-1385.

[120] Essex, D.W. The role of thiols and disulfides in platelet function. *Antioxid. Redox Signal.*, **2004**, *6*, 736-746.

[121] Essex, D.W.; Li, M. Redox modification of platelet glycoproteins. *Curr. Drug. Targets*, **2006**, *7*, 1233-1241.

[122] Jurk, K.; Lahav, J.; Van Aken, H.; Brodde, M.F.; Nofer, J.R.; Kehrel, B.E. Extracellular protein disulfide isomerase regulates feedback activation of platelet thrombin generation *via* modulation of coagulation factor binding. *J Thromb Haemost.* **2011**, *9*, 2278-2290.

[123] Janiszewski, M.; Lopes, L.R.; Carmo, A.O.; Pedro, M.A.; Brandes, R.P.; Santos, C.X.; Laurindo, F.R. Regulation of NAD(P)H oxidase by associated protein disulfide isomerase in vascular smooth muscle cells. *J. Biol. Chem.*, **2005**, *280*, 40813-40819.

[124] Amer, J.; Ghoti, H.; Rachmilewitz, E.; Koren, A.; Levin, C.; Fibach, E. Red blood cells, platelets and polymorphonuclear neutrophils of patients with sickle cell disease exhibit oxidative stress that can be ameliorated by antioxidants. *Br. J. Haematol.*, **2006**, *132*, 108-113.

[125] Amer, J.; Fibach, E. Oxidative status of platelets in normal and thalassemic blood. *Thromb. Haemost.*, **2004**, *92*, 1052-1059.

[126] Avigliano, L.; Savini, I.; Catani, M.V.; Del Principe, D. Trans-plasma membrane electron transport in human blood platelets. *Mini Rev. Med. Chem.*, **2008**, *8*, 555-563.

[127] Basili, S.; Raparelli, V.; Riggio, O.; Merli, M.; Carnevale, R.; Angelico, F.; Tellan, G.; Pignatelli, P.; Violi, F.; CALC Group. NADPH oxidase-mediated platelet isoprostane over-production in cirrhotic patients: implication for platelet activation. *Liver Int.* **2011**, *31*, 1533-1540.

[128] Savini, I.; Catani, M. V.; Arnone, R.; Rossi, A.; Frega, G.; Del Principe, D.; Avigliano, L. Translational control of the ascorbic acid transporter SVCT2 in human platelets. *Free Radic. Biol. Med.*, **2007**, *42*, 60-616.

[129] Essex, D.W.; Li, M. Redox control of platelet aggregation. *Biochemistry*, **2003**, *42*, 129-136.

[130] Pignatelli, P.; Sanguigni, V.; Paola, S.G.; Lo Coco, E.; Lenti, L.; Violi, F. Vitamin C inhibits platelet expression of CD40 ligand. *Free Radic. Biol. Med.,* **2005**, *38*, 1662-1666.

[131] Morel, O.; Jesel, L.; Hugel, B.; Douchet, M.P.; Zupan, M.; Chauvin, M.; Freyssinet, J.M.; Toti, F. Protective effects of vitamin C on endothelium damage and platelet activation during myocardial infarction in patients with sustained generation of circulating microparticles. *J. Thromb. Haemost.*, **2003**, *1*, 171-177.

[132] Amer, J.; Fibach, E. Oxidative status of platelets in normal and thalassemic blood. *Thromb. Haemost.*, **2004**, *92*, 1052-1059.

[133] Bar, J.; Ben-Haroush, A.; Feldberg, D.; Hod, M. The pharmacologic approach to the prevention of preeclampsia: from antiplatelet, antithrombosis and antioxidant therapy to anticonvulsants. *Curr. Med. Chem. Cardiovasc. Hematol. Agents*, **2005**, *3*, 181-185.

[134] McVeigh, G.E.; Hamilton, P.; Wilson, M.; Hanratty, C.G.; Leahey, W.J.; Devine, A.B.; Morgan, D.G.; Dixon, L.J.; McGrath, L.T. Platelet nitric oxide and superoxide release during the development of nitrate tolerance: effect of supplemental ascorbate. *Circulation*, **2002**, *106*, 208-213.

[135] Takajo, Y.; Ikeda, H.; Haramaki, N.; Murohara, T.; Imaizumi, T. Augmented oxidative stress of platelets in chronic smokers. Mechanisms of impaired platelet-derived nitric oxide bioactivity and augmented platelet aggregability. *J. Am. Coll. Cardiol.*, **2001**, *38*, 1320-1327.

[136] 138. Loffredo, L.; Carnevale, R.; Perri, L.; Catasca, E.; Augelletti, T.; Cangemi, R.; Albanese, F.; Piccheri, C.; Nocella, C.; Pignatelli, P.; Violi, F. NOX2-mediated arterial dysfunction in smokers: acute effect of dark chocolate. *Heart* **2011**, *97*, 1776-1781.

[137] Azzi, A.; Ricciarelli, R.; Zingg, J.M. Non-antioxidant molecular functions of alpha-tocopherol (vitamin E). *FEBS Lett.*, **2002**, *519*, 8-10.

[138] Taccone-Gallucci, M.; Lubrano, R.; Del Principe, D.; Menichelli, A.; Giordani, M.; Citti, G.; Morsetti, M.; Meloni, C.; Mozzarella, V.; Meschini, L. Platelet lipid peroxidation in haemodialysis patients: effects of vitamin E supplementation. *Nephrol. Dial. Transplant.*, **1989**, *4*, 975-978.

[139] Watanabe, J.; Umeda, F.; Wakasugi, H.; Ibayashi, H. Effect of vitamin E on platelet aggregation in diabetes mellitus. *J. Exp. Med.*, **1984**, *143*, 161-169.

[140] Unchern, S.; Laoharuangpanya, N.; Phumala, N.; Sipankapracha, P.; Pootrakul, P.; Fucharoen, S.; Wanachivanawin, W.; Chantharaksri, U. The effects of vitamin E on platelet activity in beta-thalassaemia patients. *Br. J. Haematol.*, **2003**, *123*, 738-744.

[141] Murohara, T.; Ikeda, H.; Katoh, A.; Takajo, Y.; Otsuka, Y.; Haramaki, N.; Imaizumi, T. Vitamin E inhibits lysophosphatidylcholine-induced endothelial dysfunction and platelet activation. *Antioxid. Redox. Signal.*, **2002**, *4*, 791-798.

[142] Freedman, J.E.; Farhat, J.H.; Loscalzo, J.; Keaney, J.F.Jr. alpha-tocopherol inhibits aggregation of human platelets by a protein kinase C-dependent mechanism. *Circulation*, **1996**, *94*, 2434-2440.

[143] Liu, M.; Wallmon, A.; Olsson-Mortlock, C.; Wallin, R.; Saldeen, T. Mixed tocopherols inhibit platelet aggregation in humans: potential mechanisms. *Am. J. Clin. Nutr.* **2003**, 77, 700-706.

[144] Ulker, S.; McKeown, P.P.; Bayraktutan, U. Vitamins reverse endothelial dysfunction through regulation of eNOS and NAD(P)H oxidase activities. *Hypertension*, **2003**, *41*, 534-539.

[145] Li, F.; Chong, Z.Z.; Maiese, K. Navigating novel mechanisms of cellular plasticity with the NAD+ precursor and nutrient nicotinamide. *Front. Biosci.*, **2004**, *9*, 2500-2520.

[146] Li, F.; Chong, Z.Z.; Maiese, K. ell Life *versus* cell longevity: the mysteries surrounding the NAD+ precursor nicotinamide. *Curr. Med. Chem.*, **2006**, *13*, 883-895.

[147] Ramaschi, G.; Torti, M.; Festetics, E.T.; Sinigaglia, F.; Malavasi, F.; Balduini, C. Expression of cyclic ADP-ribose-synthetizing CD38 molecule on human platelet membrane. *Blood*, **1996**, *87*, 2308-2313.

[148] Torti, M.; Festetics, E.T.; Bertoni, A.; Sinigaglia, F.; Balduini, C. Thrombin induces the association of cyclic ADP-ribose-synthesizing CD38 with the platelet cytoskeleton. *FEBS Lett.*, **1998**, *428*, 200-204.

[149] Brune, B.; Molina y Vedia, L.; Lapetina, E.G. Agonist-induced ADP-ribosylation of a cytosolic protein in human platelets. *Proc. Natl. Acad. Sci.*, **1990**, *87*, 3304-3308.

[150] Del Principe, D.; Menichelli, A.; Casini, A.; Di Giulio, S.; Mancuso, G.; Finazzi-Agrò, A. A surface NAD-glycohydrolase of human platelets may influence their aggregation. *FEBS Lett.* **1986**, *205*, 66-70.

[151] Coxon, C.H.; Lewis, A.M.; Sadler, A.J.; Vasudevan, S.R.; Thomas, A.; Dundas, K.A.; Taylor, L.; Campbell, R.D.; Gibbins, J.M.; Churchill, G.C.; Tucker, K.L. NAADP regulates human platelet function. *Biochem J.* **2012**, *441*, 435-442.

Send Orders for Reprints to reprints@benthamscience.net
Recent Advances in Medicinal Chemistry, Vol. 1, 2014, 433-455

CHAPTER 14

Role of Intermediate States in Protein Folding and Misfolding

Roberto Santucci[1,*], Fabio Polticelli[2,3], Federica Sinibaldi[1] and Laura Fiorucci[1]

[1]Department of Clinical Sciences and Translational Medicine, University of Rome 'Tor Vergata', Via Montpellier 1, 00133 Rome, Italy; [2]Department of Sciences, University Roma Tre, V.le Marconi 446, 00146 Rome, Italy and [3]National Institute of Nuclear Physics, 'Roma Tre' Section, 00146 Rome, Italy

Abstract: Most proteins fold into their native structure through defined pathways which involve a limited number of transient intermediates. Intermediates play a relevant role in the folding process; many diseases of genetic nature are in fact coupled with protein misfolding, which favours formation of stable inactive intermediate species of a protein. This review describes a number of diseases originated from protein misfolding and briefly discusses the mechanism(s) responsible, at molecular level, for these pathologies. It is also envisaged the *native* ⇄ *molten globule* transition since sometimes the conversion of the native form into a compact intermediate state permits a protein to carry out distinct physiological functions inside the cell. A non-native compact form of cyt *c*, for example, is involved in the programmed cell death (apoptosis) after that the protein is released from the mitochondrion; in addition, non-native forms of the protein are involved in some of the disorders attributed to amyloid formation.

Keywords: Alzheimer's disease, amyloid fibrils, apoptosis, conformational diseases, cystic fibrosis, cytochrome *c*, energy landscapes, folding pathways, intermediate states, Levinthal paradox, misfolding, molten globule, neurodegenerative diseases, phospholipids, protein folding.

INTRODUCTION

Defining the mechanism(s) governing protein folding remains a central point in biophysics and molecular biology. To explain how a protein synthesized as a linear unfolded polypeptide rapidly folds into its biologically active structure even though there are a large number of potential conformational states, remains an intriguing and still unsolved puzzle that has stimulated, over the last decades, the interest of researchers.

***Address correspondence to Roberto Santucci:** Department of Clinical Sciences and Translational Medicine, University of Rome 'Tor Vergata', Via Montpellier 1, 00133 Rome, Italy; Tel: + 39 06 72596364; Fax: + 39 06 72596353; E-mail: santucci@med.uniroma2.it

Atta-ur-Rahman, Muhammad Iqbal Choudhary and George Perry (Eds)

While small proteins are thought to fold through a two-state (*unfolded* ⇄ *native*) mechanism, larger proteins (*i.e.*, those with more than 100 residues) fold into the unique native structure through well defined folding pathways which involve a limited number of intermediate species. Therefore, big effort has been lavished to characterize the intermediates that form during the folding process.

In the last years, a large body of kinetic and equilibrium data have provided extensive information on the protein folding pathway(s) and on the structural properties of intermediate states, thus providing a relevant contribution to better understanding the process [1-8]. Its high complexity, however, renders hard a deep comprehension of the mechanism(s) governing protein folding [5, 9, 10]. For example, the characterization of kinetic intermediates forming when a protein folds is difficult to achieve, due in part to the high rapidity of the process and in part to the limitations of the experimental approach [11-13]. A valid alternative strategy is thus represented by the study of partially folded states at equilibrium which reveal how a non-native state of a protein is structurally organized.

Studies on proteins have outlined the strict correlation between equilibrium and kinetic intermediates and elucidated the properties of the 'molten globule', a non-native compact state considered a major intermediate in protein folding [14-17].

A deep knowledge of the properties of intermediate species is important; formation of non-native forms of a protein owing to improper folding often causes the onset of serious pathologies, generally indicated as conformational diseases. Examples include cystic fibrosis and neurodegenerative diseases such as Alzheimer's disease, Parkinson's disease, or Creutzfeldt-Jakob disease (see [18] and references therein). To date, the view that protein misfolding is mainly provoked by genetic mutations is widely accepted, although many aspects of the problem still remain unknown. Therefore it is not surprising that a big effort is still devoted to investigate protein folding/misfolding, an intriguing but not well understood process that, when fully defined, will provide a relevant contribution to the development of efficient therapeutic approaches *vs.* genetic diseases.

Unfolding / Refolding Kinetics

Under physiological-like conditions, a protein folds spontaneously into its native biologically active state following the information encoded in the amino acidic sequence of the polypeptide chain. Most proteins fold through a non-two-state process, which implies the formation of a limited number of intermediate states, as described by the following scheme:

$$U \rightleftarrows n\,(I) \rightleftarrows N$$

where U is the unfolded state, I the intermediate state(s), and N is the native state.

The kinetic investigation of protein folding is generally complicated by the uncertainty concerning the structural characteristics of partly folded states and the role they play during the folding process. An important, not fully clarified point concerns the role played by kinetic intermediates in the route towards formation of the native state. It is not clear yet, for example, whether an intermediate state I is to be treated as an original on-pathway transient species or, instead, as an accumulating off-pathway misfolded form produced by a side-reaction. Many reports (see below) hypothesize that transient intermediates accumulating as off-pathway forms in protein folding may induce diseases in humans [19-21]; thus, identifying the kinetic mechanism(s) governing protein folding and determining the properties of intermediates forming during the process remains a priority.

Protein folding is solely investigated under native-like conditions which favour the rapid formation of the biologically active structure. From a kinetic point of view, intermediates have been classified as 'early' or 'late' intermediates. While the 'early' intermediates form rapidly (within a few milliseconds), the 'late' intermediates form just prior to the lowest, rate-limiting step of the folding reaction (see [22] for a detailed description). Being short lived the 'early' intermediates are very difficult to study; conversely, the 'late' intermediates can be "kinetically trapped". Trapping facilitates the kinetic characterization of intermediate steps because it decreases the rate of folding.

The Molten Globule

The molten globule (MG), considered the major intermediate of protein folding, is a non-native compact state of a protein characterized by native-like secondary structure but fluctuating tertiary conformation. This implies that during folding the just synthesized protein collapses into a flexible compact species whose tertiary architecture lacks the tight packing typical of the native state, as illustrated in Fig. (**1**).

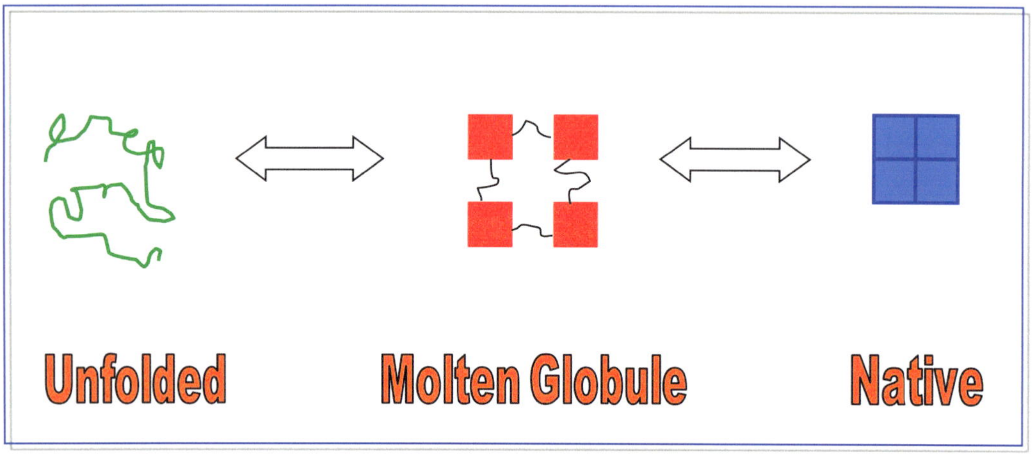

Figure 1: Molten globule model in the folding pathway of a protein.

The MG of several proteins, such as carbonic anhydrase, α-lactalbumin, apomyoglobin, cytochrome *c*, has been investigated in high detail during the last decade of the past century [23-26]. These studies have proved that the MG possesses a more expanded (less compact) structure with respect to the native state. The hypothesis asserting that the MG is stabilized mainly by hydrophobic interactions is supported by the fact that, although less packed, the non polar groups maintain in the MG their attraction in water. As observed for apomyoglobin the MG is stabilized mainly by hydrophobic interactions, while the native state is instead stabilized by a specific tight packing.

About twenty years ago, Ptitsyn and collaborators introduced the "non-uniform model" theory to describe the structural properties of the MG (see [27] and references therein). The theory is based on the hypothesis that in the MG macromolecules undergo a non uniform expansion. This implies that some regions of the macromolecule (those constituting the core of the protein, including α- and β-regions) remain fairly preserved (the core becomes a little less rigid), while other regions, as the loops and the ends of α- and β-regions, result less folded. Thus, although water molecules cannot penetrate relevantly inside the MG core because the non polar groups of α- and β-regions have a very limited capability to expand, the more unfolded regions can be easily penetrated by water. From this viewpoint, the MG can be assumed as a "liquid-like" system since the native arrangement of ordered secondary regions is stabilized by ("liquid-like") hydrophobic contacts.

The reason for which loop regions undergo conformational changes as the protein partially unfolds into the MG is due to the fact that in the native conformation some loops and the protein core are attached one to the other by interactions involving hydrophobic residues. Formation of the MG is coupled with a slight increase of the macromolecule volume, which facilitates release of these hydrophobic amino acids; as a consequence, loops unfold because their stability is significantly reduced without a proper hydrophobic support [27]. Very recent results have shown that, at the first stage of the unfolding process, the native protein expands to a form in which residues are unlocked while the van der Waals interactions persist. This intermediate, which has been called "Dry MG" (DMG), is thought to form in the first step of the unfolding process [16, 17].

The MG plays a primary role in a variety of biological processes. In some cases, proteins have been found to undergo conformational changes to perform their physiological (sometimes pathological) functions. Such conformational changes, of tertiary nature, are now becoming recognized as a mechanism potentially used by a protein to achieve structural diversity and, consequently, diversity of functions. To this issue, the MG state is ideally suited to provide the required conformational flexibility. In the case of apolipoprotein E4 (apoE4), a protein that plays an important role in lipid transport in the plasma and in the central nervous system, the enhanced propensity of a common human apoE4 isoform to stabilize the MG state is considered a major risk factor for Alzheimer's disease and atherosclerosis [28]. Also, oleic acid-bound α-lactalbumin, which acquires the properties typical of the MG, is thought to induce apoptosis in cancer cells [29].

Protein Folding: 'Classical View' *vs.* 'New View'

The mechanism(s) governing protein folding is (are) still far from being fully understood due to the high complexity of the process. Small proteins (those with less than one hundred aminoacidic residues) fold in two-state transitions; in this case, the unfolded polypeptide directly collapses into the compact, biologically active native conformation. Conversely, large proteins (those with more than one hundred aminoacidic residues) fold into the unique, native structure through a limited number of intermediate species. The intermediate states sometimes participate to productive folding (on-pathway intermediates), in other cases they accumulate as compact forms distinct from the native, trapped by non-native intramolecular interactions (off-pathway intermediates).

The "classic view" was initially introduced to tentatively explain the "Levinthal paradox" (proteins fold in a few seconds, although finding the native state among all the possible protein configurations theoretically requires a very long time) [30]. The idea is that a protein collapses into the stable, native conformation through well defined folding pathways involving a limited number of intermediate species. Much interest was thus devoted to understand how proteins find the correct folding pathway, avoiding the others. In the years, a growing number of kinetic and equilibrium studies have detected an appreciable number of folding intermediates (see [8, 12, 31-35] for recent reviews). The characterization of these partially folded forms has provided a precious contribution to a better understanding of the sequence of events carrying the protein from the unfolded to the compact native conformation.

More recently, the introduction of the so-called 'new view' has opened new perspectives. The models on which the 'new view' is based replace single specific 'folding pathways' with more complex multidimensional 'energy landscapes' (where an 'energy landscape' represents the free energy of a single protein conformation, strictly depending on the degrees of freedom) [36-40]. On the basis of Anfinsen studies [30] the 'new view' establishes that folding is path-independent, and envisages molecule populations and multiple folding pathways rather than specific structures and pathways [37, 41, 42]. According to this concept, terms like on- and off-pathways and intermediate states are now viewed as distributions of single chain conformations.

The 'new view' assumes that protein folding to the native state proceeds through multiple routes on funnel-like energy landscapes. Although energy landscapes may be rough, *i.e.* characterized by non-native minima in which partially folded molecules are trapped, the idea is that the whole protein population finally reaches the native conformation, corresponding to a free energy minimum under the solution conditions [37, 38, 42]. In other words, although single macromolecules follow an own distinct trajectory, the molecule population finally reaches the conformation corresponding to a minimum in free energy, *i.e.* the native state.

Studies carried out over the last two decades have significantly improved our understanding of the process, and led to the formulation of the 'new view' hypothesis. However, further studies are necessary for defining some still unclear aspects of the process. In a recent future the use of sophisticated techniques, such as atomic force microscopy (utilized to unfold proteins mechanically [43, 44]) or single-molecule fluorescence (able to detect the behaviour of single macromolecules, otherwise masked when the whole molecule population is considered) [45, 46] may provide, together with computational approaches, a remarkable contribution for better defining the energy landscapes and probing the 'new view' concept in protein folding.

Folding, Misfolding and Diseases: Role of Intermediate Conformers

Point mutations in proteins sometimes block the folding pathway at level of stable intermediate states. When this happens, the protein cannot adopt its native conformation and its biological activity is altered or lost. In many cases, the altered function of the mutated protein causes genetic diseases.

Cystic fibrosis is a disorder characterized by chronic lung diseases and caused by mutations of the cystic fibrosis trans-membrane conductance regulator (CFTMR) gene [47]. The protein expressed by the gene is the CFTMR protein, a member of the ATP-binding cassette transporters superfamily, which functions as a chloride channel and a regulatory protein [48]. Wild type (wt) CFTMR is synthesized and folded inside the cell endoplasmic reticulum (ER) as an immature precursor; in the Golgi apparatus the protein then folds into its 'native', biologically active structure following a complex glycosylation process which involves molecular chaperones [49]. The 70 kDa heat-shock protein (Hsp70), a major chaperone present in human cell cytosol, participates to the folding process of CFTMR; it binds to the newly synthesized, unfolded CFTMR and holds the protein in a state able to fold, thus permitting its translocation into different cell compartments [50]. The deletion of a single amino acid, Phe508, is the most common mutation occurring in CFTMR protein [48]. Unlike the wt protein, the ΔPhe508 mutant is an incompletely glycolsylated protein which folds into a relatively stable intermediate that requires a longer time to convert into the native conformation. Consequently, mutated CFTMR does not reach the cell membrane; once it folds, it is retained in the ER and degraded by the (protein-degrading complex) proteasome [51].

Chronic liver and lung diseases are caused by events of similar nature. In this case, the disease is caused by an insufficient amount of α-l-antitrypsin (AT), a 52 kDa serine protease inhibitor (see [52, 53] for recent reviews). This 394-residues protein, which is synthesized in hepatocytes, is secreted into circulation; its main function is to protect the lower respiratory tract of lungs from the proteolytic attack of neutrophil eleastase (NE). The AT structure includes 3 β-sheets and an exposed loop containing the residues interacting with the proteinase [54, 55]. The conformational changes induced in the loop by mutations convert the rigid native structure into a more flexible conformation, which causes protein inactivation. Genetic AT deficiency causes early lung and liver diseases since the delicate equilibrium between NE and AT (which is crucial for health) is lost. Of the approx 90 known variants of human AT, the S mutant (Glu264→Val) [53], and the Z mutant (Glu342→Lys) [52, 53] are the most common ones. Whereas the S variant is not associated with critical clinical diseases (the Glu264Val mutation does not provoke significant alterations of the protein structure), in the Z variant the mutation alters the protein conformation significantly, promoting the formation of protein aggregates. Formation of polymers generates insoluble inclusion bodies in the endoplasmic reticulum of the liver [56]; their accumulation in the hepatocytes provokes neonatal hepatitis and adult liver carcinoma. Further, the secretory defect leads to AT deficiency; this exposes lungs to the proteolytic attack of NE and to consequent alveolar destruction. Big effort is actually devoted to find strategies able to counteract the defects associated with AT deficiency [57, 58].

The polymerization phenomenon is not restricted to AT; other antiproteinases undergo similar processes. In antithrombin and α-antichimotrypsin, mutations sometimes favour the formation of partly folded states characterized by a high tendency to polymerize; this the cause of thrombosis or obstructive pulmonary diseases of genetic origin [59-62].

On the whole, the examples above described highlight the importance of studies devoted to define the properties of protein intermediates, of interest in biochemistry and experimental medicine.

As said above, cystic fibrosis and chronic liver diseases arise when the misfolded protein cannot reach its final cellular destination. However, a large body of misfolding-induced diseases is originated by the conversion of specific proteins from their soluble state to highly ordered aggregates, called amyloid fibrils (see [62-64] for recent reviews). Amyloid fibrils are correlated with a number of pathologies which include neurodegenerative disorders, including Alzheimer's disease and Parkinson's disease [65-68], or amyloidoses, pathological states due to formation of extracellular amyloid deposits [69, 70]. The fibrils-induced diseases have been divided into distinct groups: (i) neurodegenerative diseases (aggregation takes place in the brain), (ii) non-neuropathic localized amyloidoses (aggregation occurs in a single tissue distinct from the brain), and (iii) non-neuropathic systemic amyloidoses (aggregation occurs in more than one tissue). Amyloid fibrils are composed by a number (generally 2-to-6) of filaments, called protofilaments, twisted together. In each filament, the molecules arrangement favours formation of β-sheets running along the fibril main axis [71]. For some proteins, as amylin [72, 73] or the yeast prion protein Sup35p [74], the structural properties of fibrils have been determined in good detail.

In the last years studies have significantly improved our understanding of properties common to fibrils; these include the cross-β structure and the frequent hydrophobic (or polar) interactions along the fibril main axis [62, 70]. Further, in fibrils the β-sheets are less twisted than what expected from analysis of β-structure of globular proteins [75]. Protein conversion into fibrils follows a 'nucleated growth' biphasic mechanism, characterized by a lag phase during which the nucleus is formed, and by a rapid phase during which the oligomers, that still miss a fibrillar shape, associate with the nucleus itself [76-78]. The oligomers, unstructured and constituted by 3-6 protein molecules, are in equilibrium with monomeric forms and precede protofibrils formation. Once the critical mass is reached, the oligomers associate and give rise to protofibrils formation. As illustrated in Fig. (**2**), protofibrils act either as on-pathway [76, 79] or off-pathway [80, 81] intermediates in the kinetics of formation of fibrils. The aggregation process may be inhibited by molecular chaperones or degradation processes, while it is favoured by side chains hydrophobicity, low net charge of the polypeptide and chain propensity to form β-sheet regions.

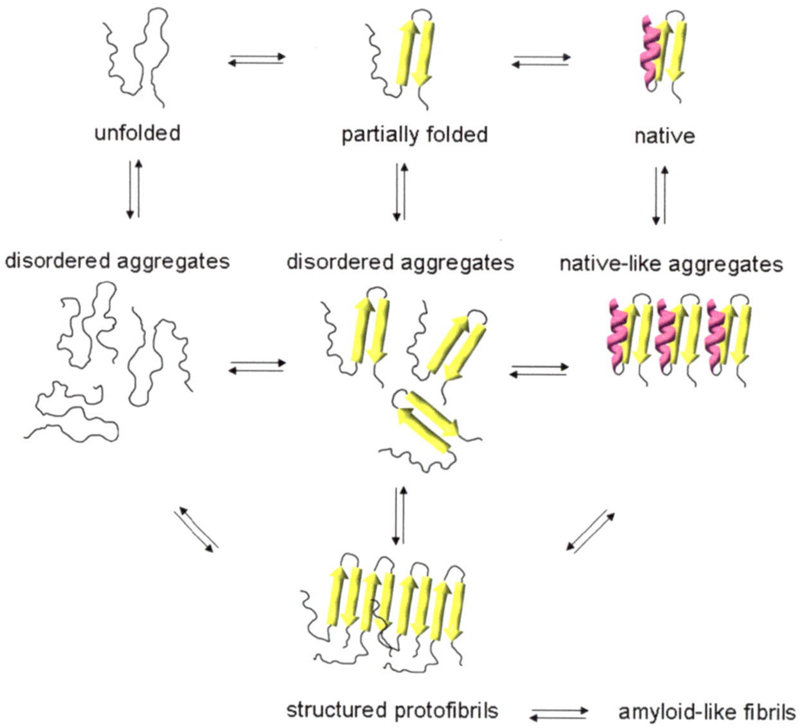

Figure 2: Schematic representation of the formation of amyloid-like fibrils with unfolded and partially folded monomeric states in equilibrium with the native state and misfolded aggregated states.

Cytochrome *c*: Non-Native Conformations and Distinct Biological Functions

Some authors have addressed the role of cofactors that may enable protein-folding variants to attain new functions. This is the case of α-lactalbumin bound to oleic acid that possesses spectroscopic properties typical of the molten globule and is able to induce apoptosis in cancer cells [82]. Recent findings have demonstrated that this same lipid cofactor (*i.e.*, oleic acid) is able to induce structural changes also in cytochrome *c* (cyt *c*), with formation of a molten globule-like state [83]. Indeed, in the last years several non-native conformers of cyt *c* have been characterized and different biological functions, strictly depending on the conformation of the hemoprotein, have been proposed [84-88].

Cyt *c*, whose structure is shown in Fig. (**3**), is a single chain hemoprotein of 104 amino acids containing three major and two minor α-helices in the structure, with the prosthetic group lying within a crevice lined with hydrophobic residues. The heme is covalently attached to the polypeptide chain by two thioether bridges with residues Cys14 and Cys17, while His18 and Met80 are the axial ligands of the six-coordinated low spin heme iron in the native state [89]. As mitochondrial peripheral membrane protein, it functions in between the inner and outer membrane, mediating electron transfer (eT) between different proteins of the respiratory chain. Studies on the interaction between cyt *c* and various membrane systems indicate that cyt *c* mediates eT between cyt *c* reductase and cyt *c* oxidase as unbound or membrane-bound protein showing a limited number of non-native exchangeable compact conformations [90, 91].

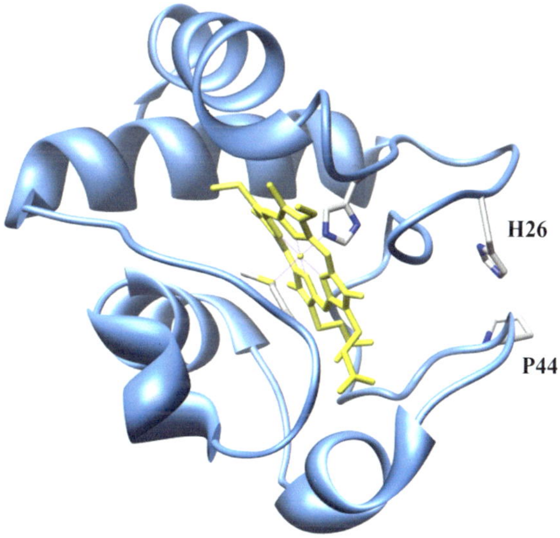

Figure 3: Ribbon structure of horse cyt *c*. The heme group, bound to His18 (right) and Met80 (left), is represented in yellow. On the right side, residues His26 and Pro44 are shown. The H-bond between His26 and the carbonyl group of the residue occupying position 44 (a proline in equine cyt *c*, a glutamic acid in yeast cyt *c* [117, 118]) keeps two Ω-loops bound one to the other. The protein structure was visualized UCSF Chimera software [119].

The finding that cyt *c* plays, as illustrated in Fig. (**4**), a role in the programmed cell death (*i.e.*, in cell apoptosis) after its release from the mitochondrion, has renewed interest in this protein (see [92, 93] and references therein). About 15% of mitochondrial cyt c remains tightly bound to the inner mitochondrial membrane (IMM); this protein fraction is thought to be involved in peroxidase activity, an event considered crucial for the initiation of the apoptotic process [94]. Cardiolipin (CL) (which constitutes approx the 20% of the total lipids composition of the mitochondrial membrane) is the phospholipid that binds to cyt *c*, in view of its unique structure that contains four, instead of two, fatty acid tails. The cyt c-CL complex plays a critical role in modulating the function of the protein and, thus, the cell fate. At the early stage of the apoptotic process, the protein-phospholipid interaction decreases and, after complex dissociation, the free protein is released into the cytosol [95].

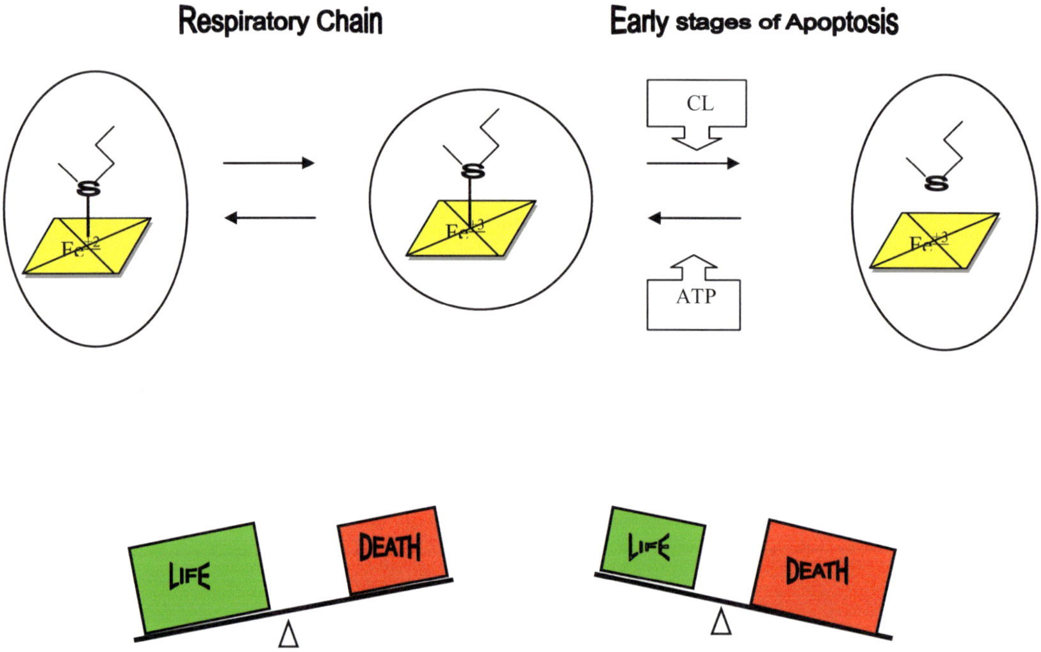

Figure 4: Role of cyt *c* in the mitochondrion (left) as electron carrier, and in the cytosol (right) where it regulates cell death and survival.

CL interaction with the protein is thought to occur at two distinct protein sites characterized by different affinity for the phospholipid [96, 97]. To date, the entry region of the acyl chain(s) into the protein is matter of debate. Two different regions of cyt *c* have been proposed to act as entry sites; one is the hydrophobic channel located close to the Asn52 residue [96], whereas the other is the region constituted by two parallel strands containing residues 67-71 and 82-85, forming a cleft that might host the acyl chain of the phospholipid [98]. A recent investigation suggests that both regions may be involved in the interaction of cyt *c* with CL to form the cyt *c*-CL complex [99]. To date, this point is still unclear and requires further investigation. What it is ascertained is that formation of the cyt *c*-CL complex is associated with tertiary conformational changes in the protein which provoke alteration of the heme pocket region and favour the displacement of Met80 from the sixth coordination position of the heme iron [95-101].

It has been proposed that in cytoplasm cyt *c* binds to the apoptosis protease activation factor (APAf-1) to form a complex, the apoptosome, activating pro-caspase 9. The originated enzymatic reaction cascade leads to the execution of apoptosis in cells. For complex formation, the presence of ATP or dATP is required [102]. That membrane-bound cyt *c* is involved in apoptotic activity has been proved; this attributes to the cyt *c*-CL binding a key role for caspase activation [92]. The conformation of the membrane-bound cyt *c* which, as said, differs from that of the native protein, has been associated with that of the protein in the complex formed with CL [95-101].

The dissociation of Met80 from the sixth coordination position of the heme iron is a peculiar characteristic of this non-native state of cyt *c* [103]. The partial unfolding of the CL-bound protein together with a significant weakening of the Met80-Fe(III) coordination bond, facilitate the access of small molecules, as H_2O_2, into the heme site of cyt *c* [104]. The peroxidase activity shown by membrane-bound cyt *c* is critical in the early stages of apoptosis. Some authors have demonstrated that the cyt *c*/CL complex acts as powerful CL-specific peroxidase, and generates CL hydroperoxides that play a role in cyt *c* (and other pro-apoptotic factors) release from mitochondrial membrane [95, 105, 106]. Thus, whereas the native fold as well as the low-spin hexacoordination of the heme iron are important for cyt *c* to function as electron carrier, the formation of a non-native compact conformation induces peroxidase activity in the protein, of relevance for the execution of the apoptotic program. ATP acts as an allosteric effector regulating structural transitions among different conformations and different oxidation states of cyt *c* endowed or not with apoptotic activity [107].

An additional role for the peroxidase activity of cyt c has been proposed in the study of the oxidative stress-induced aggregation of α-synuclein, a "natively unfolded" protein constituting the major component of intracellular inclusions in several neurodegenerative disorders, such as Parkinson's disease [108, 109]. In this regard, some authors have proposed that the peroxidase form of cyt *c* may be responsible for the damage that leads to neuronal death in the parkinsonian brain [110]. Furthermore, formation of millimeter-length fibers of phospholipid vesicles-bound cyt c displaying amyloid (ß-sheet) characteristics has been reported [111]; this implies that the cyt c-phospholipids interaction may play a role in some of the disorders attributed to amyloid formation, such as AA-amyloidosis and Alzheimer's disease. As a whole, the capability of cyt c to respond to different environments by changing its fold and exerting different biological functions may evoke its potential role in pathophysiological conditions such as neurodegenerative processes.

Table 1: Biological effects of altered folding mechanism and related diseases

Protein	**Structural Mechanism**	**Biological Effect and Related Disease**
Apolipoprotein E4 [a]	Molten globule stabilization	Alzheimer's disease
Alpha-lactalbumin [b]	Molten globule stabilization by interaction with lipids	Apoptosis in cancer cells
Cystic fibrosis Transmembrane regulator (CFTR) [c]	Phe508 deletion, misfolding, retention in the endoplasmic reticulum and degradation	Cystic fibrosis
Alpha-1-antitrypsin [d]	Aminoacid mutations conformational changes and aggregation	Alpha-1-antitrypsin deficiency
Prion protein [e]	Misfolding and aggregation in brain	Creutzfeldt-Jacob disease
Beta-amyloid [f]	Misfolding and aggregation in brain	Alzheimer's disease
Cytochrome *c* [g]	Interaction with phospholipids and non-native intermediates with peroxidase activity	Apoptosis, Parkinson's disease and amyloidosis

[a]Ref. 28; [b]Refs. 73, 82; [c]Refs. 48-51; [d]Refs.52-58; [e]Refs. 71, 74; [f]Refs. 71-73, 79; [g]Refs. 92, 95, 104, 106, 109, 112, 113

Perspectives

The present review has pointed out the important role played by intermediate states in protein folding, and indicated how dramatic consequences may be generated by protein misfolding (see Table **1**, for a schematic view). This justifies the wide body of studies devoted to determine the structural and functional properties of forms having structure intermediate between the native and fully unfolded state, and aiming to clarify how non-native compact states may influence protein folding and misfolding.

Most of the human diseases described in this review are of genetic nature, being generated by point mutations which favour accumulation of non-native compact forms of the protein. This outlines how dramatic consequences may derive from a point mutation occurring in a protein.

Studies on mutants have significantly improved our knowledge on the role of side chains (in particular, the invariant ones) in terms of structural stabilization, folding and functionality. However, such studies are not limited to elucidate specific aspects of protein misfolding and relative diseases; they also provide a precious contribution to better understanding the route of processes in which a native protein converts into the more flexible MG state. As described above, a protein acquiring a non-native compact conformation may accomplish different functions when involved in distinct processes. In the case of cyt *c*, the protein was proposed to be released from the mitochondrion as a MG state, during cell apoptosis. Work from this laboratory has provided a significant contribution to better elucidate this point, showing that oleic acid- and CL-bound cyt *c* acquires a MG character and is able to bind ATP, a molecule which actively participates to the apoptotic process [83, 97]. ATP behaves as an allosteric effector modulating structural transitions of cyt *c*, its peroxidase activity and apoptogenic properties. [112]. Studies have also revealed that the residues His26 and Asn52 are critical for stabilization of cyt *c*. The His26Tyr mutant shows a MG character, with flexibility higher than the native form, decreased stability, and altered heme crevice where Met80 is displaced by a lysine from the sixth coordination position of the heme-iron [113]. The Asn52→Ile mutation does not allow folding of equine cyt *c*; by contrast, the Asn52Ile mutant of yeast cyt *c* is able to fold, however this variant does not bind CL firmly [99]. This suggests that the mutation alters the region(s) of the protein interacting with the phospholipid. Studies on the His26Tyr and Asn52Ile variants of cyt *c* (as well as those on the ferric Lys72Asn and Lys73Asn mutants of the equine protein, actually under investigation in our laboratory) may provide in a next future useful information on the role played by specific residues in influencing the folding process of cyt *c* and in modulating the cyt *c*-CL interaction, a process critical for cell apoptosis [95].

At present, research on cyt *c* is mainly aimed to define the interaction between the protein and CL and to determine the molecular mechanism(s) favouring, and inhibiting, the peroxidase activity of cyt *c* [110, 114, 115]. This may be of relevance in the discovery of new antiapoptotic drugs. To this issue, experiments were recently carried out in this laboratory to determine if cyt *c* acts as suitable target for minocycline (a second-generation tetracycline able to cross the blood brain barrier with anti-inflammatory and neuroprotective effects). Results obtained show that minocycline may have a twofold effect on the apoptotic role of cyt *c*: (i) an initial effect due to its potency in blocking, or decreasing, the peroxidase activity of the protein at the early stages of the apoptotic process, and (ii) a subsequent effect in the cytosol, where it competes *vs.* cyt *c* for binding to Apaf-1 during formation of the apoptosome [116].

ACKNOWLEDGEMENTS

Declared none.

CONFLICT OF INTEREST

The authors confirm that this chapter contents have no conflict of interest.

DISCLOSURE

The chapter submitted for eBook Series entitled: "**Recent Advances in Medicinal Chemistry, Volume 1**" is an update of our article published in **Mini-Reviews in Medicinal Chemistry, Volume 8, Number 1, pp. 57 to 62**, with additional text and references.

REFERENCES

[1] Ewbank, J.J.; Creighton, T.E., Hayer-Hartle, M.K.; Hartle, F.U. What is the molten globule? *Struct. Biol.* **1995,** *2*, 10-11.

[2] Yeh, S-R; Rousseau, D.L. Folding intermediates in cytochrome c. *Nature Struct. Biol.* **1998,** *5*, 222-228.

[3] Dyson, H.J.; Wright, P.E. Unfolded proteins and protein folding studied by NMR. *Chem. Rev.* **2004,** *104*, 3607-3622.

[4] Englander, S.W.; Mayne, L.; Krishna, M.M. Protein folding and misfolding: mechanism and principles. *Q. Rev. Biophys.* **2007,** *40*, 287-326.

[5] Dill, K.A.; Ozkan, S.B.; Shell, M.S.; Weikl, T.R. The protein folding problem. *Annu. Rev.Biophys.* **2008,** *37*, 289-316.

[6] Baldwin, R.L. The search for folding intermediates and the mechanism of protein folding. *Annu. Rev. Biophys.* **2008,** *37*, 1-21.

[7] Bartlett, A.I.; Radford, S.E. An expanding arsenal of experimental methods yields an explosion of insights into protein folding mechanisms. *Nat. Struct. Mol. Biol.* **2009,** *16*, 582-587.

[8] Chang, J-Y. Diverse pathways of oxidative folding of disulfide proteins: underlying causes and folding models. *Biochemistry* **2011,** *50*, 3414-3431.

[9] Dill, K.A.; Ozkan, S.B.; Weikl, T.R.; Chodera, J.D.; Voelz, V.A. The protein folding problem: when will it be solved? *Curr. Opin. Struct. Biol.* **2007,** *17*, 342-346.

[10] Fersht, A.R. From the first protein structures to our current knowledge of protein folding: delights and scepticisms. *Nat. Rev. Mol. Cell. Biol.* **2008,** *9*, 650-654.

[11] Konermann, L.; Simmons, D.A. Protein-folding kinetics and mechanisms studied by pulse-labeling and mass spectrometry. Mass. Spectrom. Rev. **2003,** *22*, 1-26.

[12] Gianni, S., Ivarsson, Y., Jemth, P., Brunori, M., Travaglini-Allocatelli, C. Identification and characterization of protein folding intermediates. Biophys. Chem. **2007,** *128*, 105-113.

[13] Kathuria, S.V.; Guo, L.; Graceffa, R.; Barrea, R.; Nobrega, R.P.; Matthews, C.R.; Irving, T.C.; Bilsel, O. Minireview: structural insights into early folding events using continuous-flow time-resolved small-angle X-ray scattering. Biopolymers **2011,** *95*, 550-558.

[14] Kuwajima, K. The molten globule state as a clue for understanding the folding and cooperativity of globular-protein structure. *Proteins: Struct. Funct. Genet.* **1989,** *6*, 87-103.

[15] Ptitsyn, O.B. How the molten globule became. Trends Biochem. Sci. **1995,** *20*, 376-379.

[16] Jha, S.K.; Udgaonkar, J.B. Direct evidence for a dry molten globule intermediate during the unfolding of a small protein. *Proc. Natl. Acad. Sci. USA* **2009,** *106*, 12289-12294.

[17] Baldwin, R.L.; Frieden, C.; Rose, G.D. Dry molten globule intermediates and the mechanism of protein unfolding. *Proteins* **2010,** *78*, 2725-2737.

[18] Santucci, R.; Sinibaldi, F.; Patriarca, A.; Santucci, D.; Fiorucci L. Misfolded proteins and neurodegeneration: role of non-native cytochrome c in cell death. *Exp. Rev. Proteom.* **2010,** *7*, 507-517.

[19] Foguel, D.; Silva, J.L. New insights into the mechanisms of protein misfolding and aggregation in amyloidogenic diseases derived from pressure studies. Biochemistry **2004,** *43*, 11361-11370.

[20] Ladiwala, A.R., Lin, J.C.; Bale, S.S.; Marcelino-Cruz, A.M.; Bhattacharya, M.; Dordick, J.S.; Tessier, P.M. Resveratrol selectively remodels soluble oligomers and fibrils of amyloid Abeta into off-pathway conformers. J. Biol. Chem. **2010,** *285*, 24228-24237.

[21] Yu, H.; Liu, X:; Neupane, K.; Gupta, A.N.; Brigley, A.M.; Solanki, A.; Sosova, I.; Woodside, M.T. Direct observation of multiple misfolding pathways in a single prion protein molecule. *Proc. Natl. Acad. Sci. USA* **2012,** *109*, 5283-5288.

[22] Schmid, F.X. In: *Protein Folding*; Creighton, T.E. Ed.; Freeman and Company: New York; pp. 197-242. **1992**.

[23] Semisotnov, G.V.; Kutyshenko, V.P.; Ptitsyn, O.B. Intramolecular mobility of a protein in a "molten globule" state. A study of carbonic anhydrase by 1H-NMR. *Mol. Biol. USSR* **1989,** *23*, 808-815.

[24] Hughson, F.M.; Wright, P.E.; Baldwin, R.L. Structural characterization of a partly folded apomyoglobin intermediate. *Science* **1990,** *249*, 1544-1548.

[25] Dobson, C.M.; Hanley, C.; Radford, S.E.; Baum, J.A.; Evans, P.A. In: *Conformations and Forces in Protein Folding*; Nall, B.T.; Dill, K.A. Eds; AAAS: Washington DC; pp. 175-181. **1991**.

[26] Jeng, M-F; Englander, S.W. Stable submolecular folding units in a non-compact form of cytochrome c. *J. Mol. Biol.* **1991,** *221*, 1045-1061.

[27] Ptitsyn, O.B. Structures of folding intermediates. Curr. Opin. Struct. Biol. **1995,** *5*, 74-78.

[28] Hatters, D.M.; Peters-Libeu, C.A.; Weisgraber, K.H. Apolipoprotein E structure: insights into function. *Trends Biochem Sci.* **2006,** *31*, 445-454.

[29] Hallgren, O.; Aits, S.; Brest, P.; Gustafsson, L.; Mossberg, A.K.; Wullt, B.; Svanborg, C. Apoptosis and tumor cell death in response to HAMLET (human alpha-lactalbumin made lethal to tumor cells). *Adv. Exp. Med. Biol.* **2008,** *606*, 217-240.

[30] Anfinsen, C.B. Principles that govern the folding of protein chains. *Science* **1973,** *181*, 223-230.

[31] Englander, S.W. Protein folding intermediates and pathways studied by hydrogen exchange. Annu. *Rev. Biophys. Biomol. Struct.* **2000,** *29*, 213-238.

[32] Englander, S.W.; Mayne, L.; Krishna, M.M. Protein folding and misfolding: mechanism and principles. *Q. Rev. Biophys.* **2007,** *40*, 287-326.

[33] Baldwin, R.L. The search for folding intermediates and the mechanism of protein folding. Annu. Rev. Biophys. **2008,** *37*, 1-21.

[34] Neudecker, P.; Lundström, P.; Kay, L.E. Relaxation dispersion NMR spectroscopy as a tool for detailed studies of protein folding. *Biophys J.* **2009,** *96*, 2045-2054.

[35] Khan, M.K.; Rahaman, H.; Ahmad, F. Conformation and thermodynamic stability of pre-molten and molten globule states of mammalian cytochromes-c. *Metallomics* **2011,** *3*, 327-38.

[36] Bryngelson, J.D.; Onuchic, J.N.; Socci, N.D.; Wolynes, P.G. Funnels, pathways, and the energy landscape of protein folding: a synthesis. *Proteins* **1995,** *21*, 167-195.

[37] Dill, K.A.; Chan, H.S. From Levinthal to pathways to funnels. *Nature Struct. Biol.* **1997,** *4*, 10-19.

[38] Alm, E.; Baker, D. Prediction of protein-folding mechanisms from free-energy landscapes derived from native structures. *Proc. Natl. Acad. Sci. USA* **1999,** *96*, 11305-11310.

[39] Gianni, S.; Brunori, M.; Travaglini-Allocatelli, C. Plasticity of the protein folding landscape: switching between on- and off-pathway intermediates. *Arch. Biochem. Biophys.* **2007,** *466*, 172-176.

[40] Weinkam, P.; Zimmermann, J.; Romesberg, F.E.; Wolynes, P.G. The folding energy landscape and free energy excitations of cytochrome c. *Acc. Chem. Res.* **2010,** *43*, 652-660.

[41] Radford, S.E. Protein folding: progress made and promises ahead. *Trends Biochem. Sci.* **2000,** *25*, 611-618.

[42] Jahn, T.R.; Radford, S.E. The Yin and Yang of protein folding. *FEBS J.* **2005,** *272*, 5962-5970.

[43] Sato, T.; Esaki, M.; Fernandez, J.M.; Endo, T. Comparison of the protein-unfolding pathways between mitochondrial protein import and atomic-force microscopy measurements. *Proc. Natl. Acad. Sci. USA* **2005,** *102*, 7999-8004.

[44] Lee, H.; Kirchmeier, M., Mach, H. Monoclonal antibody aggregation intermediates visualized by atomic force microscopy. *J. Pharm. Sci.* **2011,** *100*, 416-423.

[45] Pirchi, M.; Ziv, G.; Riven, I.; Cohen, S.S.; Zohar, N., Barak, Y., Haran, G. Single-molecule fluorescence spectroscopy maps the folding landscape of a large protein. Nat. Commun. **2011,** *2*, 493.

[46] Ferreon, A.C.; Deniz, A.A. Protein folding at single-molecule resolution. Biochim. Biophys. Acta **2011,** *1814*, 1021-1029.

[47] Lee, T.W.; Matthews, D.A.; Blair, G.E. Novel molecular approaches to cystic fibrosis gene therapy. *Biochem J.* **2005,** *387*, 1-15.

[48] Amaral, M.D. Therapy through chaperones: sense or antisense? Cystic fibrosis as a model disease. *J. Inherit. Metab. Dis.* **2006,** *29*, 477-487.

[49] Amaral, M.D. Targeting CFTR: how to treat cystic fibrosis by CFTR-repairing therapies. Curr. Drug Targets 2011 *12*, 683-693.

[50] Farinah, C.M.; Nogueira, P.; Mendes, F.; Penque, D.; Amaral, M.D. The human DnaJ homologue (Hdj)-1/heat-shock protein (Hsp) 40 co-chaperone is required for the *in vivo* stabilization of the cystic fibrosis transmembrane conductance regulator by Hsp70. *Biochem. J.* **2002,** *366*, 797-806.

[51] Mendes, F.; Farinha, CM.; Roxo-Rosa, M.; Fanen, P.; Edelman, A.; Dormer, R.; McPherson, M.; Davidson, H.; Puchelle, E.; De Jonge, H.; Heda, G.D.; Gentzsch, M., Lukacs, G.; Penque, D.; Amaral, M.D. Antibodies for CFTR studies. *J. Cyst. Fibros.* **2004,** *3*, 69-72.

[52] Parfrey, H.; Mahadeva, R.; Lomas, D.A. Alpha(1)-antitrypsin deficiency, liver disease and emphysema. *Int. J. Biochem. Cell Biol.* **2003,** *35*, 1009-1014.

[53] Salahuddin, P. Genetic variants of alpha1-antitrypsin. *Curr. Protein Pept. Sci.* **2010,** *11*, 101-117.

[54] Elliott, P.R.; Lomas, D.A.; Carrel, R.W.; Abrahams, J.P. Inhibitory conformation of the reactive loop of alpha 1-antitrypsin. *Nat. Struct. Biol.* **1996,** *3*, 676-681.

[55] Kim, S-J.; Woo, J-R.; Seo, E.J.; Yu, M-H.; Ryu, S-E. A 2.1 Å resolution structure of an uncleaved alpha(1)-antitrypsin shows variability of the reactive center and other loops. *J. Mol. Biol.* **2001,** *306*, 109-119.

[56] Sivasothy, P.; Dafforn, T.R.; Gettins, P.G.; Lomas, D.A. Pathogenic alpha 1-antitrypsin polymers are formed by reactive loop-beta-sheet A linkage. *J. Biol. Chem.* **2000,** *275*, 33663-33668.

[57] Stockley, R.A. Emerging drugs for alpha-1-antitrypsin deficiency. *Expert Opin. Emerg. Drugs.* **2010,** *15*, 685-694.

[58] Rashid, S.T.; Lomas D.A. Stem cell-based therapy for α1-antitrypsin deficiency. *Stem Cell Res. Ther.* **2012,** *3*, 4.

[59] Pike, R.N.; Potempa, J.; Skinner, R.; Fitton, H.L.; McGraw, W.T.; Travis, J.; Owen, M.; Jin, L.; Carrell, R.W. Heparin-dependent modification of the reactive center arginine of antithrombin and consequent increase in heparin binding affinity. *J. Biol. Chem.* **1997,** *272*, 19652-19655.

[60] Patnaik, M.M.; Moll, S. Inherited antithrombin deficiency: a review. Haemophilia **2008,** *14*, 1229-1239.

[61] Silverman, G.A.; Bird, P.I.; Carrell, R.W.; Church, F.C.; Coughlin, P.B.; Gettins, P.G.; Irving, J.A.; Lomas, D.A.; Luke, C.J.; Moyer, R.W.; Pemberton, P.A.; Remold-O'Donnell, E.; Salvesen, G.S.; Travis, J.; Whisstock, J.C. The serpins are an expanding superfamily of structurally similar but functionally diverse proteins. Evolution, mechanism of inhibition, novel functions, and a revised nomenclature. *J. Biol. Chem.* **2001,** *276*, 33293-33296.

[62] Chiti, F.; Dobson, C.M. Amyloid formation by globular proteins under native conditions. *Nat. Chem. Biol.* **2009,** *5*, 15-22.

[63] Luheshi, L.M.; Dobson, C.M. Bridging the gap: from protein misfolding to protein misfolding diseases. FEBS Lett. **2009,** *583*, 2581-2586.

[64] Friedman, R. Aggregation of amyloids in a cellular context: modelling and experiment. *Biochem. J.* **2011,** *438*, 415-426.

[65] Van Broeck, B.; Van Broeckhoven, C.; Kumar-Singh, S. *Neurodegener. Dis.* **2007,** 4, 349-365.

[66] Uversky, V.N. Amyloidogenesis of natively unfolded proteins. *Curr. Alzheimer Res.* **2008,** *5*, 260-287.

[67] Brorsson, A.C.; Kumita, J.R.; MacLeod, I.; Bolognesi, B.; Speretta, E.; Luheshi, L.M.; Knowles, T.P.; Dobson, C.M.; Crowther, D.C. Methods and models in neurodegenerative and systemic protein aggregation diseases. *Front. Biosci.* **2010,** *15*, 373-396.

[68] Breydo, L.; Wu, J.W.; Uversky, V.N. A-synuclein misfolding and Parkinson's disease. *Biochim. Biophys. Acta* **2012,** *1822*, 261-285.

[69] Marcu, C.B.; Niessen, H.W.; Beek, A.M.; Brouwer, W.P.; Robbers, L.F.; Van Rossum, A.C. Cardiac involvement with amyloidosis: mechanisms of disease, diagnosis and management. Conn. Med. **2011,** *75*, 581-590.

[70] Kapoor, P.; Thenappan, T.; Singh, E.; Kumar, S.; Greipp, P.R. Cardiac amyloidosis: a practical approach to diagnosis and management. *Am. J. Med.* **2011,** *124*, 1006-1015.

[71] Chiti, F.; Dobson, C.M. Protein misfolding, functional amyloid, and human disease. *Annu. Rev. Biochem.* **2006,** *75*, 333-366.

[72] Kajava, A.V.; Aebi, U.; Steven, A.C. The parallel superpleated beta-structure as a model for amyloid fibrils of human amylin. *J. Mol. Biol.* **2005,** *348*, 247-252.

[73] Makin, O.; Serpell, L.C. Structural characterisation of islet amyloid polypeptide fibrils. *J. Mol. Biol.* **2004,** *335*, 1279-1288.[74] Krishnan, R.; Lindquist, S.L. Structural insights into a yeast prion illuminate nucleation and strain diversity. *Nature* **2005,** *435*, 765-772.

[75] Zandomeneghi, G.; Krebs, M.R.; McCammon, M.G.; Fandrich, M. FTIR reveals structural differences between native beta-sheet proteins and amyloid fibrils. *Protein Sci.* **2004,** *13*, 3314-3321.

[76] Serio, T.R.; Cashikar, A.G.; Kowal, A.S.; Sawicki, G.J.; Moslehi, J.J.; Serpell, L.; Arnsdorf, M.F.; Lindquist, S.L. Nucleated conformational conversion and the replication of conformational information by a prion determinant. *Science* **2000,** *289*, 1317-1321.

[77] Uversky, V.N.; Li, J.; Souillac, P.; Millett, I.S.; Doniach, S.; Jakes, R.; Goedert, M.; Fink, A.L. Biophysical properties of the synucleins and their propensities to fibrillate: inhibition of alpha-synuclein assembly by beta- and gamma-synucleins. *J. Biol. Chem.* **2002,** *277*, 11970-11978.

[78] Pedersen, J.S.; Christensen, G.; Otzen, D.E. Modulation of S6 fibrillation by unfolding rates and gatekeeper residues. *J. Mol. Biol.* **2004,** *341*, 575-588.

[79] Harper, J.D.; Lieber, C.M.; Lansbury, P.T.jr Atomic force microscopic imaging of seeded fibril formation and fibril branching by the Alzheimer's disease amyloid-beta protein. *Chem. Biol.* **1997,** *4*, 951-959.

[80] Morozova-Roche, L.A.; Zamotin, V.; Malisauskas, M.; Ohman, A., Chertkova, R.; Lavrikova, M.A.; Kostanyan, I.A.; Dolgikh, D.A.; Kirpichnikov, M.P. Fibrillation of carrier protein albebetin and its biologically active constructs. Multiple oligomeric intermediates and pathways. *Biochemistry* **2004,** *43*, 9610-9619.

[81] Gosal, W.S.; Morten, I.J.; Hewitt, E.W.; Smith, D.A.; Thomson, N.H.; Radford, S.E. Competing pathways determine fibril morphology in the self-assembly of beta2-microglobulin into amyloid. *J. Mol. Biol.* **2005,** *351*, 850-864.

[82] Mossberg, A.K.; Hun Mok K.; Morozova-Roche, L.A.; Svanborg, C. Structure and function of human α-lactalbumin made lethal to tumor cells (HAMLET)-type complexes. *Febs J.* **2010,** *277*, 4614-4625.

[83] Sinibaldi, F.; Mei, G.; Ponticelli, F.; Piro, M.C.; Howes, B.D.; Smulevich, G.; Santucci, R.; Ascoli, F.; Fiorucci, L. ATP specifically drives refolding of non-native conformations of cytochrome *c*. *Protein Sci.* **2005,** *14*, 1049-1058.

[84] Nantes, I.L.; Zucchi, M.R.; Nascimento, O.R.; Faljoni-Alario, A. Effect of heme iron valence state on the conformation of cytochrome *c* and its association with membrane interfaces. A CD and EPR investigation. *J. Biol. Chem.* **2001,** *276,* 153-158.

[85] Tuominem, E.K.; Wallace, C.J.; Clark-Lewis, I.; Craig, D.B.; Rytomaa, M.; Kinnunen, P.K. Phospholipid-cytochrome *c* interaction: evidence for the extended lipid anchorage. *J. Biol. Chem.* **2002,** *277,* 8822-8826.

[86] Sivakolundu, S.G.; Mabrouk, P.A. Structure-function relationship of reduced cytochrome *c* probed by complete solution structure determination in 30% acetonitrile/water solution. *J. Biol. Inorg. Chem.* **2003,** *8,* 527-539.

[87] Basu, S.; Keszler, A.; Azarova N.A.; Nwanze, N.; Perlegas, A.; Shiva, S.; Broniowska, K.A.; Hogg, N.; Kim-Shapiro, D.B. A novel role for cytochrome *c*: efficient catalysis of S-nitrosothiol formation. *Free Radic. Biol. Med.* **2010,** *48*, 255-263.

[88] Nantes, I.L., Kawai, C., Pessoto, F.S., Mugnol, K.C. Study of respiratory cytochromes in liposomes. *Methods Mol. Biol.* **2010,** *606*, 147-165.

[89] Banci, L.; Bertini, I.; Gray, H.B.; Luchinat, C.; Reddig, T.; Rosato, A.; Turano, P. Solution structure of oxidized horse heart cytochrome c. Biochemistry **1997,** 36, 9867-9877.

[90] Bayir, H.; Fadeel, B.; Palladino, M.J.; Witasp, E.; Kurnikov, I.V.; Tyurina, Y.Y., Tyurin, V.A.; Amoscato, A.A.; Jiang, J.; Kochanek, P.M.; DeKosky, S.T.; Greenberger, J.S. Shvedova, A.A.; Kagan, V.E. Apoptotic interactions of cytochrome *c*: redox flirting with anionic phospholipids within and outside of mitochondria. *Biochim. Biophys. Acta* **2006,** *1757*, 648-659.

[91] Berezhna, S.; Wohlrab, H.; Champion, P.M. Resonance Raman investigations of cytochrome *c* conformational change upon interaction with the membranes of intact and Ca^{2+}-exposed mitochondria. *Biochemistry* **2003,** *42*, 6149-6158.

[92] Garrido, C.; Galluzzi, L.; Brunet, M.; Puig, P.E.; Didelot, C.; Kroemer, G. Mechanisms of cytochrome *c* release from mitochondria. *Cell Death Differ.* **2006,** *13*, 1423-1433.

[93] Caroppi, P.; Sinibaldi, F.; Fiorucci L.; Santucci, R. Apoptosis and human diseases: mitochondrion damage and lethal role of released cytochrome *c* as proapoptotic protein. *Curr. Med. Chem.* **2009,** *16*, 4058-4065.

[94] Kagan, V.E.; Borisenko, G.G.; Tyurina, Y.Y.; Tyurin, V.A.; Jiang, J.; Potapovich, A.I.; Kini, V.; Amoscato, A.A.; Fujii, Y. Oxidative lipidomics of apoptosis: redox catalytic interactions of cytochrome *c* with cardiolipin and phosphatidylserine. Free Radic. Biol. Med. **2004,** *37*, 1963-1985.

[95] Kagan, V.E.; Tyurin, V.A.; Jiang, J.; Tyurina, Y.Y.; Ritov, V.B.; Amoscato, A.A.; Osipov, A.N.; Belikova, N.A.; Kapralov, A.A.; Kini, V.; Vlasova, I.I.; Zhao, Q.; Zou, M.; Di, P.; Svistunenko, D.A.; Kurnikov, I.V.; Borisenko, G.G. Cytochrome *c* acts as a cardiolipin oxygenase required for release of proapoptotic factors. *Nat. Chem. Biol.* **2005,** *1*, 223-232.

[96] Rytömaa, M.; Kinnunen, P.K. Reversibility of the binding of cytochrome *c* to liposomes. Implications for lipid-protein interactions. *J. Biol. Chem.* **1995,** *270*, 3197-31202.

[97] Sinibaldi, F.; Fiorucci, L.; Patriarca, A.; Lauceri, R.; Ferri, T.; Coletta, M.; Santucci, R. Insights into cytochrome *c*-cardiolipin interaction. Role played by ionic strength. *Biochemistry* **2008,** *47*, 6928-6935.

[98] Kalanxhi, E.; Wallace, C.J.A. Cytochrome *c* impaled: investigation of the extended lipid anchorage of a soluble protein to mitochondrial membrane models. *Biochem. J.* **2007,** *407*, 179-187.

[99] Sinibaldi, F.; Howes, B.D.; Piro, M.C.; Polticelli, F.; Bombelli, C.; Ferri, T.; Coletta, M.; Smulevich, G.; Santucci R. Extended cardiolipin anchorage to cytochrome *c*: a model for protein-mitochondrial membrane binding. *J. Biol. Inorg. Chem.* **2010,** *15*, 689-700.

[100] Schug, Z.T.; Gottlieb, E. Cardiolipin acts as a mitochondrial signaling platform to launch apoptosis. *Biochim. Biophys. Acta* **2009,** *1788*, 2022-2031.

[101] Rajagopal, B.S.; Silkstone, G.G.; Nicholls, P.; Wilson, M.T.; Worrall, J.A. An investigation into a cardiolipin acyl chain insertion site in cytochrome *c*. Biochim Biophys Acta **2012,** *1817*, 780-791.

[102] Green, D.R.; Reed, J.C. Mitochondria and apoptosis. *Science* **1998,** *281*, 1309-1312.

[103] Godoy, L.C.; Muñoz-Pinedo, C.; Castro, L.; Cardaci, S.; Schonhoff, C.M.; King, M.; Tórtora, V.; Marín, M., Miao, Q.; Jiang, J.F.; Kapralov, A.; Jemmerson, R.; Silkstone, G.G.; Patel, J.N.; Evans, J.E.; Wilson, M.T.; Green, D.R.; Kagan, V.E.; Radi, R.; Mannick, J.B. Disruption of the M80-Fe ligation stimulates the translocation of cytochrome *c* to the cytoplasm and nucleus in nonapoptotic cells. *Proc. Natl. Acad. Sci. USA* **2009,** *106*, 2653-2658.

[104] Vladimirov, Y.A.; Proskurnina, E.V.; Izmailov, D.Y.; Novikov, A.A.; Brusnichkin, A.V.; Osipov, A.N.; Kagan, V.E. Cardiolipin activates cytochrome *c* peroxidase activity since it facilitates H_2O_2 access to heme. *Biochemistry (Moscow)* **2006,** *71*, 998-1005.

[105] Belikova, N.A.; Vladimirov, Y.A.; Osipov, A.N.; Kapralov, A.A.; Tyurin, V.A.; Potapovich, M.V.; Basova, L.V.; Peterson, J.; Kurnikov, I.V.; Kagan, V.E. Peroxidase activity and structural transitions of cytochrome *c* bound to cardiolipin-containing membranes. *Biochemistry* **2006,** *45*, 4998-5009.

[106] Hüttemann, M.; Pecina, P.; Rainbolt, M.; Sanderson, T.H.; Kagan, V.E.; Samavati, L.; Doan, J.W.; Lee, I. The multiple functions of cytochrome *c* and their regulation in life and death decisions of the mammalian cell: From respiration to apoptosis. *Mitochondrion* **2011,** *11*, 369-381.

[107] Patriarca, A.; Eliseo, T.; Sinibaldi, F.; Piro, M.C.; Melis, R.; Paci, M.; Cicero, D.O.; Polticelli F.; Santucci, R.; Fiorucci, L. ATP acts as a regulatory effector in modulating structural transitions of cytochrome *c*: implications for apoptotic activity. *Biochemistry* **2009,** *48*, 3279-3287.

[108] Olteanu, A.; Pielak, G.J Peroxidative aggregation of alpha-synuclein requires tyrosines. *Protein Sci.* **2004,** *13*, 2852-2856.

[109] Bayir, H.; Kapralov, A.A.; Jiang, J.; Huang, Z.; Taurina, Y.Y.; Tyurin, V.A.; Zhao, Q.; Belikova, N.A.; Vlasova, I.I.; Maeda, A.; Zhu, J.; Na, H.M.; Mastroberardino, P.G.; Sparvero, L.J.; Amoscato, A.A.; Chu, C.T.; Greenamyre, J.T.; Kagan, V.E. Peroxidase mechanism of lipid-dependent cross-linking of synuclein with cytochrome *c*: protection against apoptosis *versus* delayed oxidative stress in Parkinson disease. *J. Biol. Chem.* **2009,** *284*, 15951-15969.

[110] Everse, J.; Liu, C.J.; Coates, P.W. Physical and catalytic properties of a peroxidase derived from cytochrome *c*. *Biochim. Biophys. Acta* **2011,** *1812*, 1138-1145.

[111] Alakoskela, J.M.; Jutila, A.; Simonsen, A.C.; Pirneskoski, J.; Pyhajoki, S.; Turunen, R.; Marttila, S.; Mouritsen, O.G.; Goormaghtigh, E.; Kinnunen, P.K.J. Characteristics of fibers formed by cytochrome *c* and induced by anionic phospholipids. *Biochemistry* **2006,** *45*, 13447-13453.

[112] Sinibaldi, F.; Mei, G.; Ponticelli, F.; Piro, M.C.; Howes, B.D.; Smulevich, G.; Santucci, R.; Ascoli, F.; Fiorucci, L. ATP specifically drives refolding of nonnative conformations of cytochrome *c*. *Protein Sci.* **2005,** *14*, 1049-1058.

[113] Sinibaldi, F.; Piro, M.C.; Howes, B.D.; Smulevich, G.; Ascoli, F.; Santucci, R. Rupture of the H-bond linking two omega-loops induces the molten globule state at neutral pH in cytochrome *c*. *Biochemistry* **2003,** *42,* 7604-7610.

[114] Samhan-Arias, A.K.; Ji, J.; Demidova, O.M.; Sparvero, L.J.; Feng, W.; Tyurin, V.; Taurina, Y.Y.; Epperly, M.W.; Shvedova, A.A.; Greenberger, J.S.; Bayır, H.; Kagan, V.E.; Amoscato, A.A. Oxidized phospholipids as biomarkers of tissue and cell damage with a focus on cardiolipin. *Biochim. Biophys. Acta* **2012,** 1818, 2413-2423.

[115] Abe, M.; Niibayashi, R.; Koubori, S.; Moriyama, I.; Miyoshi, H. Molecular mechanisms for the induction of peroxidase activity of the cytochrome *c*-cardiolipin complex. *Biochemistry*. **2011,** *50*, 8383-8391.

[116] Patriarca, A.; Polticelli, F.; Piro, M.C.; Sinibaldi, F.; Mei, G.; Bari, M.; Santucci, R.; Fiorucci, L. Conversion of cytochrome *c* into a peroxidase: inhibitory mechanisms and implication for neurodegenerative diseases. *Arch. Biochem. Biophys*., **2012,** 522, 62-69.

[117] Bushnell, G.W.; Louie, G.V.; Brayer, G.D. High-resolution three-dimensional structure of horse heart cytochrome *c*. J. Mol. Biol. **1990,** *214*, 585-595.

[118] Louie, G.V.; Brayer, G.D. High-resolution refinement of yeast iso-1-cytochrome *c* and comparisons with other eukaryotic cytochromes c. *J. Mol. Biol.* **1990,** *214*, 527-555.

[119] Pettersen, E.F.; Goddard, T.D.; Huang, C.C.; Couch, G.S.; Greenblatt, D.M.; Meng, E.C.; Ferrin, T.E. UCSF Chimera - A Visualization System for Exploratory Research and Analysis. *J. Comput. Chem.* **2004,** *25*, 1605-1612.

Index

A

B

C

D

E

F

G

J

L

M

N

O

P

Q

R

S

T

U

V

W

Made in the USA
Las Vegas, NV
04 November 2026

57423851R00288